Optimization and Control in Energy Management: Mathematical Modeling and Simulation

Optimization and Control in Energy Management: Mathematical Modeling and Simulation

Editors

Atanda Kamoru Raji
Khaled M. Abo-Al-Ez

MDPI

Basel • Beijing • Wuhan • Barcelona • Belgrade • Novi Sad • Cluj • Manchester

Editors

Atanda Kamoru Raji
Cape Peninsula University of Technology
Bellville
South Africa

Khaled M. Abo-Al-Ez
Postgraduate School of Engineering Management,
University of Johannesburg
Johannesburg
South Africa

Editorial Office
MDPI
St. Alban-Anlage 66
4052 Basel, Switzerland

This is a reprint of articles from the Special Issue published online in the open access journal *Mathematics* (ISSN 2227-7390) (available at: https://www.mdpi.com/journal/mathematics/special_issues/YUZTG35P22).

For citation purposes, cite each article independently as indicated on the article page online and as indicated below:

Lastname, A.A.; Lastname, B.B. Article Title. *Journal Name* **Year**, *Volume Number*, Page Range.

ISBN 978-3-7258-0775-8 (Hbk)
ISBN 978-3-7258-0776-5 (PDF)
doi.org/10.3390/books978-3-7258-0776-5

Contents

About the Editors

Atanda Kamoru Raji

Atanda Kamoru Raji is an Associate Professor of Microgrid and Integrated Energy Systems at the Department of Electrical Electronic and Computer Engineering, Faculty of Engineering and the Built Environment, Cape Peninsula University of Technology (CPUT), South Africa. He holds a Doctor of Technology degree in Electrical Engineering with specialization in grid-connected interconnection technology of renewables. He is the deputy head of the Energy Institute at CPUT. He is currently the Head of the Programme of Postgraduate Study and Research. He has published over 200 peer-reviewed journals and conference proceedings nationally and internationally (Google Scholar 2024: Citation 1022, h-index 15, i10-index 25). His research areas include but are not limited to modern power system design, development and optimization, hybrid renewable energy system, AI and ML applications in energy systems, energy management system, and engineering education research.

Khaled M. Abo-Al-Ez

Khaled M. Abo-Al-Ez is an Associate Professor of Engineering Management at the Postgraduate School of Engineering Management, Faculty of Engineering, and the Built Environment at the University of Johannesburg (UJ), South Africa. He was an Associate Professor of Electrical Power Systems at the Department of Electrical Electronic and Computer Engineering, Faculty of Engineering and the Built Environment, Cape Peninsula University of Technology (CPUT), South > Africa. He holds a PhD in Electrical Engineering with specialization in Electrical Power and Machines Engineering. He is an established researcher with NRF-C3 rating from the National Research Foundation (NRF) of South Africa. He is currently the Head of the Centre for Research in Power Systems at CPUT. He has published over 100 peer-reviewed journals and conference proceedings nationally and internationally (Google Scholar 2024: Citation 829, h-index 16, i10-index 20). His research areas include but are not limited to optimal operation and planning of power systems, renewable energy system, cyber-physical energy systems, AI and ML applications in energy systems, energy management system, and engineering education and management research.

Preface

The global energy crisis due to the depletion of fossil fuel resources, the need to reduce greenhouse emissions, the climate change phenomenon, and the unexpected increase in fuel prices due to global conflicts have increased the need to focus on using new and renewable energy systems. These systems can work autonomously and in a hybrid manner with other energy generation systems. The operation of these single or hybrid energy systems can be in the form of islanded or grid-connected mini or microgrids. Microgrid power fluctuation due to renewable energy systems requires energy storage systems to balance the energy and provides a continuous flow of energy even when energy fluctuates from renewable sources. Energy management strategies are essential in such systems for reliability, good power quality, and optimizing the operation of different energy and storage distributed systems in the microgrid. Energy management systems (EMS) are control techniques for managing the power flow in response to supply, demand, power quality, and storage conditions. The research work is currently focusing on different topics related to the optimization and control design of energy management systems using the mathematical modeling and simulation of different operation scenarios of these systems under different environmental and grid conditions.

The present book contains the 14 articles accepted for publication among the 30 in total manuscripts submitted to the Special Issue "Optimization and Control in Energy Management: Mathematical Modeling and Simulation" of the MDPI journal *Mathematics*.

The idea for this Special Issue arose from the interest of the Guest Editors in energy management topics, which have been a large part of their research experience over the past 20 years. From this standpoint, the researchers specialized in these topics were invited to contribute their research results to enrich this book.

The 15 articles, which appear in the present book, cover important and specialized research topics in the field of energy management systems, including Real-Time Management for an EV Hybrid Storage System Based on Fuzzy Control, Efficient Red Kite Optimization Algorithm for Integrating the Renewable Sources and Electric Vehicle Fast Charging Stations in Radial Distribution Networks, Optimizing Power Exchange Cost Considering Behavioral Intervention in Local Energy Community, Performance Evaluation of Grid-Connected DFIG-Based WECS with Battery Energy Storage System under Wind Alterations Using FOPID Controller for RSC, Strategic Electricity Production Planning of Turkey via Mixed Integer Programming Based on Time Series Forecasting, Reinforcement Learning-Enabled Electric Vehicle Load Forecasting for Grid Energy Management, Coordinated Economic Operation of Hydrothermal Units with HVDC Link Based on Lagrange Multipliers, Optimal Integration of Battery Systems in Grid-Connected Networks for Reducing Energy Losses and CO2 Emissions, Blockchain-Driven Real-Time Incentive Approach for Energy Management System, An improved Fractional MPPT Method by Using a Small Circle Approximation of the P–V Characteristic Curve, Optimal Location and Operation of PV Sources in DC Grids to Reduce Annual Operating Costs While Considering Variable Power Demand and Generation, Robust Flatness-Based Tracking Control for a "Full-Bridge Buck Inverter–DC Motor" System, Optimal Location and Sizing of PV Generation Units in Electrical Networks to Reduce the Total Annual Operating Costs: An Application of the Crow Search Algorithm, Design of Space Efficient Electric Vehicle Charging Infrastructure Integration Impact on Power Grid Network, and a systematic literature review on the research area of Variable-Speed Wind Turbines for Grid Frequency Support.

It is anticipated that the book will become interesting and useful for those working in the area of energy management systems, renewable energy systems, and hybrid electric vehicles, as well as for those having the proper mathematical background and the willingness to become familiar with the recent advances in the mathematical modeling and simulations of these systems and the design of their optimal controllers for a reliable and stable operation of smart electrical grids.

As the Guest Editors of this Special Issue, we are grateful to the authors of the papers for their quality contributions, to the reviewers for their valuable comments towards the improvement of the submitted studies, and to the administrative staff of MDPI publications for their support to complete this project.

Atanda Kamoru Raji and Khaled M. Abo-Al-Ez
Editors

mathematics

MDPI

rticle

)esign of Space Efficient Electric Vehicle Charging nfrastructure Integration Impact on Power Grid Network

uresh Chavhan [1,2,*], Subhi R. M. Zeebaree [3], Ahmed Alkhayyat [4] and Sachin Kumar [2,*]

1 Department of Electronics and Communication Engineering, Indian Institute of Information Technology Kottayam, Kerala 686635, India
2 Big Data and Machine Learning Lab, South Ural State University, 454080 Chelyabinsk, Russia
3 Energy Engineering Department, Technical College of Engineering, Duhok Polytechnic University, Duhok 42001, Iraq
4 College of Technical Engineering, The Islamic University, Najaf 54001, Iraq
* Correspondence: suresh.chavhan046@gmail.com (S.C.); kumars@susu.ru (S.K.)

Abstract: With an ever-increasing number of electric vehicles (EVs) on the roads, there is a high demand for EV charging infrastructure. The present charging infrastructure in the market requires a lot of space and sometimes leads to traffic congestion, increasing the risk of accidents and obstruction of emergency vehicles. As the current infrastructure requires ample space, the cost of setting up this charging infrastructure becomes very high in metropolitan cities. In addition, there are a lot of adverse effects on the power grid due to the integration of EVs. This paper discusses a space-efficient charging infrastructure and multi-agent system-based power grid balance to overcome these issues. The proposed multi-level EV charging station can save a lot of space and reduce traffic congestion as more vehicles can be accommodated in the space. Depending on the size, capacity, and type of multi-level vehicle charging system, it can serve as a reliable charging solution at sites with medium and high daily footfall. We integrated the EV charging station with IEEE 33 bus test system and analyzed the grid and charging stations. The proposed scheme is exhaustively tested by simulation in a discrete-time event simulator in MATLAB and analyzed with varying EV arrival rates, time periods, etc.

Keywords: multi-level charging stations; power grid network; multi-agent system; stability

MSC: 93C95

itation: Chavhan, S.; Zeebaree, R.M.; Alkhayyat, A.; Kumar, S. esign of Space Efficient Electric ehicle Charging Infrastructure tegration Impact on Power Grid etwork. *Mathematics* **2022**, *10*, 3450. tps://doi.org/10.3390/ ath10193450

cademic Editors: Atanda Raji and haled M. Abo-Al-Ez

eceived: 26 July 2022
ccepted: 12 September 2022
ublished: 22 September 2022

1. Introduction

Energy management is one of the major challenging tasks in this fast-changing world. The concept of the smart grid has brought a new change in the utilization of technology. However, on the other hand, the massive increase in the demand for production and managing the optimized energy within the grid structure is one of the significant issues with which the R&Ds are currently engaged. Not only within the intelligent grid network but also with the other minor ancillaries attached to the grid structure that accurately measure the energy utilization within the grid. Thus, the increase in Demand for Energy is one of the significant issues that is needed to be addressed [1]. In addition, EV charging stations help the industry optimize its grids' load management. It can also be used to aid the utilities in managing peak demand.

On the other hand, the smart grid (SG) also handles the demand response of the industrial and residential load. Thus, the management facility deployed in the SG Network should be strong enough to provide constant optimized power output. It should also be capable of monitoring and rectifying itself in case of any malfunction.

Recently, integrating electric vehicles with the smart grid network has been one of the standard techniques used to conserve energy. The SG network [1,2] is a combination of various distributed systems and functions such as Advanced Metering Infrastructure (AMI), Distribution Automation (DA), and Distributed Management System (DMS), which all integrate to produce the most optimized power flow. Load flow management is also another vital management technique. Especially when the EVs are integrated with the grid structure, the demand for electricity reaches its peak demand. Thus, monitoring the load at various times during the day is necessary, ensuring the overload condition is not reached. Similar load management techniques should be well implemented at various stages of the grid for constant load monitoring. The conventional load management techniques in the SG network are not well suited. Advanced load management techniques such as distributive load response (DLR) and advanced metering infrastructure (AMI) are integrated for the more intelligent operation of the network [3,4]. Thus, the SG greatly depends on the design, development, and integration of advanced devices and techniques for its efficient performance. Integrating the charging stations with the smart grid contributes to load management techniques and provides ample opportunity for research in this domain.

One of the biggest issues in crowded metropolitan areas is the space due to the increased population. There is no extra space for creating the charging infrastructure, and it may become complicated to create one in the future. Therefore, this paper proposes on-demand space-efficient multi-level EV charging station infrastructure for metropolitan cities. We have integrated the novel structure of the infrastructure with the 33-bus power grid for power transaction, load management, and power grid network balance. A novel methodology, that is, a multi-agent system, is deployed on each bus and charging station to collect, share and analyze various power system parameters. The analyzed parameters are used to control and maintain the stability of the power grid. The multi-level charging infrastructure's stress and displacement analysis has been done, and the cost analysis of the proposed infrastructure has been discussed. Developed the analytical model of the proposed method, optimization problem is formulated and is solved using the Genetic algorithm, and simulation has been conducted with varying parameters such as arrival rates, service rates, and the number of charging points available at the charging station.

The rest of the paper is organized as follows: Section 2 presents the literature review in which a discussion on existing studies is given. Section 3 provides an overview of the charging stations. Section 4 provides the detail of the proposed approach. Section 5 provides a discussion on the cost analysis of multi-level charging stations. Section 6 provides a brief discussion on the stress analysis of the designed structure. Section 7 provides a discussion on the mathematical modeling of the proposed multi-level EV charging station infrastructure. Section 8 details the simulation study and result analysis. Section 9 concludes the article.

2. Related Works

This section presents the smart grid network and EV charging station as follows:

2.1. Smart Grid Network

A traditional grid generally works as a medium and distribution hub for electricity transmission from the distribution station to the end-users. The grid model is based on the principle of electromagnetism, with few sensors and a manual degree of control. The efficiency of the system thus becomes low during the high demand. Moreover, the system reliability and security are very low compared to that of the other system [5]. The major drawback of the system is the manual distributing mechanism. The flexibility and the extension of the system thus become much more difficult.

However, the Smart Grid solves all the drawbacks mentioned above. It is an intelligent distributed electrical network that operates in an automated mode. The reliability, flexibility, security, sustainability, and efficiency are much more as compared to the conventional grid [6]. Moreover, the Smart Grid provides two-way communication, numerous sensors, self-diagnostics capabilities, and remote monitoring facilities.

The application of power electronics in the intelligent grid is off to a vast extent. Hence, the issue of the EMI in the current flow increases rapidly. The EMI issues in the smart grid have been reported to depend upon the magnitude of fault currents, their flowing paths, and harmonic contents. This interference may degrade the devices' efficiency and the system [7]. Moreover, the EMI sources coupling into a nearby device may also cause EMI radiations and lead to cascading effects. Thus, minimizing the EMI effect is one of the most critical issues to be handled in the SG network. When the circuit space becomes preferable, the size of the EMI filters can be reduced in proportional order and can be conducted simply by increasing the switching frequency of the filters. The smart meter is an important content of the Smart Grid Network. It works on the principle of the communication infrastructure to monitor real-time energy consumption and either distribute or optimize the energy per consumption. It consists of the micro-controller unit, metering chips, voltage converter, current converter, and peripheral circuit. The intelligent meters generally operate at the end-user level, primarily via LAN. Thus, a significant source of EMI issues is reported via the transmission line that causes transmission error, loss of message, and delay in the communication channel. The power Line Communication (PLC) system is the unique technology that makes the smart grid better than the others. It offers a modeling solution by importing bit error rates into the system via various architectures. The high-speed transmission with real-time bill generation, data processing, fault diagnostics, and reporting makes the PLC in the Smart grid a most crucial component. The PLC operates at a very high frequency, which might lead to EMI issues. As a result, the signal propagating generates multiple impulsive and stationary noises [5,8]. The recent Ultra-wideband (UWB) pulse transmission technique shows lower power spectral density lowers the EMI in the channel approximately by 10 dB as compared to other modulation schemes. Another important electromagnetic issue concerns the intelligent grid's AC transmission system. The current switching is one of the major sources of the generation of interferences and noises in the channel. The AC transmission consists of components such as a DC-AC inverter, setup transformer, harmonic blocking transformer, and associated hardware control. Operating at a high frequency with constantly switching with the demand response gives rise to high electromagnetic disturbances. Thus, some techniques such as a thyristor-switched capacitor, static synchronous compensators along with the use of passive filters, and active filters can be used to reduce the EMI issues. The intelligent grid network plays a significant role in the electric vehicle market. EVs and PHEVs have entered the market and are gaining significant importance. The charging infrastructure of the battery of the electric vehicles and meeting the energy demand is an essential aspect of the smart grid. The intelligent meter fixed at various levels determined the grid's capacity. Integrating the smart grid with the battery charging system without any compatibility issues is one of the significant challenges. The public charging station must be deployed in optimum numbers and be integrated with the grid network. The cost of the operation should also be optimized to meet the daily needs of the standard public [9,10]. The cost of operation is the last factor that should be considered. The policy formation and the implementation should be conducted in a secure manner that provides the optimum operating cost [11,12]. The detailed related works reflect the major issues of the intelligent grid network, which are to be addressed. This work implements and analyses the power flow results through the smart grid. The various algorithms, network structure, and operating technology are being modeled, and the results are discussed and analyzed.

2.2. EV Charging Stations

The paper [8] talks about the various charging connector types and presents the comparison between the American and European standards. In addition, the paper discusses the various energy storage systems and how to incorporate them with charging stations. A useful ESS for linking fast EV charging stations might be considered a system that includes batteries and ultra-capacitors: the first for their high energy densities and the second for their high power density. The paper [9] presents a novel design strategy for rapid

EV charging stations on a round motorway that takes into account spatial and tempora mobility habits. The planning approach is divided into three parts: an EV charging poir spatial-temporal model, a placement determination model, and a capacity determinatio model. The established planning approach not only precisely determines the most idea sites for EV charging stations while taking into account the travel convenience of EV user but also minimizes the total cost. The paper [10] presents a data-driven robust EV chargin infrastructure design strategy for residential car parks that takes into account the unce tainty of future EV charging behavior. For more realistic modeling of individual billin behavior, the suggested data-driven system makes use of real-world identity unit dat and a queuing mechanism. The non-parametric estimation of charging power probabilit density may be derived by aggregating EV individuals. Following that, a distributionall robust optimization strategy for planning EV infrastructure under stochastic chargin demand and maximum queue time is devised. The paper [11] proposes a business case fo a battery switching station as well as an optimization model. Customers' complaints abou long charging periods and range anxiety can be alleviated with the BSS. The BSS not onl benefits users financially, but it also benefits the electrical system by participating in electri ity markets and avoiding or postponing costly infrastructure changes. To be profitable, th BSS must ensure that the fees it collects, the risk it takes of failing to satisfy its clients, an the discounts it offers are all correctly constructed. The author of the paper [12] implement a periodic fluid model to represent charging operations at a BSS with time-varying deman for battery swap and time-varying prices for charging empty batteries, with the goal o determining the best battery purchasing and charging policy that best trades off batter investment cost and operating cost, including charging cost and customer waiting for cos The authors of the [13] focus on a design concept and methodology for promoting EV adop tion by automated battery pack switching at battery sharing stations as part of a batter sharing network, which would become an integral element of the smart grid. In additior the advantages and disadvantages of traditional battery swapping were also discussec The authors in [14] focus to address the issue of optimally locating charging stations i metropolitan environments. There are two optimization criteria used: maximization o reachable households and minimization of overall e-transportation energy cost. Mixec integer programming with linear and nonlinear energy-aware constraints is utilized t make decisions in both scenarios. A multi-objective optimization model is also offered which addresses both criteria (number of reachable homes and transportation energy) a the same time. The author of the paper [15] describes the development of a universa inductive charger (UIC) for electric vehicles. The suggested UIC can provide a constant o regulated charging voltage to various EVs thanks to a wide range of magnetic couplin between the transmitting and receiving coils. Zero-voltage switching of the primary dc-a inverter is universally realized in every charging cycle with a series-connected LC circui To automatically pick the best frequency in varied coupling situations and modify the frequency during the charging process, a simple and very effective control mechanism based on a variety of frequencies is used. The proposed system has advantages: Universa application, adaptive frequency, effective V and I sensing, and high efficiency. The author in the paper [16] presented a cost-effective vehicle charging technique particularly designec for smart homes/buildings with a PV(solar panels) system is presented and developed They suggested a smart algorithm for electric vehicle charging which can be used fo smart homes/buildings and is divided into two stages: the photovoltaic output and powe consumption prediction, and scheduling of EV charging. The prototype application for the home/building that has been created can give EV charging schedules based on custome preferences. This paper [17] discusses the challenges of electric vehicle charging infrastruc ture and also discusses the Indian Government policies. The paper also discusses the basi guidelines and standards to install public charging stations as described by the Ministry o Power and the Ministry of housing and urban affairs. Table 1 provides a brief comparisor of relevant existing works with the current work.

Table 1. Comparisons of Related works and Proposed work.

Paper ID	Approach	SG Ecosystem, Network Power Distribution	V2G, MAS	Power Flow Monitoring	Integration of Micro-Grids	Optimization Techniques
[11,18–21]	Distributed Power flow network model with energy management systems for EV Infrastructure	✓	✓	×	×	×
[5,8,16,22–26]	Power Flow in microgrids in SG Network	✓	✓	✓	×	×
[27–30]	Cyber Physical Protection in SG Network	✓	✓	✓	×	✓
[6,13,14,21,31–33]	Efficient Management Algorithms and optimised power flow mechanism	✓	×	✓	✓	×
Proposed Work	A comprehensive power flow analysis of the SG Network Infrastructure	✓	✓	✓	✓	×

3. Overview of Charging Stations

EVs have battery capacity with them, which, depending on the size and capacity of the battery pack, require charging from time to time. The charging need is determined by the type of vehicle (two-wheeler, three-wheeler, four-wheeler, and bus), as well as the utilitarian purpose (passenger or commercial). By 2030, the Indian Government wants to electrify 30 percent of private automobiles. This goal necessitates the simultaneous deployment of charging stations across India. Table 2 provides a comparison of charging stations.

Electric vehicles in India are currently separated into two-wheelers, passenger cars, and commercial vehicles (Buses and three-wheelers). Two-wheelers have compact batteries that can be removed and carried to homes, businesses, or stores to be charged using standard wall sockets; alternatively, they may be linked to any public charging station. Three-wheelers are excellent candidates for battery replacement. In this concept, batteries are charged in a big industrial-scale facility and trucked to three-wheeler concentration sites, where a 3-wheeler driver can exchange the used battery with a fully charged one. Swap stations can also be installed at a PCS by technology owners. Three-wheelers may be taxed at any PCS. Buses with batteries larger than 100 kWh would be sold with the battery manufacturer's approved proprietary charging standards. These batteries will set you back many million rupees, and charging them from any PCS is not recommended. The charging devices supplied (or recommended) by the bus manufacturer will be installed at bus depots and bus depots as needed by bus operators. Currently available electric car batteries range in size from 11 kWh to 40 kWh.Fast charging is required for these EVs. The issue of EVSE interoperability between different EV types is basically limited to automobiles. In this paper, we are mainly focusing on the charging infrastructure for electric cars.

Table 2. Comparison of Charging Station.

Parameters	On-Board Charging Station	Off-Board Charging Station	Fast Charging Station	Wireless Charging Station	Smart Charging Station	Battery Swapping Station	Multi-Level Charging Station
Energy transfer (in kW)	Less	High	Different ports for multiple levels	Bi-directional	Depends upon the distance between coils	Bidirectional	Bidirectional, Safe and High
Level of battery heating issue	Low	Very high	Medium	Medium	Low	Low	Very low
Battery weight on EV	Added	Removed	Removed	Removed	Moderate	Constant	Removed
Battery charging time	More	Depends on the controller of EV	Depends on the controller of EV	Depends on user control	Depends on power transmission coils	More	No delay
Flexibility	Anywhere charge	No flexibility	Anywhere charge	Anywhere charge	More flexible	More flexible	No flexibility
Cost and complexity	Low cost and complexity	High cost and complexity	High cost and low complexity	High cost and complexity	High cost and medium complexity	Medicum cost and high complexity	Low cost and low complexity

3.1. *Electric Vehicle Charging Infrastructure in India*

Currently, in India, the electric vehicle charging infrastructure is mainly divide into Battery Swapping and Charging the Battery which is further divided into AC (slow Charging and DC (fast) Charging. In the case of alternating current charging, alternatin current is provided to the electric car's onboard charger, which converts AC to DC and i then used to charge the vehicle battery. DC charging, on the other hand, converts AC t DC at the charge point and feeds DC straight to the car battery. Based on the

1. Home Charging: Home chargers commonly use a 230 V/15 A single-phase socket wit a maximum output capacity of 2.5 kW. Home charging is clearly an AC (Alternatin Current) charging method. The amount of electricity used is factored into the hom metering system. The time it takes to charge an electric vehicle is determined b the charging rate and the quantity of charge necessary (which is determined by th battery's usable capacity). Electric scooters can be charged in 2–3 h and electric cars i 6–7 h using home charging.
2. Public Charging: Public charging means charging the electric vehicle outside th home. For example charging the vehicle either in the supermarket, cinema hall, reta parks, etc.
3. Battery Swapping: The electric vehicles swap their drained batteries with full recharged batteries. The Battery Swapping concept decouples battery charging fror vehicle use, which benefits both the swapping station operator and the power systen No rapid charging is necessary, and electrical grid management is straightforward [11

3.2. *Types of Chargers*

1. Type 1 AC Charger: This is the most basic EV charger, which is used by some entr level electric vehicles. It may charge a vehicle slowly using an AC outlet or a hom charging system. It has a charging power of up to 220 volts and a maximum currer of 16 amps. It can handle up to 3 kW of single-phase input electricity. When usin these types of chargers, the vehicle must convert AC electricity to DC, which is time-consuming procedure.

2. Type 2 Charger: The Type 2 Charger can charge at a quicker rate and works with both AC and DC charging methods. These chargers are designed to work with three-phase power systems. In European charging stations, it's fairly frequent. With a 400 volt AC supply, it can handle input power ranging from 7.4 kW to 43 kW. These chargers are also commonly installed in EV owners' houses for faster charging periods because they are compatible with vehicles that use CCS connections.
3. CCS or Combined Charging System: With new-generation electric vehicles, a combination charging system plug, often known as a CCS type plug (or CCS Type 2), is becoming more widespread. These charging systems are capable of offering DC fast-charging for cars from commercial charging stations as well as standard charging from home charging stations. For DC rapid charging, the plug contains two additional contact points. Input power for most DC fast chargers is 50 kW, however, this type of socket can handle charge power of up to 350 kW.
4. CHAdeMo Charger: It was developed by Nissan, Tokyo Electric Power Company (TEPCO), Mitsubishi, Subaru, and Toyota and was first deployed in Japan [34]. This was one of the first fast-charging systems to be created, and it is now used in over 70 nations across the world. It can handle up to 50 kW of DC fast charging [35]. Newer automobiles, on the other hand, are converting to the CCS system since it is more versatile.
5. GB/T Charger Under the Bharat DC 001 standard, the Indian Government suggested the GB/T type charger for EVs. These chargers, which were erected by the government's Energy Efficiency Services Limited (EESL), are capable of DC fast-charging with a 10–15 kW output for low-power EVs. On the other hand, this sort of connector can handle capacities of up to 230 kW.

4. Proposed Space Efficient Multi-Level Charging Station Infrastructure Method

In the past decade, there has been a significant upward trend in the use of electric vehicles or EVs mainly because of the idea, that as this number goes higher we need infrastructure in place that can deliver the necessary power and electricity and also at the same time decreasing the operating costs of the said infrastructure. In metropolitan areas, creating extra infrastructure there is a necessity of space and it has become one of the biggest issues. Therefore, to provide a temporary charging station, especially at the shopping mall, big shops, etc. Hence, this can be achieved with a multi-agent system (MAS) and multi-level charging system, especially in metropolitan cities all over the world.

We developed MAS-based multi-level charging system architecture as shown in Figure 1.

Multi-level charging will help in charging more vehicles in the smaller area. This system involves stacking vehicles layer by layer thus requiring less space and greatly helping in reducing traffic congestion. Making the system automated would require less manual intervention, thus reducing damage and cost. Figure 2 shows that the EV or battery is connected as a load to each of the busses in the 33-bus power grid network. Each bus is also connected with a Multi-Agent System (MAS) and this MAS with the help of IoT Technology transmits the data from the bus and over to the cloud and it is further routed to the main/central MAS moves the data to the edge computing technology involved. When the data are transmitted to the central MAS it is processed through edge computing, this way we can save on computation costs and also use minimal hardware, this can make the whole system much cheaper. In this case, there are a few conditions that must be met to declare the bus stable or unstable, that decision is taken here.

Figure 3 shows the different components used while designing the architecture of the proposed system, such as socket, charging kiosk with cable and connector gun, and charging kiosk connected with the socket.

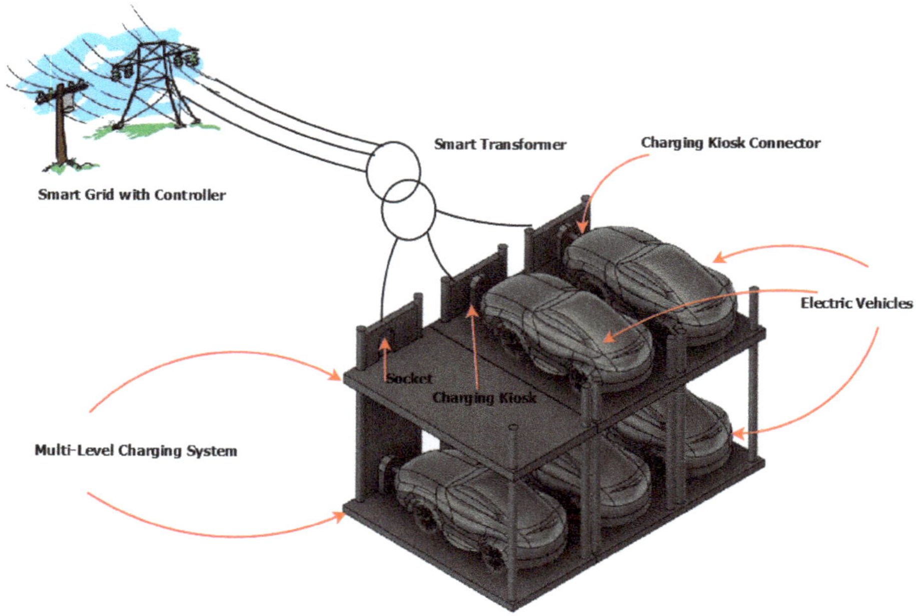

Figure 1. Multi-Level EVs Charging Design CAD Model.

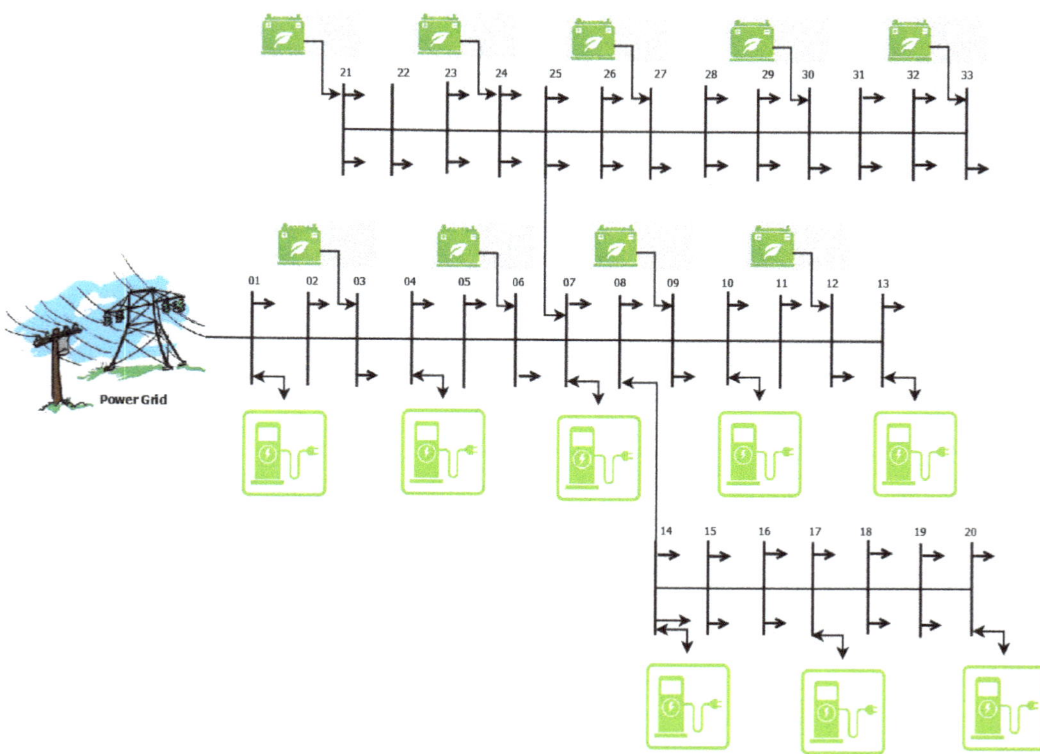

Figure 2. EV Charging Stations with 33-bus Distribution Network.

(a) (b) (c)

Figure 3. Charging System Entities. (**a**) Socket. (**b**) Charging Kiosk with cable and connector gun. (**c**) Charging kiosk connected with socket.

Implementation of an EV Charging Station with 33-Bus Distribution Network

Figure 2 is designed and implemented in Matlab/Simulink environment by adhering to the following parameters:

- Each bus is connected to the agent, and this agent, in turn, connects to the charging station. Therefore, the agent analysis its respective bus and shares the information with the charging station.
- Every EV/EV battery connected to the station has 40% of the initial state of charge (SoC).
- Every EV has the same battery parameters.
- Each charging station is equipped with similar chargers in order to linearize the distributed observations.
- The simulation assumes a charging station to be operating at 100% capacity in order to make distinct observations.
- Residential/Industrial loads are always connected and running at full power.
- A single three-phase power source powers the entirety of the grid.

Observations and Inferences from IEEE 33-Bus Distribution Network

The following observations made for the above system simulation and their results obtained are presented are as follows:

- Utility grid simulation is conducted with only residential/industrial loads and the charging station kept off. For this scenario, we have observed a stable 3-phase voltage of 9.7 kV as shown in Figure 4a in the grid.
  Figure 4b shows the reactive power of 11 MW in the system. It is absorbed by harmonic filters and some residential/industrial loads.
- When all CSs are operated at total capacity and no DC Fast Chargers are operated, there is a voltage drop to 8.88 kV as shown in Figure 4c along with the injection of reactive power as shown in Figure 4d in the system which scaled up to −8.5 MW.
  Total Harmonic Distortion (THD) has been observed that is 1.8% as shown in Figure 5a using the Powergui FFT analysis tool.
- When the simulated charging station's power draw crossed the 50% capacity threshold, the charger switched to energy stored in battery banks by enabling DCFCs at half time of simulation, and there is a significant drop of reactive power to −9.6 MW has been observed as shown in Figure 5b and a considerable increase in voltage of 9.25 kV has been observed as shown in Figure 5c. This increased the grid's power quality significantly. A substantial decline in THD that is, 1.17%. This decline is due to the disconnection of chargers at the instance of time as shown in Figure 5d.

- To reduce harmonics, we developed two advanced doubly-tuned passive harmoni filters in the grid. These filters consist of a circuit formed by inductance, capacitanc and resistances. The intended design was shown to observe an optimal amount reactive power from the grid which decreases the harmonics distortion. By usir this, we observed a drop to 0.07% in THD as shown in Figure 6 as compared to 1.8 THD without a filter. This falls within the range of acceptable limits of both THD ar reactive power as shown in Figure 5a.

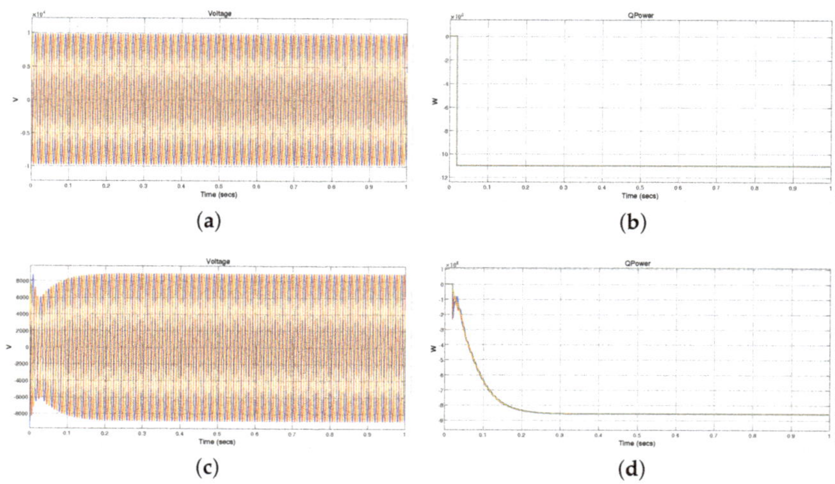

Figure 4. Voltage and Reactive power in utility grid when operated with all chargers and withou chargers. (**a**) Voltage in utility grid when operated without chargers. (**b**) Reactive power in utili grid when operated without chargers. (**c**) Voltage in utility grid when operated with all charger (**d**) Reactive power in utility grid when operated with all chargers.

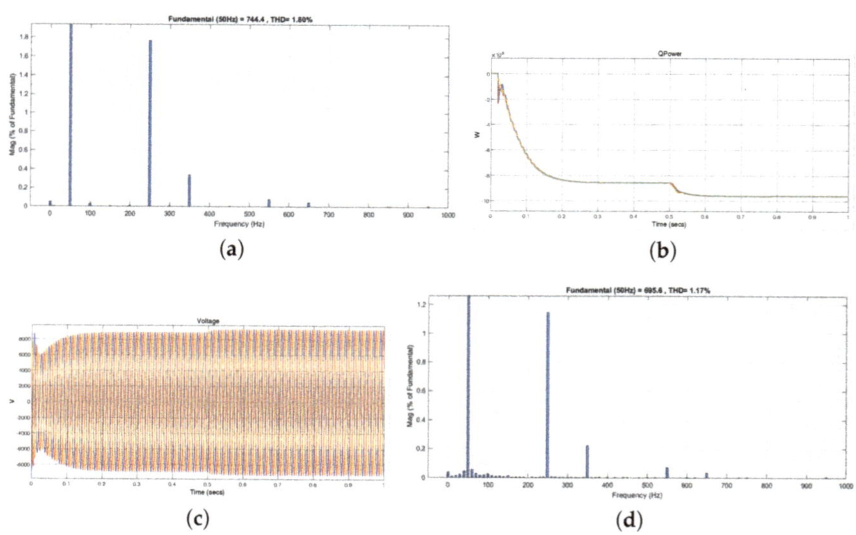

Figure 5. Harmonics, Reactive power and voltage in utility grid. (**a**) Harmonics when all chargers ar operated. (**b**) Reactive power in utility grid when CSs operated with DCFCs. (**c**) Voltage in utilit grid when CSs operated with DCFCs. (**d**) Harmonics in utility grid when CSs operated with DCFC

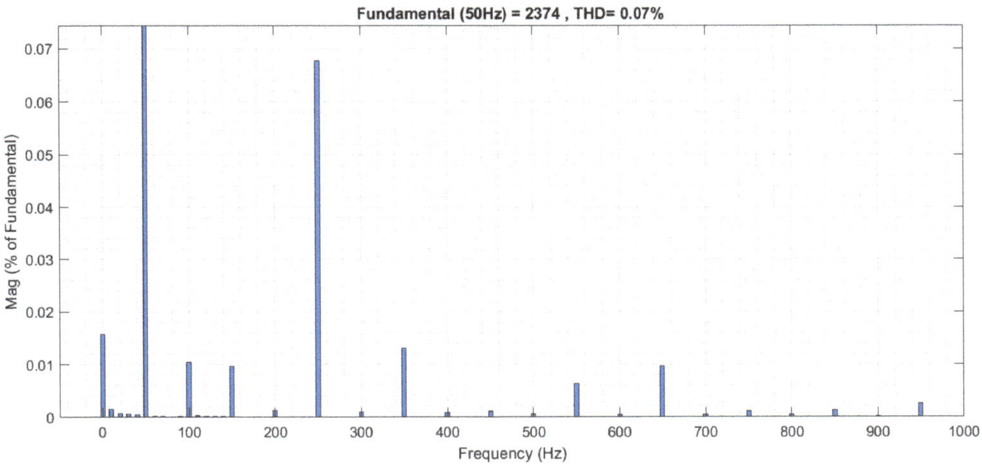

Figure 6. Harmonics in utility grid when operated with all chargers and harmonic filters.

Figure 7a shows the simulation results when there is no multiagent system (MAS) in the power distribution or grid network which is resulted in the instability at bus 20.

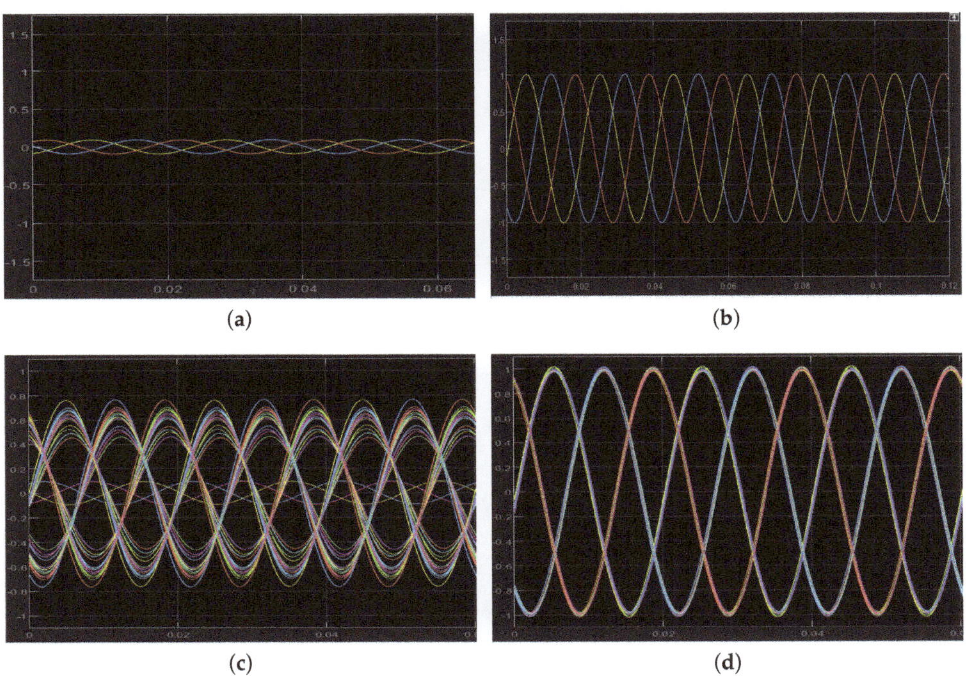

Figure 7. Stability and instability of 33-busses with and without multiagent system. (**a**) Unstable bus 20: Without MAS deployment in it. (**b**) Stable bus 20 after MAS deployment in it. (**c**) When one of the buses is unstable: Without-MAS deployment in the system. (**d**) 33-busses are stable after deployment of MAS in the system.

Figure 7b shows a stable bus that is due to the deployment of MAS at bus 20. Figure 7c shows when we consider all 33-bus in the simulation, one of the bus instability results in the whole power distribution network instability. Figure 7d shows the simulation results when we deploy the MAS to all 33-bus in the distribution network resulting in the stabilit of the system.

5. Cost Analysis of Multi-Level Charging Station Infrastructure

The cost estimation of the system depends on various factors for example the set-u city location, government norms in that area, the sizing of the system, etc. In the below tabl we have taken the location as Jaipur city and we have taken our base case as the minimur infrastructure (charger connectors) suggested by the government of India (Table 3).

From Tables 3 and 4, the total cost of setting up the charging station can be divide into two parts. The first is the capital cost (CAPEX) as shown in Table 3 which is th one-time fixed cost and the second one is the operational cost (OPEX) as shown in Table The CAPEX comes out to be around Rs. 2,955,000 and the OPEX comes out to be aroun Rs. 972,000.

Table 3. CAPEX: Cost estimation of the proposed multi-level charging infrastructure.

Type of Charger	Number of Chargers in PCS	Power Output	Approx. Cost in Indian Rupees	Number of EVs That Can Be Charged Simulataneously	Maximum Power Sold to EVs per Day (24 h/day) kWh
CCS	1	50 kW	72,500	1	1200
CHAdeMO	1	50 kW	72,500	1	1200
Type 2 AC	1	22 kW	12,500	1	528
Bharat DC-001	1	15 kW	24,000	1	360
Bharat AC-001	1	9.9 kW	7000	3	237.6
Swap station	-	-	-	-	360
New electricity connection (2SO KVA), Transformer, Cabling, Panels, Breakers, and Energy meter	-	-	75,000	-	-
Civil works (Flooring, painting, Boards, Branding, Shed/covers, etc.	-	-	750,000	-	-
EVSE Management Software-integration with chargers and payment gateway	-	-	40,000	-	-
CCTV Camera Setup	-	-	30,000	-	-
Total CAPEX	-	-	2,955,000	-	3885.6

Table 4. OPEX: Cost estimation of the proposed multi-level charging infrastructure.

Type of Service	Cost in Indian Rupees
Technician's charges	150,000 for 6 months
Site maintenance staff	180,000 per year
Land lease rental (50,000 per month)	600,000 per year
Advertising (3000 per month)	36,000 per year
Total cost	972,000 + EVSE software fees for 1st year. 822,000 + EVSE software fees for 2nd year

6. Stress Analysis of the Designed Structure of Multi-Level Charging Station Infrastructre

The structure is made up of mild steel of which Young's modulus is 220,000 MPa, Yield Strength is 207 MPa and Ultimate Tensile Strength is 345 MPaThe bottom-most surface (1st level) of the system is assumed to be a constraint to the ground and the sidewalls are also assumed to be fixed. The kerb weights of the available electric vehicles range from 1235 N (Tata Tigore EV) to 2595 N (Audi e-tron). So for the stress analysis, we have taken the weight of the car (or load) to be 3000 N.

Numerical analysis shows that the maximum stress in the structure is 68.53 MPa as shown in Figure 8a and the maximum displacement is 0.06716 mm as shown in Figure 8b.

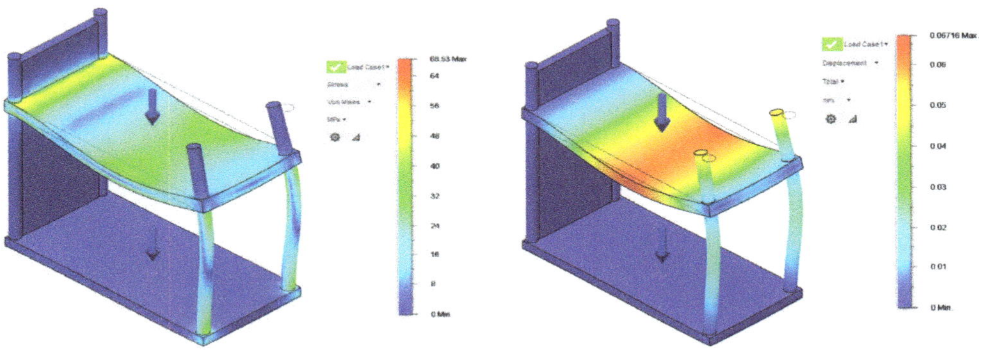

(a) (b)

Figure 8. Stress Analysis of the multilevel charging station. (**a**) Von-Mises Stress Distribution of the designed structure. (**b**) Maximum Displacement of the designed structure.

7. Multi-Level EV Charging Station Infrastructure Model

In this section, we discuss the mathematical modeling of the proposed multi-level EV charging station infrastructure, the optimization model, and to solve the optimization problem GA algorithm is developed.

7.1. Analytical Model

The conceptual framework of EVs queuing model at multi-level EV charging station infrastructure is shown in Figure 9. The state transition diagram of EVs at the charging station is shown in Figure 10. The stochastic model is based on the "$M/M/S/K$" queuing model, where the first M indicates the Poisson distribution for EV arrivals, the second M indicates the Exponential distribution for service time at the charging station, S represents a maximum number of charging sockets at charging stations, and K represents the length of the queue at charging station. In this queuing model, all EVs charging stations have plug-in sockets with identical characteristics, queue access or charging socket service is based on first come first service (FCFS), and the EVs service time depends upon the number of EVs waiting and the amount of time the station takes to charge the vehicle.

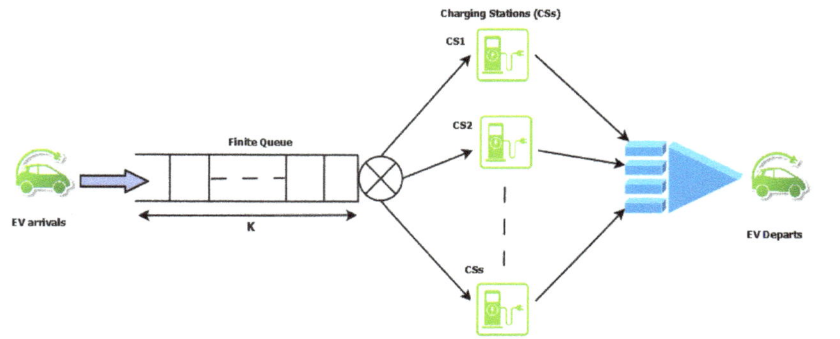

Figure 9. Conceptual Framework of EVs queuing model at Multi-level EV Charging Statio
Infrastructure.

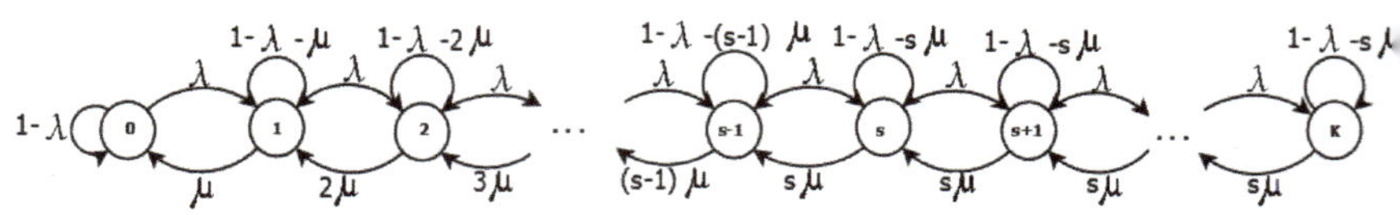

Figure 10. Stochastic EVs queuing model at Multi-level Charging Station Infrastructure: Discre
Time Markov Chain $M/M/S/K$ Model.

In the transition diagram as shown in Figure 10, there are two major functions th
will be carried out

1. Arrival rate (λ): $\lambda_i = \lambda \forall i \epsilon K$.
2. Service rate (μ): $\mu_i = \begin{cases} i\mu & 0 \leq i \leq S \\ S\mu & S < i \leq K \end{cases}$

The steady-state probabilities of EVs at multi-level charging stations are denoted as P
and are given as follows:

$$P_n = \begin{cases} \frac{(\rho)^n}{n!} P_0, 0 \leq n < s \\ \\ \frac{(\rho)^n}{S^{(n-s)} n!}, s \leq n \leq K \end{cases} \quad (1)$$

where $\rho = \lambda/\mu$ is the utilization of charging points at the charging station. The basi
condition in the Little's theorem and Markov chain is the sum of all the probabilities i
the system is equal to one, that is, $\sum_{n=0}^{K} P_n = 1$. Using the above condition we can get P
as follows:

$$P_0 = \left[\sum_{n=0}^{s-1} \frac{(\rho)^n}{n!} + \sum_{n=s}^{K} \frac{(\rho)^n}{S^{(n-s)} S!} \right]^{-1} \quad (2)$$

The average number of EVs in the system is given as

$$EV^{avg} = \frac{P_0}{S!} \left(\sum_{n=0}^{s-1} n\rho^n + S^s \sum_{n=s}^{K} n \frac{\rho^n}{s^n} \right) \quad (3)$$

The probability of the charging station is full, that is, the probability of blocking th
EVs to get charged at the charging station which is an equal probability that the chargin
station is in the state K, and is given as

$$P_B = P_m = \frac{\rho^K}{K!} \cdot P_0 \quad (4)$$

$$P_B = \frac{\rho^K}{K!}\left[\sum_{n=0}^{s-1}\frac{(\rho)^n}{n!}\right]^{-1} \quad (5)$$

$$P_B = \frac{\rho^K/K!}{\sum_{n=0}^{K}\frac{(\rho)^n}{n!}} \quad (6)$$

The above Equation is called as Erlang-B formula used for blocking the EVs from charging at the multi-level charging station.

7.2. Optimization Model

The multilevel EV charging station infrastructure planning optimization model depends upon the factors, such as initial construction cost, vehicle retention, EV charging cost, characteristics of the battery, etc.

C_{CS}^{ML} is the objective function that minimizes the total cost of multi-level charging station infrastructure, that is initial infrastructure cost and EV users' charging cost.

$$C_{CS}^{ML} = minimize \sum_i \sum_j C_j^{init} Y_{ij} + C' \sum_k \sum_i D_k X_{ki} dist_{ki} \quad (7)$$

subject to:

$$\sum_i X_{ki} = 1, \forall k \epsilon K \quad (8)$$

$$X_{ki} \leq Y_{ij}, \forall k \epsilon K, i \epsilon I, j \epsilon J \quad (9)$$

$$\sum_k D_k X_{ki} \leq \sum_j S_j Y_{ij}, \forall i \epsilon I \quad (10)$$

$$\sum_j Y_{ij} \leq 1, \forall i \epsilon I \quad (11)$$

$$\sum_i \sum_j Y_{ij} = Q \quad (12)$$

$$Y_{ij} \epsilon \{0,1\}, i \epsilon I, j \epsilon J \quad (13)$$

$$X_{ki} \epsilon \{0,1\}, \forall k \epsilon K, i \epsilon I. \quad (14)$$

where K is the set of demands of charging points, $k\epsilon K$ refers to a charging point, I is the candidate charging point set, $i\epsilon I$ refers to a candidate charging point, D_k is the demand quantity at charging point k, C_j^{init} is the EV charging station initial construction cost of level j, C' is the EV charging station's unit cost of user, S_j is the mth level serviceability of the charging station, $dist_{ki}$ is the distance between charging point demand k and candidate charging point i, Q is the number of charging station to be constructed, Y_{ij}, it is 1 when the charging point i of charging station's level j, otherwise it is 0, X_{ki} its value is 1 when the users at charging point k and receives service at candidate charging point i, otherwise 0. In the above optimization model, Equation (8) indicates the respective EV users arrive at their respective charging points in the charging station at a particular time. Equation (9) is used to make a candidate's charging point into the function using prerequisite conditions. Equation (10) is the condition of the charging demands at one station should not exceed its maximum service rate. Equation (11) is developed for constructing only one of the EV charging station levels. Equation (12) is the constraint of the total amount of charging stations that need to be developed. Equations (13) and (14) are variables required for the possible zones.

7.3. GA for Multi-Level Charging Station Infrastructure

To solve the problems described in the previous subsection in Equations (7)–(14), a genetic algorithm (GA) is proposed. The solution of the GA includes the following steps:

1. Initialization of population: We have randomly generated the initial population. Here, binary coding is employed, that is 1 means the point is selected to develop the charging station, otherwise 0.
2. Estimation of each individual fitness metric. During each scheme performance evaluation, extra work needs to be taken in order to complete the solution, that is all the charging station's charging demand points should be allocated to the potential station to end the evaluation process.
3. Estimating the next generation. For generating qualified offspring, especially the designed crossover and mutation operators are employed.
4. Convergence: There are two convergences, either the generated the best fitness out of 50 generations or the limit has reached, the developed algorithm will produce the best individual of that generation as the final output.

8. Simulation and Results Analysis

In this section, we present the simulation results of the proposed system's $M/M/S/K$ queuing model and objective function using a genetic algorithm.

We have developed and simulated the queuing model in MATLAB. In the simulation the EVs arrival flow fluctuation is modeled using a Poisson distribution. We presume that all EVPSS have a level 3 plug-in, which is the fastest type of EV charger. We have assumed that the time it takes for the smart grid to respond to each EV is zero and all vehicles must be satisfied once the total charge time has expired. In addition, the charging station's number of charging kiosks (servers) is expected to be 20.

The simulation demonstrates the performance of the queuing model with varying performance parameters. The blocking probability of the EVs with varying arrival rates service rates, and the number of charging points available. The number of charging points = NC = {3, 4, 5} for DC fast charging infrastructure (DFCI), On-Road Charging Infrastructure (ORCI), and Multi-level Charging Infrastructure (MLCI) with arrival rates $\lambda = \{7,5,3\}$ and service rates $\mu = \{1,1.5,2\}$. As shown in Figure 11, increasing the arrival of DFCI leads to an increase in the probability of blocking and vice versa. We can also notice that the variations are more in the ORCI than in the MLCI. Figures 12 and 13 shows the similar effects. We have solved the objective function using the Genetic Algorithm, the result is shown in Figure 14. The Figure 14 shows that convergence is quick and the solution is optimal.

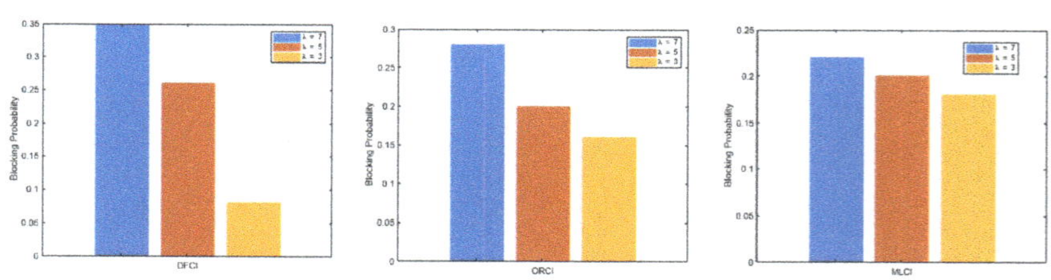

(a) (b) (c)

Figure 11. EVs Blocking Probability at Charging Station with Varying Arrival Rates. (**a**) DC Fast Charing Infrastructure. (**b**) On Road Charging Infrastructure. (**c**) Multi-level Charging Infrastructure.

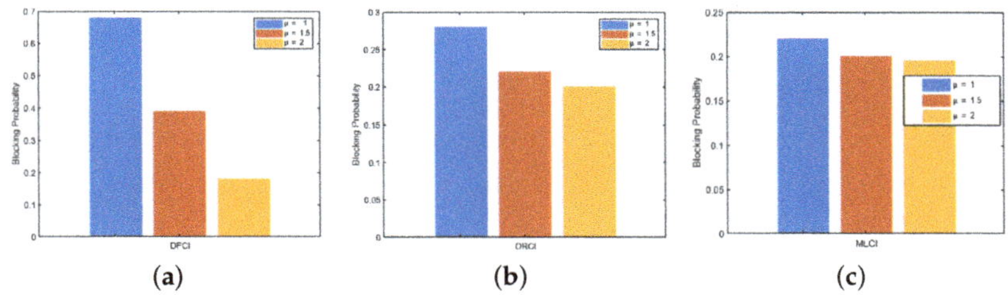

Figure 12. EVs Blocking Probability at Charging Station with Varying Service Rates. (**a**) DC Fast Charing Infrastructure. (**b**) On Road Charging Infrastructure. (**c**) Multi-level Charging Infrastructure.

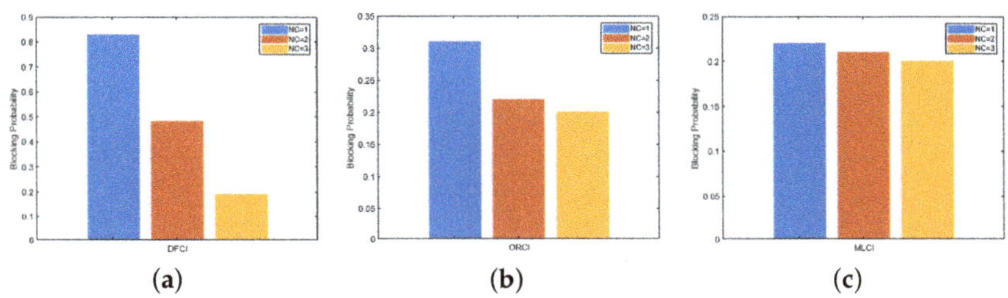

Figure 13. EVs Blocking Probability at Charging Station with Varying Number of chargers. (**a**) DC Fast Charing Infrastructure. (**b**) On Road Charging Infrastructure. (**c**) Multi-level Charging Infrastructure.

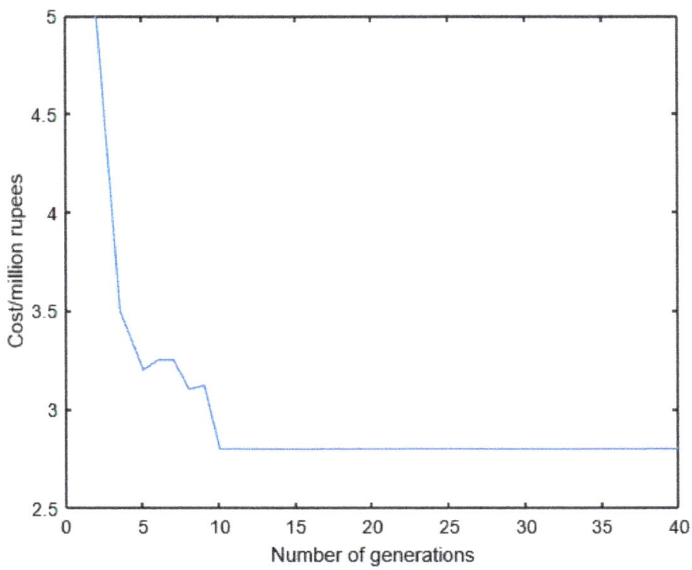

Figure 14. GA-based optimal solution.

Table 5 shows the comparisons between existing charging stations (such as on-board, off-board, battery swapping station) and multilevel charging station.

Table 5. Comparisons between Existing charging stations and Multilevel charging station.

Type of Charging Station	Safety	Traffic	Waiting Time	Cost	Complexity
On-board charging station	Less	High	More	Low	Low
Off-board charging station	Medium	High	More	High	High
Battery swapping station	Less	Low	Medium	Medium	High
Multi-level charging station	High	Low	Low	Low	Low

9. Conclusions

This paper developed the space-efficient multi-level charging station infrastructu method for metropolitan cities. A novel design of the multi-level charging station infra tructure has been discussed. The developed design is integrated with the smart grid. 33-b simulation has been conducted, and a MAS is deployed for each bus to control the grid stability. Discussed the observations and inferences of 33-bus with MAS and proved th the system's stability is maintained with MAS. The cost analysis has been conducted for th new design of the multi-level charging station infrastructure. In addition, the stress analys of the designed structure is evaluated and verified the maximum stress in the structu is 68.53 MPa and the maximum displacement is 0.06716 mm. Moreover, developing th mathematical model of multi-level EV charging station infrastructure using the $M/M/S/$ queuing model and critical results analysis have also been carried out. The limitation this study is that the proposed approach is designed for dense metropolitan areas wit high populations, such as Indian metropolitan cities. In the future, the new approach ca be designed for less densely populated areas.

Author Contributions: Conceptualization, S.C. and S.K.; methodology, S.C. and S.R.M.Z.; validatic S.C. and A.A.; writing—original draft preparation, S.C. and S.K.; writing—review and editin S.C. and S.K.; supervision, S.K. All authors have read and agreed to the published version the manuscript.

Funding: The authors received no external funding.

Institutional Review Board Statement: Not applicable.

Informed Consent Statement: Not applicable.

Data Availability Statement: Data can be made available of request.

Acknowledgments: This work was supported by the Ministry of Science and Higher Education the Russian Federation (Government Order FENU-2020–0022).

Conflicts of Interest: The authors declare no conflict of interest.

Abbreviations

EV	Electric Vehicle
SG	Smart Grid
AMI	Advanced Metering Infrastructure
DMS	Distributed Management System
PLC	Power Line Communication
UWB	Ultra-Wide Band
UIC	Universal Inductive Charger
MAS	Multi-Agent System

References

1. Colak, I.; Sagiroglu, S.; Fulli, G.; Yesilbudak, M.; Covrig, C.F. A survey on the critical issues in smart grid technologies. *Renew. Sustain. Energy Rev.* **2016**, *54*, 396–405. [CrossRef]
2. Sagiroglu, S.; Terzi, R.; Canbay, Y.; Colak, I. Big data issues in smart grid systems. In Proceedings of the 2016 IEEE International Conference on Renewable Energy Research and Applications (ICRERA), Birmingham, UK, 20–23 November 2016; pp. 1007–1012.
3. Shadare, A.E.; Sadiku, M.N.; Musa, S.M. BElectromagnetic compatibility issues in critical smart grid infrastructure. *IEEE Electromagn. Compat. Mag.* **2017**, *6*, 63–70. [CrossRef]
4. Young, J.R. Smart Grid Technology in the Developing World, 2017, Honors Projects 68. Available online: https://digitalcommons.spu.edu/honorsprojects/68 (accessed on 25 July 2022).
5. Ellis, M. Smart grid: The components and integrating communication. In Proceedings of the 2012 IEEE Green Technologies Conference, Tulsa, OK, USA, 19–20 April 2012; pp. 1–6.
6. Hu, J.; Saleem, A.; You, S.; Nordström, L.; Lind, M.; Østergaard, J. A multi-agent system for distribution grid congestion management with electric vehicles. *Eng. Appl. Artif. Intell.* **2015**, *38*, 45–58. [CrossRef]
7. Vallejo, D.; Albusac, J.; Glez-Morcillo, C.; Castro-Schez, J.J.; Jiménez, L. A multi-agent approach to intelligent monitoring in smart grids. *Int. J. Syst. Sci.* **2014**, *45*, 756–777. [CrossRef]
8. Bonaiuto, V.; Sargeni, F. A Matlab Simulink model for the study of smart grid—Grid-integrated vehicles interactions. In Proceedings of the 2017 IEEE 3rd International Forum on Research and Technologies for Society and Industry (RTSI), Modena, Italy, 11–13 September 2017; pp. 1–6.
9. Falvo, M.C.; Foiadelli, F.E.D.E.R.I.C.A. Preliminary analysis for the design of an energy-efficient and environmental sustainable integrated mobility system. In Proceedings of the IEEE PES General Meeting, Minneapolis, MN, USA, 25–29 July 2017; pp. 1–7.
10. Roscia, M.; Longo, M.; Lazaroiu, G.C. Smart City by multi-agent systems. In Proceedings of the 2013 International Conference on Renewable Energy Research and Applications (ICRERA), Madrid, Spain, 20–23 October 2018; pp. 371–376.
11. Marra, F.; Sacchetti, D.; Træholt, C.; Larsen, E. Electric vehicle requirements for operation in smart grids. In Proceedings of the 2nd IEEE PES International Conference and Exhibition on Innovative Smart Grid Technologies, Manchester, UK, 5–7 December 2011; pp. 1–7.
12. Vermesan, O.; Blystad, L.C.; Hank, P.; Bahr, R.; John, R.; Moscatelli, A. Smart, connected and mobile: Architecting future electric mobility ecosystems. In Proceedings of the 2013 Design, Automation & Test in Europe Conference & Exhibition (DATE), Grenoble, France, 18–22 March 2018; pp. 1740–1744.
13. Lin, S.Y.; Chen, J.F. Distributed optimal power flow for smart grid transmission system with renewable energy sources. *Energy* **2013**, *56*, 184–192. [CrossRef]
14. Li, Z.; Wang, J.; Sun, H.; Guo, Q. Transmission contingency analysis based on integrated transmission and distribution power flow in smart grid. *IEEE Trans. Power Syst.* **2015**, *30*, 3356–3367. [CrossRef]
15. Yang, Q.; Li, D.; Yu, W.; Liu, Y.; An, D.; Yang, X.; Lin, J. Toward data integrity attacks against optimal power flow in smart grid. *IEEE Internet Things J.* **2017**, *4*, 1726–1738. [CrossRef]
16. Liu, G.; Liu, K.; Shi, D.; Zhu, W.; Wang, Z.; Chen, X. Graph computation and its applications in smart grid. In Proceedings of the 2017 IEEE International Congress on Big Data (BigData Congress), Honolulu, HI, USA, 25–30 June 2017; pp. 507–510.
17. Owerko, D.; Gama, F.; Ribeiro, A. Optimal power flow using graph neural networks. In Proceedings of the ICASSP 2020–2020 IEEE International Conference on Acoustics, Speech and Signal Processing (ICASSP), Barcelona, Spain, 4–8 May 2020; pp. 5930–5934.
18. Sureshkumar, K.; Ponnusamy, V. Hybrid renewable energy systems for power flow management in smart grid using an efficient hybrid technique. *Trans. Inst. Meas. Control* **2020**, *42*, 2068–2087. [CrossRef]
19. Vogt, M.; Marten, F.; Braun, M. A survey and statistical analysis of smart grid co-simulations. *Appl. Energy* **2018**, *222*, 67–78. [CrossRef]
20. Mohagheghi, E.; Alramlawi, M.; Gabash, A.; Li, P. A survey of real-time optimal power flow. *Energies* **2018**, *11*, 3142. [CrossRef]
21. Sureshkumar, K.; Ponnusamy, V. Power flow management in micro grid through renewable energy sources using a hybrid modified dragonfly algorithm with bat search algorithm. *Energy* **2019**, *181*, 1166–1178. [CrossRef]
22. Montoya, O.D.; Garrido, V.M.; Gil-González, W.; Grisales-Noreña, L.F. Power flow analysis in DC grids: Two alternative numerical methods. *IEEE Trans. Circuits Syst. II Express Briefs* **2019**, *66*, 1865–1869. [CrossRef]
23. Ustun, T.S.; Aoto, Y. Analysis of smart inverter's impact on the distribution network operation. *IEEE Access* **2019**, *7*, 9790–9804. [CrossRef]
24. Ni, Z.; Paul, S. A multistage game in smart grid security: A reinforcement learning solution. *IEEE Trans. Neural Netw. Learn. Syst.* **2019**, *30*, 2684–2695. [CrossRef]
25. Pohl, O.; Rewald, F.; Dalhues, S.; Jörke, P.; Rehtanz, C.; Wietfeld, C.; Kubis, A.; Tamgue, R.K.; Kirsten, D. Advancements in distributed power flow control. In Proceedings of the 2018 53rd International Universities Power Engineering Conference (UPEC), Glasgow, UK, 4–7 September 2018; pp. 1–6.
26. Hossain, E.; Khan, I.; Un-Noor, F.; Sikander, S.S.; Sunny, M.S.H. Application of big data and machine learning in smart grid, and associated security concerns: A review. *IEEE Access* **2019**, *7*, 13960–13988. [CrossRef]
27. Kongjeen, Y.; Bhumkittipich, K. Impact of plug-in electric vehicles integrated into power distribution system based on voltage-dependent power flow analysis. *Energies* **2018**, *11*, 1571. [CrossRef]

28. Kappagantu, R.; Daniel, S.A. Challenges and issues of smart grid implementation: A case of Indian scenario. *J. Electr. Syst. In* *Technol.* **2018**, *5*, 453–467. [CrossRef]
29. Faheem, M.; Shah, S.B.H.; Butt, R.A.; Raza, B.; Anwar, M.; Ashraf, M.W.; Ngadi, M.A.; Gungor, V.C. Smart grid communicatio and information technologies in the perspective of Industry 4.0: Opportunities and challenges. *Comput. Sci. Rev.* **2018**, *30*, 1–3 [CrossRef]
30. Abdel-Nasser, M.; Mahmoud, K.; Kashef, H. A novel smart grid state estimation method based on neural networks. *IJIMAI* **201** *5*, 92–100. [CrossRef]
31. Gupta, S.; Kazi, F.; Wagh, S.; Singh, N. Analysis and prediction of vulnerability in smart power transmission system: A geometrica approach. *Int. J. Electr. Power Energy Syst.* **2018**, *94*, 77–87. [CrossRef]
32. Artale, G.; Cataliotti, A.; Cosentino, V.; Di Cara, D.; Fiorelli, R.; Guaiana, S.; Panzavecchia, N.; Tinè, G. A new low cost power lin communication solution for smart grid monitoring and management. *IEEE Instrum. Meas. Mag.* **2018**, *21*, 29–33. [CrossRef]
33. Agundis-Tinajero, G.; Segundo-Ramírez, J.; Visairo-Cruz, N.; Savaghebi, M.; Guerrero, J.M.; Barocio, E. Power flow modeling c islanded AC microgrids with hierarchical control. *Int. J. Electr. Power Energy Syst.* **2019**, *105*, 28–36. [CrossRef]
34. Lei, X.; Yang, Z.; Yu, J.; Zhao, J.; Gao, Q.; Yu, H. Data-driven optimal power flow: A physics-informed machine learning approac *IEEE Trans. Power Syst.* **2020**, *36*, 346–354. [CrossRef]
35. Muhammad, Y.; Khan, R.; Raja, M.A.Z.; Ullah, F.; Chaudhary, N.I.; He, Y. Design of fractional swarm intelligent computing wit entropy evolution for optimal power flow problems. *IEEE Access* **2020**, *8*, 111401–111419. [CrossRef]

mathematics

MDPI

eview

/ariable-Speed Wind Turbines for Grid Frequency Support: \ Systematic Literature Review

ksher Bhowon [1], Khaled M. Abo-Al-Ez [2,*] and Marco Adonis [3,*]

1 Department of Electrical, Electronic, and Computer Engineering, Cape Peninsula University of Technology, Cape Town 7535, South Africa
2 Department of Electrical, Electronic, and Computer Engineering, Centre for Power Systems Research (CPSR), Cape Peninsula University of Technology, Cape Town 7535, South Africa
3 Department of Electrical, Electronic, and Computer Engineering, Centre for Distributed Power and Electronics Systems (CDPES), Cape Peninsula University of Technology, Cape Town 7535, South Africa
* Correspondence: aboalezk@cput.ac.za (K.M.A.-A.-E.); adonism@cput.ac.za (M.A.)

Abstract: As the finite nature of non-renewable energy resources is realised and climate change concerns become more prevalent, the need to shift to more sustainable forms of energy such as the adoption of renewable energy has seen an increase. More specifically, wind energy conversion systems (WECS) have become increasingly important as a contribution to grid frequency support, to maintain power at the nominal frequency and mitigate power failures or supply shortages against demand. Therefore, limiting deviations in frequency is imperative and, thus, the control methods of WECS are called to be investigated. The systematic literature review methodology was used and aimed at investigating these control methods used by WECS, more specifically variable-speed wind turbines (VSWT), in supporting grid frequency as well as the limitations of such methods. The paper identifies these to be de-loading, energy storage systems and emulated inertial response. Further classification of these is presented regarding these control methods, which are supported by literature within period of 2015–2022. The literature indicated a persistent interest in this field; however, a few limitations of VSWTS were identified. The emulated inertial response, specifically using a droop control-based frequency support scheme, was the primary means of providing frequency support. This systematic literature review may be limited by the number of papers selected for the study. Results and conclusions will not only be useful for WECS development but also in assisting with the security of the transmission grid's frequency stability. Future work will focus on further studying the limitations of WECS providing frequency support.

Keywords: variable-speed wind turbine (VSWT); frequency support; frequency regulation; systematic literature review (SLR); wind energy conversion system (WECS); renewable energy

MSC: 37M05

itation: Bhowon, A.; Abo-Al-Ez, .M.; Adonis, M. Variable-Speed 'ind Turbines for Grid Frequency ıpport: A Systematic Literature eview. *Mathematics* **2022**, *10*, 3586. :tps://doi.org/10.3390/ ath10193586

cademic Editor: Xavier Blasco

eceived: 29 August 2022
ccepted: 27 September 2022
ublished: 1 October 2022

1. Introduction

The transmission network, commonly referred to as the grid, is the high-voltage system that interconnects and transfers electrical energy from the generation network of power plants to the distribution network. The nominal frequency of the transmission network in South Africa is 50 Hz [1]. The frequency of a transmission network or grid is maintained at its nominal value by the balance between generation and consumption. In other words, a generator is required to either increase or decrease the active power produced in reaction to deviations around the nominal system frequency [2]. A significant deviation in the frequency of the transmission network may lead to the instability of the system or damage to connected devices [3]. More specific to the topic of this paper, a deviation too far below the nominal frequency can cause generation units to fall out of step and cause out-of-step protection relays, disconnecting the generator units from the grid.

This will place further demand on the remaining generators and will potentially lead grid collapse if this frequency event continues.

Conventional rotating synchronous generators that are committed to the grid ad rotational inertia to the grid. This is due to their large rotating masses that are couple to the grid via an electro-mechanical interaction between the rotor and stator [4]. Th electro-mechanical interaction enables the synchronous generator to exchange (both abso and release) its kinetic energy with the grid, proportional to the rate of change of frequen (ROCOF) [5]. This is known as an inertial response and is described by the swing equ tion [6]. It acts to overcome the immediate imbalance between supply and demand of th network. The generator, at the instant of the disturbance, will convert the kinetic energ of the rotor to electrical energy, limiting the rate of change of frequency and frequenc nadir as the rotor slows down in the process. In power systems, a loss of generators, loa shedding or a 3–5% load change is considered a large disturbance [7]. Following the inerti response, the generator's governor will adjust the setpoint and bring the machine up speed again within a period in the order of seconds, thus providing primary frequenc regulation. Secondary frequency regulation mechanisms, such as generation redispatch an automatic generation control, occur within an order of tens of seconds to a few minutes.

Conversely, variable-speed wind turbines with back-to-back power electronic co verters provide no inertia to the power system, since the power electronic converte decouple the rotating mass of the variable-speed wind turbine (VSWT) from the grid [4 Consequently, the effective inertia of the power system begins to decline with the increas in wind power penetration. VSWTs, therefore, do not naturally respond to a system fr quency change. The operation of a VSWT is governed by its MPPT (maximum pow point tracking) algorithm so as to extract the maximum possible power from the win to convert to electrical power [8]. The decline in rotating inertia leads to an increase i speed of grid frequency dynamics and may lead to situations where traditional frequenc controllers become too slow, relative to the disturbance, to limit large frequency deviatio The problem of low grid inertia is more prevalent in isolated systems or systems with hig wind power penetration. When a frequency drop occurs in such a system, the system ca experience a large ROCOF and frequency nadir. This concern is further asserted by th stochastic nature of wind energy which can cause an appreciable imbalance between suppl and demand [9]. This is of concern to the frequency stability of such electrical networks.

To overcome the challenges associated with frequency stability, wind turbine gene ators (WTGs) need to implement frequency control systems to allow them to partake i the regulation of the power system frequency [10]. Modern technologies and new contr systems aid in the feasibility for wind power plants (WPPs) to achieve this. Escalate by the finite nature of fossil fuels, in conjunction with climate change, the reliance o such renewable energy systems is increasing globally. More specifically, South Africa installed wind turbine generating capacity has increased from 790 MW in 2015 to 1468 M in 2017 [11]. As a country that has signed the Paris Climate Accord (and is still an activ member at the time of writing), an agreement that endorses a limit of a 1.5 °C increase i global temperature, South Africa may be turning to renewable energy sources such as, bu not limited to, WTGs to decrease the dependency of the power system on fossil fuels [10]. is of importance to address the frequency stability concerns that may arise with an increas in wind power penetration.

This systematic literature review aims to evaluate the recent work, within the period c 2015–2022, of wind energy conversion systems for grid frequency support. More specificall it aims to review the methods in which this is achieved as well as provide an analytic view based on the frequency of research papers published annually pertaining to this fiel Thereafter, a conclusion can be formulated to determine whether wind energy conversio systems for grid frequency support have gained interest and relevance.

Structed as follows—Section 2 sets out the methodology used to conduct this stud which is followed by the findings thereof, presented in Section 3. Based on the finding

Section 4 identifies areas to be considered for future works within this field. Finally, Section 5 concludes this study.

2. Materials and Methods

A research methodology is the practical framework used to identify, select, process, and analyse information within a research study [12]. In the context of a research paper, this section allows the reader to critically evaluate the study's validity and reliability [12].

Specifically, this review adopts the basic systematic literature review methodology, as described by Kitchenham [13]. This method of review is often used in software engineering research with success due to its success and easy implementation into other fields such as economics and education [14]. The following actions were followed when conducting the research for this study:

1. The formulation of research questions;
2. Search process;
3. Inclusion criteria;
4. Exclusion criteria;
5. Quality assessment;
6. Data collection;
7. Data analysis.

As stipulated by the methodology, the first process involves the formulation of research questions from which the basis and focus of the study are determined. The research questions are rooted in identifying solutions and limitations to VSWTs providing grid frequency support. The two research questions investigated in this study include:

RQ1: What control methods are VSWTs using to support the grid frequency?

RQ2: What are the limitations of VSWTs in supporting the grid frequency?

The search process included the manual entry of simple search strings based on title, keywords and abstract. Given the relevance of IEEE to this research topic, IEEE Xplore Digital Library was the primary library used, whilst ScienceDirect and Wiley Online Library were included to supplement the search process.

The inclusion and exclusion criteria for this study are criteria used to refine the search. The inclusion criteria included all papers relevant to the keywords searched and period set by this study: 2015–2022.

Conversely, all papers outside this period were excluded. The exclusion criteria further excluded papers which were not written in English and/or were unpublished, including conference papers, reviews and case studies. In addition, duplicate papers and/or those not relevant to the topic were excluded. Furthermore, papers related to frequency support but not in the context of wind energy conversion systems or wind turbine generators were excluded. Table 1 summarizes the search criteria and filters applied to each of the digital libraries searched.

A secondary screening was applied to each paper that satisfied the keywords used during the search to determine if it addressed the research questions. Those papers that did not were excluded. Based on the refinement and screening criteria, a sum of 56 papers were included in this review. Figure 1 shows the distribution and number of the articles across libraries included in this study.

A consistent data collection and analysis process for each of the 56 papers was applied, with respect to the research questions posed. In this way, the data collected were classified and categorized based on the control methods used by VSWTs to support grid frequency and further evaluated by their limitations thereof.

Table 1. Search Criteria Applied.

Digital Library	Search Criteria
IEEE Xplore Digital Library	1. WECS Grid Frequency support a. Journals b. 2015–2022 2. Wind turbine Frequency support a. Journals b. 2015–2022 3. Variable-speed wind turbine frequency support a. Journals b. 2015–2022
ScienceDirect (Elsevier)	1. WECS Grid Frequency support a. Article b. Engineering c. 2015–2022 2. Wind turbine frequency support a. Article b. Engineering c. 2015–2022 3. Variable-speed wind turbine frequency support a. Article b. Engineering c. 2015–2022
Wiley Online Library	1. WECS Grid Frequency support a. Journals b. Energy c. Electrical and Electronics Engineering d. 2015–2022 2. Wind turbine Frequency support a. Journals b. Energy c. Electrical and Electronics Engineering d. 2015–2022 3. Variable-speed wind turbine frequency support a. Journals b. Energy c. Electrical and Electronics Engineering d. 2015–2022

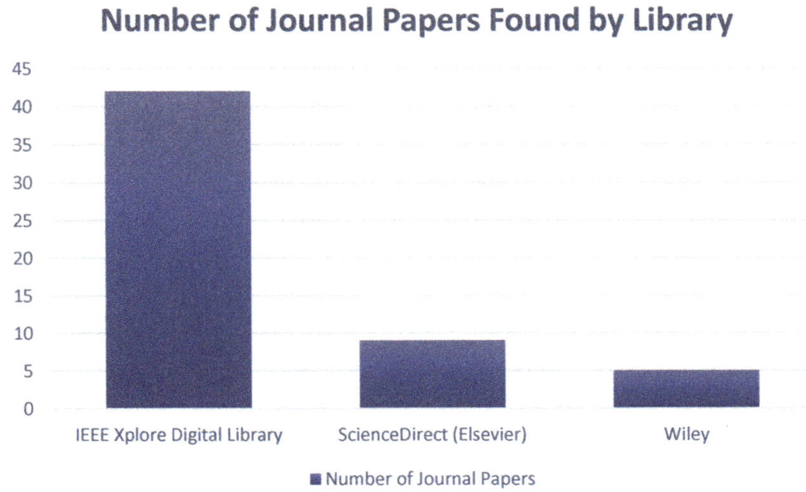

Figure 1. Article Sources and Number of Articles Found.

3. Results

The following section discusses the results of the literature review conducted in supporting the two research questions posed by this paper. The results associated to each research question are sub-categorized and each paper's contribution is discussed thereunder.

3.1. RQ1: How Are VSWTs Supporting the Grid Frequency?

Wind energy conversion systems differ from conventional synchronous generators since they do not have an inertial response to naturally suppress frequency disturbances by exchanging the kinetic energy of the rotor with the grid. The rate at which the grid frequency will change in an imbalance between supply and demand is related to system inertia [15]. As the number of wind turbine generators connected to the grid increases, the effective system inertia will decrease and, by extension, grid frequency dynamics will increase. The literature suggests that it is indeed possible for VSWTs to provide grid frequency support through either an inertial response or by partaking in grid frequency regulation. This research question (RQ1) is aimed at investigating the various methods and implementations of WECS supporting the grid frequency.

Recent works within the literature that were surveyed identified three prevailing methods of supporting the grid frequency. These include VSWTs which operate in a de-loaded state, implement an embedded energy storage and those which emulate an inertial response. Systems which emulate an inertial response either rely on the kinetic energy of the rotor or the electrostatic energy stored within the DC-link capacitor of the VSWT WTG itself. These classifications are illustrated by a hierarchical diagram in Figure 2. The surveyed works from the literature are classified by these three prevailing methods.

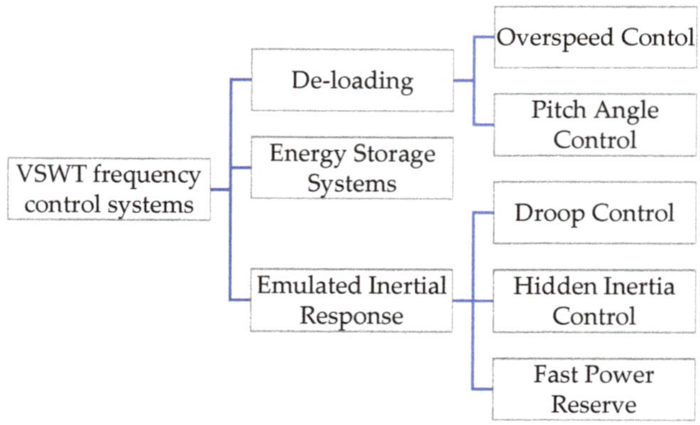

Figure 2. General Classification of VSWTs Frequency Control Systems [16].

3.1.1. De-Loading

Wind turbines operate along an MPPT curve to extract the maximum energy fro the wind. However, operating along this MPPT curve leaves no additional power reserv for frequency support [17]. To provide additional power reserve for frequency suppor a de-loading control system can be implemented. A de-loading control system operate a WTG at a suboptimal point, where the WTG deviates from the MPPT curve to provid additional active power for grid frequency support functions [18]. There are two differe methods to implement de-loading, namely overspeed control and pitch angle control [17 A summary of the papers classified as de-loading control, grouped by research area, shown in Table 2.

Table 2. Publications in De-loading Control for VSWTs.

Research Area	Reference	Year of Publication
Overspeed Control	[19]	2018
	[20]	2016
	[21]	2017
	[22]	2018
	[23]	2021
Pitch Angle Control	[24]	2020
	[25]	2017
	[26]	2015
	[27]	2016
	[28]	2021

The paper [19] evaluates pitch de-loading, kinetic energy recovery and WTG ove speeding, which enable WTGs to provide short-term frequency support, from an electric and mechanical perspective. The authors found that WTG overspeeding showed a fast response to a frequency event, accompanied by a higher initial power surge, while pitc de-loading provided a more sustainable support followed by a smoother recovery. WT overspeeding could have implications on the mechanical stability of the WTG; howeve the method wasted less energy when compared to pitch de-loading. An important aspe of kinetic energy recovery is that it does not deviate from the MPPT curve at norm operation. However, the system is most likely to suffer a second frequency dip as the roto of the WTGs recover to the nominal rotor speed. The simulation environment used wa MATLAB/Simulink.

The paper [20] presents an analytical model for short-term grid frequency report to evaluate the contribution of inertial and droop responses from a wind farm. The available mechanical power of a VSWT was approximated by a second-order polynomial which quantified the kinetic energy and wind power reserve over a wide range of operating points. Since the VSWT operating characteristics and conditions were now quantified, the frequency controller gains could be adjusted accordingly, thereby ensuring stable performance of the wind farm during frequency transients. To stimulate the frequency response of a wind farm during wind power fluctuations, a modified system frequency response model which considered the inertial and droop responses was developed. The efficacy of the model was verified by comparisons of the results with those obtained empirically.

Two virtual inertia control schemes, an equation-based scheme and an adaptive fuzzy logic-based scheme, are proposed in [21]. The proposed control schemes dynamically modulate the gains of the inertia control schemes based on system events to improve the primary frequency response of the WECS. The efficacy of the proposed schemes was validated in MATLAB/Simulink, while additional hardware-in-the-loop simulations on the OPAL-RT real-time simulator platform were presented to further substantiate the proposed schemes. It was concluded that the adaptive fuzzy logic-based scheme provided better frequency regulation when compared to the dynamic equation-based scheme.

The paper [22] proposes an active power control scheme which enables DFIGs (doubly fed induction generators) to partake in grid frequency support. Herein, a power surge-based co-ordination strategy provides an inertial response, while a power reserve control method assists with primary frequency control. The DFIG is, therefore, designed to provide both inertial and primary frequency support by adjusting the reserve amount whilst considering under- and over-frequency events. The effectiveness of this proposed control scheme was validated through case studies on a 181-bus WECC system with 50% wind penetration. The paper concludes by stating that future work will see the inclusion of secondary frequency control within the control scheme, which will make use of a co-ordinated strategy between DFIGs and conventional synchronous generators.

A droop control scheme for WTGs that uses rotor speed control for frequency support is proposed in [23]. The proposed control scheme does not add frequency-droop control signals to the WTG's power reference, but instead relies on the method of power tracking by adjusting the wind turbine's power tracking curve for primary frequency regulation. The efficacy of the proposed control scheme was validated through simulations in DIgSILENT Power Factory on a modified IEEE 39-bus power system. The results showed that the proposed scheme produces a linear frequency-droop response independent of the power tracking method used. This is an improvement over conventional frequency-droop controllers whose frequency-regulating responses are dependent on the method of power tracking used.

In [24], the authors propose an optimisation function which determines the operating point of de-loaded WTGs to reduce the magnitude of speed deviation and settling time caused by the inertial response and participation in frequency control. The optimisation function was integrated into a combined inertial and frequency control strategy for type-4 (fully rated power converter) WTGs, based on the concept of a virtual synchronous generator. The paper presents an analysis of the impact of the inertial response and frequency control on the dynamics of VSWTs. The results showed that the kinetic energy and mechanical energy variations in the WTG depend on its operating point. The control strategy and optimisation process were evaluated by non-linear time-domain simulations using the ODE23tb solver in MATLAB. The validation of the optimisation function included wind speed and load variations.

In [25], an integrated controller to provide both an inertial response and primary frequency regulation is proposed. More specifically, this includes a de-loaded pitch control system, which is proposed alongside an optimised MPPT controller to reserve capacity for frequency regulation and provide an inertial response while under de-loaded operation. The de-loading controller can estimate the proper pitch angle or regulate the tip ratio to attain the required de-loading. Under the de-loading option, the MPPT controller will switch from the MPPT curve to a virtual inertial control curve depending on the frequency variation. In this way, the WTG may provide frequency support to the grid by shifting its active power reference.

The authors in [26] present a method which estimates the grid frequency response as a result of a generator tripping. This method applies to both conventional synchronous machines and wind turbine generators for grid frequency support. In this way, system operators can continuously evaluate the inertia and headroom produced and adjust the amount of WTG inertia and active power control required for reliable system operation accordingly. The authors' full proposed model was validated by simulation on PSS/E. As concluded by the authors, determining the optimal amount of WTG inertia and active power support from conventional synchronous generators will be the focus of their future work.

The authors in [27] present a co-ordinated, distributed control scheme which allows offshore wind power plants (WPPs), connected through an HVDC system, to support the primary frequency control efforts of AC grids on land. The control scheme is designed to account for AC areas and WPPs which may be operated by different operators. In addition the control scheme considers the limited power reserve of the wind turbine generators in the WPPs and will adjust accordingly to maintain a suitable frequency regulation. In the case of a large change in the power demand, the control scheme will permit all stations to share their power reserves so that the frequency in the AC areas converges to the nominal frequency. Small changes to the power demand result in the control scheme restoring the frequency in the AC areas to the nominal frequency value. This control scheme was validated through transient simulations in a modified version of Cigré DC grid benchmark which included a five-terminal HVDC grid, comprised of two WPPs and three AC networks. MATLAB/SimPowerSystem was the simulation environment used in this paper.

A dynamic de-loading control scheme for a DFIG to provide additional active power for grid frequency support is proposed in [28]. The proposed control scheme co-ordinates the rotor acceleration control and pitch angle control while considering the frequency regulation demands of the grid and the prediction error of wind power. This control scheme was simulated in MATLAB. The results show that despite a small amount of wind abandonment, this control scheme can provide frequency support to the grid and mitigate the secondary frequency drop following frequency regulation.

3.1.2. Energy Storage Systems

Energy storage systems can partake in frequency regulation and work with WTGs to improve the system inertia or augment the frequency response of WTGs [17]. Specifically, in the context of frequency regulation, energy storage systems can mitigate the secondary frequency drop of WTGs during the rotor recovery phase of the WTG [17]. A summary of the papers classified as regarding ESS is shown in Table 3.

Table 3. Publications in ESS for VSWTs.

Research Area	Reference	Year of Publication
Energy Storage Systems	[15]	2015
	[29]	2021
	[30]	2016
	[31]	2015
	[32]	2021
	[33]	2019
	[34]	2019
	[35]	2020
	[36]	2022

The authors in [15] propose a strategy that provides better performance of temporary frequency support and addresses the problem of system frequency oscillation and secondary frequency drop. The solution is based on a co-ordinated control of WTGs in conjunction with an energy storage system (ESS). The proposed strategy was simulated on PSCAD/EMTDC to verify its effectiveness. The authors concluded that a secondary frequency drop may be avoided by an ESS rated to only 10% of the WTG.

A specified-time consensus (STC) control for ESS-assisted DFIG WTGs to assist in frequency regulation is proposed in [29]. The efficacy of the proposed control scheme is validated by simulations conducted in MATLAB/Simulink. The results showed that the proposed strategy can provide a smooth power output of the WTG and improve the frequency regulation capability.

The paper [30] presents a hybrid operation strategy for a WECS using an ESS for grid frequency support. This operating strategy provides a reserve power margin by de-loaded operation in addition to relying on the kinetic energy of the rotor. The ESS is used concurrently with the kinetic energy discharge to provide additional energy to the grid during a frequency deviation. Furthermore, the ESS is used to maintain the power balance between generation and consumption, thereby mitigating the stochastic nature of wind. The effectiveness of the proposed hybrid operation strategy was verified using PSCAD/EMTDC.

In [31], an algorithm is proposed to integrate hydro-electric pumped storage (HEPS) station with WECS, providing grid frequency support. This strategy embeds the idea that excess energy from the WECS is stored in a HEPS, providing energy during frequency drops. The major benefit in this is that the WTG will always follow its MPPT curve since no frequency support methods are applied to the WTG allowing for deviation. This ensures that the maximum energy will always be extracted from the wind. The major HEPS aspects of this system were estimated, while case studies examined the impact of the algorithm on frequency recovery at 40% wind power penetration. The simulation environment used was MATLAB/Simulink.

A co-operative control strategy of a WECS and compressed air energy storage for frequency regulation is proposed in [32]. The effectiveness of the proposed co-ordinated control strategy was evaluated under various scenarios and load profiles using MATLAB/Simulink simulations.

The authors in [33] present an embedded solution that uses a hierarchal controller on a microgrid comprising wind turbines and battery units. In this way, the solution provides a co-ordinated frequency support to a weak grid by adjusting the active power flow through the tie-line in accordance with grid frequency requirements. To deal with the interactions between the impedance of the weak grid and output impedance of the microgrid, a stability analysis model was developed. This proposed approach was validated through simulation using MATLAB/SimPowerSystem toolbox.

The paper [34] proposes a multilevel embedded ESS consisting of super-capacitors and lead–acid batteries inside a PMSG (permanent magnet synchronous generator) to provide frequency support. The super-capacitors are used to emulate an inertial response while the lead–acid batteries are used to provide the primary frequency response. The secondary frequency response is provided by the mechanical power reserved in the wind turbine by using a suboptimal MPPT strategy. In addition, a supplementary control strategy is proposed, which makes use of the super-capacitors and lead–acid batteries to provide a primary and secondary frequency response, respectively. Simulations were performed using MATLAB/SimPowerSystem and laboratory tests were conducted to validate the effectiveness of the proposed control strategy. By utilising the complimentary characteristics of lead–acid batteries and super-capacitors, the solution eliminates frequent cycling of the lead–acid batteries while also relieving mechanical stress from WECS due to abrupt electromagnetic changes when partaking in frequency regulation.

An inertial response control scheme with a super-capacitor ESS is proposed in [35]. The inertial response is designed using the generator torque limits. The energy released from the ESS is used to augment the active power deficit during the recovery process of th turbine rotor. In addition, a damping controller is added to the inertial control to suppre mechanical oscillations in the shaft and tower of the turbine during frequency support. Th mechanical system of the wind turbine generator is modelled and simulated using FAS by the NREL, whilst the electrical system of the wind turbine generator is developed i MATLAB/Simulink. The simulation results show that the proposed method can improv the frequency nadir, mitigate the secondary frequency dip and reduce the magnitude in th mechanical subsystem.

To mitigate the impact of the wake effect in grid frequency support, the paper [3 proposes an optimal ESS allocation (OEA) scheme for DFIG-based wind turbines. To realis the OEA scheme, wind turbines are placed in a cluster, considering their received win speed. Each wind turbine within a cluster will share the same ESS. The OEA scheme trie to optimise the coherency of all the clusters' frequency support margins so that all win turbines maintain the same frequency stability. In the OEA scheme, the ESSs do not direct provide frequency support services to the grid, rather, they serve to improve the win turbine's frequency support capability and the system security. In this implementatio the required capacity of the ESSs need not be so large as in the case of when ESSs directl provide frequency support, thereby reducing the level of investment needed to deploy a ESS. The effectiveness of the OEA is verified by simulation studies conducted on a modifie 29-bus Hydro-Quebec transmission system with one DFIG wind farm.

3.1.3. Emulated Inertial Response

The energy contained within a WTG system can be used when the frequency deviatio exceeds the allowable frequency range [16]. Droop control provides additional power i proportion to the frequency deviation [17]. When a wind turbine operates at its maximur power, the additional power required is obtained from the kinetic energy of the rotatin mass [16]. Similar to droop control, hidden inertia emulation also releases kinetic energ from the wind turbine generator, but the inertial response is instead based on the respons of traditional synchronous generators [16]. Kinetic energy from the wind turbine can als be released by the fast power reserve controller, which acts on the rotor speed signal [16 A summary of the papers classified as regarding emulated inertial control, grouped b research area, is shown in Table 4.

Table 4. Publications in Emulated Inertial Control for VSWTs.

Research Area	Reference	Year of Publication
Droop Control	[37]	2020
	[38]	2017
	[39]	2016
	[40]	2017
	[41]	2020
	[42]	2015
	[43]	2016
	[44]	2016
	[45]	2016
	[46]	2017
	[47]	2017
	[48]	2017

Table 4. *Cont.*

Research Area	Reference	Year of Publication
Droop Control	[37]	2020
	[7]	2017
	[49]	2017
	[50]	2020
	[51]	2021
	[52]	2021
	[53]	2020
	[54]	2018
	[55]	2021
	[56]	2018
	[57]	2020
	[58]	2022
Hidden Inertia Control	[59]	2019
	[60]	2018
	[61]	2021
	[62]	2020
	[63]	2015
	[64]	2018
	[65]	2021
Fast Power Reserve	[66]	2020
	[67]	2017
	[68]	2019
	[69]	2019
	[70]	2018
	[71]	2016
	[72]	2017

To improve the primary frequency response of wind power plants, the authors in [37] propose a distributed synchronised control technique. The droop control uses an optimum power share ratio and the frequency of each WPP at the point of common coupling (PCC) to calculate the droop. The control technique finds the optimum power share ratio through an interactive algorithm. The synchronised droop characteristics of the WPPs are functions of the frequency variation at the PCC and are inversely proportional to the respective power share ratios. The droop is, therefore, changed with frequency variation followed by a load-generation imbalance due to the disturbance. The synchronised droop varies with increasing communications delay between nodes, which mitigates the effects of communications delay on the proposed method. The effectiveness of the proposed control technique was tested on a WPP-integrated 39-bus New England system and simulated in PSCAD/EMTDC. The simulation results show that the primary frequency response of the proposed control technique is better than that of the distributed Newton method. However, the performance of the distributed synchronised droop control deteriorates in the event of a false data-injection attack on the communications system.

The paper [38] presents a communication-free alternative co-ordinated control scheme that prioritises frequency versus active power droop, fitted onto onshore VSCs. This scheme aims to transfer the wind turbine recovery power to undisturbed AC grids and allow for the correct control operation of multi-terminal high-voltage direct current (MTDC) systems during multiple power imbalances on different AC grids. The control scheme presented in the paper is compared with another co-ordinated control scheme which uses frequency versus DC voltage droop; the MTDC frequency support capability when wind farms do not provide extra power is also evaluated using a four-terminal HVDC system. The results presented in this paper show that during a single power imbalance in one AC grid, the fast frequency response from MTDC-connected wind farms limits the RoCoF on disturbed AC grids and transfers additional active power from another AC system, containing the system frequency deviation. The MTDC-connected windfarms were equipped with both the alternative co-ordinated control scheme and the co-ordinated control scheme. However, the co-ordinated control scheme was shown to have larger power oscillations when compared to the alternative co-ordinated control scheme under certain conditions. A three-terminal HVDC system was used to test the effectiveness of the co-ordinated control scheme and the alternative co-ordinated control scheme, both of which were modelled using PSCAD and experimental results were obtained using the RTDS tool of RSCAD. The MTDC frequency support capability when find farms do not provide extra power was verified using a four-terminal system and modelled using MATLAB/Simulink.

In [39], a control scheme is proposed for DFIG-based wind turbines for improved transient response and participation in grid frequency support. The proposed control scheme consists of a main controller with two auxiliary controllers. The main controller is a fuzzy-based controller whose parameters are optimised using the genetic algorithm to achieve an optimal transient response. The two auxiliary controllers, frequency deviation and wind speed oscillation controllers, enable the DFIG to provide frequency to the grid and mitigate the impacts of wind speed fluctuations on the WTG output power by using the kinetic energy of the WTG. The performance of the controllers is highly dependent on the operating point of the WECS. Simulations were performed considering various scenarios to prove that the proposed control scheme can enhance the power system's frequency performance after disturbances.

The authors in [40] propose a non-linear dynamic model for the DFIG's output power integrated into a dynamic model of the power grid. In addition to this, a state feedback controller is proposed by considering if DFIGs participate in grid frequency regulation or not. The stability of the entire system is considered using the input-to-state stability theory. The controller was embedded in the DFIG's detailed model and simulations were conducted to evaluate its performance. In comparison to a conventional controller presented in the paper, the proposed controller delivered more output power during grid frequency support and a negated transient recovery period.

A fast frequency support scheme for wind turbine systems to raise the frequency nadir close to the settling frequency and eliminate the secondary frequency dip is proposed in [41]. The frequency support scheme uses the kinetic energy of the wind turbine to raise the frequency nadir close to the settling frequency. In addition to raising the frequency nadir, an adaptive gain function of the real-time rotor speed and wind power penetration level is proposed to provide frequency nadir improvements under various wind speed and wind penetration levels. The last aspect of the proposed frequency support scheme is a new speed recovery strategy to mitigate the secondary frequency deviation associated with the rotor recovering its speed. The speed recovery strategy of the proposed frequency support scheme does not require the wind turbine to recover to the MPPT operating point during the primary frequency support phase. This allows the release of kinetic energy to be larger and no secondary frequency deviation will occur as no energy is extracted from the grid to recover the rotor speed. The three aspects of the proposed frequency support scheme have been implemented together, and the performance and stability of the wind turbines have been verified by comparisons of simulation results using the two-are power system for DFIG-based wind turbines and an IEEE 39-bus power system for PMSG-based wind turbines. The simulations were conducted using RTDS and Dymola.

In [42], an active power control method for VSWTs to enhance the inertial response and damping capability during transient events is proposed. The control method implements an optimised power point tracking controller, which shifts the wind turbine operating point from the MPPT curve to the virtual inertia control curves in relation to the frequency deviation. This allows the wind turbine to use its kinetic energy to provide grid frequency support. In addition to frequency support, the authors also theoretically evaluate the effects of the virtual inertia curves on damping power oscillations. When compared with a supplementary derivative control, the proposed optimised power point tracking control method avoids an interaction between the inertial response and the MPPT controller. The optimised power point tracking control method can also contribute to the power system damping and provides a smoother recovery of the rotor speed. The control method was prototyped on a three-machine system comprising two synchronous generators and a PMSG WTG with a wind penetration level of 31% to validate the proposed method.

The authors in [43] propose a control strategy that shifts the MPPT curve of the WTG to a virtual inertial control curve in relation to the frequency deviation so as to recover the kinetic energy of the wind turbine to provide grid frequency support. When compared to a PD-based inertial controller, the proposed virtual inertia control scheme provides a rapid response in the event of a sudden change in power and a smoother recovery to the MPPT operation. Comparative studies of the network frequency responses with and without the proposed virtual inertial control curves following a sudden system load and wind power change were conducted using MATLAB/Simulink. The simulation results show that the proposed control scheme can provide rapid dynamic frequency support to the grid with a reduction in frequency variations during fluctuations in both load and generation.

In [44], a control strategy for primary and inertial responses for high-wind-integrated power systems is proposed. The proposed method couples the pitch and power control loops and provides control at both sub- and super-synchronous operation. This method can be used for either a WTG or can be used to dispatch an entire wind farm for a primary frequency response and avoid individual turbine control. The proposed control scheme was evaluated on a 39-bus dynamic IEEE New England test system with 39 buses and 10 generators. Two aggregated windfarms were placed at bus 38 and bus 32. The dynamic models of the turbine and converter dynamics were modelled in DIgSILENT Power Factory.

The authors in [45] propose an inertial control scheme for a DFIG-based wind power plant. The proposed scheme aims to improve the frequency nadir and stable operation of the DFIG, especially when wind speed decreases during inertial control, by using adaptive gains set to be proportional to the kinetic energy stored in the DFIG, and the gains decrease with the declining kinetic energy. The results presented in the paper indicate that the proposed scheme improves the frequency nadir and prevents over-deceleration under certain wind and system conditions. The performance of the proposed adaptive gain scheme was verified through simulations of case studies using EMTP-RV.

The paper [46] proposes two novel control strategies that enable an inertial response from PMSG wind turbines during transient events. The first strategy aims to provide inertia to the system by simultaneously using the energy of the DC-link capacitor and kinetic energy of the wind turbine rotor. The second strategy aims to provide inertia to the system by first using the energy of the DC-link capacitor and then the kinetic energy of the rotor in a cascading control scheme. The two control strategies were validated using a case study of one PSMG-based wind turbine subject to sudden load variations and then compared. Subject to the same disturbance event, it is shown that both control strategies can provide similar performance in stabilising the system frequency, permitting that the control parameters are set correctly in advance. However, the cascaded control scheme characterises itself by enabling a better energy harvest during a frequency disturbance. The simulation environment used in [46] was not mentioned.

A non-linear system model of a VSWT may experience large deviations from its operating point during times of frequency support when linear control techniques are used. In [47], a novel non-linear controller used to enable an inertial response from a VSWT is presented. A regular controller based on the input–output feedback linearisation approac was first designed, and then replaced by the proposed controller, where the output power the wind turbine is expressed in terms of state variables and its Taylor series expansion use in the design of the control system. The proposed controller was embedded in a detaile model of a VSWT in MATLAB/Simulink and the various effects of different controll parameters were investigated. It is also emphasized that an appropriate co-ordinatio between conventional generating units and VSWTs equipped with the proposed controll results in a negligible transient recovery period. In addition to this, the behaviour the controller was also investigated in the presence of fluctuating wind power input an was shown to provide a smoother frequency response. The advantages of the propose controller over that of conventional PI control were also verified.

A co-ordinated control strategy to provide system inertial support for an offsho windfarm connected via an HVDC transmission line to an onshore main grid is discusse in [48]. The authors compare two proposed strategies: one where the energy of th HVDC capacitors and the kinetic energy of the wind turbine are used simultaneously provide inertial support without the installation of remote communication between the tw terminals of the HVDC transmission line, while the other strategy employs a sequenti release of energy starting with the HVDC capacitors and then using the kinetic energy the wind turbine, with the aid of communication between the onshore and offshore grid A detailed design and case study of the two control strategies have been conducted t compare and demonstrate the effectiveness of the control strategies in DIgSILENT Pow Factory. It was shown from the test system that when subject to the same disturbanc event, both strategies exhibit similar performance in stabilising the system frequenc permitting that the control parameters are set correctly in advance. However, the cascade control scheme has been shown to have better wind energy harvest during a frequenc disturbance event. The impact of the time delay, introduced by the communication the two converters in the cascaded control scheme, may have some limited impact on th control system performance and overall system stability.

A distributed virtual inertia scheme that can be implemented by grid-side-connecte power converters without any modification to the system hardware is proposed in [7]. Th virtual inertia provided to the grid uses the energy stored in the DC-link capacitors of th grid-side-connected power converters, and by regulating the DC-link voltages in proportio to the grid frequency, the DC-link capacitors are aggregated into a large equivalent capacit The limitations of the virtual inertia along with the design parameters such as DC-lin capacitance, DC-link voltage and maximum DC-link voltage deviation were identified i this paper. The concept was verified by MATLAB/Simulink simulations and addition experimental results were also obtained to verify the efficacy of the proposed concept. Th proposed concept indicates improvements in both the reduction in frequency deviatio and the improvement in the RoCoF.

In [49], the system frequency dynamics during inertia emulation and primary fr quency support from wind turbine generators are studied and a mode-switching schem of a wind turbine generator for frequency support is proposed. A proposed concept, th region of safety, which is the initial set of safe trajectories, is used to determine switchin instances. The barrier certificate methodology was used to derive a new algorithm t enlarge the region of safety for given desired safety limits and worst-case disturbanc scenarios that lead to finding the critical switching instants and a safe recovery procedur Furthermore, the inertial response and load-damping effects are derived in the respectiv time frames of both inertial and primary frequency responses, respectively. The paper [50 presents theoretical results under critical cases.

The paper [50] proposes a torque limit-based method to emulate an inertial respons The efficacy of two torque limit-based methods is investigated through a power syster with varying degrees of wind penetration levels in DIgSILENT Power Factory 2018. Th first torque limit-based method considers a definite ramp rate for the inertial power of th wind turbines, while the second torque limit-based method varies the key parameters t obtain a linear relation between its characteristics and the operating point of the VSWT. It was shown that the second torque limit-based method places lower stress on the mechanical parts of the wind turbine, particularly at low rotor speeds, compared to the first torque limit-based method. While the results show a more stable system operation due to the emulated inertia, the results do, however, also reveal a deteriorated frequency nadir for both torque limit-based methods, while the degree of wind penetration is high due to the incoordination of the VSWT inertial response and the governor response of fossil-fuelled generating units. To mitigate this, it is suggested that the inertial power of the VSWT is multiplied by a frequency-dependent gain.

Based on a frequency response model derived in [51], an estimation method to calculate the virtual moment of inertia provided by a DFIG-based wind farm is proposed. By using the Routh approximation method, an expression for the virtual moment of inertia for the grid-connected DFIG system is derived. To augment the availability of the expression, an estimation method based on the matrix pencil method and least squared algorithm for estimating the virtual moment of inertia provided by the wind farm is proposed. The effectiveness of the proposed method and the derived expression are tested on a DFIG grid-connected system and a modified IEEE 30-bus system in MATLAB/Simulink. Based on the results, the derived expression of the virtual moment of inertia directly expresses the inertial behaviour of a VSWT with additional frequency control and the estimation can efficiently calculate the virtual moment of inertia provided by the wind farm, regardless of whether VSWTs are involved in the frequency regulation of the grid or not. The ability to distinguish whether VSWTs are involved or not has value in power systems with high wind power penetration since it avoids the complex processing of parametric derivation and frequency response integration.

The frequency nadir and RoCoF are two metrics that temporary grid frequency support schemes strive to improve. The paper [52] proposes a rotor speed-based inertia control scheme aimed at improving the performance of the system's frequency regulation. Several case studies were performed, by varying wind speed conditions and wind power penetration levels, to investigate the performance of the proposed control scheme. The proposed control scheme was tested on a seven-machine power system, with an aggregated wind farm on MATLAB/Simulink. The results showed that the proposed scheme contributes to reducing the frequency nadir and RoCoF. In addition, the rotor speed-based inertia control scheme is decoupled from the system frequency which, in turn, decouples the inertial response speed from the inertia response amplitude, allowing an appropriate response to be realised, ensuring the stable operation of the WTG and the system.

In [53], a frequency support scheme for DFIG wind turbines that implements a two-stage switching control scheme is proposed. The first stage of the control scheme uses a variable proportion coefficient designed to emulate an inertial response from the DFIG wind turbine, while the second stage of the control scheme uses a fuzzy logic control scheme to determine the variable proportion coefficient to both quickly restore the DFIG wind turbine to MPPT operation and avoid a secondary frequency dip in the system. The proposed control scheme was simulated in MATLAB/Simulink and case studies were performed on a WSCC 9-bus and IEEE 39-bus power system to verify the effectiveness of the proposed control scheme. The results show that the proposed control scheme can provide a satisfactory performance for frequency support despite varying operating conductions. The paper concludes by stating that in future work, the feasibility of implementing the control scheme in large-scale power systems with integrated wind turbines will be verified by hardware-in-the-loop simulation.

A time-variable droop control method for WTGs to provide grid frequency support is proposed in [54]. In the proposed method, a positive droop characteristic provides frequency support and is followed by a period of negative droop to restore the kinetic energy of the rotor. The proposed method is simulated in DIgSILENT Power Factory, and the performance of the proposed control method is studied when subjected to various operating conditions. The results show that the frequency nadir can be improved while WTGs regain their kinetic energy and continue MPPT operation.

An adaptive droop control scheme for DFIGs to provide frequency support is proposed in [55]. The initial value of the adaptive droop gain is determined in accordance with the wind speed and decreases with time. The efficacy of the proposed control scheme was simulated on a modified IEEE 14-bus system in EMTP-RV. The results show that the proposed method prevents a DFIG from stalling and reduces the second frequency drop during the rotor recovery phase.

A RoCoF droop control strategy to mitigate the problems of frequency deviation and high RoCoF is proposed in [56]. In response to a high-frequency event, power is contributed to the grid from the reserve margin stored in the DC-link capacitor to mitigate the RoCoF. Once a new steady state has been reached, the reserve energy expended from the DC-link capacitor is recovered. An experimental comparison between the proposed RoCoF droop control strategy and conventional droop control strategy is conducted on the RT-LAB platform. The results of the comparison show that by adjusting control parameters such as increasing the droop coefficient or decreasing the bandwidth, a larger virtual inertia can be obtained than that contributed by the capacitor alone.

The paper [57] proposes a strategy to co-ordinate the use of the kinetic energy of a VSWT and the energy stored in the DC-link capacitor from a VSC-HVDC-connected offshore wind farm to provide grid frequency support. The minimum rotor speed limit for the VSWT is determined by the rate of change of the mechanical power with rotor speed which reduces the reduction in mechanical power caused by the low rotor speed, while the capacitor energy of the VSC-HVDC is used to contribute to the frequency nadir when the output power of the VSWT is less than the initial value. A detailed selection process of the controller parameters to improve the efficacy of the frequency control strategy is discussed and the proposed strategy was verified using a power system model in MATLAB/Simulink. The simulation results show that the proposed strategy can significantly reduce the grid frequency variation and maintain frequency stability in grids with high penetration of renewable energy.

In [58], a time-sharing frequency co-ordinated control scheme is proposed. The implementation of the proposed scheme uses a frequency dead-band to co-ordinate the priority of the rotor virtual inertia control, DC-link virtual inertia control and a new adaptive frequency droop control. In the proposed control scheme, the electrostatic energy stored within the DC-link capacitor is always used first for frequency support, while the kinetic energy of the rotor is only used once the energy stored within the DC-link capacitor has been depleted. The performance of the proposed control scheme is simulated and studied in PSCAD/EMTDC. The results indicate that the time-sharing frequency co-ordinated control scheme has better performance in wind energy harvesting and system stability when compared to either traditional cascaded control or co-ordinated control strategies. In addition, the proposed scheme can also effectively avoid power oscillations.

In [59], a co-ordinated control strategy for PMSG-based VSWTs to provide frequency support services is proposed. An inertial response is emulated by using the energy stored in the DC-link capacitor. However, since the available energy is relatively small, it is supplemented by a virtual capacitor control strategy which uses the rotor-side converter to provide a virtual capacitance that is larger than the actual DC-link capacitance by using the wind turbine's kinetic energy in a similar way to the inertial response of a synchronous generator. Lastly, a power–frequency droop control is used to simulate the primary frequency control of a synchronous generator. The efficacy of the proposed control strategy was verified by simulation results in PSCAD/EMTDC and indicated that the RoCoF is mitigated and the frequency nadir is improved by adopting the proposed co-ordinated control strategy.

The authors in [60] present a control approach based on the VSM concept. The approach presented in [60], called an enhanced virtual synchronous machine (eVSM), uses the existing inertia of the DC-link component and does not emulate the rotating inertia based on an assumption of unlimited energy. The eVSM concept aims to reduce the need of a large DC-link component or battery storage while providing the same inertial response of an equivalent synchronous machine.

The paper [61] proposes a two-level combined control strategy for offshore wind farms to provide frequency support for an onshore system. On the wind turbine level, each wind turbine implements inertial and droop control with adaptive coefficients, which allow wind turbines with high rotor speeds to release more kinetic energy, while working at MPPT mode. To mitigate the second frequency dip that occurs after frequency support, the wind turbines are divided into clusters based on their rotor speed and a step start-up control scheme is implemented for the wind turbine clusters to provide frequency support sequentially. The sequential frequency support allows one cluster to provide frequency support, while another recovers their rotor speed. At the system level, a communication-free allocation control strategy is proposed using the local frequency signal of onshore voltage-source converter stations to share the active power among the onshore stations. Case studies were conducted on a three-area four-terminal voltage-source converter-based multi-terminal high-voltage direct current offshore windfarm, in MATLAB/Simulink, to determine effectiveness of the two-level combined control strategy. The simulation results show the efficacy of the proposed control scheme under various scenarios. The paper concludes by stating that future work will consider optimising the number of clusters and asynchronous interconnection to the onshore system.

The authors in [62] propose a control strategy for offshore HVDC-connected windfarms to provide an inertial response and primary frequency support. The proposed control strategy uses the HVDC converters to map the onshore frequency variation onto a voltage variation in the offshore grid. The authors have identified various limitations of conventional frequency support strategies from offshore-connected windfarms including large frequency deviations and high RoCoF in the offshore gird. In contrast, the proposed control strategy achieves the frequency requirements of the regional grid code without stressing the offshore grid as the frequency deviations and RoCoF are kept low. The proposed control strategy was tested on two test systems, one built entirely in the DIgSILENT Power Factory environment and the other on an IEEE 39-bus system. However, the proposed control strategy relies on using existing local communication systems to monitor the HVDC link voltage to distinguish between internal disturbances of the offshore AC grid and disturbances that arise due to a frequency disturbance of the main onshore AC grid. The authors are developing an adaptive control strategy with minimal reliance on local communication systems that can discern between internal and external disturbances.

The paper [63] expands on a communication-less approach previously proposed in the literature for offshore wind turbines connected though a DC grid to multiple onshore AC grids. To extend this concept, the authors instead adopt a strategy where the onshore frequency variations are communicated to the offshore converters by using a fibre optic link embedded within the sub-sea DC cables. In the paper, a case study on a four-terminal DC grid connecting an offshore wind farm and two onshore AC systems illustrates the inertial support from the offshore wind farms with the proposed strategy. An aggregated model of the WTGs within the wind farm was used in the simulations and the parameters were obtained from a simulation model in DIgSILENT Power Factory. It is shown that the proposed weighted frequency scheme strategy can improve the transient frequency deviation in AC grids experiencing an under-frequency event.

The authors of [64] propose a clustering-based co-ordinated control scheme for large-scale wind farms to provide frequency support. The proposed control scheme considers the distributed layout of WTGs and the wake effects inside the wind farm. WTGs are grouped by their wind profiles and the same control commands are dispatched to the same group. This effectively realises a group of WTGs as one single WTG and greatly reduces the control variables and complexity of the optimization problem. Simulations were conducted in MATLAB/Simulink to verify the effectiveness of the proposed control strategy.

A hierarchical non-linear model predictive control (MPC) for frequency support proposed in [65]. The proposed hierarchical non-linear MPC reduces the computational burden of the central controller when compared to a centralised non-linear MPC. The efficac efficiency and robustness of the proposed hierarchical non-linear MPC is validated by numerical simulations in MATLAB/Simulink. The results show that the non-linear MPC ca reduce the computation time by as much as 50% when compared to a centralised non-linea MPC. This makes the proposed MPC method more favourable for large-scale wind far implementations by avoiding the requirement of higher performance computation faciliti

The authors of [66] present a short-term frequency support method for doubly fe induction generators (DFIGs) to improve the frequency nadir of the system using less kinet energy and negate the second frequency drop caused by the rotor speed recovery. Th proposed method adds a time-varying constant to the power reference for the maximu power point tracking (MPPT) function and allows the rotor to recover its speed alon with the MPPT curve. The proposed method was analysed and validated on a modifie IEEE 14-bus test system; however, the simulation environment was not disclosed. Th simulation results indicate that the proposed method could provide a frequency nad improvement with a lower kinetic energy cost for the DFIG in various scenarios and reduc the secondary frequency nadir. The performance of the proposed method is affected b wind speed, as the winds speed affects the levels of kinetic energy available. The propose method was only tested against two wind speed conditions: 8 m/s and 10 m/s. The pap concludes by stating that due to the inherent variability and uncertainty of wind power, th frequency support capacity of DFIGs may be unable to meet the instantaneous frequenc system requirements and may be required to operate with ESS systems. The autho state that future work will focus on a co-ordinated frequency control scheme betwee DFIGs and ESSs for frequency control will be studied in power systems with high win power penetration.

In [67], the effect of different power curves in releasing kinetic energy for grid fr quency support is considered and two solutions are proposed. The first of the two propose solutions aims at reducing the rate of change of frequency (RoCoF) while the second metho is aimed at improving the RoCoF and the frequency nadir. The amount of kinetic energ used for frequency support was considered and the minimum wind turbine rotor speec at various wind conditions were defined to avoid a large reduction in mechanical pow while the wind turbine supports the grid frequency. The proposed solutions were validate by simulations using MATLAB/Simulink and indicate that the proposed solutions ca reduce the RoCoF and frequency nadir while minimising the power imbalance when th rotor is reaccelerated. Given that wind turbines can regulate their power quicker tha conventional generators and only have a limited and temporary kinetic energy reserve th can be used to support the grid frequency (that has to return to the wind turbine syste after frequency control), it might be beneficial to use wind turbines to improve the RoCo while using conventional generators to reduce the frequency nadir.

A power reference model that operates reliably during uncertain wind conditio while providing grid frequency support is proposed in [69]. The proposed frequenc support scheme provides an emulated inertial response to a disturbance by releasing portion of the rotor's available kinetic energy and decreases linearly with rotor speed. T determine the effectiveness of the proposed scheme, the proposed scheme was studied—alon with other methods—on large-scale integrated wind farms on IEEE 9-bus and New Englan 39-bus systems simulated in DIgSILENT Power Factory and MATLAB.

In [69], a dynamic demand control strategy that co-ordinates with the DFIG control i presented. This co-ordinated control strategy for DFIGs can provide an inertial respons and primary frequency support. In addition, it can mitigate the secondary frequenc dip following frequency support and improve the performance of the primary frequenc support. For the demand control to work optimally, controllable loads with high pow ratings that can accept a changing connection state and have minimal user impact ar suggested. The proposed control strategy was simulated in MATLAB/Simulink, while th

DFIG unit model and aggregated controllable load model were developed in OpenModelica and linked to ePHASORsim, a real-time power system transient simulation software.

To improve the frequency nadir while ensuring a rapid frequency stabilisation for high wind power penetration levels, a temporary frequency support scheme of a DFIG is proposed in [70]. At the onset of a frequency disturbance, the power reference is incremented by the incremental power, which is a value that varies with rotor speed and wind power penetration level and is maintained for a pre-set period. The reference will decrease during this deceleration phase with the reduction in speed and release less kinetic energy. This, however, will aid in the rapid recovery of the rotor speed and, during the acceleration phase, the scheme will decrease the output of the DFIG with both rotor speed and time until the reference intersects the MPPT curve again. The results are obtained by running various scenarios on an IEEE 14-bus system and indicate that the scheme arrests the frequency nadir at a higher level, except for Scenario 2. The proposed control scheme was simulated using EMTP-RV.

The paper [71] proposes a method to control variable-speed wind turbines to provide a frequency response through temporary over-production. The control method aims to determine the optimal power extraction profile between multiple generators to minimize the total loss of efficiency while allowing for an increase in generation. In case of frequency events, the cumulative generation is increased by a specified amount, while the power profile of individual turbines is determined by resolving an optimal control problem to minimise the loss of efficiency. The results of this method were evaluated through simulations and the concept was also extended to recover the VSWTs to their initial state of maximum efficiency in minimum time.

In [72], a stable stepwise short-term frequency support scheme based on a DFIG wind farm is proposed and aims to improve the frequency nadir while quickly recovering the rotor speed. When a frequency event occurs, the output power is raised by increasing the power reference prior to the event and sustaining it for some predetermined period determined by the frequency nadir. The power reference then linearly decreases with the rotor speed until it converges with rotor speed to some pre-set value. To then recover the rotor speed, at the pre-set rotor speed, the power reference decreases with the rotor speed and time until it intersects with the MPPT reference. The proposed scheme was deployed on an IEEE 14-bus system with one DFIG-based wind farm and simulated on EMTP-RV. The simulation results of the four case studies showed that the proposed stepwise short-term frequency support scheme can raise the frequency nadir while recovering the rotor quicker than other conventional schemes in the case studies.

3.2. RQ2: What Are the Limitations of VSWTs in Supporting the Grid Frequency

The rate of change of power of wind turbines could limit the effective implementation of frequency support methods. High rate of change of power in wind turbines leads to an increase in maintenance cost. Simply applying a rate limiter to the power function of the wind turbine generator may reduce the efficacy of frequency support strategies. In [73], a small-signal analysis is performed to study the effects of frequency regulation methods on the rate of change of power of wind turbines. Both DFIG and PMSG wind turbine generators are considered. A detailed model of a wind turbine generator was developed in the paper. The model considered the double-mass nature of the mechanical system, active damping controller and frequency support mechanisms, namely droop and virtual inertia. The model in the paper allowed for a small-signal analysis of the frequency support dynamics in the wind turbine generator and the characterisation of the associated mechanical stresses imposed on the generator. The paper has shown that droop and virtual inertia frequency support methods subject the mechanical system to high rates of change of torque and power which accelerate the aging process. The paper has investigated the use of DC-link capacitors to respond to the fast transient portion of the frequency support dynamics while the kinetic energy stored within the rotating mass of the turbine is still used in the slower but high-energy portion of the frequency support. Furthermore, the paper also studies the management of these two sources and verifies all analytical results using time-domain simulations in MATLAB/Simulink based on detailed non-linear models.

The paper [74] considers the potential issues associated with system-wide integration of emulated inertia technologies and issues associated with implementing large-scale frequency support controls, namely the potential delayed frequency recovery and the variation in the optimal response with different frequency conditions, are identified in this paper. Two differently sized systems with wind integration and emulated inertial response capabilities that represent simplified dynamic power system models are developed in MATLAB/Simulink to investigate the impact of varying system conditions. The findings of the paper show that there are implications for the development of ancillary service markets and grid code requirements for systems with a high penetration level of non-synchronous generation. The differences between the emulated inertial response from wind power plants and the inertial response from conventional synchronous generator-based plants should be considered if the integrity and stable operation of the grid are to be maintained. System planners and operators should ensure that the way in which energy is delivered and recovered from wind turbines is realised as a function of the system demand, wind penetration level, geographical wind distribution and reserve level. Failure to consider this may lead to an undesirable frequency response from wind generation and impact the grid stability.

In [75], the frequency support capabilities of VSWTs are studied. The impact of inertial control is analysed, and it has been determined that at a large scale, VSWTs may lead to a potential delayed grid frequency recovery and a variation in the optimal response under varying system conditions. In addition, the paper states that should inertial control or emulated inertia technology be adopted at scale, further development will be required. Two simplified dynamic power system models, one small- and one medium–large-sized system were developed in MATLAB/Simulink with inertial control capabilities to investigate the impact of varying system conditions on the optimal tuning of the inertial control parameters. The results highlight the importance of resource tuning and issues relating to what is defined as an optimal response, while implementation approaches for system operators considering the use of inertial control to emulate an inertial response are proposed.

4. Discussion

With the relevance of VSWTs for frequency support owing to the increase in wind power penetration and concerns about the frequency of the grid [16], research in this field can make a great impact in addressing practical concerns. The categories of VSWT frequency support are based on the prevailing methods used, since each method has advantages and practical limitations.

The period from which surveyed works were drawn showed a marked increase in publications from 2017 to 2019, followed by a period of consistent publications except for the current year, as seen in Figure 3. The absence of exponential growth and, instead, a consistent rate of publication could be attributed to the well-understood problem of increasing wind power penetration and grid frequency stability without ancillary support from WTGs. Recent publications surveyed revealed that the focus is in improving the frequency support characteristics of VSWTs by modifying already understood frequency support methods, combining methods or proposing novel methods.

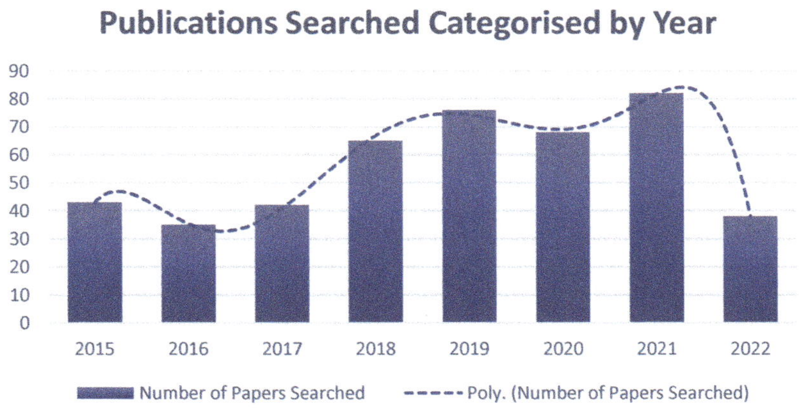

Figure 3. Cumulative Search Strings SCOPUS Look up.

Figure 4 shows a graph of the number of journal articles admitted to this study, based on the inclusion and exclusion criteria, grouped by the year of publication. From this graph, it can be noted that there is a persistent interest in this field, with a peak in publications in 2017 and a reduction in publications in 2019 but recovering in 2020. The year 2022 is excluded from comment as this is present year, and some journals may not have published papers at the time of conducting this study. This indicates a consistent interest in this research area. This may be attributed to the periodic publishing of journals and the extensive review process resulting in a more even distribution of published papers. Nevertheless, this does indicate relevance in this area of study. Furthermore, the trend of the graph in Figure 4 likely deviates from the general trend (Figure 3) due to the exclusion of conference papers and not necessarily because of a decline in interest in this research area.

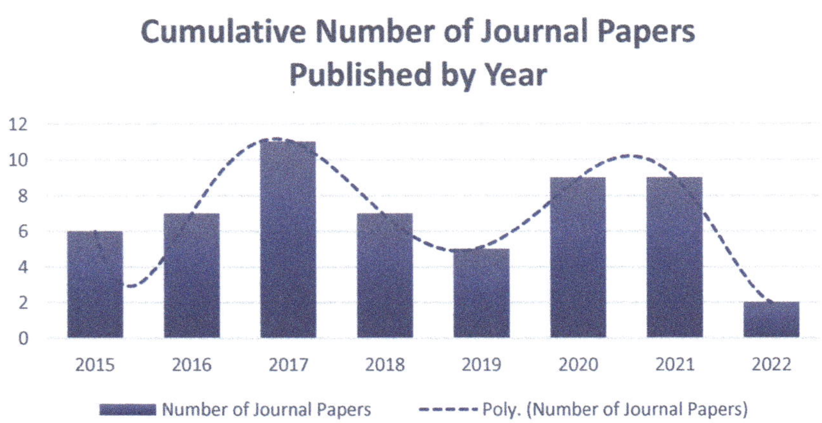

Figure 4. Cumulative Number of Journal Papers Grouped by Year of Publication.

Of the journal papers studied in this systematic literature review, 11 papers were classified as those that used de-loading as a primary means of providing frequency support, 9 relied on ESS and 45 implemented an emulated inertial response. The research area that gained the most focus identified in this review was droop control-based, with 23 journal publications. The distribution of the papers surveyed in this study classified by research focus is shown in Figure 5.

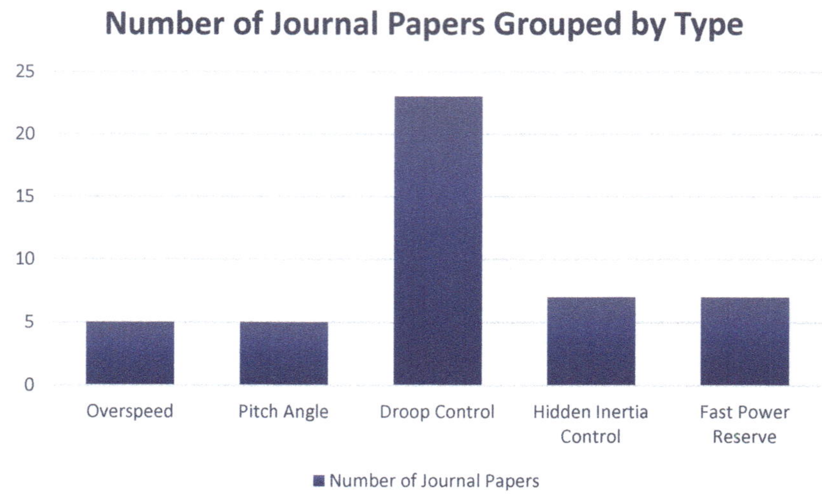

Figure 5. Distribution of Journal Papers surveyed.

When screening the literature, it was observed that a limited number of publications identified and studied the limitations of VSWTs providing grid frequency support. Since the impact of low inertia has been studied, this area of research will continue to be relevant as the prevalence of WTGs increases. However, a better understanding of the impact of VSWTs providing grid frequency support on the VSWT structure and the small-signal stability of the grid would greatly benefit this field of research. Future works will focus on understanding the limitations of VSWTs providing grid frequency support and small-signal stability of the grid.

5. Conclusions

VSWTs for grid frequency support is a research area that can provide solutions to grids with high wind power penetration in a world where grids seek to decarbonise their generating fleet. The main objective of this paper is to present relevant information in a modern timeframe on the current implementations and limitations for VSWTs to provide grid frequency support to provide the reader with a comprehensive view on the topic. The study was conducted using the systematic literature review methodology, which facilitates the systematic process of searching, locating, assessing, analysing and categorising existing papers in the literature. An extensive review was conducted from various databases from 2015 to mid-August 2022. From the results, it is seen that there is a consistent interest in the research at the present time.

The main limitation of this work is that number of papers selected for the study and the number of studies identifying the limitations of VSWTs providing grid frequency support were limited. This will be examined in future works.

Author Contributions: Conceptualization, A.B., K.M.A.-A.-E. and M.A.; methodology, A.B., K.M.A.-A.-E. and M.A.; formal analysis, A.B., K.M.A.-A.-E. and M.A.; investigation, A.B.; resources, A.B. and K.M.A.-A.-E.; data curation, A.B.; writing—original draft preparation, A.B.; writing—review and editing, A.B., K.M.A.-A.-E. and M.A.; visualization, A.B.; supervision, K.M.A.-A.-E. and M.A.; project administration, K.M.A.-A.-E.; funding acquisition, K.M.A.-A.-E. and M.A. All authors have read and agreed to the published version of the manuscript.

Funding: This research received no external funding.

Data Availability Statement: Not applicable.

Conflicts of Interest: The authors declare no conflict of interest. The funders had no role in the design of the study; in the collection, analyses, or interpretation of data; in the writing of the manuscript; or in the decision to publish the results.

References

1. National Energy Regulator of South Africa. The South African Grid Code: The Network Code. 2010; pp. 1–59. Available online: http://www.nersa.org.za/ (accessed on 10 May 2019).
2. Anaya-Lara, O.; Jenkins, N.; Ekanayake, J.; Cartwright, P.; Hughes, M. *Wind Energy Generation: Modelling and Control*; John Wiley & Sons: Hoboken, NJ, USA, 2009; Volume 54.
3. Ma, H.; Li, H. Analysis of frequency dynamics in power grid: A Bayesian structure learning approach. *IEEE Trans. Smart Grid* **2013**, *4*, 457–466. [CrossRef]
4. Ulbig, A.; Borsche, T.S.; Andersson, G. *Impact of Low Rotational Inertia on Power System Stability and Operation*; IFAC: New York, NY, USA, 2014; Volume 19. [CrossRef]
5. Boyle, J.; Littler, T.; Foley, A. Review of frequency stability services for grid balancing with wind generation. *J. Eng.* **2018**, *2018*, 1061–1065. [CrossRef]
6. Okedu, K.E. Improving grid frequency dynamics of synchronous generators considering wind energy penetration grid dynamics and operation. In Proceedings of the 2017 IEEE International Electric Machines and Drives Conference, IEMDC 2017, Miami, FL, USA, 21–24 May 2017; pp. 1–6. [CrossRef]
7. Fang, J.; Li, H.; Tang, Y.; Blaabjerg, F. Distributed Power System Virtual Inertia Implemented by Grid-Connected Power Converters. *IEEE Trans. Power Electron.* **2018**, *33*, 8488–8499. [CrossRef]
8. Oshnoei, A.; Khezri, R.; Muyeen, S.; Blaabjerg, F. On the Contribution of Wind Farms in Automatic Generation Control: Review and New Control Approach. *Appl. Sci.* **2018**, *8*, 1848. [CrossRef]
9. Munteanu, I.; Cutululis, N.-A.; Bratcu, A.I.; CeangĂ, E. *Optimal Control of Wind Energy Systems*; Springer: London, UK, 2008.
10. Dreidy, M.; Mokhlis, H.; Mekhilef, S. Inertia response and frequency control techniques for renewable energy sources: A review. *Renew. Sustain. Energy Rev.* **2017**, *69*, 144–155. [CrossRef]
11. DoE. State of Renewable Energy in South Africa—2017. 2018. Available online: http://www.energy.gov.za/files/media/Pub/2017-State-of-Renewable-Energy-in-South-Africa.pdf (accessed on 11 May 2019).
12. Research Methodology—Research Support—LibGuides at University of the Witwatersrand. Available online: https://libguides.wits.ac.za/c.php?g=693518&p=4914913 (accessed on 1 May 2022).
13. Kitchenham, B.; Pretorius, R.; Budgen, D.; Brereton, O.P.; Turner, M.; Niazi, M.; Linkman, S. Systematic literature reviews in software engineering—A tertiary study. *Inf. Softw. Technol.* **2010**, *52*, 792–805. [CrossRef]
14. Torreglosa, J.P.; García-Triviño, P.; Fernández-Ramirez, L.M.; Jurado, F. Control strategies for DC networks: A systematic literature review. *Renew. Sustain. Energy Rev.* **2016**, *58*, 319–330. [CrossRef]
15. Miao, L.; Wen, J.; Xie, H.; Yue, C.; Lee, W.-J. Coordinated Control Strategy of Wind Turbine Generator and Energy Storage Equipment for Frequency Support. *IEEE Trans. Ind. Appl.* **2015**, *51*, 2732–2742. [CrossRef]
16. Fernández-Bustamante, P.; Barambones, O.; Calvo, I.; Napole, C.; Derbeli, M. Provision of frequency response from wind farms: A review. *Energies* **2021**, *14*, 6689. [CrossRef]
17. Lin, C.H.; Wu, Y.K. Overview of Frequency-Control Technologies for a VSC-HVDC-Integrated Wind Farm. *IEEE Access* **2021**, *9*, 112893–112921. [CrossRef]
18. Fernández-Guillamón, A.; Gómez-Lázaro, E.; Muljadi, E.; Molina-Garcia, Á. A Review of Virtual Inertia Techniques for Renewable Energy-Based Generators. In *Renewable Energy–Technologies and Applications*; IntechOpen: London, UK, 2020. [CrossRef]
19. Attya, A.B.T.; Dominguez-Garcia, J.L. Insights on the Provision of Frequency Support by Wind Power and the Impact on Energy Systems. *IEEE Trans. Sustain. Energy* **2018**, *9*, 719–728. [CrossRef]
20. Ye, H.; Pei, W.; Qi, Z. Analytical modeling of inertial and droop responses from a wind farm for short-term frequency regulation in power systems. *IEEE Trans. Power Syst.* **2016**, *31*, 3414–3423. [CrossRef]
21. Pradhan, C.; Bhende, C.N.; Samanta, A.K. Adaptive virtual inertia-based frequency regulation in wind power systems. *Renew. Energy* **2018**, *115*, 558–574. [CrossRef]
22. Wang, S.; Tomsovic, K. A Novel Active Power Control Framework for Wind Turbine Generators to Improve Frequency Response. *IEEE Trans. Power Syst.* **2018**, *33*, 6579–6589. [CrossRef]
23. Boyle, J.; Littler, T.; Muyeen, S.M.; Foley, A.M. An alternative frequency-droop scheme for wind turbines that provide primary frequency regulation via rotor speed control. *Int. J. Electr. Power Energy Syst.* **2021**, *133*, 107219. [CrossRef]
24. Tessaro, H.J.; de Oliveira, R.V. Operating point optimization of wind turbine generators for provision of inertial response and frequency control. *Int. Trans. Electr. Energy Syst.* **2020**, *30*, e12382. [CrossRef]
25. Fu, Y.; Wang, Y.; Zhang, X. Integrated wind turbine controller with virtual inertia and primary frequency responses for grid dynamic frequency support. *IET Renew. Power Gener.* **2017**, *11*, 1129–1137. [CrossRef]
26. Yan, R.; Saha, T.K. Frequency response estimation method for high wind penetration considering wind turbine frequency support functions. *IET Renew. Power Gener.* **2015**, *9*, 775–782. [CrossRef]

27. Bianchi, F.D.; Dominguez-Garcia, J.L. Coordinated Frequency Control Using MT-HVDC Grids With Wind Power Plants. *IEE Trans. Sustain. Energy* **2016**, *7*, 213–220. [CrossRef]
28. Ouyang, J.; Pang, M.; Li, M.; Zheng, D.; Tang, T.; Wang, W. Frequency control method based on the dynamic deloading of DFIG for power systems with high-proportion wind energy. *Int. J. Electr. Power Energy Syst.* **2021**, *128*, 106764. [CrossRef]
29. Xiong, L.; Yang, S.; He, Y.; Li, P.; Huang, S.; Wang, Z.; Wang, J. Specified time consensus control of ESSs for frequency suppo with DFIG wind turbines. *Int. J. Electr. Power Energy Syst.* **2022**, *135*, 107520. [CrossRef]
30. Choi, J.W.; Heo, S.Y.; Kim, M.K. Hybrid operation strategy of wind energy storage system for power grid frequency regulatio *IET Gener. Transm. Distrib.* **2016**, *10*, 736–749. [CrossRef]
31. Attya, A.B.T.; Hartkopf, T. Utilising stored wind energy by hydro-pumped storage to provide frequency support at high levels c wind energy penetration. *IET Gener. Transm. Distrib.* **2015**, *9*, 1485–1497. [CrossRef]
32. Abouzeid, S.I.; Guo, Y.; Zhang, H.C. Cooperative control framework of the wind turbine generators and the compressed ai energy storage system for efficient frequency regulation support. *Int. J. Electr. Power Energy Syst.* **2021**, *130*, 106844. [CrossRef]
33. Zhao-Xia, X.; Mingke, Z.; Yu, H.; Guerrero, J.M.; Vasquez, J.C. Coordinated Primary and Secondary Frequency Support Betwee Microgrid and Weak Grid. *IEEE Trans. Sustain. Energy* **2019**, *10*, 1718–1730. [CrossRef]
34. Tan, Y.; Muttaqi, K.M.; Ciufo, P.; Meegahapola, L.; Guo, X.; Chen, B.; Chen, H. Enhanced Frequency Regulation Using Multileve Energy Storage in Remote Area Power Supply Systems. *IEEE Trans. Power Syst.* **2019**, *34*, 163–170. [CrossRef]
35. Yan, W.; Wang, X.; Gao, W.; Gevorgian, V. Electro-mechanical modeling of wind turbine and energy storage systems wit enhanced inertial response. *J. Mod. Power Syst. Clean Energy* **2020**, *8*, 820–830. [CrossRef]
36. Xiong, L.; Yang, S.; Huang, S.; He, D.; Li, P.; Khan, M.W.; Wang, J. Optimal Allocation of Energy Storage System in DFIG Win Farms for Frequency Support Considering Wake Effect. *IEEE Trans. Power Syst.* **2022**, *37*, 2097–2112. [CrossRef]
37. Mahish, P.; Pradhan, A.K. Distributed Synchronized Control in Grid Integrated Wind Farms to Improve Primary Frequenc Regulation. *IEEE Trans. Power Syst.* **2020**, *35*, 362–373. [CrossRef]
38. OAdeuyi, D.; Cheah-Mane, M.; Liang, J.; Jenkins, N. Fast Frequency Response from Offshore Multiterminal VSC–HVDC Scheme *IEEE Trans. Power Deliv.* **2017**, *32*, 2442–2452. [CrossRef]
39. Ashouri-Zadeh, A.; Toulabi, M.; Ranjbar, A.M. Coordinated design of fuzzy-based speed controller and auxiliary controllers in variable speed wind turbine to enhance frequency control. *IET Renew. Power Gener.* **2016**, *10*, 1298–1308. [CrossRef]
40. Toulabi, M.; Bahrami, S.; Ranjbar, A.M. An Input-to-State Stability Approach to Inertial Frequency Response Analysis o Doubly-Fed Induction Generator-Based Wind Turbines. *IEEE Trans. Energy Convers.* **2017**, *32*, 1418–1431. [CrossRef]
41. Zhao, X.; Xue, Y.; Zhang, X.-P. Fast Frequency Support from Wind Turbine Systems by Arresting Frequency Nadir Close t Settling Frequency. *IEEE Open Access J. Power Energy* **2020**, *7*, 191–202. [CrossRef]
42. Wang, Y.; Meng, J.; Zhang, X.; Xu, L. Control of PMSG-Based Wind Turbines for System Inertial Response and Power Oscillatio Damping. *IEEE Trans. Sustain. Energy* **2015**, *6*, 565–574. [CrossRef]
43. Zhang, X.; Wang, Y.; Fu, Y.; Xu, L. A novel method for obtaining virtual inertial response of DFIG-based wind turbines. *Win Energy* **2016**, *19*, 313–328. [CrossRef]
44. Ghosh, S.; Kamalasadan, S.; Senroy, N.; Enslin, J. Doubly Fed Induction Generator (DFIG)-Based Wind Farm Control Framewor for Primary Frequency and Inertial Response Application. *IEEE Trans. Power Syst.* **2016**, *31*, 1861–1871. [CrossRef]
45. Lee, J.; Jang, G.; Muljadi, E.; Blaabjerg, F.; Chen, Z.; Kang, Y.C. Stable Short-Term Frequency Support Using Adaptive Gains for DFIG-Based Wind Power Plant. *IEEE Trans. Energy Convers.* **2016**, *31*, 1068–1079. [CrossRef]
46. Li, Y.; Xu, Z.; Wong, K.P. Advanced Control Strategies of PMSG-Based Wind Turbines for System Inertia Support. *IEEE Trans Power Syst.* **2017**, *32*, 3027–3037. [CrossRef]
47. Toulabi, M.; Dobakhshari, A.S.; Ranjbar, A.M. An Adaptive Feedback Linearization Approach to Inertial Frequency Response o Wind Turbines. *IEEE Trans. Sustain. Energy* **2017**, *8*, 916–926. [CrossRef]
48. Li, Y.; Xu, Z.; Ostergaard, J.; Hill, D.J. Coordinated Control Strategies for Offshore Wind Farm Integration via VSC-HVDC fo System Frequency Support. *IEEE Trans. Energy Convers.* **2017**, *32*, 843–856. [CrossRef]
49. Zhang, Y.; Tomsovic, K.; Djouadi, S.M.; Pulgar-Painemal, H. Hybrid Controller for Wind Turbine Generators to Ensure Adequat Frequency Response in Power Networks. *IEEE J. Emerg. Sel. Top. Circuits Syst.* **2017**, *7*, 359–370. [CrossRef]
50. Azizipanah-Abarghooee, R.; Malekpour, M.; Dragicevic, T.; Blaabjerg, F.; Terzija, V. A Linear Inertial Response Emulation fo Variable Speed Wind Turbines. *IEEE Trans. Power Syst.* **2020**, *35*, 1198–1208. [CrossRef]
51. Chen, P.; Qi, C.; Chen, X. Virtual Inertia Estimation Method of DFIG-based Wind Farm with Additional Frequency Control. *J Mod. Power Syst. Clean Energy* **2021**, *9*, 1076–1087. [CrossRef]
52. Lao, H.; Zhang, L.; Zhao, T.; Zou, L. Novel insight on temporary frequency support of variable speed wind energy conversio system. *Wind Energy* **2021**, *24*, 1353–1367. [CrossRef]
53. Peng, X.; Yao, W.; Yan, C.; Wen, J.; Cheng, S. Two-Stage Variable Proportion Coefficient Based Frequency Support of Grid Connected DFIG-WTs. *IEEE Trans. Power Syst.* **2020**, *35*, 962–974. [CrossRef]
54. Garmroodi, M.; Verbic, G.; Hill, D.J. Frequency Support from Wind Turbine Generators with a Time-Variable Droop Characteristic *IEEE Trans. Sustain. Energy* **2018**, *9*, 676–684. [CrossRef]
55. Yang, D.; Jin, Z.; Zheng, T.; Jin, E. An adaptive droop control strategy with smooth rotor speed recovery capability for type II wind turbine generators. *Int. J. Electr. Power Energy Syst.* **2022**, *135*, 107532. [CrossRef]

6. Binbing, W.; Abuduwayiti, X.; Yuxi, C.; Yizhi, T. RoCof droop control of PMSG-based wind turbines for system inertia response rapidly. *IEEE Access* **2020**, *8*, 181154–181162. [CrossRef]
7. Guo, X.; Yan, J.; Xu, G.; Hu, J.; Bi, T. Control of VSC-HVDC connected wind farms for system frequency support. *Int. Trans. Electr. Energy Syst.* **2020**, *30*, e12352. [CrossRef]
8. Jiang, Q.; Zeng, X.; Li, B.; Wang, S.; Liu, T.; Chen, Z.; Wang, T.; Zhang, M. Time-Sharing Frequency Coordinated Control Strategy for PMSG-Based Wind Turbine. *IEEE J. Emerg. Sel. Top. Circuits Syst.* **2022**, *12*, 268–278. [CrossRef]
9. Zeng, X.; Liu, T.; Wang, S.; Dong, Y.; Chen, Z. Comprehensive Coordinated Control Strategy of PMSG-Based Wind Turbine for Providing Frequency Regulation Services. *IEEE Access* **2019**, *7*, 63944–63953. [CrossRef]
0. Khajehoddin, S.A.; Karimi-Ghartemani, M.; Ebrahimi, M. Grid-Supporting Inverters with Improved Dynamics. *IEEE Trans. Ind. Electron.* **2019**, *66*, 3655–3667. [CrossRef]
1. Xiong, Y.; Yao, W.; Wen, J.; Lin, S.; Ai, X.; Fang, J.; Wen, J.; Cheng, S. Two-Level Combined Control Scheme of VSC-MTDC Integrated Offshore Wind Farms for Onshore System Frequency Support. *IEEE Trans. Power Syst.* **2021**, *36*, 781–792. [CrossRef]
2. Kabsha, M.M.; Rather, Z.H. A New Control Scheme for Fast Frequency Support from HVDC Connected Offshore Wind Farm in Low-Inertia System. *IEEE Trans. Sustain. Energy* **2020**, *11*, 1829–1837. [CrossRef]
3. Sanz, I.M.; Chaudhuri, B.; Strbac, G. Inertial Response from Offshore Wind Farms Connected Through DC Grids. *IEEE Trans. Power Syst.* **2015**, *30*, 1518–1527. [CrossRef]
4. Ma, S.; Geng, H.; Yang, G.; Pal, B.C. Clustering-Based Coordinated Control of Large-Scale Wind Farm for Power System Frequency Support. *IEEE Trans. Sustain. Energy* **2018**, *9*, 1555–1564. [CrossRef]
5. Li, G.; Ye, H. Hierarchical nonlinear model predictive control for frequency support of wind farm. *Int. J. Electr. Power Energy Syst.* **2021**, *129*, 106820. [CrossRef]
6. Yang, D.; Gao, H.-C.; Zhang, L.; Zheng, T.; Hua, L.; Zhang, X. Short-term frequency support of a doubly-fed induction generator based on an adaptive power reference function. *Int. J. Electr. Power Energy Syst.* **2020**, *119*, 105955. [CrossRef]
7. Xu, G.; Xu, L. Improved use of WT kinetic energy for system frequency support. *IET Renew. Power Gener.* **2017**, *11*, 1094–1100. [CrossRef]
8. Kheshti, M.; Ding, L.; Nayeripour, M.; Wang, X.; Terzija, V. Active power support of wind turbines for grid frequency events using a reliable power reference scheme. *Renew. Energy* **2019**, *139*, 1241–1254. [CrossRef]
9. Wang, S.; Tomsovic, K. Fast frequency support from wind turbine generators with auxiliary dynamic demand control. *IEEE Trans. Power Syst.* **2019**, *34*, 3340–3348. [CrossRef]
0. Yang, D.; Kim, J.; Kang, Y.C.; Muljadi, E.; Zhang, N.; Hong, J.; Song, S.-H.; Zheng, T. Temporary Frequency Support of a DFIG for High Wind Power Penetration. *IEEE Trans. Power Syst.* **2018**, *33*, 3428–3437. [CrossRef]
1. de Paola, A.; Angeli, D.; Strbac, G. Scheduling of Wind Farms for Optimal Frequency Response and Energy Recovery. *IEEE Trans. Control. Syst. Technol.* **2016**, *24*, 1764–1778. [CrossRef]
2. Yang, D.; Kang, M.; Kim, J.; Hong, J.; Kang, Y.C. Stable stepwise short-term frequency support of a DFIG-based wind farm. *Int. Trans. Electr. Energy Syst.* **2018**, *28*, e2495. [CrossRef]
3. Arani, M.F.M.; Mohamed, Y.A.-R.I. Analysis and Mitigation of Undesirable Impacts of Implementing Frequency Support Controllers in Wind Power Generation. *IEEE Trans. Energy Convers.* **2016**, *31*, 174–186. [CrossRef]
4. Ruttledge, L.; Flynn, D. Emulated Inertial Response from Wind Turbines: Gain Scheduling and Resource Coordination. *IEEE Trans. Power Syst.* **2016**, *31*, 3747–3755. [CrossRef]
5. Wang, X.; Gao, W.; Scholbrock, A.; Muljadi, E.; Gevorgian, V.; Wang, J.; Yan, W.; Zhang, H. Evaluation of different inertial control methods for variable-speed wind turbines simulated by fatigue, aerodynamic, structures and turbulence (FAST). *IET Renew. Power Gener.* **2017**, *11*, 1534–1544. [CrossRef]

mathematics

MDPI

Article

Optimal Location and Sizing of PV Generation Units in Electrical Networks to Reduce the Total Annual Operating Costs: An Application of the Crow Search Algorithm

Brandon Cortés-Caicedo [1], Luis Fernando Grisales-Noreña [1,2,*], Oscar Danilo Montoya [3,4], Miguel-Angel Perea-Moreno [5] and Alberto-Jesus Perea-Moreno [6,*]

1 Departamento de Mecatrónica y Electromecánica, Facultad de Ingeniería, Instituto Tecnológico Metropolitar Medellín 050036, Colombia
2 Estudiante de Doctorado, Departamento de Ingeniería Eléctrica, Campus Lagunillas s/n, University of Jaé Edificio A3, 23071 Jaén, Spain
3 Grupo de Compatibilidad e Interferencia Electromagnética, Facultad de Ingeniería, Universidad Distrital Francisco José de Caldas, Bogotá 110231, Colombia
4 Laboratorio Inteligente de Energía, Universidad Tecnológica de Bolívar, Cartagena 131001, Colombia
5 Grupo de investigación TEP178 Nuevas Tecnologías Aplicadas a la Agricultura y el Medioambiente, Universidad de Córdoba, 14071 Córdoba, Spain
6 Departamento de Física Aplicada, Radiología y Medicina Física, Universidad de Córdoba, Campus de Rabanales, 14071 Córdoba, Spain
* Correspondence: luisgrisales@itm.edu.co or lfgn0001@red.ujaen.es (L.F.G.-N.); aperea@uco.es (A.-J.P.-M.)

Abstract: This study presents a master–slave methodology to solve the problem of optimally locatir and sizing photovoltaic (PV) generation units in electrical networks. This problem is represente by means of a Mixed-Integer Nonlinear Programming (MINLP) model, whose objective function to reduce the total annual operating costs of a network for a 20-year planning period. Such cos include (i) the costs of purchasing energy at the conventional generators (the main supply noc in this particular case), (ii) the investment in the PV generation units, and (iii) their correspondin operation and maintenance costs. In the proposed master–slave method, the master stage uses th Discrete–Continuous version of the Crow Search Algorithm (DCCSA) to define the set of nodes whe the PV generation units will be installed (location), as well as their nominal power (sizing), and th slave stage employs the successive approximation power flow technique to find the value of th objective function of each individual provided by the master stage. The numerical results obtaine in the 33- and 69-node test systems demonstrated its applicability, efficiency, and robustness whe compared to other methods reported in the specialized literature, such as the vortex search algorith the generalized normal distribution optimizer, and the particle swarm optimization algorithm. A simulations were performed in MATLAB using our own scripts.

Keywords: crow search algorithm; discrete–continuous codification; master–slave strategy; locatio and sizing of photovoltaic generation units; reduction in total annual operating costs; alternatin current networks

MSC: 65K05; 65K10; 68N99; 90C26; 90C59

Citation: Cortés-Caicedo, B.; Grisales-Noreña, L.F.; Montoya, O.D.; Perea-Moreno, M.-A.; Perea-Moreno, A.-J. Optimal Location and Sizing of PV Generation Units in Electrical Networks to Reduce the Total Annual Operating Costs: An Application of the Crow Search Algorithm. *Mathematics* **2022**, *10*, 3774. https://doi.org/10.3390/math10203774

Academic Editors: Atanda Raji and Khaled M. Abo-Al-Ez

Received: 7 September 2022
Accepted: 10 October 2022
Published: 13 October 2022

1. Introduction

1.1. General Context

In recent years, the rapid modernization of countries, the fast advancement of tecl nology, and the ongoing population growth have led to a significant increase in electric energy consumption and, therefore, to a looming global energy crisis. As a result, conve tional energy resources—often used to meet the energy demand—have begun to run out [1 In fact, energy sources based on, for instance, coal, natural gas, and oil are unable to mee the current energy demands [2]. Additionally, these sources have a negative impact on the environment. This issue has attracted the attention of researchers, who are constantly looking for ways to transform the current energy system (based on conventional energy resources) into a cleaner one (based on renewable energy resources) in order to meet the energy demand while also protecting the environment [1,3].

Distributed energy resources, such as photovoltaic (PV) generation units and wind turbines, are becoming increasingly popular and widely employed because they are considered clean and limitless energy sources. Likewise, better and more advanced technologies are being developed to use these resources in a practical and cost-effective manner [1,4]. In Colombia, thanks to the abundance of the solar resource in the Caribbean and Pacific regions, PV systems are the most extensively employed technology to produce electricity and displace fossil fuel generation [5]. The country, thus, has a huge potential for integrating PV generation sources, which will allow it to propose solutions that are both energetically and environmentally sustainable in order to meet the energy demand while reducing the use of fossil fuels [6].

Optimally designing PV generation units and installing them in electrical systems, however, is a challenging task because an improper planning may result in overvoltages and overcurrents in the nodes and lines that make up the electrical system. This, in turn, can cause a variety of problems [7], including an increase in energy losses and a decline in energy quality, which affect not only the network's operational capabilities but also its financial viability by failing to meet the standards enforced by regulatory bodies as the Commission for the Regulation of Energy and Gas in Colombia (CREG, by its Spanish acronym) [8]. This poses a challenge for engineers in charge of the design, planning, and operation of electrical networks, as effective strategies for the proper integration of PV generation units must be devised. These strategies must be able to ensure the project's financial viability over a specific time horizon as well as the safe and reliable provision of the service while adhering to the standards set out by regulatory bodies.

A lot of research has been completed, from a technical approach, on the optimal location and sizing of PV generation units in electrical systems to reduce power losses [9], improve voltage profiles [10], and enhance voltage stability [11]. Such an approach, however, does not analyze the financial viability of the proposed solutions, as the costs associated with the investment in PV generation units and their related operation and maintenance costs, as well as the planning horizon, are not taken into account in the calculation of the objective function. Considering these two aspects as well would ensure that the solution is both technically and economically feasible [12,13].

1.2. State of the Art

In recent years, different combinatorial optimization techniques mostly based on master–slave methodologies that use a discrete–continuous codification have been developed to solve the problem of optimally locating and sizing PV generation units in electrical systems. These techniques consider financial aspects that respect the technical and operating conditions of the network.

A discrete–continuous codification allows optimal location and sizing problems to be solved jointly. For example, the authors of [14] presented a master–slave methodology that combines the Discrete–Continuous version of the Chu and Beasley Genetic Algorithm (DCCBGA) and the successive approximation power flow method. Their main goal was to reduce the total annual operating costs of electrical networks, including the costs associated with the investment in PV systems, as well as their corresponding maintenance and operation costs. In order to assess the applicability and effectiveness of their proposed methodology, the 33- and 69-node test systems were used. In addition, they compared the results obtained by their proposed methodology with the exact solution to the Mixed-Integer Nonlinear Programming (MINLP) model (which represented the problem being addressed) produced by the BONMIN solver of the General Algebraic Modeling System (GAMS). Additionally, the authors performed a statistical analysis and examined processing times in order to assess the repeatability and robustness of the algorithm. Importantly, the mathematical model developed in their study has been used as the basis to design new optimization techniques that employ a discrete–continuous codification to solve the problem of optimally integrating PV generation units in electrical networks.

In [15], the authors employed the Newton Metaheuristic Algorithm (NMA) for solving the problem of optimally siting and sizing PV generation units in the IEEE 34- and 85-node test systems. Their main goal was also to minimize the annual operating costs of electrical networks. When compared to the results obtained by the BONMIN solver and the DCCBGA, their proposed methodology was found to be effective. The authors did not perform a statistical analysis or examine processing times to evaluate the repeatability and robustness of their algorithm. For their part, the authors of [12] proposed using the Discrete–Continuous Vortex Search Algorithm (DCVSA) to solve the problem of optimally integrating PV generation units in alternating and direct current networks. In fact, this is the first study to assess the reduction in total annual operating costs in both types of networks. According to the results obtained in the 33- and 69-node test systems, the proposed methodology was capable of finding an optimal solution while observing the voltage and current constraints established in the MINLP model. The authors also compared their proposed methodology with the BONMIN solver and the DCCBGA, conducted statistical analyses, and evaluated the processing times with the purpose of evaluating the repeatability and robustness of the algorithm.

In [16], as in the previous study, the authors tested a Discrete–Continuous version of the Generalized Normal Distribution Optimizer (DCGNDO) in the 33- and 69-node test systems. When compared to the DCVSA, their proposed methodology achieved significant reductions in the total annual operating costs. The authors, however, did not perform a statistical analysis or examine processing times, which does not guarantee the repeatability or the robustness of the algorithm. Finally, the work by [17] employed a Discrete–Continuous version of the Parallel Particle Swarm Optimization algorithm (DCPPSO) algorithm. When compared to other optimization methodologies that also use a discrete–continuous codification, their proposed methodology obtained the best results in terms of best solution, processing time, and standard deviation in the 33- and 69-node test systems. Moreover, the authors conducted a statistical analysis and examined processing times, thus proving that their proposed methodology is a robust and reliable tool to solve the problem of optimally integrating D-STATCOMs units in electrical systems.

1.3. Motivations, Contributions, and Scope

From the literature review, the importance of considering an objective function focused on financial aspects was identified. This is because the primary goals are to (i) provide an efficient and cost-effective service, (ii) reduce the costs associated with the investment in PV generation units and their operation while adhering to the standards of energy quality, voltage, and service, and (iii) meet the energy demand in compliance with the regulations in force. In addition, it was observed that the Crow Search Algorithm (CSA) has not been yet used to solve the optimization problem addressed in this study.

Therefore, this paper presents a master–slave method whose master stage uses a Discrete–Continuous Crow Search Algorithm (DCCSA) to solve the problem of optimally locating and sizing PV generation units in electrical systems, and the slave stage employs the successive approximation power flow technique to find the value of the objective function, which is the reduction in the total annual operating costs of an electrical system over a 20-year planning and operation horizon. These costs include (i) the annual costs of purchasing energy at the conventonal generators (Slack node in this particular case); (ii) the annual investment in PV generation units; and (iii) their corresponding annual operation and maintenance costs. The following are the main contributions of this study to solving the optimization problem under analysis:

- A thorough description of the mathematical model that represents the problem of optimally locating and sizing PV generation units in electrical systems. This model has, as the objective function, the reduction in the total annual operating costs and observes the set of constraints that represent the behavior of electrical networks in a DG scenario.
- A new master–slave methodology to solve the MINLP model that represents the problem under study. In this methodology, the master stage uses a Discrete–Continuous version of the Crow Search algorithm (DCCSA) to define the set of nodes where the PV generation units will be installed (location), as well as their corresponding nominal power (sizing). Meanwhile, the slave stage employs the successive approximation power flow method to evaluate the total annual operating costs of the network.
- A new master–slave methodology that finds a global optimal solution to the problem of optimally locating and sizing PV generation units in electrical systems and produces the best results in terms of solution quality and repeatability.

Figure 1 shows the graphical abstract that summarizes the contents of the article. The main idea of this research paper is to use a CSA-based master–slave methodology that employs the successive approximations power flow method for solving the problem of siting and sizing PV generation units to minimize the total annual operating costs in electrical distribution systems.

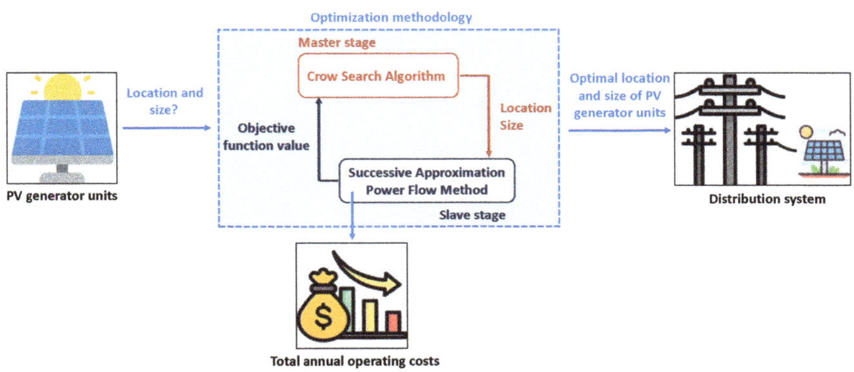

Figure 1. Graphic proposed methodology.

1.4. Structure of the Paper

This paper is organized as follows. Section 2 introduces the mathematical formulation of the problem regarding the optimal location and sizing of PV generation units in electrical systems, with the objective function being the reduction in the total annual operating costs. Section 3 presents the proposed methodology, which combines the DCCSA and the successive approximation power flow technique. Section 4 presents the main characteristics of the 33- and 69-node test systems, the generation and demand curves employed, and the parametric information necessary to find the value of the fitness function. Section 5 presents a discussion on the results obtained for the problem under analysis as well as the total annual operating costs. Finally, Section 6 outlines the conclusions and future lines of research.

2. Mathematical Formulation

The problem of optimally locating and sizing PV generation units in electrical systems can be expressed and solved using an MINLP model. In this model, the decision variables (i.e., those of a binary nature) are associated with the selection of the nodes where the PV generation units will be installed [18], whereas its nonlinearities appear in the formulation of the power flow due to the nonlinear nature of its equations [19,20].

The next subsections present the objective function and the set of constraints that represent the problem under analysis.

2.1. Formulation of the Objective Function

The main focus of power system planning projects is to recover the initial investment made by the grid operator. In this case study, in order to recover the investment made to integrate the PV generation units into electricity systems, the aim is the minimization of the total power purchase costs at the node that interconnects the distribution system with the transmission/subtransmission grid. Note that the installation and maintenance costs of the PV generation units are charged to the distribution system operators (DSO), as they are the body in charge of distributing and managing the power required to supply the demand of the users together with the system losses, in order to provide a service with high power quality that is as economical as possible.

In this context, the objective function considered is the reduction in the total annual operating costs of an electrical network, which will allow the initial investment to be recovered. Such costs include (i) the costs of purchasing energy at the main supply node (also known as the slack node or substation node), (ii) the investment in PV generation units, and (iii) their corresponding maintenance and operation costs. Such an objective function is presented in (1)–(4).

$$\min \ A_{cost} = f_1 + f_2 + f_3, \tag{1}$$

$$f_1 = C_{kWh} T f_a f_c \left(\sum_{h \in \mathcal{H}} \sum_{k \in \mathcal{N}} p_{k,h}^{cg} \Delta h \right), \tag{2}$$

$$f_2 = C_{pv} f_a \left(\sum_{k \in \mathcal{N}} p_k^{pv} \right), \tag{3}$$

$$f_3 = C_{O\&M} T \left(\sum_{h \in \mathcal{H}} \sum_{k \in \mathcal{N}} p_{k,h}^{pv} \Delta h \right), \tag{4}$$

with

$$f_a = \left(\frac{t_a}{1 - (1 + t_a)^{-N_t}} \right),$$

$$f_c = \left(\sum_{t \in \mathcal{T}} \left(\frac{1 + t_e}{1 + t_a} \right)^t \right),$$

where A_{cost} is the objective function value and represents the total annual operating costs of the system. f_1 denotes the annual costs of purchasing energy at the main supply node. f_2 is the annual investment in PV generation units; and f_3 is their corresponding operation and maintenance costs. C_{kWh} represents the average energy purchasing cost at the substation node. T is the number of days in an ordinary year (365). f_a denotes the annuity factor that can be used to calculate the regular payments that the network operator must make, which are dependent on the expected internal return rate (t_a) and the planning horizon (years, N_t). f_c represents the factor associated with the increase in the cost of energy during the planning period, which depends on the annual percentage rise in the cost of energy expected by the network operator (t_e). $p_{k,h}^{cg}$ is the active power produced by each conventional generator connected to node k in time period h. Δh denotes the time during which the electrical variables are assumed constant (i.e., 1 h for a one-day operation scenario). C_{pv} represents the average installation cost for 1 kW of PV power. p_k^{pv} is the nominal power of each PV generation unit connected to node k. $C_{O\&M}$ denotes the maintenance and operation costs associated with each PV generation unit. $p_{k,h}^{pv}$ represents the active power produced by each PV generation unit connected to node k in time period h. Finally, $\mathcal{N}$, $\mathcal{H}$, and $\mathcal{T}$ are the sets containing all the network nodes, the time periods in a one-day operation scenario, and the number of years in the planning horizon, respectively.

2.2. Set of Constraints

The set of constraints of the problem regarding the optimal placement and sizing of PV generation units in electrical systems represents the different operational limits found in such systems, such as active and reactive power balance, maximum and minimum capacities of each device, and voltage regulation bounds at each node in the system. Such constraints are presented in (5)–(13).

$$p^{cg}_{k,h} + p^{pv}_{k,h} - P^{d}_{k,h} = v_{k,h} \sum_{j \in \mathcal{N}} Y_{kj} v_{j,h} \cos\left(\theta_{k,h} - \theta_{j,h} - \varphi_{kj}\right), \{\forall k \in \mathcal{N}, \ \ \forall h \in \mathcal{H}\}, \quad (5)$$

$$q^{cg}_{k,h} - Q^{d}_{k,h} = v_{k,h} \sum_{j \in \mathcal{N}} Y_{kj} v_{j,h} \sin\left(\theta_{k,h} - \theta_{j,h} - \varphi_{kj}\right), \{\forall k \in \mathcal{N}, \ \ \forall h \in \mathcal{H}\}, \quad (6)$$

$$p^{pv}_{k,h} = p^{pv}_{k} G^{pv}_{h}, \{\forall k \in \mathcal{N}, \ \ \forall h \in \mathcal{H}\}, \quad (7)$$

$$P^{cg,\min}_{k} \le p^{cg}_{k,h} \le P^{cg,\max}_{k}, \{\forall k \in \mathcal{N}, \ \ \forall h \in \mathcal{H}\}, \quad (8)$$

$$Q^{cg,\min}_{k} \le q^{cg}_{k,h} \le Q^{cg,\max}_{k}, \{\forall k \in \mathcal{N}, \ \ \forall h \in \mathcal{H}\}, \quad (9)$$

$$x_k P^{pv,\min}_{k} \le p^{pv}_{k} \le x_k P^{pv,\max}_{k}, \{\forall k \in \mathcal{N}\}, \quad (10)$$

$$v^{\min}_{k} \le v_{k,h} \le v^{\max}_{k}, \{\forall k \in \mathcal{N}, \ \ \forall h \in \mathcal{H}\}, \quad (11)$$

$$\sum_{k \in \mathcal{N}} x_k \le N^{avail}_{pv}, \quad (12)$$

$$x_k \in \{0,1\}, \{\forall k \in \mathcal{N}\}, \quad (13)$$

where $P^{d}_{k,h}$ and $Q^{d}_{k,h}$ are the active and reactive power demanded at node k in time period h, respectively. $q^{gc}_{k,h}$ denotes the reactive power produced by each conventional generator connected to node k in time period h. $v_{k,h}$ and $v_{j,h}$ denote the voltages at nodes k and j in time h, respectively. Y_{kj} is the admittance relating nodes k and j and whose angle is $\varphi_{k,j}$. $\theta_{k,h}$ and $\theta_{j,h}$ denote the voltage angles at nodes k and j in time period h, respectively. G^{pv}_{h} represents the expected PV generation curve in the area of influence of the electrical system. $P^{cg,\min}_{k}$ and $P^{cg,\max}_{k}$ are the active power bounds for each conventional generator connected to the node k, whereas $Q^{cg,\min}_{k}$ and $Q^{cg,\max}_{k}$ denote the reactive power bounds for each conventional generator connected to node k. $P^{fv,\min}_{k}$ and $P^{fv,\max}_{k}$ represent the active power bounds for the PV generation units connected to node k. x_k is the binary variable in charge of locating each PV generation unit in a node k of the system. $v^{\min}_{k}$ and $v^{\max}_{k}$ denote the minimum and maximum voltage regulation values allowed at each node that makes up the electrical system. Finally, N^{avail}_{fv} represents a constant parameter related to the maximum number of PV generation units available for installation along the network.

2.3. Model Interpretation

The model presented in (1)–(13), which represents the problem of optimally locating and sizing PV generation units in electrical systems, can be interpreted as follows: Equation (1) defines the objective function of the problem, which is the sum of (i) the annual energy purchasing costs at the node that connects the electrical system with a transmission/distribution network, as shown in Equation (2); (ii) the annual investment in PV systems, as shown in Equation (3); and (iii) their corresponding maintenance and operation costs, as shown in Equation (4). Equality Equations (5) and (6) represent the active and reactive power balance at each system node in each time period, respectively. These are the most complex constraints in the problem under analysis, and, due to their nonlinear and nonconvex nature, they require numerical methods to be properly solved [21]. Equation (7) establishes that the active power produced by the PV generation units varies depending on their nominal power and the expected generation curve in the area of influence of the network. Inequality Equations (8) and (9) define the minimum and maximum active and reactive power injected by the conventional generators. Inequality Equation (10) determines the minimum and maximum active power produced by the PV generation units that will be installed along the system. Box-type Equation (11) presents the lower and upper voltage regulation bounds for all nodes and time periods. Equation (12) defines the maximum number of PV generation units available for installation along the network. Equation (13) shows the binary nature of the decision variable x_k.

Note that the MINLP model in Equations (1)–(13) is a general representation of the problem under study. The two main drawbacks of this model are (i) its nonlinearities and nonconvexities in the active and reactive power balance equations and (ii) the fact that it combines binary and integer variables. Consequently, as there might be multiple solutions to this model, which will be local optima, several authors have proposed using master–slave methods to solve it because they enable separating the continuous optimization problem from its discrete part [22].

Therefore, to solve the problem under study, this research presents a master–slave methodology that combines the DCCSA and the successive approximations method. This technique has not been reported in the specialized literature and belongs to the main contributions of this work.

3. Proposed Solution Methodology

To solve the problem of optimally locating and sizing PV generation units in electrical systems, which was described above, this study proposes using a master–slave methodology that employs a Discrete–Continuous version of the Crow Search Algorithm (DCCSA) [23] in the master stage and the successive approximations power flow method [24] (slave stage). In this methodology, the master stage defines the set of nodes where the PV generation units will be installed, as well as the size of such units, and the slave stage evaluates the objective function and constraints associated with network operation, which were presented in (5)–(13).

The next subsections present the codification used to represent the problem under analysis as well as each component of the proposed methodology (i.e., master stage and slave stage).

3.1. Proposed Codification

The DCCSA is the cornerstone of our proposed solution methodology because it is responsible for determining the optimal locations and sizes of PV generation units in electrical networks. To that end, each individual in the metaheuristic algorithm uses a discrete–continuous codification of the form

$$C_i^t = \left[2,\ z,\ ...,\ n\ |0.0000,\ p_z^{pv},\ ...,\ 2.4000\right]; i = 1, 2, ..., N_i, \quad (14)$$

where C_i^t is an individual i from population C at iteration t, whose size is $1 \times (2N_{pv}^{avail})$. z is a random number that defines the node where each PV generation unit will be installed. This number can take a value between 2 and the number of nodes in the system (i.e., n), which means that the PV generation units are only placed in the demand nodes. Finally, N_i is the number of individuals in the population.

As can be seen in Equation (14), each individual in the population has two components: (i) the first N_{pv}^{avail} parameters of the solution vector, which define the demand nodes where the PV generation units are to be installed, and (ii) the subsequent N_{pv}^{avail} parameters of the solution vector, which determine the optimal sizes of each PV generation unit to be installed in the system.

The main advantage of this codification is that it allows the optimal location and sizing problem to be solved in a single stage by transforming the MINLP model defined from (1)–(13) into a nonlinear programming model. Consequently, the solution space can be efficiently explored and exploited in shorter processing times [16].

3.2. Master Stage: The Discrete–Continuous Version of the Crow Search Algorithm (DCCSA)

The DCCSA is a bio-inspired optimization metaheuristic algorithm, which is based on the intelligent behavior of crow flocks [23]. In the animal kingdom, crows are considered the smartest birds because they are capable of memorizing and remembering faces, using tools, communicating with one another, and properly feeding throughout the year [23]. They are known to be ambitious birds because they compete with one another for better food sources and pay attention to where other birds hide their food to steal it [25]. After stealing food, crows take the necessary precautions to avoid becoming victims, such as changing their hiding places and course [26].

This behavior can be modeled mathematically by considering the following simple principles in order to properly explore and exploit the solution space [23]:

- ✓ Crows live in flocks.
- ✓ Crows remember where they hide their food.
- ✓ Crows follow other crows to steal their food.
- ✓ Crows protect their hiding places from theft via stochastic processes.

3.2.1. Initial Population

The DCCSA is a population-based algorithm. The population in this algorithm consists of crows that are randomly located in the environment, which allows it to start exploring and exploiting the solution space. The structure of the initial population of crows is as shown below:

$$C^t = \begin{bmatrix} C^t_{11} & C^t_{12} & \cdots & C^t_{1N_v} \\ C^t_{21} & C^t_{22} & \cdots & C^t_{2N_v} \\ \vdots & \vdots & \ddots & \vdots \\ C^t_{N_i1} & C^t_{N_i2} & \cdots & C^t_{N_i,N_v} \end{bmatrix}, \tag{15}$$

where C^t is the population of crows at iteration t, and N_v is the number of variables or the dimension of the solution space, that is, the number of PV generation units to be installed in the electrical system and their sizes, i.e., $2N_{pv}^{ava}$.

To generate an initial population of crows that respects the structure shown in (14), Equation (16) is used. This equation creates a matrix of random numbers (within the lower and upper limits) containing all possible solutions.

$$C^0 = y_{\min} ones(N_i, N_v) + (y_{\max} - y_{\min}) rand(N_i, N_v) \tag{16}$$

In (16), $ones(N_i, N_v) \in \mathbb{R}^{N_i \times N_v}$ is a matrix filled with ones. $rand(N_i, N_v) \in \mathbb{R}^{N_i \times N_v}$ is a matrix filled with random numbers between 0 and 1, which are generated with a uniform distribution. Finally, $y_{\min} \in \mathbb{R}^{N_v \times 1}$ and $y_{\max} \in \mathbb{R}^{N_v \times 1}$ are vectors that represent the lower and upper bounds of the solution space, respectively:

$$y_{\min} = \begin{bmatrix} y_{1,\min} \\ y_{2,\min} \end{bmatrix}, y_{\max} = \begin{bmatrix} y_{1,\max} \\ y_{2,\max} \end{bmatrix},$$

where $y_{1,\min} \in \mathbb{R}^{N_{pv}^{ava} \times 1}$ and $y_{1,\max} \in \mathbb{R}^{N_{pv}^{ava} \times 1}$ denote the lower and upper bounds, respectively, of the decision variables related to the locations of the PV generation units in the demand nodes, and $y_{2,\min} \in \mathbb{R}^{N_{pv}^{ava} \times 1}$ and $y_{2,\max} \in \mathbb{R}N_{pv}^{ava} \times 1$ are the lower and upper bounds, respectively, of the decision variables associated with the sizes of the PV generation units.

Finally, at each iteration (t), each crow (i) in the population memorizes the position of its hiding place, as shown in Equation (17). This equation thus stores the position of the best food source found thus far by each crow. Note that crows memorize their best experience thanks to the fact that they move around in their environment in search of the best food source.

$$M^t = \begin{bmatrix} C_{11}^t & C_{12}^t & \cdots & C_{1N_v}^t \\ C_{21}^t & C_{22}^t & \cdots & C_{2N_v}^t \\ \vdots & \vdots & \ddots & \vdots \\ C_{N_i1}^t & C_{N_i2}^t & \cdots & C_{N_i,N_v}^t \end{bmatrix} \tag{17}$$

3.2.2. Crows' Movement

One could say that in iteration t, crow j wants to visit its hideout (i.e., best food sourc found thus far), which is at position M_j^t, and crow i decides to follow crow j to be near i cache. At this point, there are two possible scenarios:

1. Scenario 1: Search
   In this scenario, crow j is unaware that crow i is following it. Hence, i can get close t the cache of crow j and updates its position in the solution space. This new positio can be modeled mathematically as follows:

$$C_i^{t+1} = C_i^t + rand\, fl\, (M_j^t - C_i^t), \tag{18}$$

   where *rand* is a random number between 0 and 1, generated with a uniform distrib tion, and fl is the flight length of crow i. As per [23], small values of fl allow for local exploration of the solution space (close to C_i^t), whereas large values of fl allo for a global exploration of the solution space (far from C_i^t).
2. Scenario 2: Evasion
   In this scenario, crow j is aware that crow i is following it. Hence, to prevent i hidden food from being stolen, it tries to fool crow i by moving to a random positio in the solution space.

These two scenarios can be summarized as follows:

$$C_i^{t+1} = \begin{cases} C_i^t + rand\, fl\, (M_j^t - C_i^t) & \text{If } r_j \geq A_p \\ \text{a random position} & \text{otherwise} \end{cases}, \tag{19}$$

where r_j is a random number between 0 and 1, which is generated by a uniform distributio and A_p is the probability that crow j finds out that crow i is following it.

3.2.3. Memory Updating

Once the position of the crows is updated considering the two scenarios describe above, the new position of the food source must be memorized based on its quality. Thu if the fitness function value of the new food source is better than the fitness function value c the previously memorized food source, crows update their memory with the new positio

$$M_i^{t+1} = \begin{cases} C_i^{t+1} & \text{If } F_F(C_i^{t+1}) < F(M_i^t) \\ M_i^t & \text{otherwise} \end{cases}, \tag{20}$$

where $F_f(\cdot)$ represents the fitness function to minimize.

3.2.4. General Implementation of the DCCSA

Algorithm 1 shows how the DCCSA is implemented to solve the problem addresse in this study.

3.3. Slave Stage: Successive Approximation Power Flow Method

The successive approximation method used to solve the power flow in electrica systems was first introduced by Montoya and Gil-González in [24]. The active and reactiv power balance equations given by (5) and (6), respectively, can be solved iteratively usin this method. It enables the slave stage to evaluate the fitness function value for eac individual in the population of crows while ensuring that the constraints specified in th

MINLP model (described in Section 2) are respected. Likewise, this method was selected because it requires short processing times and rapidly converges toward the solution.

Algorithm 1: Crow search algorithm used to solve optimization problems

Define parameters N_i, Nv, A_p, fl, $t_{\max}$, $y^{\min}$, and $y^{\max}$;
Generate the initial population using Equation (16);
Calculate the fitness function value (see Equation (24)) of each individual;
Initialize the memory (M_i^0) of each crow (i);
for $t \leq t_{\max}$ **do**
  for $j = 1 : N_i$ **do**
    Randomly select a crow (i);
    if $rand_j \geq A_P$ **then**
      Generate the new position of crow i using Equation (18);
    else
      Generate a random position for crow i;
  Evaluate the fitness function value of crows' new position (see Equation (24));
  Update crows' memory (M^{t+1}) using Equation (20);
Result: The best solution is found for $C_i^{t_{\max}}$, and its fitness function is $F(C_i^{t_{\max}})$.

The recursive formula that can be employed to solve the power flow presented in (5) and (6) is given by

$$\mathbb{V}_{d,h}^{t+1} = -\mathbf{Y}_{dd}^{-1}\left[\mathbf{diag}^{-1}(\mathbb{V}_{d,h}^{t,*})(\mathbb{S}_{d,h}^{*} - \mathbb{S}_{pv,h}^{*}) + \mathbf{Y}_{ds}\mathbb{V}_{s,h}\right], \tag{21}$$

where t is the iteration counter. $\mathbb{V}_{d,h}$ denotes the vector that contains the voltage at the demand nodes for each time period h, i.e., the variables of interest. $\mathbf{Y}_{ds}$ is the component of the admittance matrix that associates the slack node with the demand nodes, whereas $\mathbf{Y}_{dd}$ denotes the component of the admittance matrix that relates the demand nodes to each other. $\mathbb{S}_{d,h}$ represents the vector in the complex domain that contains the active and reactive power demanded at the load nodes for each time period h. $\mathbb{S}_{pv,h}$ is the vector in the complex domain that contains the active power produced by each PV unit for each time period h. $\mathbb{V}_{s,h}$ denotes the vector that contains the voltage at the terminals of the substation node for each time period h, which is a known parameter in the solution of the power flow. Finally, $\mathbf{diag}(z)$ represents a diagonal matrix made up of the elements of vector z.

To assess the convergence of the iterative process, the criterion shown in (22) is used. According to this criterion, the maximum difference between the demand voltages (i.e., $\mathbb{V}_{d,h}$) for each time period h in two consecutive iterations should be below a predefined threshold.

$$\max_{h}\left\{||\mathbb{V}_{d,h}^{t+1}| - |\mathbb{V}_{d,h}^{t}||\right\} \leq \zeta \tag{22}$$

In (22), ζ represents the convergence error, which, for the purposes of this study, will be 1×10^{-10}, as recommended by the authors of [24].

Once the power flow is solved for all time periods h using the successive approximation method, the power produced at the terminals of the substation node must then be calculated, as follows:

$$\mathbb{S}_{s,h}^{*} = \mathbf{diag}(\mathbb{V}_{s,h}^{*})(\mathbf{Y}_{ss}\mathbb{V}_{s,h} + \mathbf{Y}_{sd}\mathbb{V}_{d,h}), \tag{23}$$

where $\mathbb{S}_{s,h}$ denotes the vector in the complex domain that contains the active and reactive power produced at the slack node for each time period h. $\mathbf{Y}_{ss}$ is the component of the admittance matrix associated with the slack node, while $\mathbf{Y}_{sd}$ is the component of the admittance matrix that associates the slack node to the demand ones.

Note that the value of f_1 can be obtained by solving (23). Similarly, the solution provided by each individual in the master stage, which respect the codification established in (14), can be used to obtain the value of f_2 and f_3. However, to rule out possible infeasible solutions, which violate the boundaries of the solution space, the objective function described in (1) is replaced by the fitness function shown in (24) [27,28].

$$F_f = A_{cost} + \beta_1 \max_h\{0, |\mathbb{V}_{d,h}| - v^{\max}\} - \beta_2 \min_h\left\{0, |\mathbb{V}_{d,h}| - v^{\min}\right\} - \beta_3 \min_h\left\{0, \mathrm{real}(\mathbb{S}_{s,h} - P_k^{gc,\min})\right\} \quad (24)$$

In (24), F_f is the value of the fitness function, and β_1, β_2, and β_3 denote the penalty factors applied to the objective function. These penalty factors come into play when the solutions provided by the master stage violate the voltage regulation constraints or power generation capacities at the substation node. In this study, the value of such penalty factors is 1×10^6, and each penalty factor has its corresponding unit.

One of the main advantages of using a fitness function is that it helps the metaheuristic optimization algorithm to efficiently explore and exploit the solution space. If all the constraints presented in (5)–(13) are met, the final value of F_f equals the original value of the objective function (A_{cost}). If not, the solution is discarded as a possible optimal solution [29].

4. Test Systems

To validate the master–slave methodology proposed in this paper to solve the problem of optimally locating and sizing PV generation units in electrical systems, the 33- and 69-node test systems were used, both of which have a radial topology [30]. These test systems are selected for the sake of comparison, as they have been previously used in the literature to solve the problem of locating and sizing PV generation units. This allows evaluating and comparing the best response, repeatability, and processing times of the proposed master–slave methodology. The next subsections present the main characteristics of each test system.

4.1. First Test Feeder: 33-Node Test System

This system consists of 33 nodes and 32 distribution lines, as shown in Figure 2. It operates at a base voltage of 12.66 kV and a base power of 100 kVA. In the peak power consumption scenario, the loads of this system demand $(3715 + j2300)$ kVA. Its parametric information can be found in [31].

4.2. Second Test Feeder: 69-Node Test System

This system consists of 69 nodes and 68 distribution lines, as illustrated in Figure 2. It operates at a base voltage of 12.66 kV and a base power of 100 kVA. In the peak power consumption scenario, the loads of this system demand $(3890.7 + j2693.6)$ kVA. Its parametric information can be found in [31].

4.3. Calculation of the Objective Function

To calculate the value of the fitness function defined in (24), the parametric data shown in Table 1 were used [32,33].

To assess the impact of integrating PV generation units in the systems described above, typical generation and demand curves reported for Medellín (Colombia) were used, which are illustrated in Figure 3.

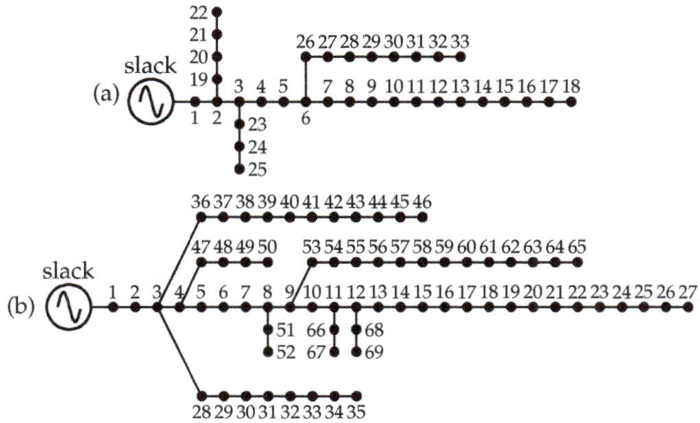

Figure 2. Single-line diagram of the two test feeders used in this study: (**a**) 33-node test system and (**b**) 69-node test system.

Table 1. Parameters used to calculate the objective function.

Parameter	Value	Unit	Parameter	Value	Unit
C_{kWh}	0.1390	USD/kWh	T	365	days
t_a	10	%	N_t	20	years
Δh	1	h	t_e	2	%
C_{pv}	1036.49	USD/kWp	$C_{O\&M}$	0.0019	USD/kWh
N_{pv}^{avail}	3	-	ΔV	± 10	%
$P_k^{pv,\min}$	0	kW	$P_k^{pv,\max}$	2400	kW
β_1	1×10^6	USD/V	β_2	1×10^6	USD/V
β_3	1×10^6	USD/W	-	-	-

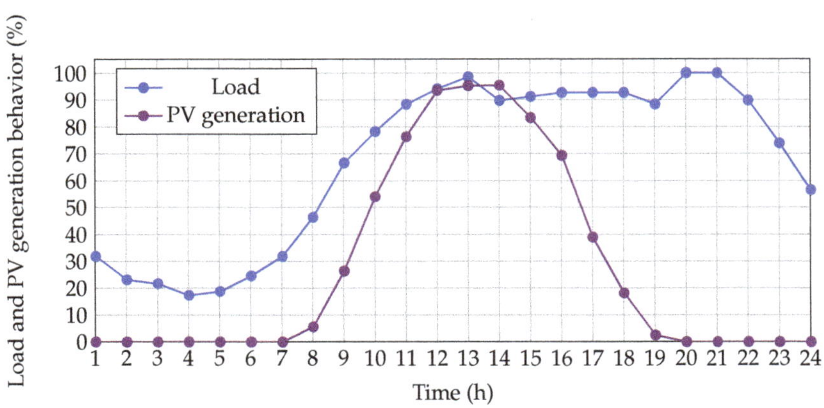

Figure 3. Typical behavior of the generation and demand curves reported for Medellín (Colombia).

5. Numerical Results and Discussion

This section discusses the numerical results obtained by the DCCSA in solving the problem of optimally locating and sizing PV generation units in the two test systems under analysis. To show the efficiency of the proposed metaheuristic algorithm, it was compared against the following six methods, which have also been used to solve the same problem: (i) the BONMIN solver of the GAMS (exact solution to the MINLP model) [14], (ii) the Discrete–Continuous version of the Chu and Beasley Genetic Algorithm (DC-CBGA) [14], (iii) the Discrete–Continuous version of the Newton Metaheuristic Algorithm

(DCNMA) [15], (iv) the Discrete–Continuous version of the Vortex Search Algorithm (DCVSA) [12], (v) the Discrete–Continuous version of the Generalized Normal Distribution Optimizer (DCGNDO) [16], and (vi) the Discrete–Continuous version of the Parallel Particle Swarm Optimization (DCPPSO) algorithm [17].

For both test systems, this study considered installing three PV generation units, each with a maximum capacity of 2400 kW. All simulations were performed in MATLAB (version 2022*a*) using our own scripts on a desktop computer with an Intel(R) Core(TM) i9-119(CPU@2.50Ghz processor and 64.0 GB RAM, which was running 64-bit Windows 10 Pro.

5.1. DCCSA Parameters

The information presented in Table 2 was used to implement the master–slave methodology proposed in this study to solve the problem of optimally locating and sizing PV generation units in electrical systems.

Table 2. Parameters of the DCCSA employed in the master stage.

Parameter	DCCSA
Number of individuals (N_i)	87
Maximum iterations ($t_{\max}$)	816
Flight length (fl)	2.8741
Awareness probability (A_p)	0.0046

To define the parameters shown in Table 2, the DCCSA was tuned using the Chu and Beasley genetic algorithm [34], with an initial population of 50 individuals and a maximum number of iterations of 350 for the 69-node test system because it is the largest of the two systems used to validate the proposed methodology. The tuning parameters were: (i) a population size (N_i) in the [1, 100] range, (ii) a maximum number of iterations ($t_{\max}$) in the [1, 1000] range, (iii) a flight length (fl) in the [0, 3.5] range, and (iv) an awareness probability (A_p) in the [0, 1] range. Moreover, the proposed methodology was evaluated 100 consecutive times to find the best, average, and worst values for the objective function. Additionally, the standard deviation and average time required by the algorithm to determine the optimal locations and sizes of the PV generation units were calculated for the two test systems under analysis.

5.2. Results Obtained in the First Test System under Analysis

5.2.1. Numerical Results

Table 3 shows the numerical results of the proposed methods and of those used as comparison in the 33-node test system. From left to right, this table specifies the methodology implemented, the nodes where the PV generation units were installed and their nominal power, the annual operating costs provided by each solution methodology, the reduction percentage obtained by each methodology with respect to the base case (values reported in the second row), the average processing time, and the standard deviation.

According to the information in Table 3, the solution provided by each metaheuristic algorithm outperformed that by the BONMIN solver (i.e., the exact solution to the MINLP model), which confirms that the presence of binary variables causes conventional optimization techniques to get stuck in local optima. Additionally, the proposed DCCSA, like the DCGNDO and the DCPPSO, managed to reduce the total annual operating costs by 1,000,783.62 USD/year when compared to the base case. This suggests that the global optimal solution for this test system is 2,699,671.76 USD/year, which is found by placing the PV generation units at nodes 10, 16, and 31, for a total installed capacity of 3647.65 kW. Finally, all the methods allowed a reduction of more than 26.95% with respect to the base case, with the DCCSA allowing the highest reduction (27.0449%). When compared to the other methods in terms of reduction in total annual operating costs, the DCCSA ou performed the BONMIN solver by 0.0581%, the DCNMA by 0.0151%, the DCCBGA by 0.0071%, the DCVSA by 0.0025%, and the DCGNDO and the DCPPSO by 0.0013%.

Table 3. Numerical results obtained in the 33-node test system.

Method	Location (Node)/Power (MW)	A_{cost} (USD/Year)	Reduction (%)	Time (s)	STD (%)
Base case	- - -	3,700,455.38	0	-	-
BONMIN	17/1.3539 18/0.2105 33/2.1452	2,701,824.14	26.9867	3.64	0
DCNMA	8/2.0961 16/1.2688 30/0.2770	2,700,227.33	27.0298	20.21	0.0812
DCCBGA	11/0.7605 15/0.9690 30/1.9060	2,699,932.29	27.0378	5.30	0.0452
DCVSA	11/0.7606 14/1.0852 31/1.8030	2,699,761.71	27.0424	170.23	0.0427
DCGNDO	10/1.0083 16/0.9137 31/1.7257	2,699,671.76	27.0436	268.69	0.0700
DCPPSO	10/1.0092 16/0.9137 31/1.7245	2,699,671.76	27.0436	8.32	0.0246
DCCSA	10/1.0093 16/0.9138 31/1.7246	2,699,671.76	27.0449	77.00	0.0037

5.2.2. Statistical Analysis

To show the effectiveness and robustness of the DCCSA in solving the problem of optimally locating and sizing PV generation units in electrical systems, it was run 100 consecutive times in the 33-node test system. The results of such validation are illustrated in Figure 4, which shows the improvements obtained by the DCCSA in terms of best solution, processing time, and standard deviation when compared to the other solution methodologies. The numbers in red indicate that the method used for comparison outperformed the DCCSA.

As observed in Figure 4, the DCCSA produced the best results in terms of reduction in annual operating costs when compared to the other methods. It outperformed the BONMIN solver by 0.0797%, the DCNMA by 0.0206%, the DCCBGA by 0.0097%, the DCVSA by 0.0053%, and the DCNGDO and the DCPPSO by 1×10^{-7}%.

Regarding processing times, the BONMIN solver, the DCNMA, the DCCBGA, and the DCPPSO were faster than the proposed solution methodology. When compared to the DCCSA, they reduced processing times by 95.2727%, 73.7529%, 93.1168%, and 89.1947%, respectively. Importantly, these differences in processing times are attributed to the fact that the population size employed for the proposed DCCSA included 77 more individuals than those used for the other methods. This means that at each iteration, the proposed algorithm had to evaluate 1848 power flows more than the other techniques. The DCCSA, however, was faster than the DCVSA and the DCNGDO; it reduced processing times by 121.0809% and 248.6963% when compared to the DCVSA and the DCNGDO, respectively. The processing times obtained by the DCCSA can be considered negligible when compared to the planning horizon chosen for this study (i.e., 20 years).

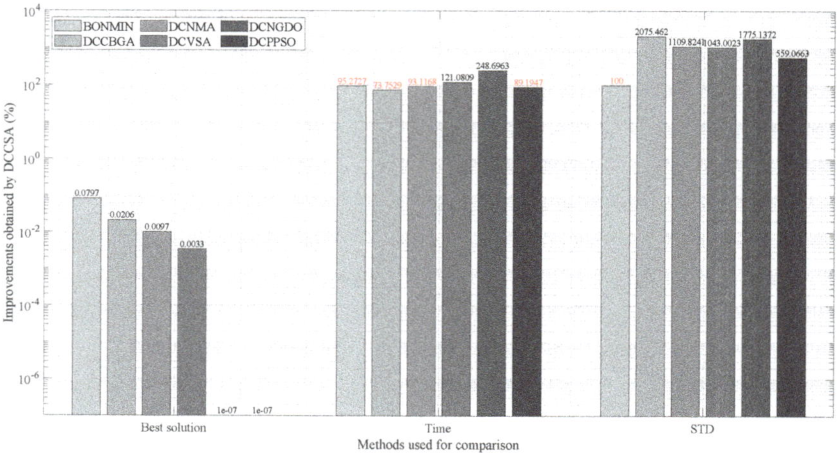

Figure 4. Improvements obtained by the DCCSA in the 33-node test system.

As for the standard deviation, the proposed DCCSA was superior to the other method as it achieved an improvement of 2075.4620% with respect to the DCNMA, of 1109.8241% with respect to the DCCBGA, of 1043.0023% with respect to the DCVSA, of 1775.1372% with respect to the DCNGDO, and of 559.0663% with respect to the DCPPSO. Note tha in this case, the DCCSA was not compared to the BONMIN solver. The reason for thi is that the solution of the BONMIN solver will always be the same because it is an exac solution to the MINLP model, so even if it is run 100 times, its standard deviation wil always be 0.

The results mentioned above confirm the effectiveness and reliability of the DCCSA as when solving the problem of optimally locating and sizing PV generation units i electrical networks to reduce the annual operating costs, it produced the best results i terms of solution quality and repeatability. Hence, the proposed methodology is regarde as the best option to solve such a problem in the 33-node test system.

5.2.3. Feasibility Check

To verify that the optimal solution yielded by the DCCSA is feasible, i.e., it satisfie the electrical constraints proposed by the mathematical model presented in (5)–(13) an considered in the formulation of the fitness function given by (24), the active powe generation at the main supply node was analyzed before and after implementing th solution obtained by the proposed methodology (see Figure 5).

When the solution provided by the DCCSA was implemented in the 33-node tes system, the power produced by the slack node was inversely proportional to the powe produced by the PV units. This means that as the power produced by the PV units increase from hour 7 to 14 (see Figure 3), the power produced at the substation node decreased unti it hit zero (right when the PV power reached its maximum value). Similarly, as the powe produced by the PV units decreased from hour 15 to 20, the power produced at the slac node increased. This proves that power generation respected the capacity constraint, as i yielded positive or zero values.

Finally, to confirm that the voltage profiles were within the regulation bounds (i.e $\pm 10\%$), the behavior of the minimum and maximum voltages was examined for all tim periods once the solution provided by the DCCSA was implemented (see Figure 6).

As may be concluded from Figure 6, the minimum and maximum voltage values i all time periods respected the voltage regulation bounds, as they remained within $\pm 10\%$ Additionally, the maximum voltage (i.e., 1.0322 pu) was recorded at node 16 when the PV units injected 100% of their nominal power. The minimum voltage (i.e., 0.9038 pu), whic coincided with the minimum voltage of the base case, was recorded at node 18 when th

PV units did not inject power and during the period of peak demand (from hour 20 to hour 21).

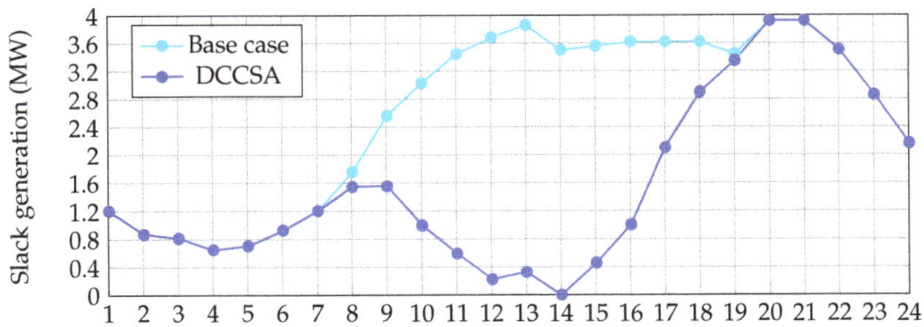

Figure 5. Impact of PV integration in the 33-node test system.

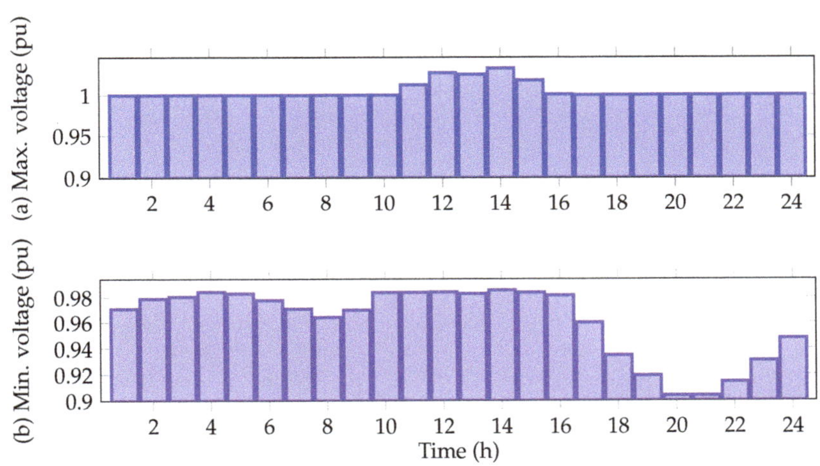

Figure 6. Voltage behavior during one day in the 33-node test system: (**a**) maximum voltage and (**b**) minimum voltage.

5.3. Results Obtained in the Second Test System under Analysis

5.3.1. Numerical Results

Table 4, which was organized the same way as Table 3, shows the numerical results of the proposed technique and the methods used for the sake of comparison in the 69-node test system. Importantly, the BONMIN solver was not employed for comparison purposes in this test system because it failed to converge to any feasible solution. This can be explained by the fact that the solution space in this test system was larger than that in the first test system.

According to the information in Table 4, the proposed DCCSA provided the best solution for the 69-node test system, with a reduction in the total annual operating costs of approximately 1,053,276.87 USD/year with respect to the base case. This means that the optimal solution for this test system is 2,824,923.05 USD/year, which is found by placing the PV generation units at nodes 21, 61, and 64, for a total installed capacity of 3807.02 kWp. Moreover, all the methods used to solve the problem addressed in this paper allowed a reduction of more than 27% with respect to the base case, with the DCGNDO, the DCPPSO, and the DCCSA allowing the highest reduction (27.1589%). When compared to the other methods in terms of reduction in the annual operating costs, the proposed methodology outperformed the DCNMA by 0.0373%, the DCCBGA by 0.0192%, and the DCVSA by 0.0087%.

Table 4. Numerical results obtained in the 69-node test system.

Method	Location (Node)/Power (MW)	A_{cost} (USD/Year)	Reduction (%)	Time (s)	STD (%)
Base case	- - -	3,878,199.93	0	-	-
DCNMA	12/0.0794 60/1.3805 61/2.3776	2,826,368.60	27.1216	91.81	0.1900
DCCBGA	24/0.5326 61/1.8954 64/1.3772	2,825,783.33	27.1397	22.36	0.0999
DCVSA	16/0.2632 61/2.2719 63/2.2934	2,825,264.56	27.1502	887.64	0.0942
DCGNDO	21/0.4812 61/2.4 64/0.9259	2,824,923.38	27.1589	1237.23	0.2558
DCPPSO	21/0.4890 61/2.4 64/0.9169	2,824,923.29	27.1589	55.15	0.0267
DCCSA	21/0.4816 61/2.4 64/0.9254	2,824,923.05	27.1589	377.49	0.0225

5.3.2. Statistical Analysis

As in the previous test system, the proposed methodology was run 100 consecutiv times in the 69-node test system to validate its efficiency and robustness in solving the pro lem of optimally locating and sizing PV generation units in electrical systems. The resul of such validation are shown in Figure 7, which shows the improvements obtained by th DCCSA in terms of best solution, processing time, and standard deviation when compare to the other solution methodologies. The numbers in red indicate that the method used fc comparison outperformed the DCCSA.

As can be seen in Figure 7, the DCSSA provided the best results in terms of reductio in annual operating costs when compared to the others methods. It outperformed th DCNMA by 0.0512%, the DCCBGA by 0.0306%, the DCVSA by 0.01198%, the DCGND by 1.1589×10^{-5}%, and the DCPPSO by 8.621×10^{-6}%.

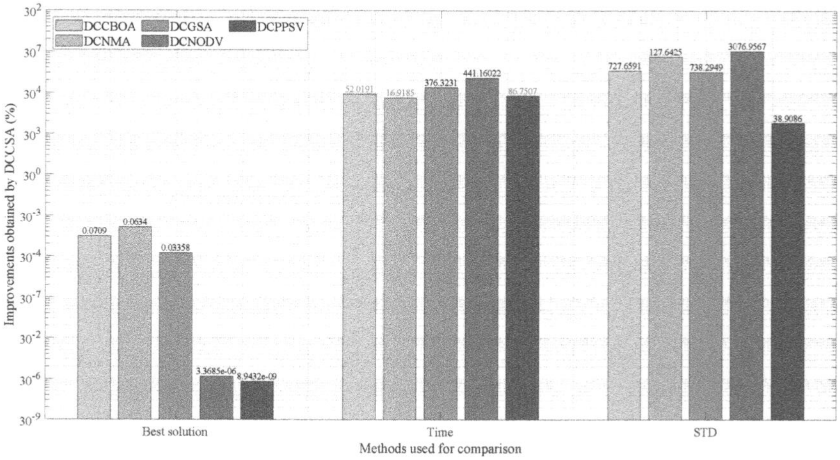

Figure 7. Improvements obtained by the DCCSA in the 69-node test system.

Regarding processing times, the DCNMA, the DCCBGA, and the DCPPSO were faster than the proposed methodology. When compared to the DCCSA, they reduced processing times by 75.6789%, 94.0767%, and 85.3903%, respectively. Importantly, these differences in processing time are attributed to the population size used for the DCCSA. The DCCSA, however, was faster than the DCVSA and the DCGNDO; it reduced processing times by 135.1417% and 227.75044% when compared to the DCVSA and the DCGNDO, respectively. The processing times obtained by the DCCSA can be considered negligible when compared to the planning horizon chosen for this study (i.e., 20 years).

As for the standard deviation, the proposed DCCSA produced the best results, as it achieved an improvement of 743.5249% with respect to the DCNMA, of 343.5967% with respect to the DCCBGA, of 318.4626% with respect to the DCVSA, of 1035.6953% with respect to the DCNGDO, and of 18.6085% with respect to the DCPPSO.

According to this, the proposed DCCSA provided the best results in terms of solution quality and repeatability, which makes it the best option for solving the problem of optimally locating and sizing PV generation units in the 69-node test system.

5.3.3. Feasibility Check

To verify whether the optimal solution yielded by the DCCSA is feasible, the active power generation at the main supply node was evaluated before and after implementing the solution obtained by the proposed methodology (see Figure 8).

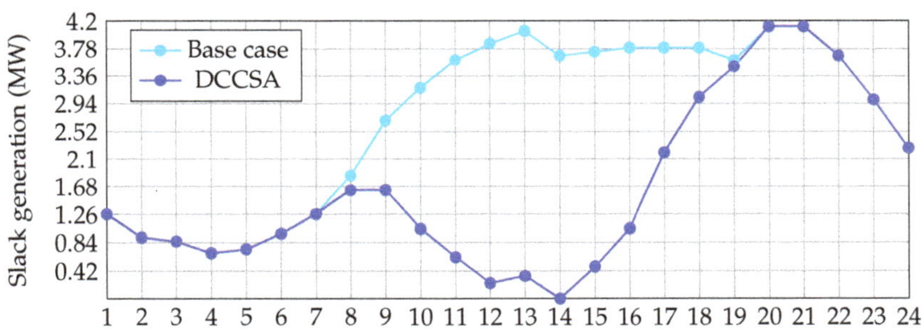

Figure 8. Impact of PV integration in the 69-node test system.

Before implementing the solution provided by the DCCSA, the power produced at the slack node followed the same behavior of the demanded active power (see Figure 3) along with the system losses. However, once the best solution delivered by the DCCSA was implemented, the power produced at the slack node significantly decreased as the power produced by the PV units increased until it hit zero in time period 14 when the PV units injected 100% of their capacity. This proves that power generation at the slack node respected the capacity constraint, as it yielded positive or zero values.

Finally, to confirm that the voltage profiles were within the regulation bounds (i.e., $\pm 10\%$), the behavior of the minimum and maximum voltages was examined for all time periods once the solution delivered by the DCCSA was implemented (see Figure 9).

As may be concluded from Figure 9, the minimum and maximum voltage values in all time periods respected the voltage regulation bounds, as they remained within $\pm 10\%$. In addition, the maximum voltage (i.e., 1.0322 pu) was recorded at node 64 when the PV units injected 100% of their nominal power. The minimum voltage (i.e., 0.9092 pu), which coincided with the minimum voltage of the base case, was recorded at node 65 when the PV units did not inject power and during the period of peak demand (from hour 20 to 21).

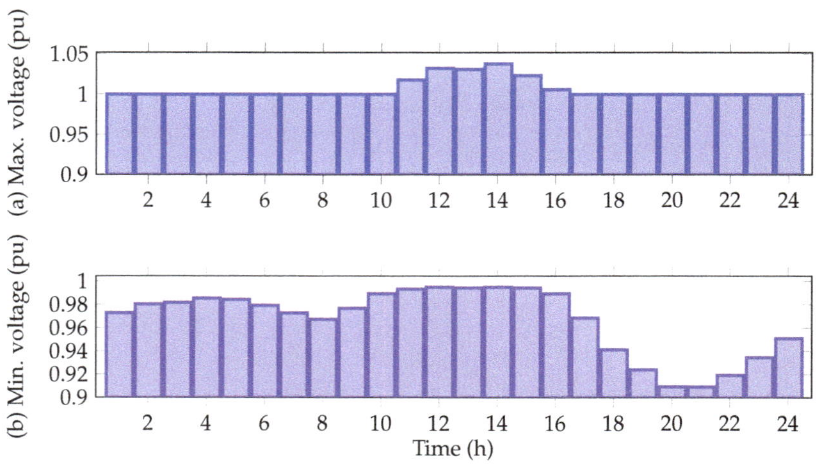

Figure 9. Voltage behavior during one day in the 69-node test system: (**a**) maximum voltage and (**b**) minimum voltage

6. Conclusions and Future Work

This study presented a master–slave method that employs a discrete–continuous version of the Crow Search Algorithm to solve the problem of optimally locating and sizing PV generation units in electrical networks. In the slave stage, the DCCSA is responsible for defining the set of nodes where the PV generation units are to be installed as well as the sizes of such units. In the slave stage, the successive approximations power flow method is in charge of finding the fitness function value. The objective function was the reduction in the total annual operating costs of a electrical network, which include (i) the energy purchasing costs at the main supply node, (ii) the investment in the PV generation units, and (iii) their corresponding operation and maintenance costs. The parameters of the proposed methodology were tuned using the CBGA.

The numerical results generated by our solution method in the 33- and 69-node test systems proved its applicability and effectiveness in comparison with other six methods reported in the specialized literature (the BONMIN solver of the GAMS, the DCCBGA, the DCNMA, the DCVSA, the DCGNDO, and the DCPPSO algorithm). The following are the key findings of this study:

- ✓ The DCCSA managed to reduce the total annual operating costs by approximately 1,000,783.62 USD/year and 1,053,276.87 USD/year in the 33- and 69-node test systems, respectively. These values represent reductions of 27.0449% and 27.1589%. These are the largest reductions found for the problem of locating and sizing PV generation units, which indicates that the overall optimal solutions to this problem for both test systems are 2,699,671.76 USD/year and 2,824,923.05 USD/year, respectively.
- ✓ After 100 consecutive evaluations, the proposed DCCSA showed the lowest standard deviation values in both test systems, with improvements of 559.0663% and 18.6085% with respect to the DCPPSO (the second method with the best results) in the 33- and 69-node test systems, respectively. These results confirm the repeatability and robustness of the DCCSA in solving the problem under study, which makes the methodology used in this study the best option (i.e., over the other methodologies used in this topic) to solve the problem regarding the location and sizing of PV generation units. Moreover, this guarantees that in each evaluation, the solutions will be close to 80 USD/year and 637 USD/year for the 33- and 69-node test systems, respectively.
- ✓ The processing times required by the proposed technique to find an optimal and feasible solution was 76.9990 s in the 33-node test system and 377.4915 s in the 69-node test system. These are good values, considering that at each iteration, the DCCSA

evaluated 1848 power flows more than the other methods. Additionally, processing times are not critical in power system planning because the quality of the solution provided by the methodology is what really matters.

✓ Due to the nonlinearities and nonconvexities of the mathematical model used to express the problem of optimally locating and sizing PV generation units in electrical systems, the complexity of the problem rises as the number of nodes increases. As a result, the BONMIN solver of the GAMS was unable to find an optimal solution in the 69-node test system. The proposed DCCSA, on the contrary, was found to be independent of the number of nodes in the electrical system because it produced the best results in terms of reductions in the total annual operating costs and standard deviation, even as the complexity of the problem increased. This allows concluding that the proposed DCCSA is the best option to solve the problem under analysis. Yet, as the number of system nodes increases, so does the size of the solution space, which implies that the time required to find an optimal solution will increase as well.

Based on our findings, future studies could reformulate the mathematical model of the problem under study, taking into account the maximum thermal current supported by the conductors in an electrical network. They could also solve the problem addressed in this paper using a multi-objective optimization approach that improves not only economic but also technical and environmental aspects that represent the operating conditions of electrical systems. Finally, the optimal conductor selection problem could be included in power system planning, and the costs associated with the investment in each conductor could be considered.

Author Contributions: Conceptualization, methodology, software, and writing (review and editing), B.C.-C., L.F.G.-N., O.D.M., M.-A.P.-M. and A.-J.P.-M. All authors read and agreed to the published version of the manuscript.

Funding: This research was funded by Minciencias through the *Fondo Nacional de Financiamiento para la Ciencia, la Tecnología y la Innovación, Fondo Francisco José de Caldas;* Instituto Tecnológico Metropolitano; Universidad Nacional de Colombia; and Universidad del Valle under the project entitled *"Estrategias de dimensionamiento, planeación y gestión inteligente de energía a partir de la integración y la optimización de las fuentes no convencionales, los sistemas de almacenamiento y cargas eléctricas, que permitan la generación de soluciones energéticas confiables para los territorios urbanos y rurales de Colombia (código: 70634)"*, which is part of the research program entitled *"Estrategias para el desarrollo de sistemas energéticos sostenibles, confiables, eficientes y accesibles para el futuro de Colombia (Código:1150-852-70378)*.

Data Availability Statement: Not applicable.

Conflicts of Interest: The authors of this paper declare no conflict of interest.

Acronyms

PV	Photovoltaic
MINLP	Mixed-Integer Nonlinear Programming
GAMS	General Algebraic Modeling System
DCCBGA	Discrete–Continuous Chu and Beasley Genetic Algorithm
DCNMA	Discrete–Continuous Newton Metaheuristic Algorithm
DCVSA	Discrete–Continuous Vortex Search Algorithm
DCGNDO	Discrete–Continuous Generalized Normal Distribution Optimizer
DCPPSO	Discrete–Continuous Parallel Particle Swarm Optimization

References

1. Saeed, M.H.; Fangzong, W.; Kalwar, B.A.; Iqbal, S. A Review on Microgrids' Challenges & Perspectives. *IEEE Access* **2021**, *9*, 166502–166517.
2. Pimm, A.J.; Palczewski, J.; Barbour, E.R.; Cockerill, T.T. Using electricity storage to reduce greenhouse gas emissions. *Appl. Energy* **2021**, *282*, 116199. [CrossRef]
3. González, D.M.L.; Rendon, J.G. Opportunities and challenges of mainstreaming distributed energy resources towards the transition to more efficient and resilient energy markets. *Renew. Sustain. Energy Rev.* **2022**, *157*, 112018. [CrossRef]

4. Valencia, A.; Hincapie, R.A.; Gallego, R.A. Optimal location, selection, and operation of battery energy storage systems ar renewable distributed generation in medium–low voltage distribution networks. *J. Energy Storage* **2021**, *34*, 102158. [CrossRef]
5. IPSE. *Boletín Datos IPSE Mayo 2022*; IPSE : Bogotá, Colombia, 2021.
6. López, A.R.; Krumm, A.; Schattenhofer, L.; Burandt, T.; Montoya, F.C.; Oberländer, N.; Oei, P.Y. Solar PV generation in Colomb A qualitative and quantitative approach to analyze the potential of solar energy market. *Renew. Energy* **2020**, *148*, 1266–12 [CrossRef]
7. Eltawil, M.A.; Zhao, Z. Grid-connected photovoltaic power systems: Technical and potential problems—A review. *Renew. Susta Energy Rev.* **2010**, *14*, 112–129. [CrossRef]
8. Cabrera, J.B.; Veiga, M.F.; Morales, D.X.; Medina, R. Reducing power losses in smart grids with cooperative game theory. *Advanced Communication and Control Methods for Future Smartgrids*; Intechopen: London, UK, 2019; p. 49.
9. Almadhor, A.; Rauf, H.T.; Khan, M.A.; Kadry, S.; Nam, Y. A hybrid algorithm (BAPSO) for capacity configuration optimization a distributed solar PV based microgrid. *Energy Rep.* **2021**, *7*, 7906–7912. [CrossRef]
10. Ngamprasert, P.; Rugthaicharoencheep, N.; Woothipatanapan, S. Application Improvement of Voltage Profile by Photovolta Farm on Distribution System. In Proceedings of the 2019 IEEE International Conference on Power, Energy and Innovatio (ICPEI), Pattaya, Thailand, 16–18 October 2019; pp. 98–101.
11. Al Abri, R.; El-Saadany, E.F.; Atwa, Y.M. Optimal placement and sizing method to improve the voltage stability margin in distribution system using distributed generation. *IEEE Trans. Power Syst.* **2012**, *28*, 326–334. [CrossRef]
12. Cortés-Caicedo, B.; Molina-Martin, F.; Grisales-Noreña, L.F.; Montoya, O.D.; Hernández, J.C. Optimal Design of PV Systems Electrical Distribution Networks by Minimizing the Annual Equivalent Operative Costs through the Discrete-Continuous Vort Search Algorithm. *Sensors* **2022**, *22*, 851. [CrossRef] [PubMed]
13. Jiang, S.; Wan, C.; Chen, C.; Cao, E.; Song, Y. Distributed photovoltaic generation in the electricity market: status, mode ar strategy. *CSEE J. Power Energy Syst.* **2018**, *4*, 263–272. [CrossRef]
14. Montoya, O.D.; Grisales-Noreña, L.F.; Perea-Moreno, A.J. Optimal Investments in PV Sources for Grid-Connected Distributic Networks: An Application of the Discrete–Continuous Genetic Algorithm. *Sustainability* **2021**, *13*, 13633. [CrossRef]
15. Montoya, O.D.; Grisales-Noreña, L.F.; Alvarado-Barrios, L.; Arias-Londoño, A.; Álvarez-Arroyo, C. Efficient reduction in th annual investment costs in AC distribution networks via optimal integration of solar PV sources using the newton metaheurist algorithm. *Appl. Sci.* **2021**, *11*, 11525. [CrossRef]
16. Montoya, O.D.; Grisales-Noreña, L.F.; Ramos-Paja, C.A. Optimal Allocation and Sizing of PV Generation Units in Distributic Networks via the Generalized Normal Distribution Optimization Approach. *Computers* **2022**, *11*, 53. [CrossRef]
17. Grisales-Noreña, L.F.; Montoya, O.D.; Hernández, J.C.; Ramos-Paja, C.A.; Perea-Moreno, A.J. A Discrete-Continuous PSO for th Optimal Integration of D-STATCOMs into Electrical Distribution Systems by Considering Annual Power Loss and Investme Costs. *Mathematics* **2022**, *10*, 2453. [CrossRef]
18. Duong, M.Q.; Pham, T.D.; Nguyen, T.T.; Doan, A.T.; Tran, H.V. Determination of optimal location and sizing of solar photovolta distribution generation units in radial distribution systems. *Energies* **2019**, *12*, 174. [CrossRef]
19. Khoso, A.H.; Shaikh, M.M.; Hashmani, A.A. A New and Efficient Nonlinear Solver for Load Flow Problems. *Eng. Technol. Ap Sci. Res.* **2020**, *10*, 5851–5856. [CrossRef]
20. Kim, Y.; Kim, K. Accelerated Computation and Tracking of AC Optimal Power Flow Solutions using GPUs. *arXiv* **202** arXiv:2110.06879.
21. Mahmoud, M.S.; Fouad, M. *Control and Optimization of Distributed Generation Systems*; Springer: New York, NY, USA, 2015.
22. Devikanniga, D.; Vetrivel, K.; Badrinath, N. Review of meta-heuristic optimization based artificial neural networks and i applications. *Proc. J. Phys. Conf. Ser.* **2019**, *1362*, 012074. [CrossRef]
23. Askarzadeh, A. A novel metaheuristic method for solving constrained engineering optimization problems: Crow search algorith *Comput. Struct.* **2016**, *169*, 1–12. [CrossRef]
24. Montoya, O.D.; Gil-González, W. On the numerical analysis based on successive approximations for power flow problems in A distribution systems. *Electr. Power Syst. Res.* **2020**, *187*, 106454. [CrossRef]
25. Jain, M.; Rani, A.; Singh, V. An improved crow search algorithm for high-dimensional problems. *J. Intell. Fuzzy Syst.* **201** *33*, 3597–3614. [CrossRef]
26. Hussien, A.G.; Amin, M.; Wang, M.; Liang, G.; Alsanad, A.; Gumaei, A.; Chen, H. Crow Search Algorithm: Theory, Rece Advances, and Applications. *IEEE Access* **2020**, *8*, 173548–173565. [CrossRef]
27. Sahoo, R.R.; Ray, M. PSO based test case generation for critical path using improved combined fitness function. *J. King Sa Univ.-Comput. Inf. Sci.* **2020**, *32*, 479–490. [CrossRef]
28. Zhang, X.; Beram, S.M.; Haq, M.A.; Wawale, S.G.; Buttar, A.M. Research on algorithms for control design of human–machi interface system using ML. *Int. J. Syst. Assur. Eng. Manag.* **2021**, *13*, 462–469. [CrossRef]
29. Harman, M.; Jia, Y.; Zhang, Y. Achievements, open problems and challenges for search based software testing. In Proceedings the 2015 IEEE 8th International Conference on Software Testing, Verification and Validation (ICST), Graz, Austria, 13–17 Apr 2015; pp. 1–12.
30. Kaur, S.; Kumbhar, G.; Sharma, J. A MINLP technique for optimal placement of multiple DG units in distribution systems. *Int. Electr. Power Energy Syst.* **2014**, *63*, 609–617. [CrossRef]

1. Sahoo, N.; Prasad, K. A fuzzy genetic approach for network reconfiguration to enhance voltage stability in radial distribution systems. *Energy Convers. Manag.* **2006**, *47*, 3288–3306. [CrossRef]
2. Castiblanco-Pérez, C.M.; Toro-Rodríguez, D.E.; Montoya, O.D.; Giral-Ramírez, D.A. Optimal Placement and Sizing of D-STATCOM in Radial and Meshed Distribution Networks Using a Discrete-Continuous Version of the Genetic Algorithm. *Electronics* **2021**, *10*, 1452. [CrossRef]
3. Wang, P.; Wang, W.; Xu, D. Optimal sizing of distributed generations in DC microgrids with comprehensive consideration of system operation modes and operation targets. *IEEE Access* **2018**, *6*, 31129–31140. [CrossRef]
4. Beasley, J.E.; Chu, P.C. A genetic algorithm for the set covering problem. *Eur. J. Oper. Res.* **1996**, *94*, 392–404. [CrossRef]

mathematics

MDPI

Article

Robust Flatness-Based Tracking Control for a "Full-Bridge Buck Inverter–DC Motor" System

Ramón Silva-Ortigoza [1,*], Magdalena Marciano-Melchor [1], Rogelio Ernesto García-Chávez [1], Alfredo Roldán-Caballero [2], Victor Manuel Hernández-Guzmán [3], Eduardo Hernández-Márquez [4], José Rafael García-Sánchez [5], Rocío García-Cortés [6] and Gilberto Silva-Ortigoza [7]

1 Laboratorio de Mecatrónica y Energía Renovable, CIDETEC, Instituto Politécnico Nacional, Mexico City 07700, Mexico
2 Unidad Profesional Interdisciplinaria de Ingeniería Campus Tlaxcala, Instituto Politécnico Nacional, Tlaxcala 90000, Mexico
3 Facultad de Ingeniería, Centro Universitario, Universidad Autónoma de Querétaro, Querétaro 76010, Mexico
4 Departamento de Ingeniería Mecatrónica, Instituto Tecnológico Superior de Poza Rica, Tecnológico Nacional de México, Veracruz 93230, Mexico
5 División de Ingeniería Mecatrónica, Tecnológico de Estudios Superiores de Huixquilucan, Tecnológico Nacional de México, Mexico City 52773, Mexico
6 Secretaría de Investigación y Posgrado, Dirección de Investigación, Instituto Politécnico Nacional, Mexico City 07738, Mexico
7 Facultad de Ciencias Físico Matemáticas, Benemérita Universidad Autónoma de Puebla, Puebla 72570, Mexico
* Correspondence: rsilvao@ipn.mx; Tel.: +52-55-5729-6000 (ext. 52530)

Citation: Silva-Ortigoza, R.; Marciano-Melchor, M.; García-Chávez, R.E.; Roldán-Caballero, A.; Hernández-Guzmán, V.M.; Hernández-Márquez, E.; García-Sánchez, J.R.; García-Cortés, R.; Silva-Ortigoza, G. Robust Flatness-Based Tracking Control for a "Full-Bridge Buck Inverter–DC Motor" System. *Mathematics* **2022**, *10*, 4110. https://doi.org/10.3390/math10214110

Academic Editors: Atanda Raji and Khaled M. Abo-Al-Ez

Received: 14 October 2022
Accepted: 31 October 2022
Published: 4 November 2022

Abstract: By developing a robust control strategy based on the differential flatness concept, this paper presents a solution for the bidirectional trajectory tracking task in the "full-bridge Buck inverter–DC motor" system. The robustness of the proposed control is achieved by taking advantage of the differential flatness property related to the mathematical model of the system. The performance of the control, designed via the flatness concept, is verified in two ways. The first is by implementing experimentally the flatness control and proposing different shapes for the desired angular velocity profiles. For this aim, a built prototype of the "full-bridge Buck inverter–DC motor" system, along with Matlab–Simulink and a DS1104 board from dSPACE are used. The second is via simulation results, i.e., by programming the system in closed-loop with the proposed control algorithm through Matlab–Simulink. The experimental and the simulation results are similar, thus demonstrating the effectiveness of the designed robust control even when abrupt electrical variations are considered in the system.

Keywords: motor drivers; power converters; full-bridge Buck inverter; DC motor; differential flatness; flatness-based control; trajectory tracking task

MSC: 37N35; 57N45; 93-05; 93B52; 93C73; 93D20

1. Introduction

During the 21st century, the progress in energy transformation has allowed expansion of the spectrum of technological applications from a purely industrial sector to a domestic one. A clear example of such a change has been the diversification of devices that require the transformation of electric energy into mechanical energy. With the aim of accomplishing the aforesaid task of energetic transformation, one of the electric machines mainly used is the DC motor. Thus, currently there is wide use of the DC motor in applications ranging from home and public services to the entertainment industry. This fact, joined with the historical necessity of DC motors in industrial applications [1], means the research community is still interested in developing drivers for DC motors. Moreover, derived from the excellent benefits arising from feeding DC motors with power electronics converters, in the last two decades, several efforts have developed new controllers for different "DC/DC power electronic converters–DC motor" connections [2–60]. In particular, for the DC/DC Buck converter, these connections can be classified into systems generating unidirectional [2–35] or bidirectional movement [36–41], both related to the shaft of the DC motor. Based on the fact that the angular velocity is one of the variables of interest to be controlled in these kinds of systems, this work focuses on the trajectory tracking problem of a "full-bridge Buck inverter–DC motor" system. Thus, the state-of-the-art associated with controller design for the DC/DC Buck power electronic converter connected with the DC motor is presented next.

1.1. Unidirectional "DC/DC Buck Converter–DC Motor" System

At the beginning of this millennium, Lyshevski proposed the first mathematical model of the DC/DC Buck converter–DC motor system and solved the regulation control task via a nonlinear PI control [2]. Afterwards in [3], Ahmad et al., designed and compared the performance of the PI, fuzzy PI, and LQR control algorithms. In the same way, Bingöl and Paçaci reported in [4] the development of software for controlling the system under study via neural networks. Meanwhile, Sira-Ramírez and Oliver-Salazar in [5] studied the concepts of active disturbance rejection control and differential flatness in two combinations of the Buck converter with DC motors. Moreover, in recent years, several active disturbance rejection control schemes have been developed for governing Buck converter-driven motor systems, e.g., [6–11], while the study of controls based on differential flatness enabling solving the trajectory tracking task have been proposed in [12–14]. On the other hand, the applications of zero average dynamics of fixed point induction control techniques to control the speed of a permanent magnet DC motor with a Buck converter were detailed by Hoyos et al., in [15–18]. In the meantime, papers based on the sliding mode control (SMC) were presented by Wei et al., Silva-Ortigoza et al., and Hernández-Guzmán et al. in [19–21], respectively. More recently, other studies based on SMC were exhibited by Rauf et al. in [22,23] and by Ravikumar and Srinivasan in [24]. Another solution was implemented by Khubalkar et al. via fractional order PID controllers, whose tuning was executed with a dynamic particle swarm optimization (dPSO) technique [25], with an improved dPSO technique [26], and by using an ant colony optimization technique [27]. Additionally, Srinivasan et al. [28,29] introduced a sensitivity analysis applied to the DC/DC Buck converter–DC motor system by exploiting the exact tracking error dynamics passive output feedback (ETEDPOF) methodology. Other recent control techniques investigated in the literature include neuroadaptive backstepping based control, intelligent nonlinear adaptive control, and neural network based intelligent control by Nizami et al. [30–32], piecewise affine PI-based control by Hanif et al. [33], and adaptive neurofuzzy H_∞-based control by Rigatos et al. [34], while Kazemi and Montazeri in [35] elaborated a fault detection control algorithm by combining a switching observer with the bond graph method.

1.2. Bidirectional "DC/DC Buck Converter–DC Motor" Systems

When the bidirectional rotation of the DC motor shaft is considered, and in the search to overcome the intrinsic dynamic limitations associated with the DC/DC Buck converter, two alternatives have been proposed. On the one hand, Silva-Ortigoza et al. [36] proposed the dynamic model and experimental validation of the "DC/DC Buck converter–inverter–DC motor" topology. In addition, a sensorless passivity-based control, via the ETEDPOF methodology, for executing the bidirectional angular velocity trajectory tracking task in this topology, was addressed in [37]. Meanwhile, two robust differential flatness-based tracking controls for the "DC/DC Buck converter–inverter–DC motor" topology were designed by Hernández-Márquez et al. [38]. Lastly, an adaptive backstepping SMC associated with Chebyshev neural network estimation for the angular velocity trajectory tracking task for such a topology was considered by Chi et al. [39]. On the other hand, Hernández-Márquez et al. [40] developed a mathematical model, experimentally validated, for a new "full-bridge Buck inverter–DC motor" system that accomplishes bidirection rotation of the DC motor shaft. Later, the angular velocity trajectory tracking proble related to this system was tackled by Silva-Ortigoza et al. [41], where a sensorless contr based on the ETEDPOF methodology was utilized.

1.3. *Other "DC/DC Power Converters–DC Motor" Systems*

Complementary to the control schemes that have been developed for the connectic between the DC/DC Buck converter and the DC motor, other topologies of DC/DC pow electronic converters to drive the DC motors include the following. For the unidirection and bidirectional "DC/DC Boost converter–DC motor" systems, several controllers ha been presented in [42–51], respectively. For the DC/DC Buck-Boost converter with a D motor, different control schemes were introduced for the unidirectional case in [52,5 and for the bidirectional case in [54,55]. On the other hand, the design of controllers f the Sepic, Luo, and Cuk converters feeding DC motors have been reported in [56–5 respectively. Finally, control algorithms for the multilevel DC/DC Buck converter an parallel DC/DC Buck converter connected with the DC motor have been studied in [59,6 respectively.

1.4. *Discussion of Related Work, Motivation, and Contribution*

Regarding the aforementioned literature on the "DC/DC Buck converter–DC moto system [2–35], it is worth noting that all these control solutions for the angular veloci regulation and tracking problems have only dealt with the unidirectional rotation of th DC motor shaft. This is because the DC/DC Buck converter only supplies electric power the form of unipolar voltage. This restriction is solved by integrating an inverter in suc systems with the aim of providing them with the capability of supplying electric pow with bipolar voltage. As a consequence, bidirectional "DC/DC Buck converter–DC moto systems have emerged, and proposals to solve the bidirectional control of the motor sha angular velocity have been introduced in [36–41].

Related to some applications of these systems, it is possible to find mechatron systems [61], robotic arms [62], and wheeled mobile robots [63] for the unidirection "DC/DC Buck converter–DC motor" system, whereas, applications of the bidirection "DC/DC Buck converter–DC motor" system have been recently presented in renewab energy by Chi et al. in [39] and wheeled mobile robots by Hernández-Guzmán et al. in [6 On the other hand, important research works recently developed and devoted to the desig of controllers for the "full-bridge Buck inverter" system have been reported in [65,66].

In the context of the bidirectional "DC/DC Buck converter–DC motor" systems, on of these bidirectional proposals is the "full-bridge Buck inverter–DC motor" system. Th system, recently reported in [40], was controlled by designing a passive tracking contr based on the ETEDPOF method in [41]. With such an approach, the control objective achieved, i.e., the angular velocity ω converges to a desired angular velocity profile ω However, when abrupt variations in load R are introduced into the system in closed-loo ω no longer tracks ω^*. This can be seen in Figure 1, which corresponds to the experiment results reported in [41] associated with abrupt changes in load R.

Motivated by the benefits of the "full-bridge Buck inverter–DC motor" system topo ogy presented in [40], the possible potential applications of this topology, and the expe imental results of the passivity-based control reported in [41], the main contribution this study is twofold: (1) to develop a robust differential flatness-based tracking contr for the "full-bridge Buck inverter–DC motor" system and (2) to experimentally valida such a proposed scheme on a built platform of the system and to corroborate the resul with the corresponding simulation results in closed-loop. It is worth noting that, compare with [41], the control algorithm designed herein is robust against parametric variations.

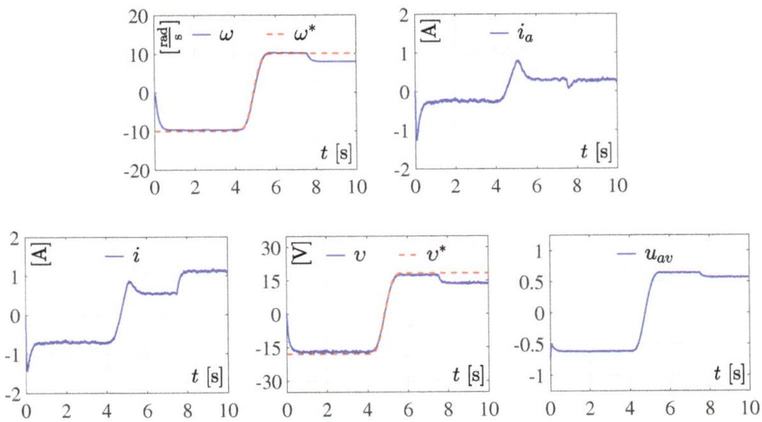

Figure 1. Experimental results of the passivity-based control reported in [41], when ω^* is of the Bézier type (22), and the abrupt variations in load R, declared in (26), are carried out in the system.

From the aforementioned discussion of the state-of-the-art, motivation, and contributions of the work, the remaining parts of this paper are organized as follows. Section 2 presents the generalities of the "full-bridge Buck inverter–DC motor" system. Section 3 provides the design of the robust flatness-based control for solving the angular velocity trajectory tracking problem on the system. In Section 4, the blocks of the built experimental platform used for the experimental tests in closed-loop are described. In Section 5, the experimental and the corresponding simulation results of the system in closed-loop are detailed and discussed. Finally, in Section 6, the conclusions of this study are presented.

2. "Full-Bridge Buck Inverter–DC Motor" System

The electric circuit of the full-bridge Buck inverter–DC motor system is depicted in Figure 2.

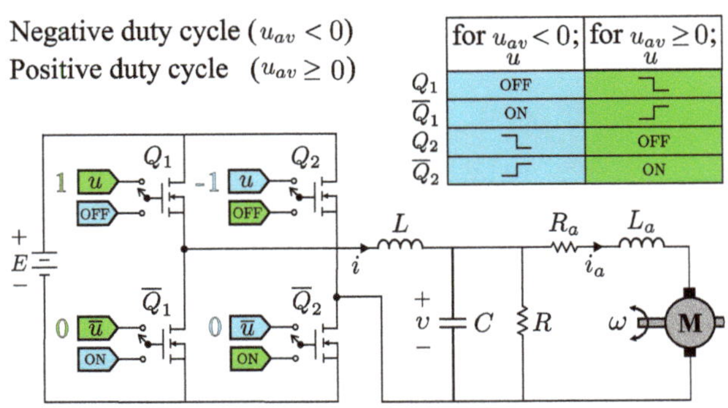

Figure 2. Full-bridge Buck inverter–DC motor system and clock signals u associated with Q_1, $\overline{Q}_1$, Q_2, and $\overline{Q}_2$, for the negative duty cycle ($u_{av} < 0$) and for the positive duty cycle ($u_{av} \geq 0$).

The circuit shown in Figure 2 can be divided into two subsystems:

- The subsystem *full-bridge Buck inverter*, which modulates and feeds with bipolar voltage v the DC motor through the input u. It is composed of a power supply E and an array of four transistors, denoted as Q_1, $\overline{Q}_1$, Q_2, and $\overline{Q}_2$, which operate in accordance with the clock cycles shown in Figure 2. This subsystem is made up of an LC filter, where i is the circulating current over the inductor L, and v is the voltage appearing across the terminals of the parallel connection between the capacitor C and the load R.

- The subsystem *DC motor*, which relates to the actuator system and comprises the armature resistance R_a, armature inductance L_a, and armature current i_a. ω corresponds to the angular velocity associated with the motor shaft. Additional parameters are the moment of inertia of the rotor and motor load J, the viscous friction coefficient of the motor b, the counterelectromotive force constant k_e, and the motor torque constant k_m.

As stated in [40], the switched model of the full-bridge Buck inverter–DC motor system shown in Figure 2 is given by

$$\begin{aligned} L\frac{di}{dt} &= -v + Eu, \\ C\frac{dv}{dt} &= i - \frac{v}{R} - i_a, \\ L_a\frac{di_a}{dt} &= v - R_a i_a - k_e\omega, \\ J\frac{d\omega}{dt} &= k_m i_a - b\omega, \end{aligned} \tag{1}$$

where $u \in \{-1, 0, 1\}$ represents the positions of the transistors Q_1, $\overline{Q}_1$, Q_2, and $\overline{Q}_2$ when operating in the negative or positive cycles according to the clock signals u that are shown in Figure 2. The rest of the variables and constants associated with the mathematical model (1) were previously defined. Moreover, according to [40], the average mathematical model of the full-bridge Buck inverter–DC motor system is determined by

$$\begin{aligned} L\frac{di}{dt} &= -v + Eu_{av}, \\ C\frac{dv}{dt} &= i - \frac{v}{R} - i_a, \\ L_a\frac{di_a}{dt} &= v - R_a i_a - k_e\omega, \\ J\frac{d\omega}{dt} &= k_m i_a - b\omega, \end{aligned} \tag{2}$$

with $u_{av} \in [-1, 1]$, where u_{av} is the duty cycle or average input signal of the system. The rest of the variables and constants concerning the mathematical model (2) were previously defined.

3. Robust Flatness-Based Tracking Control

This section presents the development of a robust differential flatness-based tracking control for the full-bridge Buck inverter–DC motor system. Such a control exploits the differential flatness property employed by the mathematical model of system (2). The following proposition summarizes the main result.

Proposition 1. *Consider the average mathematical model of the full-bridge Buck inverter–DC motor system given in (2), in closed-loop, with the following controller:*

$$\begin{aligned} u_{av} &= \left[\frac{CJLL_a}{k_m E}\right]\mu + \left[\frac{bCLL_aR + CJLRR_a + JLL_a}{k_m RE}\right]\omega^{(3)} \\ &\quad + \alpha\ddot{\omega} + \left[\frac{bLR + bLR_a + bL_aR + k_ek_mL + JRR_a}{k_m RE}\right]\dot{\omega} + \left[\frac{bR_a + k_ek_m}{k_m E}\right]\omega, \end{aligned} \tag{3}$$

where

$$\begin{aligned} \mu &= \omega^{*(4)} - k_4\left[\omega^{(3)} - \omega^{*(3)}\right] - k_3[\ddot{\omega} - \ddot{\omega}^*] - k_2[\dot{\omega} - \dot{\omega}^*] - k_1[\omega - \omega^*] \\ &\quad - k_0\int_0^t [\omega - \omega^*]d\tau, \end{aligned} \tag{4}$$

and

$$\begin{aligned} k_0 &= a\omega_n^4, \\ k_1 &= 4a\zeta\omega_n^3 + \omega_n^4, \\ k_2 &= 2a\omega_n^2 + 4a\zeta^2\omega_n^2 + 4\zeta\omega_n^3, \\ k_3 &= 4a\zeta\omega_n + 2\omega_n^2 + 4\zeta^2\omega_n^2, \\ k_4 &= a + 4\zeta\omega_n. \end{aligned} \tag{5}$$

Then, $\omega(t) \to \omega^*(t)$ exponentially, where $\omega^*(t)$ is continuously differentiable up to the fourth time derivative.

Proof. In [40], it was demonstrated that the full-bridge Buck inverter–DC motor system average model given in (2) is differentially flat, with the motor angular velocity as the flat output, i.e.,

$$\mathcal{F} = \omega. \tag{6}$$

This means that all of the state variables x as well as the input u_{av} are given as functions of the flat output and a number of its time derivatives, which is [40]

$$x = \left[\ i(\omega),\ v(\omega),\ i_a(\omega),\ \omega\ \right]^T, \quad u_{av} = u_{av}(\omega), \tag{7}$$

where

$$\begin{aligned} i(\omega) &= \left[\tfrac{CJL_a}{k_m}\right]\omega^{(3)} + \left[\tfrac{bCL_aR+CJRR_a+JL_a}{k_mR}\right]\ddot{\omega} \\ &+ \left[\tfrac{bL_a+JR+JR_a+bCRR_a+k_ek_mCR}{k_mR}\right]\dot{\omega} + \left[\tfrac{bR+bR_a+k_ek_m}{k_mR}\right]\omega, \end{aligned} \tag{8}$$

$$v(\omega) = \tfrac{JL_a}{k_m}\ddot{\omega} + \left[\tfrac{bL_a+JR_a}{k_m}\right]\dot{\omega} + \left[\tfrac{bR_a}{k_m} + k_e\right]\omega, \tag{9}$$

$$i_a(\omega) = \tfrac{J}{k_m}\dot{\omega} + \tfrac{b}{k_m}\omega, \tag{10}$$

and

$$\begin{aligned} u_{av}(\omega) &= \left[\tfrac{CJLL_a}{k_mE}\right]\omega^{(4)} + \left[\tfrac{bCLL_aR+CJLRR_a+JLL_a}{k_mRE}\right]\omega^{(3)} + \alpha\ddot{\omega} \\ &+ \left[\tfrac{bLR+bLR_a+bL_aR+k_ek_mL+JRR_a}{k_mRE}\right]\dot{\omega} + \left[\tfrac{bR_a+k_ek_m}{k_mE}\right]\omega, \end{aligned} \tag{11}$$

with

$$\alpha = \frac{bLL_a+JLR+JLR_a+bCLRR_a+k_ek_mCLR+JL_aR}{k_mRE}.$$

Notice there are four variables that can be known by direct measurements, i.e., i, v, i_a, and ω. Moreover, there are three linear Equations (8)–(10). Hence, the three unknowns $\dot{\omega}, \ddot{\omega}$, and $\omega^{(3)}$ can be computed. This means that the control law in (3) can be obtained by mimicking (11) and replacing the unknown variable $\omega^{(4)}$ by another variable, say μ. Likewise, notice that (11) represents the flatness-based model of the full-bridge Buck inverter–DC motor system. Thus, (11) and (3) can be equated to obtain the closed-loop dynamics, i.e.,

$$\omega^{(4)} = \mu. \tag{12}$$

This expression represents a linear system given by a chain of four integrators, i.e., an unstable linear system. Hence, μ must be designed in order to stabilize this chain of integrators. This motivates the definition of μ in (4) to obtain

$$e^{(4)} + k_4e^{(3)} + k_3\ddot{e} + k_2\dot{e} + k_1e + k_0\int_0^t e(\tau)d\tau = 0, \tag{13}$$

where $e(t) = \omega(t) - \omega^*(t)$. The integral component ensures zero static error in steady sta and compensates the abrupt variations that can be generated in some parameters of th system. Differentiating once the above expression with respect to time yields

$$e^{(5)} + k_4 e^{(4)} + k_3 e^{(3)} + k_2 \ddot{e} + k_1 \dot{e} + k_0 e = 0. \quad (1$$

This linear fifth-order dynamics is exponentially stable if and only if all of the fiv roots of the following polynomial have negative real parts

$$P(s) = s^5 + k_4 s^4 + k_3 s^3 + k_2 s^2 + k_1 s + k_0. \quad (1$$

This is ensured if the set of gains $\{k_0, k_1, k_2, k_3, k_4\}$ are chosen such that $P(s)$ is identic to the following Hurwitz polynomial

$$P_H(s) = (s + a)\left(s^2 + 2\zeta\omega_n s + \omega_n^2\right)^2,$$

where a, ζ, and ω_n are positive constants. This is accomplished if we choose $\{k_0, k_1, k_2, k_3, k$ as in (5). Since (14) has been ensured to be exponentially stable, we have that $\omega(t)$ - $\omega^*(t)$ exponentially.

Now, consider the expressions in (8)–(10), and suppose that ω is constant, i.e., that a velocity time derivatives are zero. Then, the following is found:

$$\bar{i}(\overline{\omega}) = \left[\frac{bR + bR_a + k_e k_m}{k_m R}\right]\overline{\omega}, \quad (1$$

$$\overline{v}(\overline{\omega}) = \left[\frac{bR_a}{k_m} + k_e\right]\overline{\omega}, \quad (1$$

$$\bar{i}_a(\overline{\omega}) = \frac{b}{k_m}\overline{\omega}. \quad (1$$

Assume that uncertainty exists in some of the system parameters appearing in (16)–(1 From these expressions, it can be realized that the steady-state value of the velocity chang if $\bar{i}, \overline{v}$, and $\bar{i}_a$, do not change. Hence, consider (13). In order to take into account that the v locity has deviated from its desired steady-state value, we replace $k_1 e$ by $k_1(\omega + \Delta\omega - \omega^*$ where $\Delta\omega$ stands for the constant velocity deviation. Then, we obtain

$$e^{(4)} + k_4 e^{(3)} + k_3 \ddot{e} + k_2 \dot{e} + k_1 e + k_0 \int_0^t e(\tau)d\tau = -k_1 \Delta\omega. \quad (1$$

After differentiating once the above expression, and since $-k_1\Delta\omega$ is a constant, (14) obtained again. Thus, it is concluded one more time that $\omega(t) \to \omega^*(t)$ exponentially. Th completes the proof of Proposition 1. □

4. Description of the Built Experimental Platform

This section describes the connections of the built prototype full-bridge Buck inverte DC motor system and the implementation of the designed robust flatness-based control.

A photograph showing the connections of the built full-bridge Buck inverter–D motor system in closed-loop is exhibited in Figure 3. The elements associated wit this photograph are the following: (1) energy power source E, (2) desktop comput (3) dSPACE DS1104 R&D controller board, (4) voltage differential probe to obtain v (via Tektronix P5200A probe), (5) current probe to find i (via a Tektronix A622 probe), (6) fu bridge Buck inverter, (7) energy power source for the instrumentation of the electroni stage, (8) current probe to acquire i_a (via a Tektronix A622 probe), (9) Omron E6B2-CWZ6 encoder to determine ω, and (10) Engel GNM 5440E-G3.1 DC motor.

Figure 3. Photograph of the hardware used for the experimental tests of the full-bridge Buck inverter–DC motor system in closed-loop.

Figure 4 shows the block diagram of the connections between the system, Simulink, and the dSPACE DS1104 controller board.

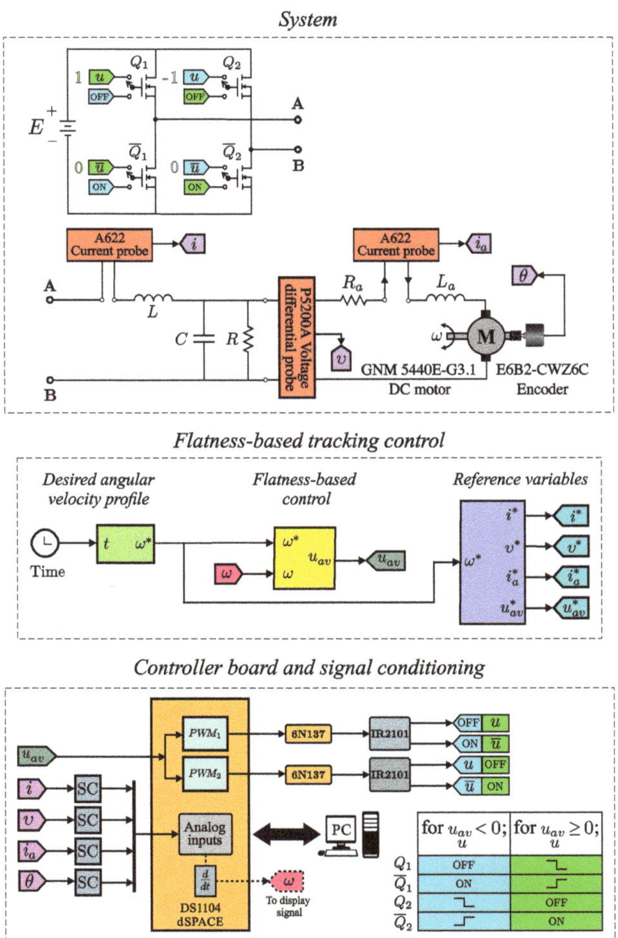

Figure 4. Connections between the software and hardware blocks of the built full-bridge Buck inverter–DC motor system in closed-loop.

The blocks composing the diagram of Figure 4 are as follows:

- *System.* This block corresponds to the built full-bridge Buck inverter–DC motor system. Here, the variables i, v, i_a, and ω were measured via a Tektronix A622 current probe, a Tektronix P5200A voltage probe, a Tektronix A622 current probe, and an Omron E6B2-CWZ6C encoder, respectively. On the other hand, the values of the parameters associated with the Buck converter and the Engel GNM 5440E-G3.1 DC motor were

$$E = 32\ \mathrm{V},\ \ L = 4.94\ \mathrm{mH},\ \ C = 4.7\ \mu\mathrm{F},\ \ R = 48\ \Omega, \tag{20}$$

and

$$\begin{aligned} L_a &= 2.22\ \mathrm{mH}, \quad R_a = 0.965\ \Omega, \quad k_m = 120.1 \times 10^{-3}\ \tfrac{\mathrm{N{\cdot}m}}{\mathrm{A}}, \\ k_e &= 120.1 \times 10^{-3}\ \tfrac{\mathrm{V{\cdot}s}}{\mathrm{rad}}, \quad J = 118.2 \times 10^{-3}\ \mathrm{kg{\cdot}m^2}, \quad b = 129.6 \times 10^{-3}\ \tfrac{\mathrm{N{\cdot}m{\cdot}s}}{\mathrm{rad}}, \end{aligned} \tag{21}$$

respectively.
- *Flatness-based tracking control.* The flatness-based control (3) was implemented in this block with Simulink. On the one hand, the determination of the angular velocity ω, required by the control, was carried out via an E6B2-CWZ6C encoder. On the other hand, the desired angular velocity profile ω^*, also required by the control, was produced from the *desired trajectory* subblock. Meanwhile, the control gains were obtained after the parameters $a = 0.2$, $\zeta = 10$ and $\omega_n = 1200$ were introduced in (5). Lastly, the reference variables i^*, v^*, i_a^*, and u_{av}^* were generated when ω^* was replaced into (8)–(10), and (11), respectively.
- *Controller board and signal conditioning.* In this block, the implementation of the switched control, u, corresponding to the obtained flatness-based tracking control u_{av} (3), was carried out via the PWM subblocks of the dSPACE DS1104 controller board. Thus, after obtaining u, the correct on/off activation of the transistors Q_1, $\overline{Q}_1$, Q_2, and $\overline{Q}_2$ of the full-bridge inverter circuit was achieved. Such a circuit was built with four IRF640N MOSFET transistors, two IR2101 drivers, and two 6N137 optocouplers. Finally, the signals i, v, i_a, and θ were correctly adjusted through signal conditioning (SC) blocks.

5. Experimental and Simulation Tests in Closed-Loop and Discussion of the Results

In order to validate the performance of the designed robust flatness-based tracking control, experimental tests were carried out on the aforementioned built experimental platform. After this, by using Matlab–Simulink, the corresponding simulation results were derived, which are presented.

5.1. Results of the System in Closed-Loop

Five different experimental tests were implemented. From the first test to the fourth one, four desired trajectories of ω^*, along with the system nominal parameters declared in (20) and (21), were considered. In the fifth experimental test, with the aim of verifying the control's robustness, some perturbations were introduced into the system. In all the plotted graphs, the variables of the system are shown; first ω and i_a and then i, v, and u_{av}. To corroborate the experimental results, the associated simulation results are also shown.

5.1.1. Experimental and Simulation Test 1

In this experiment, the desired angular velocity profile ω^* was declared by the tenth-order Bézier polynomial as follows:

$$\omega^*(t) = \overline{\omega}_i(t_i) + [\overline{\omega}_f(t_f) - \overline{\omega}_i(t_i)]\varphi\left(t, t_i, t_f\right), \tag{22}$$

being $\varphi\left(t, t_i, t_f\right)$ defined by

$$\varphi\left(t,t_i,t_f\right)=\begin{cases}0 & t\leq t_i,\\ \left(\frac{t-t_i}{t_f-t_i}\right)^5\times\left[252-1050\left(\frac{t-t_i}{t_f-t_i}\right)+1800\left(\frac{t-t_i}{t_f-t_i}\right)^2\right. & \\ \left.-1575\left(\frac{t-t_i}{t_f-t_i}\right)^3+700\left(\frac{t-t_i}{t_f-t_i}\right)^4-126\left(\frac{t-t_i}{t_f-t_i}\right)^5\right] & t\in\left(t_i,t_f\right),\\ 1 & t\geq t_f.\end{cases}$$

Here, we proposed $\overline{\omega}_i = -10\ \frac{\mathrm{rad}}{\mathrm{s}}$ and $\overline{\omega}_f = 10\ \frac{\mathrm{rad}}{\mathrm{s}}$ for times $t_i = 4$ s and $t_f = 6$ s, respectively. Figure 5 shows the experimental evaluation performed by the full-bridge Buck inverter–DC motor system in closed-loop.

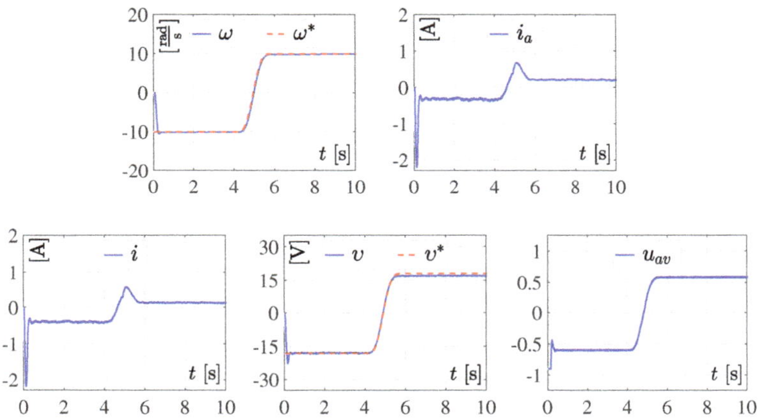

Figure 5. Experimental dynamic response of the full-bridge Buck inverter–DC motor system in closed-loop for the first desired trajectory ω^* (22).

The obtained simulation results of this test are shown in Figure 6.

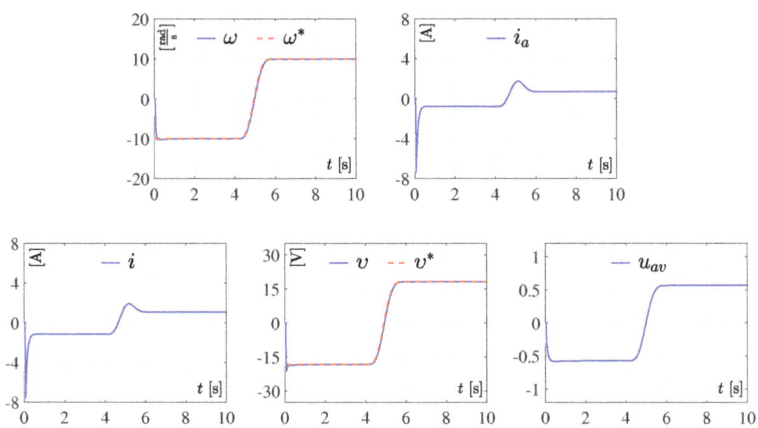

Figure 6. Simulation results of the full-bridge Buck inverter–DC motor system in closed-loop for the first desired trajectory ω^* (22).

5.1.2. Experimental and Simulation Test 2

To demonstrate the effectiveness of the developed control when periodic signals a
contemplated, sinusoidal trajectories were explored as desired angular velocity profile
Thus, in the second experiment ω^* was chosen as:

$$\omega^*(t) = 10\sin(0.8\pi t). \quad (2$$

Figure 7 exhibits the corresponding experimental results of the full-bridge Buck inverte
DC motor system in closed-loop.

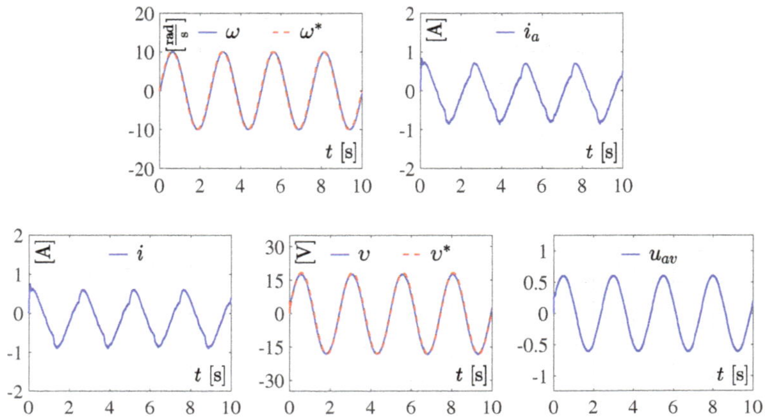

Figure 7. Experimental results when the second desired angular velocity profile, given by (23), contemplated for the system in closed-loop.

The simulation results associated with this desired trajectory are exhibited in Figure

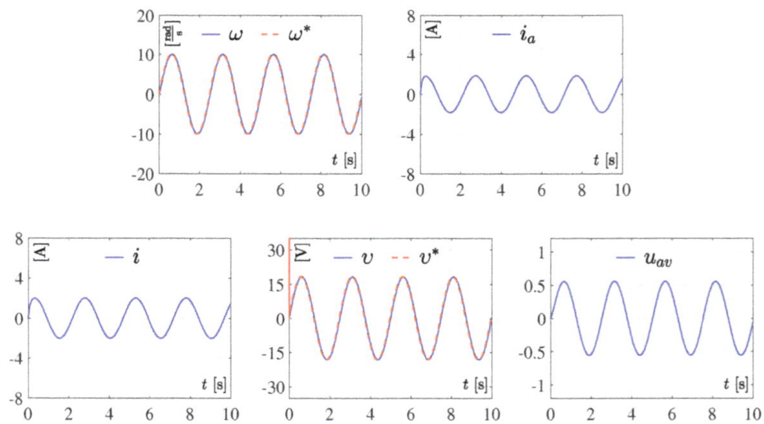

Figure 8. Simulation results when the second desired angular velocity profile, given by (23), contemplated for the system in closed-loop.

5.1.3. Experimental and Simulation Test 3

A second sinusoidal waveform was considered as the desired trajectory ω^* in th
implementation of the third experiment. Here, ω^* was given by

$$\omega^*(t) = 10\left(1 - e^{-2t^2}\right)\sin(0.8\pi t). \quad (2$$

The evolution of the system variables associated with the proposed flatness control approach is illustrated in Figure 9.

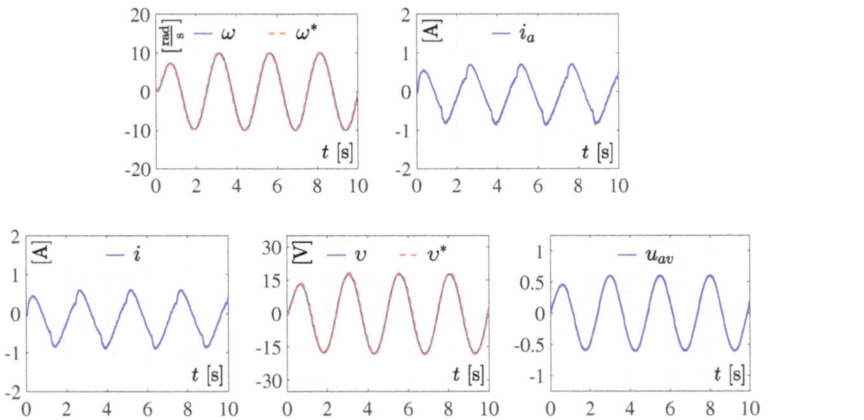

Figure 9. Experimental response of the system in closed-loop for the third desired angular velocity ω^* (24).

The simulation results associated with the experimental results depicted in Figure 9 are presented in Figure 10.

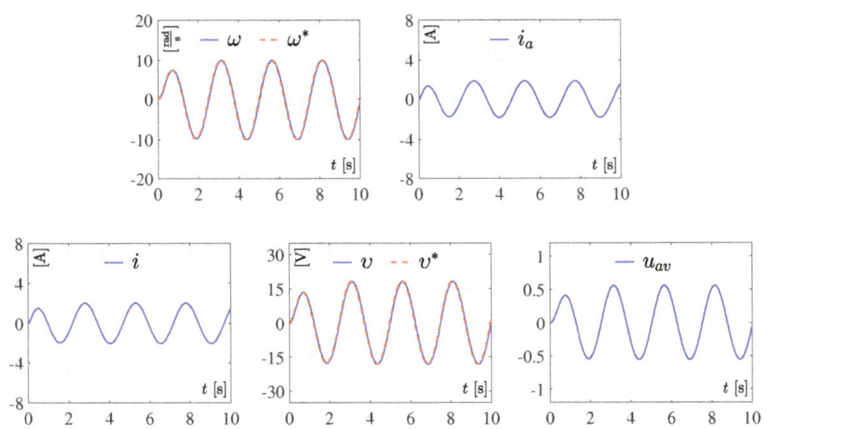

Figure 10. Simulation response of the system in closed-loop for the third desired angular velocity ω^* (24).

5.1.4. Experimental and Simulation Test 4

In the realization of the fourth experiment, the sinusoidal signal determined by (25) was introduced as the desired trajectory ω^*, i.e.,

$$\omega^*(t) = 10\sin(0.125\pi t^{\frac{3}{2}}). \tag{25}$$

For this experiment, the desired trajectory tracking performance of the system in closed-loop is presented in Figure 11. The corresponding simulation results are shown in Figure 12.

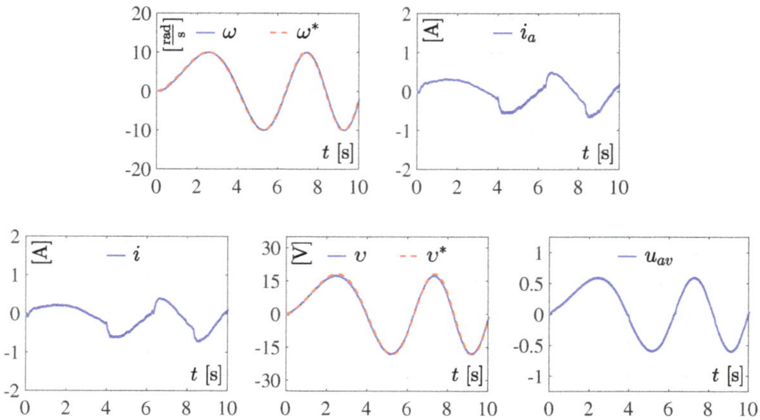

Figure 11. Experimental results of the system in closed-loop, when the fourth desired trajectory fo
ω^*, determined by (25), is taken into account.

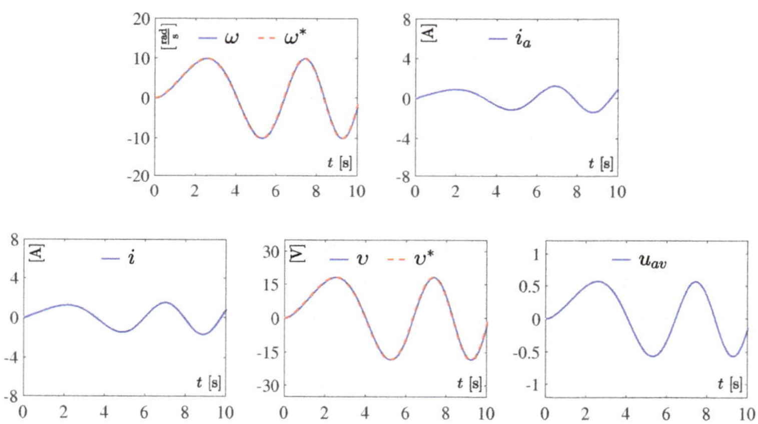

Figure 12. Simulation results of the system in closed-loop when the fourth desired trajectory for ω^*
determined by (25), is taken into account.

5.1.5. Experimental and Simulation Test 5

Lastly, the fifth experiment was carried out with abrupt perturbations in the converte load R inserted in real time into the built platform of the system. For this experiment, th desired angular velocity profile ω^* to be followed by the motor shaft was considered a in (22). Meanwhile, the abrupt perturbations in R were selected as:

$$R_m = \begin{cases} R & 0\,\mathrm{s} \leq t < 7.5\,\mathrm{s}, \\ 30\%R & 7.5\,\mathrm{s} \leq t \leq 10\,\mathrm{s}. \end{cases} \quad (26$$

The experimental performance of the flatness-based control when (26) was execute in the system is displayed in Figure 13. The simulation results are displayed in Figure 14

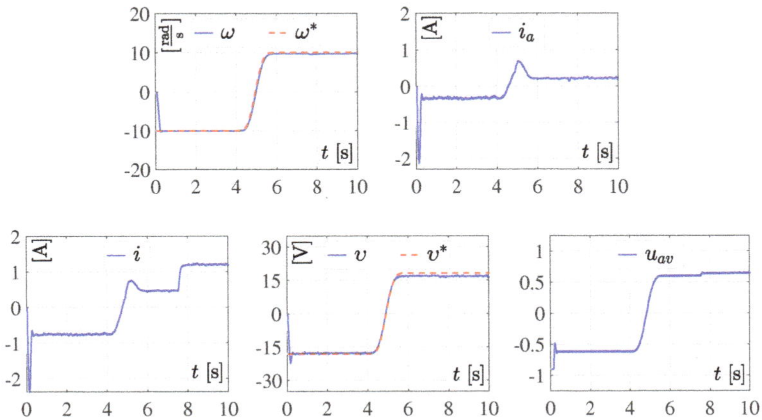

Figure 13. Experimental results associated with the designed flatness-based control (3), when ω^* is of the Bézier type (22), and the abrupt variations in load R, declared in (26), are introduced into the system.

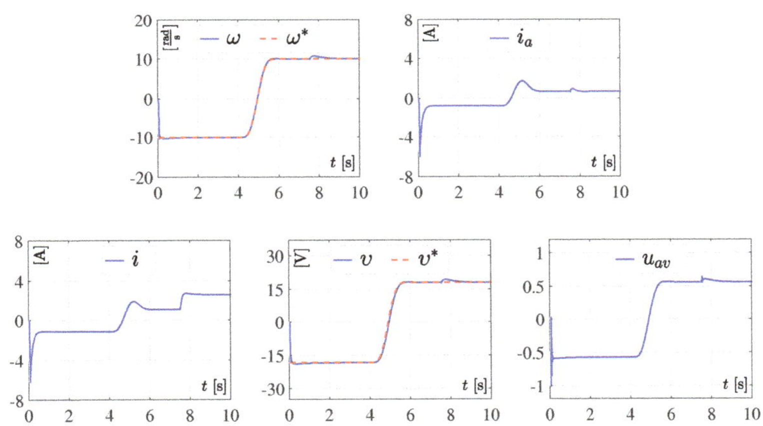

Figure 14. Simulation results associated with the designed flatness-based control (3), when ω^* is of the Bézier type (22), and the abrupt variations in load R, declared in (26), are introduced into the system.

5.2. Discussion of the Experimental and Simulation Results

The results of the experimental and simulation tests are discussed here. According to the experimental results presented in Figures 5, 7, 9, 11, and 13 along with the corresponding simulation results depicted in Figures 6, 8, 10, 12, and 14, respectively, we deduced that the flatness-based control satisfactorily accomplished the execution of the angular velocity trajectory tracking task in the full-bridge Buck inverter–DC motor system, i.e., $\omega \to \omega^*$. Thus, since we have shown that the obtained experimental results and their corresponding simulation results had a similar behavior, the remainder of this section is devoted to the comparative analysis of the experimental results obtained (i.e., by using the flatness concept) with the results reported in the literature.

To highlight the good performance of the designed flatness-based control (3) on the system, in the following, for the four different desired profiles ω^*, i.e., (22)–(25), the plotted graphics of the tracking errors corresponding to ω and v are presented for the next three cases:

1. The associated experimental tests of the system in closed-loop with the flatness-base control (studied in this paper). Here, the closed-loop tracking errors for the angul velocity ($e_{\omega\,Ej_F}$) and voltage ($e_{v\,Ej_F}$) were determined by

$$\begin{aligned} e_{\omega\,Ej_F} &= \omega - \omega^*, \\ e_{v\,Ej_F} &= v - v^*, \end{aligned} \quad (2$$

where subscript Ej_F, for $j \in \{1,2,3,4\}$, associates the experimental test from whic the tracking error was obtained. That is, for $j=1$, $j=2$, $j=3$, and $j=4$, the desire trajectories ω^* correspond to (22)–(25), respectively.

2. The experimental dynamic responses of the system in closed-loop with the passiv control based on the ETEDPOF strategy, developed in [41]. Here, the closed-loop trac ing errors of the angular velocity and voltage denoted by $e_{\omega\,Ej_P}$ and $e_{v\,Ej_P}$, respectivel were declared as

$$\begin{aligned} e_{\omega\,Ej_P} &= \omega - \omega^*, \\ e_{v\,Ej_P} &= v - v^*, \end{aligned} \quad (2$$

where subscript Ej_P, for $j \in \{1,2,3,4\}$, represents the experimental response fro which the tracking error was obtained.

3. The obtained experimental results for the system in open-loop, analyzed in [40]. Her the open-loop tracking errors of the angular velocity ($e_{\omega\,Ej_{ol}}$) and voltage ($e_{v\,Ej_{ol}}$) we defined by

$$\begin{aligned} e_{\omega\,Ej_{ol}} &= \omega - \omega^*, \\ e_{v\,Ej_{ol}} &= v - v^*, \end{aligned} \quad (2$$

where subscript Ej_{ol}, for $j \in \{1,2,3,4\}$, indicates the experimental result in open-loc from which the tracking error was obtained.

Figure 15 depicts the plotted graphics of the tracking errors related to ω and v give by (27)–(29) of the full-bridge Buck inverter–DC motor system. On the one hand, accordir to Figure 15a,c,e,g, it can be seen that the closed-loop tracking errors obtained with th flatness-based control for ω (i.e., $e_{\omega\,E\{1,2,3,4\}_F}$) were lower in magnitude than the close loop tracking errors associated with the passive control (i.e., $e_{\omega\,E\{1,2,3,4\}_P}$). In additio in accordance with Figure 15b,d,f,h, a similar evaluation was achieved for the flatne closed-loop tracking errors corresponding to v (i.e., $e_{v\,E\{1,2,3,4\}_F}$) in comparison with the associated passive closed-loop tracking errors (i.e., $e_{v\,E\{1,2,3,4\}_P}$). Meanwhile, note that th open-loop tracking errors, defined in (29) for ω and v (i.e., $e_{\omega\,E\{1,2,3,4\}_{ol}}$ and $e_{v\,E\{1,2,3,4\}_{ol}}$ are also presented in Figure 15 with the aim of exhibiting how the system profited from th proposed flatness-based control and the developed passive control reported in [41] and nc for the purpose of comparative evaluation. On the other hand, a visual comparison betwee the experimental results of the passivity-based control reported in [41] (see Figure 1) an the experimental results of the designed flatness-based control (see Figure 13) shows ho the performance of the latter control was superior. This is due to the fact that it is robu against parametric variations in comparison with the passivity-based control.

Lastly, from the aforementioned comments on the experimental results, it is inferre that the developed flatness-based control given by (3) satisfactorily carried out the aim c solving the angular velocity trajectory tracking problem in the full-bridge Buck inverter–D motor system, that is, $\omega \to \omega^*$.

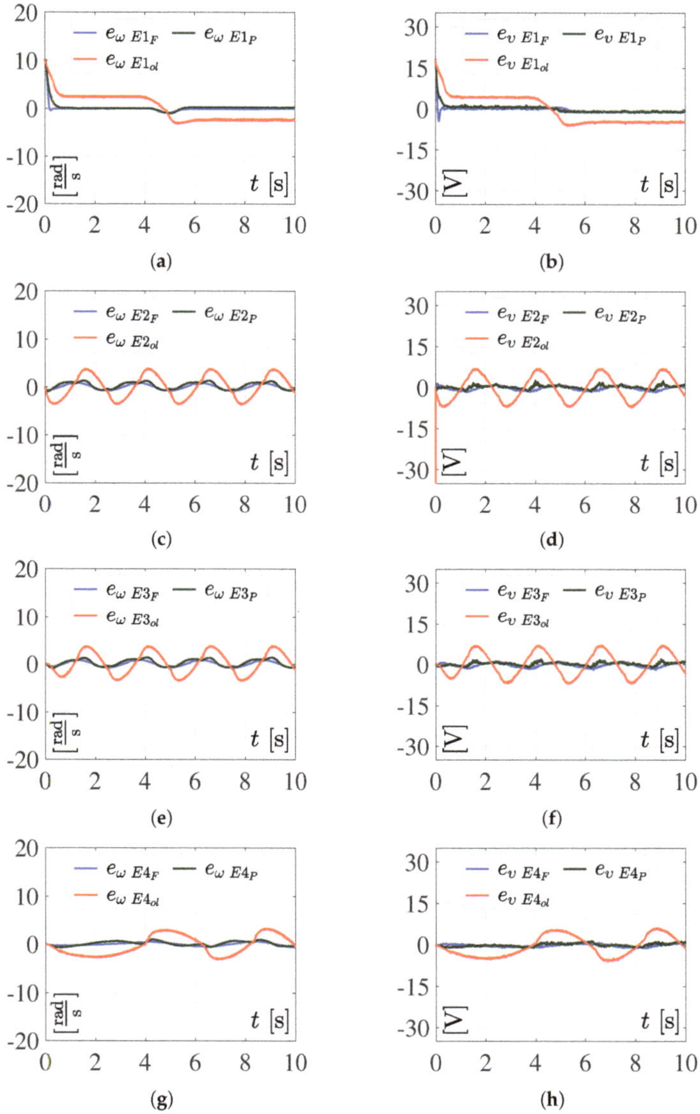

Figure 15. Plotted graphics of the tracking errors associated with variables ω and v of the full-bridge Buck inverter–DC motor system. The signal pairs $e_{\omega\,E\{1,2,3,4\}_F}$–$e_{v\,E\{1,2,3,4\}_F}$, $e_{\omega\,E\{1,2,3,4\}_P}$–$e_{v\,E\{1,2,3,4\}_P}$, and $e_{\omega\,E\{1,2,3,4\}_{ol}}$–$e_{v\,E\{1,2,3,4\}_{ol}}$ are the tracking errors obtained for ω and v, when the system considers the flatness-based control, the passivity-based control [41], and the experimental results in open-loop [40], respectively. (**a**) Tracking errors of ω, for the desired trajectory ω^* declared in (22). (**b**) Tracking errors of v, for the desired trajectory ω^* declared in (22). (**c**) Tracking errors of ω, when the desired profile ω^* is given by (23). (**d**) Tracking errors of v, when the desired profile ω^* is given by (23). (**e**) Tracking errors of ω, for ω^* defined by (24). (**f**) Tracking errors of v, for ω^* defined by (24). (**g**) Tracking errors of ω, for the desired trajectory considered in (25). (**h**) Tracking errors of v, for the desired trajectory considered in (25).

6. Conclusions

A robust differential flatness-based control for carrying out the angular velocity tra-jectory tracking task associated with the full-bridge Buck inverter–DC motor system wa developed and experimentally validated in this work. This robust control was designe by exploiting the differential flatness property of the mathematical model related to th full-bridge Buck inverter–DC motor system. Then, the experimental implementation of th robust flatness-based control was programmed by utilizing Simulink and a dSPACE DS110 controller board on a built prototype of the full-bridge Buck inverter–DC motor systen Lastly, the closed-loop system was also programmed in Matlab–Simulink with the purpos of obtaining the corresponding simulation results. After analyzing the experimental result of the closed-loop system, it was concluded that the control objective, i.e., $\omega \to \omega^*$, wa solved even when abrupt variations were added to the nominal parameters of the systen Such a conclusion was corroborated by comparison with the simulation results.

Finally, derived from the experimental results satisfactorily obtained, future wor could be developed on proposing this system as a viable electronics power stage for AC motors, renewable energy systems, electromechanical systems, robotic arms, wheele mobile robots, uncrewed underwater vehicles, and mechatronic systems.

Author Contributions: Conceptualization, R.S.-O., M.M.-M., R.E.G.-C. and A.R.-C.; methodolog R.S.-O., M.M.-M. and R.E.G.-C.; software, A.R.-C., V.M.H.-G., E.H.-M. and J.R.G.-S.; validatio A.R.-C., V.M.H.-G., E.H.-M. and J.R.G.-S.; formal analysis, R.S.-O., A.R.-C., E.H.-M. and J.R.G.-S investigation, R.S.-O., M.M.-M., R.E.G.-C. and A.R.-C.; resources, R.S.-O., M.M.-M., R.E.G.-C., A.R.-C J.R.G.-S., R.G.-C. and G.S.-O.; data curation, A.R.-C., V.M.H.-G., E.H.-M. and J.R.G.-S.; writing-original draft preparation, R.S.-O., M.M.-M., R.E.G.-C., A.R.-C., V.M.H.-G., E.H.-M. and J.R.G.-S writing—review and editing, R.S.-O., M.M.-M., R.E.G.-C., A.R.-C., J.R.G.-S., R.G.-C. and G.S.-O visualization, R.S.-O. and M.M.-M.; supervision, R.S.-O., M.M.-M., R.G.-C. and G.S.-O.; projec administration, R.S.-O. and M.M.-M.; funding acquisition, R.S.-O. and M.M.-M. All authors hav read and agreed to the published version of the manuscript.

Funding: This research was funded by the Comisión de Operación y Fomento de Actividade Académicas (COFAA) and the Secretaría de Investigación y Posgrado (SIP), both from the Institut Politécnico Nacional, México.

Institutional Review Board Statement: Not applicable.

Informed Consent Statement: Not applicable.

Data Availability Statement: Not applicable.

Acknowledgments: This work was supported by the Instituto Politécnico Nacional, México. Th work of Ramón Silva-Ortigoza and Magdalena Marciano-Melchor was supported in part by SN México, and in part by IPN programs EDI and SIBE. The work of Rogelio Ernesto García-Chávez wa supported by the CONACYT-México and BEIFI scholarships. The work of Victor Manuel Hernánde Guzmán, Eduardo Hernández-Márquez, José Rafael García-Sánchez, and Gilberto Silva-Ortigoz was supported by SNI-México.

Conflicts of Interest: The authors declare no conflicts of interest.

References

1. Gieras, J.F. *Permanent Magnet Motor Technology: Design and Applications*, 3rd ed.; CRC Press: New York, NY, USA, 2010.
2. Lyshevski, S.E. *Electromechanical Systems, Electric Machines, and Applied Mechatronics*; CRC Press: Boca Raton, FL, USA, 200 ISBN 0-8493-2275-8.
3. Ahmad, M.A.; Ismail, R.M.T.R.; Ramli, M.S. Control strategy of Buck converter driven DC motor: A comparative assessmen *Austral. J. Basic Appl. Sci.* **2010**, *4*, 4893–4903.
4. Bingöl, O.; Paçaci, S. A virtual laboratory for neural network controlled DC motors based on a DC-DC Buck converter. *Int. J. En Educ.* **2012**, *28*, 713–723.
5. Sira-Ramírez, H.; Oliver-Salazar, M.A. On the robust control of Buck-converter DC-motor combinations. *IEEE Trans. Powe Electron.* **2013**, *28*, 3912–3922. [CrossRef]
6. Yang, J.; Wu, H.; Hu, L.; Li, S. Robust predictive speed regulation of converter-driven DC motors via a discrete-time reduced-orde GPIO. *IEEE Trans. Ind. Electron.* **2019**, *66*, 7893–7903. [CrossRef]

Stanković, M.R.; Madonski, R.; Shao, S.; Mikluc, D. On dealing with harmonic uncertainties in the class of active disturbance rejection controllers. *Int. J. Control* **2021**, *94*, 2795–2810. [CrossRef]

Madonski, R.; Łakomy, K.; Stankovic, M.; Shao, S.; Yang, J.; Li, S. Robust converter-fed motor control based on active rejection of multiple disturbances. *Control Eng. Pract.* **2021**, *107*, 104696. [CrossRef]

Zhang. L.; Yang, J.; Li, S. A model-based unmatched disturbance rejection control approach for speed regulation of a converter-driven DC motor using output-feedback. *IEEE-CAA J. Autom. Sin.* **2022**, *9*, 365–376. [CrossRef]

Guerrero-Ramírez, E.; Martínez-Barbosa, A.; Guzmán-Ramírez, E.; Barahona-Ávalos, J.L. Design methodology for digital active disturbance rejection control of the DC motor drive. *e-Prime—Adv. Electr. Eng. Electron. Energy* **2022**, *2*, 100050. [CrossRef]

Guerrero-Ramírez, E.; Martínez-Barbosa, A.; Contreras-Ordaz, M.A.; Guerrero-Ramírez, G.; Guzman-Ramirez, E.; Barahona-Ávalos, J.L.; Adam-Medina, M. DC motor drive powered by solar photovoltaic energy: An FPGA-based active disturbance rejection control approach. *Energies* **2022**, *15*, 6595. [CrossRef]

Rigatos, G.; Siano, P.; Wira, P.; Sayed-Mouchaweh, M. Control of DC–DC converter and DC motor dynamics using differential flatness theory. *Intell. Ind. Syst.* **2016**, *2*, 371–380. [CrossRef]

Rigatos, G.; Siano, P.; Ademi, S.; Wira, P. Flatness-based control of DC-DC converters implemented in successive loops. *Electr. Power Compon. Syst.* **2018**, *46*, 673–687. [CrossRef]

Silva-Ortigoza, R.; Roldán-Caballero, A.; Hernández-Márquez, E.; García-Sánchez, J.R.; Marciano-Melchor, M.; Hernández-Guzmán, V.M.; Silva-Ortigoza, G. Robust flatness tracking control for the "DC/DC Buck converter-DC motor" system: Renewable energy-based power supply. *Complexity* **2021**, *2021*, 2158782. [CrossRef]

Hoyos, F.E.; Rincón, A.; Taborda, J.A.; Toro, N.; Angulo, F. Adaptive quasi-sliding mode control for permanent magnet DC motor. *Math. Probl. Eng.* **2013**, *2013*, 693685. [CrossRef]

Hoyos Velasco, F.E.; Candelo-Becerra, J.E; Rincón Santamaria, A. Dynamic analysis of a permanent magnet DC motor using a Buck converter controlled by ZAD-FPIC. *Energies* **2018**, *11*, 3388. [CrossRef]

Hoyos, F.E.; Candelo-Becerra, J.E.; Hoyos Velasco, C.I. Application of zero average dynamics and fixed point induction control techniques to control the speed of a DC motor with a Buck converter. *Appl. Sci.* **2020**, *10*, 1807. [CrossRef]

Hoyos, F.E.; Candelo-Becerra, J.E.; Rincón, A. Zero average dynamic controller for speed control of DC motor. *Appl. Sci.* **2021**, *11*, 5608. [CrossRef]

Wei, F.; Yang, P.; Li, W. Robust adaptive control of DC motor system fed by Buck converter. *Int. J. Control Autom.* **2014**, *7*, 179–190. [CrossRef]

Silva-Ortigoza, R.; Hernández-Guzmán, V.M.; Antonio-Cruz, M.; Muñoz-Carrillo, D. DC/DC Buck power converter as a smooth starter for a DC motor based on a hierarchical control. *IEEE Trans. Power Electron.* **2015**, *30*, 1076–1084. [CrossRef]

Hernández-Guzmán, V.M.; Silva-Ortigoza, R.; Muñoz-Carrillo, D. Velocity control of a brushed DC-motor driven by a DC to DC Buck power converter. *Int. J. Innov. Comp. Inf. Control* **2015**, *11*, 509–521. [CrossRef]

Rauf, A.; Li, S.; Madonski, R.; Yang, J. Continuous dynamic sliding mode control of converter-fed DC motor system with high order mismatched disturbance compensation. *Trans. Inst. Meas. Control* **2020**, *42*, 2812–2821. [CrossRef]

Rauf, A.; Zafran, M.; Khan, A.; Tariq, A.R. Finite-time nonsingular terminal sliding mode control of converter-driven DC motor system subject to unmatched disturbances. *Int. Trans. Elect. Energy Syst.* **2021**, *31*, e13070. [CrossRef]

Ravikumar, D.; Srinivasan, G.K. Implementation of higher order sliding mode control of DC–DC Buck converter fed permanent magnet DC motor with improved performance. *Automatika* **2022**. [CrossRef]

Khubalkar, S.; Chopade, A.; Junghare, A.; Aware, M.; Das, S. Design and realization of stand-alone digital fractional order PID controller for Buck converter fed DC motor. *Circuits Syst. Signal Process.* **2016**, *35*, 2189–2211. [CrossRef]

Khubalkar, S.W.; Junghare, A.S.; Aware, M.V.; Chopade, A.S.; Das, S. Demonstrative fractional order—PID controller based DC motor drive on digital platform. *ISA Trans.* **2018**, *82*, 79–93. [CrossRef] [PubMed]

Patil, M.D.;Vadirajacharya, K.; Khubalkar, S.W. Design and tuning of digital fractional-order PID controller for permanent magnet DC motor. *IETE J. Res.* **2021**. [CrossRef]

Kumar, S.G.; Thilagar, S.H. Sensorless load torque estimation and passivity based control of Buck converter fed DC motor. *Sci. World J.* **2015**, *2015*, 132843. [CrossRef]

Srinivasan, G.K.; Srinivasan, H.T.; Rivera, M. Sensitivity Analysis of Exact Tracking Error Dynamics Passive Output Control for a Flat/Partially Flat Converter Systems. *Electronics* **2020**, *9*, 1942. [CrossRef]

Nizami, T.K.; Chakravarty, A.; Mahanta, C. Design and implementation of a neuro-adaptive backstepping controller for Buck converter fed PMDC-motor. *Control Eng. Pract.* **2017**, *58*, 78–87. [CrossRef]

Nizami, T.K.; Chakravarty, A.; Mahanta, C.; Iqbal, A.; Hosseinpour, A. Enhanced dynamic performance in DC–DC converter-PMDC motor combination through an intelligent non-linear adaptive control scheme. *IET Power Electron.* **2022**. [CrossRef]

Nizami, T.K.; Gangula, S.D.; Reddy, R.; Dhiman, H.S. Legendre neural network based intelligent control of DC-DC step down converter-PMDC motor combination. *IFAC* **2022**, *55*, 162–167. [CrossRef]

Hanif, M.I.F.M.; Suid, M.H.; Ahmad, M.A. A piecewise affine PI controller for Buck converter generated DC motor. *Int. J. Power Electron. Drive Syst.* **2019**, *10*, 1419–1426. [CrossRef]

Rigatos, G.; Siano, P.; Sayed-Mouchaweh, M. Adaptive neurofuzzy H-infinity control of DC-DC voltage converters. *Neural Comput. Appl.* **2020**, *33*, 2507–2520. [CrossRef]

35. Kazemi, M.G.; Montazeri, M. Fault detection of continuous time linear switched systems using combination of bond grap method and switching observer. *ISA Trans.* **2019**, *94*, 338–351. [CrossRef] [PubMed]
36. Silva-Ortigoza, R.; Alba-Juárez, J.N.; García-Sánchez, J.R.; Antonio-Cruz, M.; Hernández-Guzmán, V.M.; Taud, H. Modeling ar experimental validation of a bidirectional DC/DC Buck power electronic converter–DC motor system. *IEEE Latin Am. Tra* **2017**, *15*, 1043–1051. [CrossRef]
37. Silva-Ortigoza, R.; Alba-Juárez, J.N.; García-Sánchez, J.R.; Hernández-Guzmán, V.M.; Sosa-Cervantes, C.Y.; Taud, H. A sensorle passivity-based control for the DC/DC Buck converter–Inverter–DC motor system. *IEEE Latin Am. Trans.* **2016**, *14*, 4227–42 [CrossRef]
38. Hernández-Márquez, E.; García-Sánchez, J.R.; Silva-Ortigoza, R.; Antonio-Cruz, M.; Hernández-Guzmán, V.M.; Taud, H Marcelino-Aranda, M. Bidirectional tracking robust controls for a DC/DC Buck converter-DC motor system. *Complexity* **20** *2018*, 1260743. [CrossRef]
39. Chi, X.; Quan, S.; Chen, J.; Wang, Y.-X.; He, H. Proton exchange membrane fuel cell-powered bidirectional DC motor control bas on adaptive sliding-mode technique with neural network estimation. *Int. J. Hydrog. Energy* **2020**, *45*, 20282–20292. [CrossRef]
40. Hernández-Márquez, E.; Avila-Rea, C.A.; García-Sánchez, J.R.; Silva-Ortigoza, R.; Marciano-Melchor, M.; Marcelino-Aranc M.; Roldán-Caballero, A.; Márquez-Sánchez, C. New "full-bridge Buck inverter—DC motor" system: Steady-state and dynam analysis and experimental validation. *Electronics* **2019**, *8*, 1216. [CrossRef]
41. Silva-Ortigoza, R.; Hernandez-Marquez, E.; Roldan-Caballero, A.; Tavera-Mosqueda, S.; Marciano-Melchor, M.; Garcia-Sanch J.R.; Hernandez-Guzman, V.M.; Silvia-Ortigoza, R. Sensorless tracking control for a "full-bridge Buck inverter—DC motor" syste Passivity and flatness-based design. *IEEE Access* **2021**, *9*, 132191–132204. [CrossRef]
42. Linares-Flores, J.; Reger, J.; Sira-Ramírez, H. Load torque estimation and passivity-based control of a Boost-converter/DC-mot combination. *IEEE Trans. Control Syst. Technol.* **2010**, *18*, 1398–1405. [CrossRef]
43. Alexandridis, A.T.; Konstantopoulos, G.C. Modified PI speed controllers for series-excited DC motors fed by DC/DC Boo converters. *Control Eng. Pract.* **2014**, *23*, 14–21. [CrossRef]
44. Malek, S. A new nonlinear controller for DC-DC Boost converter fed DC motor. *Int. J. Power Electron.* **2015**, *7*, 54–71. [CrossRe
45. Konstantopoulos, G.C.; Alexandridis, A.T. Enhanced control design of simple DC-DC Boost converter-driven DC motors: Analys and implementation. *Electr. Power Compon. Syst.* **2015**, *43*, 1946–1957. [CrossRef]
46. Mishra, P.; Banerjee, A.; Ghosh, M.; Baladhandautham, C.B. Digital pulse width modulation sampling effect embodied stead state time-domain modeling of a Boost converter driven permanent magnet DC brushed motor. *Int. Trans. Elect. Energy Sy* **2021**, *31*, e12970. [CrossRef]
47. Govindharaj, A.; Mariappan, A. Design and analysis of novel Chebyshev neural adaptive backstepping controller for Boo converter fed PMDC motor, *Int. J. Autom. Control* **2020**, *14*, 694–712. [CrossRef]
48. Alajmi, B.; Ahmed, N.A.; Al-Othman, A.K. Small-signal analysis and hardware implementation of Boost converter fed PMDC motor f electric vehicle applications. *J. Eng. Res.* **2021**, *9*, 189–208. [CrossRef]
49. Manikandan, C.T.; Sundarrajan, G.T.; Krishnan, V.G.; Ofori, I. Performance analysis of two-loop interleaved Boost converter f PMDC-motor system using FLC. *Math. Probl. Eng.* **2022**, *2022*, 1639262. [CrossRef]
50. García-Sánchez, J.R.; Hernández-Márquez, E.; Ramírez-Morales, J.; Marciano-Melchor, M.; Marcelino-Aranda, M.; Taud, H Silva-Ortigoza, R. A robust differential flatness-based tracking control for the "MIMO DC/DC Boost converter—Inverter—D motor" system: Experimental results. *IEEE Access* **2019**, *7*, 84497–84505. [CrossRef]
51. Egidio, L.N.; Deacto, G.S.; Jungers, R.M. Stabilization of rank-deficient continuous-time switched affine systems. *Automatica* **20** *143*, 110426. [CrossRef]
52. Sönmez, Y.; Dursun, M.; Güvenç, U.; Yilmaz, C. Start up current control of Buck-Boost convertor-fed serial DC motor. *Pamukk Univ. J. Eng. Sci.* **2009**, *15*, 278–283.
53. Linares-Flores, J.; Barahona-Avalos, J.L.; Sira-Ramírez, H.; Contreras-Ordaz, M.A. Robust passivity-based control of a Buc Boost-converter/DC-motor system: An active disturbance rejection approach. *IEEE Trans. Ind. Appl.* **2012**, *48*, 2362–23 [CrossRef]
54. Hernández-Márquez, E.; Avila-Rea C.A.; García-Sánchez, J.R.; Silva-Ortigoza, R.; Silva-Ortigoza, G.; Taud, H.; Marcelino-Arand M. Robust tracking controller for a DC/DC Buck-Boost converter–Inverter–DC motor system. *Energies* **2018**, *11*, 2500. [CrossRe
55. Ghazali, M.R.; Ahmad, M.A.; Ismail, R.M.T.R. Adaptive safe experimentation dynamics for data-driven neuroendocrine-PI control of MIMO systems. *IETE J. Res.* **2022**, *68*, 1611–1624. [CrossRef]
56. Linares-Flores, J.; Sira-Ramírez, H.; Cuevas-López, E.F.; Contreras-Ordaz, M.A. Sensorless passivity based control of a DC mot via a solar powered Sepic converter-full bridge combination. *J. Power Electron.* **2011**, *11*, 743–750. [CrossRef]
57. Srinivasan, G.K.; Srinivasan, H.T.; Rivera, M. Low-cost implementation of passivity-based control and estimation of load torq for a Luo converter with dynamic load. *Electronics* **2020**, *9*, 1914. [CrossRef]
58. Arshad, M.H.; Abido, M.A. Hierarchical control of DC motor coupled with Cuk converter combining differential flatness ar sliding mode control. *Arab. J. Sci. Eng.* **2021**, *46*, 9413–9422. [CrossRef]
59. Ismail, A.A.A.; Elnady, A. Advanced drive system for DC motor using multilevel DC/DC Buck converter circuit. *IEEE Acce* **2019**, *7*, 54167–54178. [CrossRef]
60. Guerrero, E.; Guzmán, E.; Linares, J.; Martínez, A.; Guerrero, G. FPGA-based active disturbance rejection velocity control for parallel DC/DC Buck converter-DC motor system. *IET Power Electron.* **2020**, *13*, 356–367. [CrossRef]

41. Erenturk, K. Hybrid control of a mechatronic system: Fuzzy logic and grey system modeling approach. *IEEE/ASME Trans. Mechatron* **2007**, *12*, 703–710. [CrossRef]
42. Swathi, K.V.R.; Nagesh-Kumar, G.V. Design of intelligent controller for reduction of chattering phenomenon in robotic arm: A rapid prototyping. *Comput. Electr. Eng.* **2019**, *74*, 483–497. [CrossRef]
43. García-Sánchez, J.R.; Tavera-Mosqueda, S.; Silva-Ortigoza, R.; Hernández-Guzmán, V.M.; Marciano-Melchor, M.; Rubio, J.J.; Ponce-Silva, M.; Hernández-Bolaños, M.; Martínez-Martínez, J. A novel dynamic three-level tracking controller for mobile robots considering actuators and power stage subsystems: Experimental assessment. *Sensors* **2020**, *20*, 4959. [CrossRef] [PubMed]
44. Hernández-Guzmán, V.M.; Silva-Ortigoza, R.; Tavera-Mosqueda, S.; Marcelino-Aranda, M.; Marciano-Melchor, M. Path-tracking of a WMR fed by inverter-DC/DC Buck power electronic converter systems. *Sensors* **2020**, *20*, 6522. [CrossRef] [PubMed]
45. Zhao, Z.; Yang, J.; Li, S.; Yu, X.; Wang, Z. Continuous output feedback TSM control for uncertain systems with a DC-AC inverter example. *IEEE Trans. Circuit Syst. II Express Briefs* **2018**, *65*, 71–75. [CrossRef]
46. Chang, E.-C.; Cheng, C.-A.; Wu, R.-C. Robust optimal tracking control of a full-bridge DC-AC converter. *Appl. Sci.* **2021**, *11*, 1211. [CrossRef]

mathematics

MDPI

Article

Optimal Location and Operation of PV Sources in DC Grids to Reduce Annual Operating Costs While Considering Variable Power Demand and Generation

Luis Fernando Grisales-Noreña [1,*], Oscar Danilo Montoya [2,3] and Carlos Andres Ramos-Paja [4]

1 Department of Electrical Engineering, Faculty of Engineering, Universidad de Talca, Campus Curicó, Curicó 3340000, Chile
2 Grupo de Compatibilidad e Interferencia Electromagnética (GCEM), Facultad de Ingeniería, Universidad Distrital Francisco José de Caldas, Bogotá 110231, Colombia
3 Laboratorio Inteligente de Energía, Universidad Tecnológica de Bolívar, Cartagena 131001, Colombia
4 Facultad de Minas, Universidad Nacional de Colombia, Medellin 050041, Colombia
* Correspondence: luiferg3190@gmail.com

Abstract: Due to the need to include renewable energy resources in electrical grids as well as th development and high implementation of PV generation and DC grids worldwide, it is necessar to propose effective optimization methodologies that guarantee that PV generators are located an sized on the DC electrical network. This will reduce the operation costs and cover the investmer and maintenance cost related to the new technologies (PV distributed generators), thus satisfyin all technical and operative constraints of the distribution grid. It is important to propose solutio methodologies that require short processing times, with the aim of exploring a large number c scenarios while planning energy projects that are to be presented in public and private contract as well as offering solutions to technical problems of electrical distribution companies within sho periods of time. Based on these needs, this paper proposes the implementation of a Discrete Continuous Parallel version of the Particle Swarm Optimization algorithm (DCPPSO) to solve th problem regarding the integration of photovoltaic (PV) distributed generators (DGs) in Direct Currer (DC) grids, with the purpose of reducing the annual costs related to energy purchasing as well a the investment and maintenance cost associated with PV sources in a scenario of variable powe demand and generation. In order to evaluate the effectiveness, repeatability, and robustness of th proposed methodology, four comparison methods were employed, i.e., a commercial software an three discrete–continuous methodologies, as well as two test systems of 33 and 69 buses. In analyzin the results obtained in terms of solution quality, it was possible to identify that the DCPPSO propose obtained the best performance in relation to the comparison methods used, with excellent result in relation to the processing times and standard deviation. The main contribution of the propose methodology is the implementation of a discrete–continuous codification with a parallel processin tool for the evaluation of the fitness function. The results obtained and the reports in the literature fo alternating current networks demonstrate that the DCPPSO is the optimization methodology wit the best performance in solving the problem of the optimal integration of PV sources in economi terms and for any kind of electrical system and size.

Keywords: DC networks; discrete–continuous metaheuristic; parallel processing tool; photovoltai generation; variable power demand; variable renewable generation

MSC: 65K05; 90C26; 90C27

Citation: Grisales-Noreña, L.F.; Montoya, O.D.; Ramos-Paja, C.A. Optimal Location and Operation of PV Sources in DC Grids to Reduce Annual Operating Costs While Considering Variable Power Demand and Generation. *Mathematics* **2022**, *10*, 4512. https://doi.org/10.3390/math10234512

Academic Editors: Atanda Raji and Khaled M. Abo-Al-Ez

Received: 27 October 2022
Accepted: 23 November 2022
Published: 29 November 2022

1. Introduction

Nowadays, DC grids are widely used around the world given their different advantage in comparison with their alternate current (AC) counterparts, i.e., their easy implementatio low investment and maintenance costs, and low operating complexity [1–4]. For these reasons, several electrical industry manufacturers and researchers have put their efforts and money into developing efficient methodologies for the operation and planning of this kind of grid, with the optimal integration of distributed energy resources currently being the topic of interest [5–8]. The optimal integration of distributed energy resources aims to locate and operate distributed generators (DGs) and energy storage systems within electrical systems in order to improve the different technical, economic, and environmental indices of DC grids [9–11]. This problem has been highly studied in alternating current networks by multiple authors, who obtained excellent results in terms of solution and processing times [12–14] and have taken advantage of different methodologies for solving problems with similar characteristics proposed in the literature [15,16], with particular adaptations to different electrical issues. However, the performance of the methodologies for AC networks reported in the literature must be evaluated in DC grids through the proposal of very new methodologies; in each of these cases, the mathematical formulations are different due to the absence of reactive or frequency components in the DC networks, which generates a completely different problem. Consequently, the optimal integration of PV sources into DC grids has become a widely studied topic in recent years [17–19]. In the specialized literature, many studies have been reported with regard to solving the problem of the optimal integration of PV sources into DC grids in order to improve technical conditions (power loss, voltage profiles, system chargeability, etc.) and reduce the environmental impact associated with fossil fuel-based generation as well as with operating and investment costs [20–22].

As for the methodologies used to improve the technical characteristics of the grids, multiple works can be found in the literature. One example is [23], which proposed the use of mixed-integer quadratic programming and the General Algebraic Modeling System (GAMS) software as a solution. In this work, the authors demonstrated the effectiveness of GAMS solvers in terms of standard deviation and processing times. However, this kind of software is often stuck in the local optima and increases the costs and complexity of the solution methodology. Other works have used integer non-linear programming methods to represent the mathematical model that could describe the optimal integration of DGs into DC grids for the reduction of power losses [24], which also requires specialized software, thus increasing the acquisition costs and the complexity of the solution. In order to minimize the implementation of specialized software, different authors have used optimization methods that were based on sequential programming and developed in free software, aiming to reduce power losses via algorithms such as the particle swarm optimization method [12] and the vortex search algorithm [25], among others [26]. The solution impact, standard deviation, and processing time from these works' simulation results were then evaluated in order to demonstrate the effectiveness and robustness of the proposed solution methodologies.

In the last decade, several works have been reported whose objective function is to reduce CO_2 emissions in DC grids by optimally sitting and sizing PV sources. These works employed optimization methods based on sequential programming and aimed to improve the quality of the solution as well as reduce processing times by avoiding the implementation of specialized software [27,28]. However, most of the works published in recent years have focused on improving economical indices such as the reduction of energy purchasing/production, investment, and maintenance costs associated with DG as these indices directly affect both users and electrical operators. In this vein, PV sources are the most developed and installed technology around the world. An example of this is the work presented in [29], which used the BONMIN solver of GAMS to solve the mathematical model that represents the problem of optimal integration of DGs in DC grids in order to reduce annual costs. This methodology remained stuck in the local optima, with short processing times and a standard deviation of zero. Similarly, in [30], a two-stage optimization process was proposed for the integration of PV and wind generators in a DC grid, with the aim of reducing production and investment costs. The results demonstrated the effectiveness of the proposed methodology in terms of its objective function. Howev the authors did not analyze the method's repeatability and processing times. In additio this work employed a master–slave methodology that is traditionally used in the literatu to solve for the optimal integration of DGs in both DC and AC grids [31]. This methodolog employs two kinds of optimization methods, i.e., discrete and continuous, thus requirir more time and increasing the complexity of the solution.

Seeking to improve the effectiveness and robustness of the solution methodologies f the problem under study, the literature has proposed modified optimization algorithms th employ discrete–continuous methodologies to solve the problem of PV sources' optim integration into DC grids. These techniques involve continuous optimization metho that force some variables within the solution to be discrete, thus offering a solution to th location problem while keeping the rest of the variables continuous in order to solve th DG sizing problem. An example of this is [29], where a discrete–continuous version of th vortex search algorithm for integrating PV sources in DC grids was proposed. Its aim w to reduce the annual costs associated with energy purchasing, investment, and maintenan In this paper, the authors compared the results obtained by the proposed methodology to discrete–continuous version of the Chu and Beasley genetic algorithm and the BONMI solver of GAMS, which are used to solve the same problem in AC grids [32]. This wo compared the average results obtained by the solution methodologies, demonstrating th effectiveness of the proposed solution, but it did not analyze the effects of all solution met ods on aspects such as standard deviation and processing times. Furthermore, within th mathematical model and its validation, the voltage and branch current limits associate with the test systems were not analyzed. By using a discrete–continuous codificatio ref. [33] proposed a modified version of the generalized normal distribution optimizer solve the studied problem. In their manuscript, the authors compared the results obtaine with a discrete–continuous version of the vortex search algorithm, a genetic algorith and the BONMIN solver of GAMS. The results obtained demonstrated the effectiveness the proposed methodology, and all constraints related to the operation of the microgri were satisfied. However, the proposed methodology reported longer times in compa son with other methods. Moreover, the authors did not include the analysis of standar deviation and the impact of processing times.

Based on the aforementioned works, it is possible to conclude that it is currently nece sary to propose new methodologies for solving the problem of integrating PV sources in DC grids, ones that reduce complexity by implementing discrete–continuous codificatio and by reducing the processing times. These must also guarantee excellent performance terms of the objective function and standard deviation, with the aim of obtaining a solutic of good quality each time that the algorithm is executed. Another objective should be achieve shorter processing times in order to explore a large number of scenarios whi planning the energy projects that are to be presented in public and private contracts, as we as to offer a solution to the technical problems in distribution electrical companies within short period of time [34].

In light of the issues mentioned above, this study implemented a discrete–continuou parallel version of the particle swarm optimization (PSO) algorithm to solve the proble regarding the integration of PV sources into DC grids. This solution methodology employ a parallel processing tool that takes advantage of all the functions of the computer, with th purpose of reducing processing times. This methodology has been used in the literature solve the problem of the optimal integration of PV sources in AC grids [13]; however, i performance has not been validated in DC grids. The reduction of annual costs associate with energy purchasing, investment, and maintenance in PV sources installed in D grids was thus used as an objective function by implementing a fitness function to ensu compliance with the technical and operative constraints that represent the operation of th DC in an environment of PV sources. Furthermore, to demonstrate the effectiveness of th proposed methodology, two test systems of 33 and 69 buses were used, and we focuse on the methodologies that consider the integration of distributed energy resources in

DC grids. In addition, four methods were employed for comparison (most of which were discrete–continuous, as identified in the state of the art), including the BONMIN solver of GAMS. This paper makes the following contributions:

i. A new approach for the discrete–continuous version of the particle swarm optimization algorithm;
ii. The implementation of parallel processing to solve the problem concerning the integration of PV sources into DC grids in order to reduce processing times and improve the exploration of the algorithm;
iii. The identification of the most efficient methodology to date via simulation results, which could solve the problem of the optimal integration of PV sources into AC and DC grids for annual cost reduction. This considers the results reported in [13] for AC networks.

This paper is structured as follows: Section 2 presents the mathematical model used to solve the problem of the optimal integration of PV sources into DC grids for annual cost reduction; Section 3 describes the proposed solution methodology; Section 4 describes the 33- and 69-node test systems as well as the generation and demand curves used and the considerations made in obtaining the simulation results; Section 5 analyzes the results obtained by the methodologies in terms of the solution, processing times, and repeatability; and Section 6 presents the conclusions and future works derived from this research.

2. Mathematical Formulation

This section presents the mathematical formulation of the problem concerning the integration of PV sources into DC grids for the reduction of energy production/purchasing costs associated with conventional generators (electrical grid, DIESEL generators, among others) as well as the initial investment and maintenance costs related to PV generators. Furthermore, this mathematical model includes all constraints related to the technical and operating constraints of DC grids in the context of PV sources.

2.1. Objective Function

The objective function is presented in Equation (1). This equation considers the minimization of the annual costs associated with energy purchasing in relation to conventional generators (f_1) as well as to investment and maintenance with regard to the installation of PV sources (f_2).

$$OF = \min Annual_{costs} = \min(f_1 + f_2) \tag{1}$$

To calculate the annualized energy purchasing costs while considering the lifetime of the PV generators and the increase in power demand, Equation (2) was used, where C_{kWh} corresponds to the cost of each kWh, T represents the number of days in a regular year (365), and F_a is the factor that annualizes the total energy purchasing/production costs by conventional generators installed in the DC grid. In Equation (3), t_a represents the fixed return rate for the investment made in the integration of the PV generators, N_t corresponds to the number of years contained inside the useful life of the PV sources, and F_c corresponds to the annual increase in power demand within the planning horizon. This factor is described in Equation (4), where t_e represents the increase in the energy purchasing cost within the analyzed time (expressed as a percentage); $p_{i,h}^{cg}$ represents the power supplied by the conventional generator located at node i in the period of time h; and Δh is the duration of the said power supply. Finally, $\Omega_{\mathcal{N}}$, $\Omega_{\mathcal{H}}$, and $\Omega_{\mathcal{T}}$ represent the set of nodes that make up the DC grid, the total periods of time considered for a day of operation, and the useful life of the PV generators (years).

$$f_1 = C_{kWh} T F_a F_c \left(\sum_{i \in \Omega_{\mathcal{H}}} \sum_{i \in \Omega_{\mathcal{N}}} p_{i,h}^{cg} \Delta h \right) \tag{2}$$

$$F_a = \left(\frac{t_a}{1-(1+t_a)^{-N_t}} \right) \quad (3)$$

$$F_c = \left(\sum_{t \in \mathcal{T}} \left(\frac{1+t_e}{1+t_a} \right)^t \right) \quad (4)$$

Equation (5) calculates the investment and maintenance costs associated with the integration of PV sources into the DC grid, where C_{pv} corresponds to the cost per kW of the PV sources, p_i^{pv} represents the total PV power installed at bus i, $C_{O\&M}^{pv}$ denotes the maintenance cost of the PV generators by kW generated, and $p_{i,h}^{pv}$ is the power supply of the PV sources located at bus i for the period of time h. In this equation, Ω_{pv} denotes the set of buses that contain PV sources.

$$f_2 = C_{pv} F_a \left(\sum_{i \in \Omega_{pv}} p_i^{pv} \right) + T \left(\sum_{i \in \Omega_{\mathcal{H}}} \sum_{i \in \Omega_{pv}} C_{O\&M}^{pv} p_{i,h}^{pv} \Delta h \right) \quad (5)$$

2.2. Constraints

The set of constraints that represent the problem of the optimal integration of PV sources into DC grids while considering variable power demand and generation is presented in Equations (6)–(11).

$$p_{i,h}^{cg} + p_i^{pv} C_h^{pv} - P_{i,h}^d = v_{i,h} \sum_{j \in \Omega_N} G_{ij} v_{j,h} \quad (6)$$

$$P_i^{cg,\min} \le p_{i,h}^{cg} \le P_i^{cg,\max} \quad (7)$$

$$x_i p_i^{pv,\min} \le p_i^{pv} \le x_i p_i^{pv,\max} \quad (8)$$

$$\sum_{i \in \Omega_{pv}} x_i \le N_{pv}^{dev} \quad (9)$$

$$V_i^{\min} \le v_{i,h} \le V_i^{\max} \quad (10)$$

$$I_{ij,h} \le I_{ij}^{\max} \quad (11)$$

Equation (6) represents the power balance of the electrical grid. Here, $P_{i,h}^d$ denotes the active power demanded at node i in the period of time h, C_h^{pv} is the factor that determines (from 0 to 100%) the power production of PV sources in relation to the radiance potential of the region where the electrical system is located, $v_{i,h}$ and $v_{j,h}$ represent the voltage profiles at buses i and j in period of time h, and G_{ij} denotes the conductance value related to the branch that connects buses i and j. Equation (7) formulates the constraint that ensures that the minimum ($p_i^{cg,\min}$) and maximum ($p_i^{cg,\max}$) power are supplied by the conventional generator located at node i. Equation (8) establishes the minimum ($p_i^{pv,\min}$) and maximum ($p_i^{pv,\max}$) power bounds for the PV sources located at node i, where x_i corresponds to a binary variable that takes a value of 1 when a PV source is located at node i and a value of 0 when it is not. In this way, Equation (8) limits the maximum number of PV sources (N_{pv}^{dev}) to be located in the DC grid. Equation (10) describes the constraints related to the voltage profile limits, where $V_i^{\min}$ and $V_i^{\max}$ correspond to the minimum and maximum voltage profiles allowed at node i. Finally, Equation (11) represents the maximum branch limit, where $I_{ij,h}$ and $I_{ij}^{\max}$ correspond to the branch and maximum current allowed for the branch that connects buses i and j.

$$FF = min(f_1 + \beta PF) \quad (12)$$

$$PF = \begin{pmatrix} max\left\{0, \sum_{i\in\mathcal{N}} (p_i^{pv} - P_i^{pv,max} x_i^{DG})\right\} \\ +\left|min\left\{0, \sum_{i\in\mathcal{N}} (p_i^{pv} - P_i^{pv,\min} x_i^{DG})\right\}\right| \\ +max\left\{0, \sum_{i\in\mathcal{N}} (v_i - V_i^{max})\right\} \\ +\left|min\left\{0, \sum_{i\in\mathcal{N}} (v_i - V_i^{min})\right\}\right| \\ +max\left\{0, \sum_{i\in\mathcal{N}} \sum_{j\in\mathcal{N}} (I_{ij} - I_{ij}^{max})\right\} \\ +max\left\{0, \left(\sum_{i\in\mathcal{N}} x_i - N_{pv}^{dev}\right)\right\} \end{pmatrix} \quad (13)$$

In order to guarantee all the previously described constraints and improve the exploration of the solution space, this paper used the fitness function presented in (12), which allows the penalization of the objective function when some constraints are violated by permitting the solution methodology to explore infeasible regions. This helps to improve the solution and reduce processing times [35]. In FF, β is entrusted with normalizing the values calculated by the penalty Factor (PF) described in Equation (13). For this manuscript, $\beta = 1000$, which is obtained through a heuristic process.

3. Proposed Solution Methodology

This paper used a parallel discrete–continuous version of the PSO method [31] to solve the problem regarding the optimal integration of PV sources into DC grids. This solution methodology resolves the discrete–continuous codification that describes the problem. It avoids the use of the traditional master–slave strategy used in the specialized literature for sitting and sizing DGs in electrical networks [34,36], which requires two solution methods: a binary/discrete optimization method to solve the location problem and a continuous optimization method to solve the sizing problem. This offers a solution to the problem, but it results in increased complexity and processing times.

The discrete–continuous codification used here is illustrated in Figure 1. It can be seen that the location and power values (sizing) of the different PV sources in the DC grid are included in the same vector of size $1X2N_{pv}^{dev}$. The discrete variables are associated with the location problem, and the continuous variables are related to the sizing of the DGs. Note that in the same figure, the DGs located at node 4 have a nominal power of 0.19 Kw, while the PV sources installed at node 41 have a nominal power of 2.4 kW.

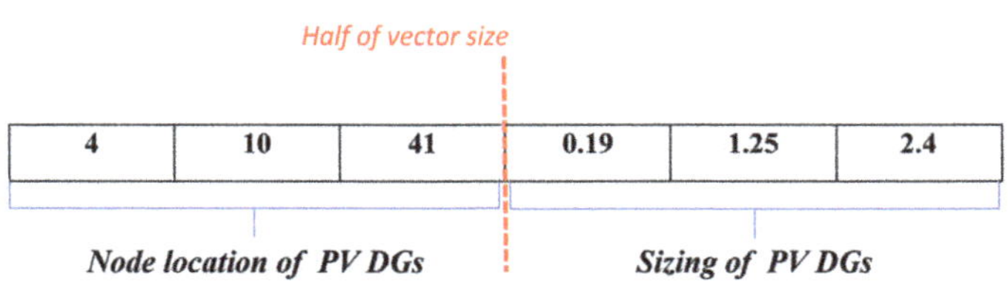

Figure 1. Codification used for the optimal integration of PV sources.

To solve the problem concerning the optimal integration of PV sources into DC grids via the aforementioned codification, this paper used the DCPPSO. This modified version of PSO discretizes the variables related to the location for each particle by using the number of candidate buses in the electrical system, and it allows for the variables associated with PV source sizing to remain continuous. Furthermore, the modified PSO uses parallel processing to evaluate the FF of each particle, which enables the reduction of processing times. It is important to highlight that in order to evaluate the FF, it is necessary to use an hourly power flow (HPF) that allows for the inclusion of variable power demand and generation related to PV sources, which is caused by the variation in the solar radiance in the region where the electrical system is located [34]. The pseudo-code that describes th DCPPSO/HPF methodology is presented below:

It can be observed that the first step of Algorithm 1 consists of reading all the D grid data and the parameters associated with DCPPSO. A maximum number of iteratio ($iter_{max}$) is set as a stopping criterion. Then, the first iteration generates the initial partic swarm via the codification presented in Figure 1 while considering random values betwe the number of buses and the power limits assigned to the DGs. Afterwards, the FF each particle is calculated using Equation (12), for which an hourly power flow based the successive approximations method is used [35]. Every hour, this method updates th power demand and PV generation during an average operation day in order to calculate th objective function described in Equation (1), thus taking variable demand and generatic into account.

Algorithm 1: Pseudo-code for the DCPPSO/HPF

```
Data: Read DC grid data and optimization parameters
for iter = 1 : iter_max do
    if iter == 1 then
        Generate the particle swarm;
        Calculate the FF for each particle by using the HPF;
        Select the FF and position obtained by each particle as the best particle
          solution and position;
        Select the best solution in the swarm and its position as the incumbent;
    else
        Calculate the velocity vector (VV);
        Update the position of the particle swarm by using the last position,
          the incumbent, and the VV;
        Calculate the FF for each particle by using the HPF;
        Update the best particle solution and position;
        Update the incumbent;
        if The stopping criterion has been met? then
            Finish the optimization process;
            Print the incumbent;
            Break;
        else
            Continue;
        end
    end
end
```

The base of the Particle Swarm Optimization algorithm (PSO) is the use of a populatic to explore the solution space in each iteration, taking advantage of the social and cognitiv knowledge to converge on a solution of good quality. In each iteration of the algorithm, it necessary to evaluate the FF of each particle that makes up the population, which requir long processing times. The proposed methodology considered the implementation of parallel processing tool that uses all Workers (W) of the computer to evaluate as individua the number of workers that exist, which reduces the processing time [35]. In order to car out this task, a highly used the tool in the literature called "parfor" of Matlab [13] wa employed, which made the parallel evaluation of the FF that comprises the populatio possible by allowing for the reduction of the processing times inside the iterative proces In this parallel processing, the PC employs all workers to evaluate the different particl that make up the swarm in groups with a size equal to W, executing as many process as necessary to evaluate all particles in the swarm. The time required for this task can b calculated by means of Equations (14) and (15). The former allows obtaining the number parallel processes (NPP) required for evaluating all particles in the swarm, while the latt calculates the total processing time required to carry out the process (PPT). Here, MPT represents the maximum processing time required for all particles to be evaluated [35].

$$NPP = CEIL(n/W) \tag{14}$$

$$PPT = NPP \cdot MTRP \tag{15}$$

After evaluating the FF of all particles, in the first iteration, the solution and position of each particle are assigned as the best values. Furthermore, the particle with the best solution is selected as the incumbent of the problem (the best solution) by storing its position and FF.

From the second iteration until the end of the process, each iteration calculates the velocity vector (VV) by using random values, as well as the position information of each particle (cognitive knowledge) and the incumbent (social knowledge). Subsequently, the particle movement is generated by using the VV and the current position. Then, the FF of the swarm is calculated, and the best particle position and solution as well as the incumbent are updated. At the end of each iteration, the stopping criterion is analyzed. In this particular case, it is verified whether the maximum number of iterations has been achieved. If this has occurred, the optimization process ends; if not, another iteration is carried out.

4. Test Scenarios and Considerations

4.1. Test Systems

In this paper, the DC versions of the 33- and 69-bus test systems were used, which are illustrated in Figure 2. These are widely used for validating planning strategies in DC grids [33].

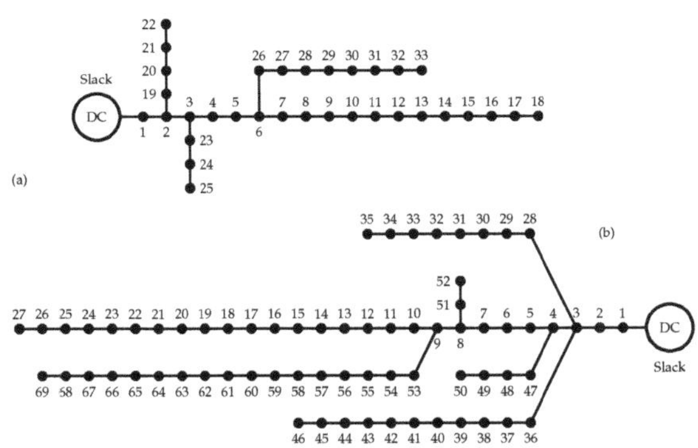

Figure 2. Electrical configuration of the DC (**a**) 33- and (**b**) 69-bus test systems [33].

Figure 2a illustrates the first test system, which comprises 33 buses and 32 branches. It employs a voltage of 12.66 kV and a power of 100 kW as base values. The parametric information of this test system is described in [37]. To obtain the branch current limits for this test system, the HPF described in the last section was used while considering the power demand and generation curves for Antioquia, Colombia (Figure 3). Consequently, the maximum current allowed for this test system was 310 A, and the electrical conductor was 350 kcmils.

Figure 2b presents the second test system employed in this research. It comprises 69 buses and 68 branches and uses the same base values as the 33-bus grid [37]. To obtain the branch current limits, the same methodology was employed, which provided a maximum allowed current of 335 A for an electrical conductor with a caliber of 400 kcmils.

4.2. Power Generation and Demand

To estimate the average impact on the annual costs of the electrical grid, this study considered the power demand and generation of PV sources in the region of Antioquia,

Colombia for an average day of operation. Figure 3 describes this behavior for a period of 24 h.

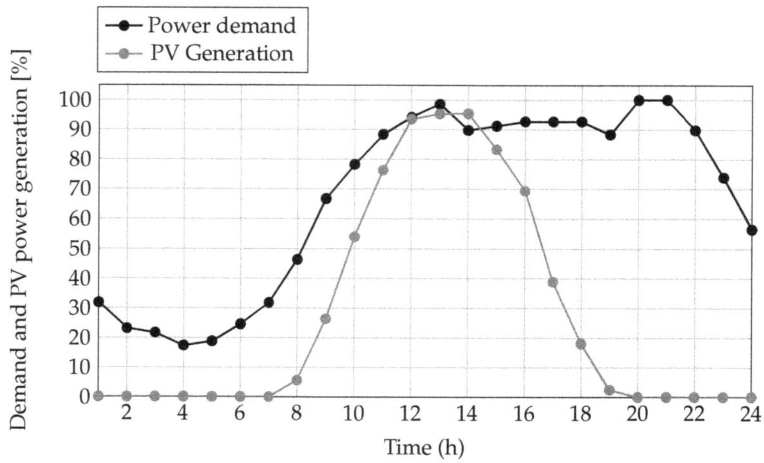

Figure 3. Typical power demand and PV generation behavior in Antioquia, Colombia [35].

4.3. *Comparison and Considerations*

One of the main objectives of this paper is to demonstrate the effectiveness of the proposed methodology in terms of the solution, repeatability, and processing times. To this effect, four comparison methods were selected from the specialized literature, which used specialized software and sequential programming optimization methods to solve the studied problem. The first solution method was the BONMIN solver of GAMS, the second was a discrete–continuous version of the Chu and Beasley genetic algorithm, the third was a discrete–continuous version of the vortex search algorithm (DCVSA), and the fourth was a discrete–continuous version of the generalized normal distribution optimizer (DCGNDO). The selection of these methods was based on the fact that they have all been used to solve the problem regarding the optimal integration of PV sources into DC grids, and that they have been evaluated in the same test systems and conditions as those used in this paper [29,33]. Furthermore, most of these comparison methods take advantage of the discrete–continuous codifications used by the proposed methodology (DCPPSO).

The main considerations and information used to validate the effectiveness and robustness of the proposed methodology in relation to the comparison methods are described below:

- All parameters used to evaluate the effects of PV source methods are presented in Table 1 [29].
- The maximum number of PV sources considered for installation was 3, and the maximum power capacity was 2.4 p.u. [29].
- The maximum allowed oscillation for the voltage profiles was $+/-$ 10% of each test system's nominal voltage.
- Both test systems considered non-telescopic grids, for which the maximum current in all branches corresponds to the maximum current allowed: 310 and 350 A for 33- and 69-bus test systems, respectively.
- The optimization parameters of the comparison methods were taken from the original paper [29,33], while the parameters of the DCPPSO are reported in Table 2.
- To evaluate the average processing time and the repeatability of the proposed methodology, each technique was executed 100 times while also analyzing the standard deviation obtained.

- The simulations were carried out in a Dell Precision T7600 Workstation with an Intel(R) Xeon(R) CPU ES-2670 @2.50 GHz and 32 GB of RAM while running the Matlab software, version 2022a.

Table 1. Parameters for the optimal integration of PV sources into electrical grids.

Param.	Value	Unit	Param.	Value	Unit
C_{kWh}	0.1390	USD/kWh	T	365	days
t_a	10	%	N_t	20	years
Δh	1	h	t_e	2	%
C_{pv}	1036.49	USD/kWp	$C_{O\&M}$	0.0019	USD/kWh
N_{pv}^{ava}	3	-	ΔV	± 10	%
$s_k^{pv,\min}$	0	kW	$s_k^{pv,\max}$	2400	kW
α_1	100×10^4	USD/V	α_2	100×10^4	USD/V
α_3	100×10^4	USD/W	α_4	100×10^4	USD/A

Table 2. DCPPSO parameters.

Parameter	33-Bus Test System
Number of particles (N_i)	100
Maximum iterations ($t_{\max}$)	1000
Maximum inertia	0.4133
Cognitive component	1.96236
Social component	2

5. Simulation Results

Tables 3 and 4 present and analyze the simulation results obtained with the proposed methodology and the comparison methods in the 33- and 69-bus test systems. This table shows, from left to right, the following information: the methodology used, the PV source location and capacity, the total annual costs, the reduction achieved with respect to the baseline case (the scenario without PV sources), the average processing time, the standard deviation, the worst voltages, and the maximum current.

Table 3. Simulation results obtained via different methodologies in the 33-bus test system in both AC and DC grids.

Methodology	Bus/Power (MVAr)	A_{cost} (USD/year)/ Reduction (%)	Time (s)	STD (%)	V_{worst} (p.u)	I_{max} (A)
Baseline case	[0–2.4]	3,644,043.01	- - -	- - -	[0.9–1.1]	310
BONMIN	18/1.4301 32/2.0611 33/1.7155	2,664,089.12/26.8919	0.77	0	0.93	304
DCCBGA	11/1.1629 14/0.9434 31/1.4827	2,662,724.82/26.9293	2.43	0.0557	0.93	304
DCVSA	9/0.5803 15/1.2913 31/1.7155	2,662,425.32/26.9375	76.86	0.0620	0.93	304
DCGNDO	10/0.9742 16/0.9202 31/1.6925	2,662,371.59/26.9390	166.15	0.0601	0.93	304
DCPPSO	10/0.9680 16/0.9189 31/1.6999	2,662,371.59/26.9390	8.52	0.0398	0.93	304

Table 4. Simulation results obtained by solution methods in the 69-bus test system.

Methodology	Bus/Power (MVAr)	A_{cost} (USD/year)/ Reduction (%)	Time (s)	STD (%)	V_{worst} (p.u)	I_{ma} (A)
Base case	[0–2.4]	3,817,420.38	- - -	- - -	[0.9–1.1]	335
BONMIN	21/0.4971 61/2.3999 65/0.8530	2,785,208.63/27.0395	2.02	0	0.93	304
DCCBGA	19/0.7908 61/1.7890) 64/1.1474	2,785,598.84/27.0292	7.74	0.1289	0.93	319
DCVSA	23/0.7720 62/2.3402 63/0.6185	2,785,538.58/27.0308	269.22	0.0974	0.93	319
DCGNDO	19/0.4969 61/2.3999 64/0.8470	2,785,011.53/27.0446	376.88	0.2384	0.93	319
DCPPSO	22/0.5310 61/2.4 64/0.8105	2,784,987.68/27.0452	28.24	0.0226	0.93	319

5.1. 33 Bus Test System

In the results presented in Table 3, it can be observed that all solution methodologies r duce the annual costs by 26.92% on average in comparison with the baseline case. Howev the DCPPSO obtained the best results. Figure 4 illustrates the improvement achieved wi this methodology, which obtained reductions of 6.44×10^{-2}%, 1.32×10^{-2}%, 2×10^{-3} and 0 when compared to the BONMIN, DCCBGA, DCVSA, and DCGNDO technique respectively. Thus, the average annual cost reduction was 2.62×10^{-2}%. It is possible conclude that DCGNDO achieved the same results as DCPPSO in this test system.

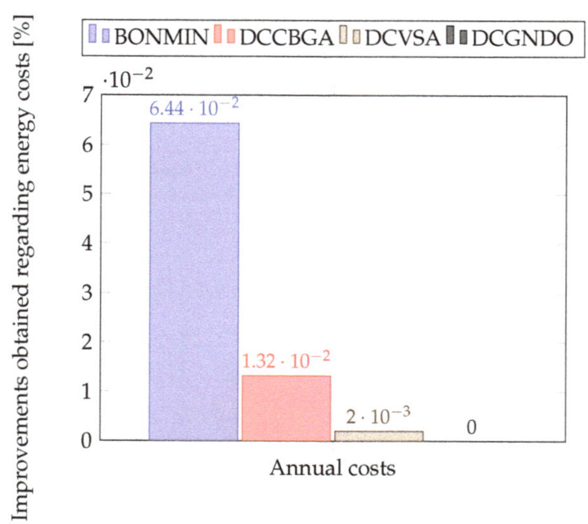

Figure 4. Improvements regarding annual cost reduction achieved by the DCPPSO with respect the comparison methods in the 33-bus test system.

As for the processing times, the fastest method was the BONMIN solver, followe by the DCCBGA (Figure 5). These methods required 95.41% and 85.54% less time tha the DCPPSO. In the 33-bus test system, the DCPPSO was in third place, with a reductio of 78.12% and 89.88% in processing times when compared to the DCVSA and DCGNDO. Despite these results, DCPSO had the best performance in terms of the solution. Figure 5 shows that the proposed methodology obtained the best results in terms of repeatability when compared to the solution methods based on sequential programming; it achieved an average standard deviation (STD) reduction of 48.80%. It is important to highlight that the BONMIN solver showed an STD of 0% since it belongs to the family of specialized software that guarantees the same solution each time that the algorithm is executed, although it has disadvantages in terms of purchase costs and implementation complexity [35,38].

Finally, columns 6 and 7 of Table 3 show that all solution methods satisfy the voltages and branch current bounds.

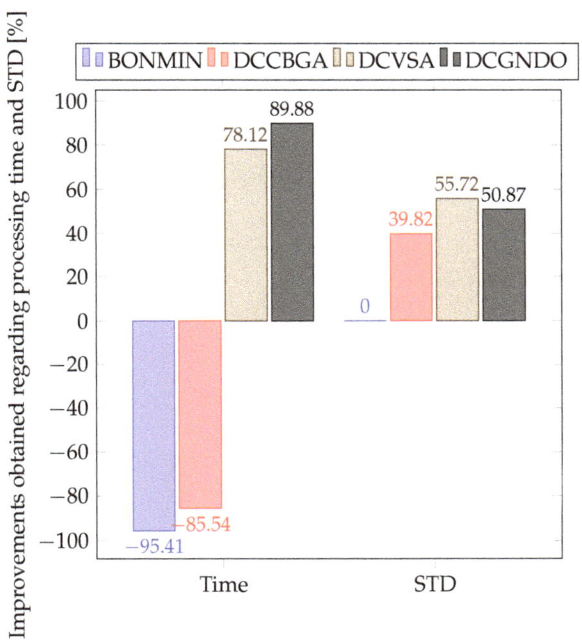

Figure 5. Improvements obtained by the DCPPSO with respect to the comparison methods in the 33-bus test system in terms of processing time and standard deviation.

5.2. 69-Bus Test System

The results obtained for the 69-node test system are shown in Table 4. Figure 6 illustrates the improvements achieved by the DCPPSO with respect to the comparison methods. Figure 6 shows that the proposed methodology obtained the best results in terms of annual cost reduction, with an average reduction of 1.24×10^{-2}%.

As for the processing times (Figure 7), DCPPSO was in third place, being outperformed by BONMIN and DCCBGA by 92.81% and 72.59%, respectively. However, DCPPSO was superior to the DCVSA and DCGNDO, reducing the processing times by 89.51% and 92.50%. When comparing these results with those obtained in the 33-bus test system, it can be observed that the differences with respect to the faster methods (BONMIN and DCCBGA) were reduced, while the quality of the solution was also the best. The proposed methodology achieved STD reductions of 82.46, 82.46, and 82.46% with respect to the DCCBGA, the DCVSA, and DCGNDO, respectively, (i.e., 83.23% on average). In this test system, the BONMIN solver reported an STD equal to 0, given its aforementioned characteristics. Finally, columns 6 and 7 of Table 4 demonstrate that all methodologies satisfied the voltage and current branch bounds established for the 69-bus test system.

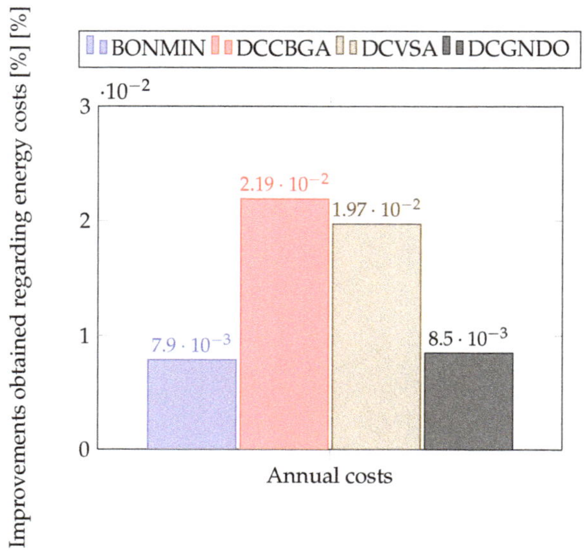

Figure 6. Improvements in the reduction of annual costs as achieved by DCPPSO with respect to the comparison methods in the 69-bus test system.

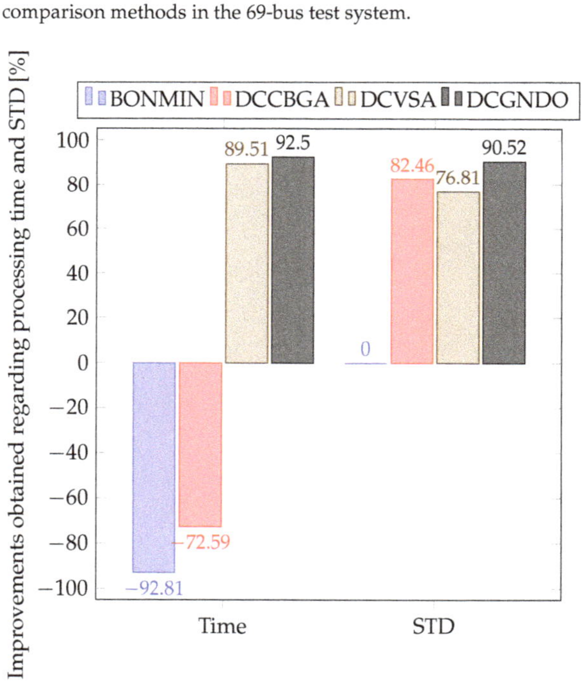

Figure 7. Improvements obtained by the DCPPSO with respect to the comparison methods in the 69-bus test system in terms of processing time and standard deviation.

6. Conclusions and Future Work

This paper proposed a discrete–continuous version of the particle swarm optimization algorithm to solve the problem regarding the optimal integration of PV sources into DC grids, with the purpose of reducing the energy purchasing/production costs associated with conventional generators as well as the investment and maintenance costs related to distributed generation. To evaluate the effectiveness, repeatability, and robustness of this solution methodology, four comparison methods and two test systems (33 and 69 buses) were used, which executed each methodology 100 times.

The advantages that the DCPPSO have are related to the implementation of a discrete–continuous codification that allows for the resolution of discrete and continuous problems at the time of its execution, which considers the discrete variables related to the location problem in this particular case as well as the continuous variables associated with the sizing problem of PV distributed generators on the DC grid. Furthermore, this methodology takes advantage of the PSO, a metaheuristic optimization technique that has been highly used in the literature to solve non-linear problems [12,35] due to its excellent effectiveness and robustness in solving non-linear and non-convex problems, such as those addressed here. Finally, inside the DCPPSO proposed, a parallel processing tool for evaluating the objective function of the population in each iteration was used, which allowed for the reduction of the processing time required for the solution methodology; thus, it was possible to present a solution of excellent quality with short processing times.

As for the results obtained in the 33-bus test system, even though DCPPSO was not the fastest method, it yielded excellent results in terms of processing times, with an average of 16.81 seconds. This method also achieved the best result in terms of the quality of the solution (annual cost reduction), with an average reduction of 0.0199% in comparison with the other methods. Furthermore, the proposed methodology was in second place in terms of the standard deviation, with a reduction of 48.80% when compared to the other discrete–continuous methodologies; it was only surpassed by the BONMIN solver of GAMS, a commercial software that increases the complexity and cost of solving the studied problem. Based on these results, the DCPPSO is considered to be an excellent methodology for the optimal integration of PV sources into small DC grids, with excellent results in terms of processing times and standard deviation.

For the 69-bus test system, the simulation results showed a similar behavior. Regarding the solution impact, DCPPSO obtained the best results, with an average reduction of 0.0126% with respect to the other methods. It also obtained the best results in terms of standard deviation; it was only surpassed by the BONMIM solver. As for the processing times, the DCPPSO was also in third place, taking longer than the BONMIN solver and the DCCBGA. This was due to the fact that BONMIN is a commercial solver and that the DCCBGA does not operate similarly to the other discrete–continuous solution methods; it does not work with population in its iterative process, which reduces the processing times and negatively impacts the quality of the solution. With respect to the other discrete–continuous methods, DCPPSO obtained an average processing time reduction of 91%. These methods were the most efficient methods after DCPPSO with regard to the quality of the solution. Based on the last results, the DCPPSO is the solution methodology with the best outcome in terms of the solution (annual cost reduction), as it showed excellent performance in terms of standard deviation and processing times for DC grids of any size. Furthermore, the results reported in [13] concluded that the DCPPSO is the most effective methodology for solving the problem of the optimal integration of PV sources in AC networks of different sizes. It can thus be concluded that to date, the DCPPSO is the solution methodology with the best trade-off in economic terms to solve the integration problem of PV sources in any kind of electrical network, with reduced processing times. The above-mentioned conclusions are based on the state-of-the-art knowledge acquired by the authors.

The proposed methodology works in other DC networks that consider the main generator, distribution system load, and PV distributed generators due to the mathematical model used, which considered all technical and operative constraints that represent the DC grids in an environment of PV distributed generations. However, in the particular case when other distributed energy resources are considered, e.g., energy storage systems, all equations that model this kind of electric devices must be included in the mathematical formulation with the aim of guaranteeing an adequate representation of the electric networks used.

Future work could consider the implementation of new solution methodologies ar other parallel processing tools that allow for the improvement of the results in terms of tl quality of the solution and processing times. In addition, the mathematical model cou include other technical, economic, and environmental indices related to the operation the DC grid, such as the power losses produced in the transport of energy, voltage stabili load chargeability, reductions in fossil fuel-based energy production and CO_2 emissior among others. By including the technical, economic, and environmental aspects of the gri it is possible to consider a multi-objective optimization algorithm in solving the proble of the optimal integration of PV sources into the DC grid. Finally, DCPPSO could l used to solve the problem concerning the optimal integration of energy storage systen into DC grids by taking advantage of the power of PV sources at times without sol production, which would improve the technical, economic, and environmental conditio of electrical networks.

Author Contributions: Conceptualization, methodology, software, and writing (review and editin L.F.G.-N., O.D.M. and C.A.R.-P. All authors have read and agreed to the published version of tl manuscript.

Funding: This research was supported by the University of Talca- Chile, Universidad Distrital Fra cisco José de Caldas, Colombia, and Universidad Nacional de Colombia under the research proje "Dimensionamiento, planeación y control de sistemas eléctricos basados en fuentes renovables r convencionales, sistemas de almacenamiento y pilas de combustible para incrementar el acce y la seguridad energética de poblaciones colombianas" (Minciencias code 70386), which belon to the research program "Estrategias para el desarrollo de sistemas energéticos sostenibles, con ables, eficientes y accesibles para el futuro de Colombia" (Minciencias code 1150-852-70378, Herm code 46771).

Data Availability Statement: Not applicable.

Conflicts of Interest: The authors of this paper declare no conflicts of interest.

Acronyms

PV	Photovoltaic
MINLP	Mixed-Integer Non-Linear Programming
GAMS	General Algebraic Modeling System
DCCBGA	Discrete–Continuous Chu and Beasley Genetic Algorithm
DCNMA	Discrete–Continuous Newton Metaheuristic Algorithm
DCVSA	Discrete–Continuous Vortex Search Algorithm
DCGNDO	Discrete–Continuous Generalized Normal Distribution Optimizer
DCPPSO	Discrete–Continuous Parallel Particle Swarm Optimization

References

1. Monteiro, V.; Oliveira, C.; Coelho, S.; Afonso, J.L. Hybrid AC/DC Electrical Power Grids in Active Buildings: A Power Electroni Perspective. In *Active Building Energy Systems*; Springer: Berlin/Heidelberg, Germany, 2022; pp. 71–97.
2. Ghiasi, M. Detailed study, multi-objective optimization, and design of an AC-DC smart microgrid with hybrid renewable energ resources. *Energy* **2019**, *169*, 496–507. [CrossRef]
3. Ghiasi, M.; Dehghani, M.; Niknam, T.; Baghaee, H.R.; Padmanaban, S.; Gharehpetian, G.B.; Aliev, H. Resiliency/cost-base optimal design of distribution network to maintain power system stability against physical attacks: A practical study case. *IEE Access* **2021**, *9*, 43862–43875. [CrossRef]
4. Vasant, P.; Weber, G.W.; Thomas, J.J.; Marmolejo-Saucedo, J.A.; Rodriguez-Aguilar, R. *Artificial Intelligence for Renewable Energ and Climate Change*; John Wiley & Sons: Hoboken, NJ, USA, 2022.
5. Ahmad, R.; Mohamed, A.A.A.; Rezk, H.; Al-Dhaifallah, M. DC Energy Hubs for Integration of Community DERs, EVs, an Subway Systems. *Sustainability* **2022**, *14*, 1558. [CrossRef]
6. Rezaeeian, S.; Bayat, N.; Rabiee, A.; Nikkhah, S.; Soroudi, A. Optimal Scheduling of Reconfigurable Microgrids in Bot Grid-Connected and Isolated Modes Considering the Uncertainty of DERs. *Energies* **2022**, *15*, 5369. [CrossRef]

7. Elsayed, A.T.; Mohamed, A.A.; Mohammed, O.A. DC microgrids and distribution systems: An overview. *Electr. Power Syst. Res.* **2015**, *119*, 407–417. [CrossRef]
8. Nigam, A.; Sharma, K.K. Planning Methodologies of Hybrid Energy System. In *Planning of Hybrid Renewable Energy Systems, Electric Vehicles and Microgrid*; Springer: Berlin/Heidelberg, Germany, 2022; pp. 9–26.
9. Mohseni, S.; Brent, A.C. Quantifying the effects of forecast uncertainty on the role of different battery technologies in grid-connected solar photovoltaic/wind/micro-hydro micro-grids: An optimal planning study. *J. Energy Storage* **2022**, *51*, 104412. [CrossRef]
10. Ghiasi, M.; Olamaei, J. Optimal capacitor placement to minimizing cost and power loss in Tehran metro power distribution system using ETAP (A case study). *Complexity* **2016**, *21*, 483–493. [CrossRef]
11. Fathi, M.; Ghiasi, M. Optimal DG placement to find optimal voltage profile considering minimum DG investment cost in smart neighborhood. *Smart Cities* **2019**, *2*, 328–344. [CrossRef]
12. Wang, P.; Wang, W.; Xu, D. Optimal sizing of distributed generations in DC microgrids with comprehensive consideration of system operation modes and operation targets. *IEEE Access* **2018**, *6*, 31129–31140. [CrossRef]
13. Grisales-Noreña, L.F.; Montoya, O.D.; Marín-García, E.J.; Ramos-Paja, C.A.; Perea-Moreno, A.J. Integration of PV Distributed Generators into Electrical Networks for Investment and Energy Purchase Costs Reduction by Using a Discrete–Continuous Parallel PSO. *Energies* **2022**, *15*, 7465. [CrossRef]
14. Ghiasi, M.; Niknam, T.; Dehghani, M.; Siano, P.; Haes Alhelou, H.; Al-Hinai, A. Optimal multi-operation energy management in smart microgrids in the presence of ress based on multi-objective improved de algorithm: Cost-emission based optimization. *Appl. Sci.* **2021**, *11*, 3661. [CrossRef]
15. Goli, A.; Ala, A.; Hajiaghaei-Keshteli, M. Efficient Multi-Objective Meta-heuristic Algorithms for Energy-aware Flexible Flow-shop Scheduling Problem. *Expert Syst. Appl.* **2022**, *213*, 119077. [CrossRef]
16. Tirkolaee, E.B.; Mahdavi, I.; Esfahani, M.M.S.; Weber, G.W. A robust green location-allocation-inventory problem to design an urban waste management system under uncertainty. *Waste Manag.* **2020**, *102*, 340–350. [CrossRef]
17. Kallel, R.; Boukettaya, G. An energy cooperative system concept of DC grid distribtion and PV system for supplying multiple regional AC smart grid connected houses. *J. Build. Eng.* **2022**, *56*, 104737. [CrossRef]
18. Shirazi, H.; Ghiasi, M.; Dehghani, M.; Niknam, T.; Garpachi, M.G.; Ramezani, A. Cost-emission control based physical-resilience oriented strategy for optimal allocation of distributed generation in smart microgrid. In Proceedings of the 2021 IEEE 7th International Conference on Control, Instrumentation and Automation (ICCIA), Tabriz, Iran, 23–24 February 2021; pp. 1–6.
19. Ishaq, S.; Khan, I.; Rahman, S.; Hussain, T.; Iqbal, A.; Elavarasan, R.M. A review on recent developments in control and optimization of micro grids. *Energy Rep.* **2022**, *8*, 4085–4103. [CrossRef]
20. Jithin, K.; Haridev, P.; Mayadevi, N.; Kumar, R.H.; Mini, V. A Review on Challenges in DC Microgrid Planning and Implementation. *J. Mod. Power Syst. Clean Energy* **2022**, 1–21. [CrossRef]
21. Anitha, D.; Premkumar, K. DC Microgrid: A Review on Issues and Control. *Smart Grids Green Energy Syst.* **2022**, 207–229. [CrossRef]
22. Kaushik, E.; Prakash, V.; Mahela, O.P.; Khan, B.; El-Shahat, A.; Abdelaziz, A.Y. Comprehensive overview of power system flexibility during the scenario of high penetration of renewable energy in utility grid. *Energies* **2022**, *15*, 516. [CrossRef]
23. Montoya, O.D.; Gil-González, W. A MIQP model for optimal location and sizing of dispatchable DGs in DC networks. *Energy Syst.* **2021**, *12*, 181–202. [CrossRef]
24. Montoya, O.D.; Gil-González, W.; Grisales-Noreña, L. Relaxed convex model for optimal location and sizing of DGs in DC grids using sequential quadratic programming and random hyperplane approaches. *Int. J. Electr. Power Energy Syst.* **2020**, *115*, 105442. [CrossRef]
25. Duman, S. A modified moth swarm algorithm based on an arithmetic crossover for constrained optimization and optimal power flow problems. *IEEE Access* **2018**, *6*, 45394–45416. [CrossRef]
26. Ajayi, O.; Nwulu, N.; Damisa, U. Application of metaheuristic algorithms in DC-optimal power flow. *Afr. J. Sci. Technol. Innov. Dev.* **2020**, *12*, 867–872. [CrossRef]
27. Ma, X.; Liu, S.; Liu, H.; Zhao, S. The Selection of Optimal Structure for Stand-Alone Micro-Grid Based on Modeling and Optimization of Distributed Generators. *IEEE Access* **2022**, *10*, 40642–40660. [CrossRef]
28. Qing, Y.; Liu, T.; He, C.; Nan, L.; Dong, G.; Gao, W.; Yu, Y. Low-carbon coordinated scheduling of integrated electricity-gas distribution system with hybrid AC/DC network. *IET Renew. Power Gener.* **2022**, *16*, 2566–2578. [CrossRef]
29. Cortés-Caicedo, B.; Molina-Martin, F.; Grisales-Noreña, L.F.; Montoya, O.D.; Hernández, J.C. Optimal Design of PV Systems in Electrical Distribution Networks by Minimizing the Annual Equivalent Operative Costs through the Discrete-Continuous Vortex Search Algorithm. *Sensors* **2022**, *22*, 851. [CrossRef]
30. Khezri, R.; Mahmoudi, A.; Haque, M.H. Two-stage optimal sizing of standalone hybrid electricity systems with time-of-use incentive demand response. In Proceedings of the 2020 IEEE Energy Conversion Congress and Exposition (ECCE), Detroit, MI, USA, 11–15 October 2020; pp. 2759–2765.
31. Grisales-Noreña, L.; Montoya-Giraldo, O.; Gil-González, W. Optimal Integration of Distributed Generators into DC Microgrids Using a Hybrid Methodology: Genetic and Vortex Search Algorithms. *Arab. J. Sci. Eng.* **2022**, *47*, 14657–14672. [CrossRef]
32. Montoya, O.D.; Grisales-Noreña, L.F.; Perea-Moreno, A.J. Optimal Investments in PV Sources for Grid-Connected Distribution Networks: An Application of the Discrete–Continuous Genetic Algorithm. *Sustainability* **2021**, *13*, 13633. [CrossRef]

33. Montoya, O.D.; Gil-González, W.; Grisales-Noreña, L.F. Solar Photovoltaic Integration in Monopolar DC Networks via the GND Algorithm. *Algorithms* **2022**, *15*, 277. [CrossRef]
34. Planas, E.; Andreu, J.; Gárate, J.I.; De Alegria, I.M.; Ibarra, E. AC and DC technology in microgrids: A review. *Renew. Sustai Energy Rev.* **2015**, *43*, 726–749. [CrossRef]
35. Grisales-Noreña, L.F.; Montoya, O.D.; Ramos-Paja, C.A. An energy management system for optimal operation of BSS in D distributed generation environments based on a parallel PSO algorithm. *J. Energy Storage* **2020**, *29*, 101488. [CrossRef]
36. Moradi, M.H.; Abedini, M. A combination of genetic algorithm and particle swarm optimization for optimal DG location an sizing in distribution systems. *Int. J. Electr. Power Energy Syst.* **2012**, *34*, 66–74. [CrossRef]
37. Kaur, S.; Kumbhar, G.; Sharma, J. A MINLP technique for optimal placement of multiple DG units in distribution systems. *Int. Electr. Power Energy Syst.* **2014**, *63*, 609–617. [CrossRef]
38. Kendrick, D.; Krishnan, R. A comparison of structured modeling and GAMS. *Comput. Sci. Econ. Manag.* **1989**, *2*, 17–36. [CrossRe

mathematics

MDPI

Article

An improved Fractional MPPT Method by Using a Small Circle Approximation of the P–V Characteristic Curve

Ernesto Bárcenas-Bárcenas *, Diego R. Espinoza-Trejo, José A. Pecina-Sánchez , Héctor A. Álvarez-Macías, Isaac Compeán-Martínez and Ángel A. Vértiz-Hernández

Academic Coordination of the Altiplano-Region Campus, Autonomous University of San Luis Potosi, Matehuala, San Luis Potosi 78700, Mexico
* Correspondence: ernesto.barcenas@uaslp.mx

Citation: Bárcenas-Bárcenas, E.; Espinoza-Trejo, D.R.; Pecina-Sánchez, J.A.; Álvarez-Macías, H.A.; Compeán-Martínez, I.; Vértiz-Hernández, Á.A. An Improved Fractional MPPT Method by Using a Small Circle Approximation of the P–V Characteristic Curve. *Mathematics* 2023, *11*, 526. https://doi.org/10.3390/math11030526

Academic Editors: Atanda Raji and Khaled M. Abo-Al-Ez

Received: 31 December 2022
Revised: 14 January 2023
Accepted: 14 January 2023
Published: 18 January 2023

Abstract: This paper presents an analytical solution to the maximum power point tracking (MPPT) problem for photovoltaic (PV) applications in the form of an improved fractional method. The proposal makes use of a mathematical function that describes the relationship between power and voltage in a PV module in a neighborhood including the maximum power point (MPP). The function is generated by using only three points of the P–V curve. Next, by using geometrical relationships, an analytical value for the MPP can be obtained. The advantage of the proposed technique is that it provides an explicit mathematical expression for calculation of the voltage at the maximum power point (v_{MPP}) with high accuracy. Even more, complex calculations, manufacturer data, the measurements of short circuit current (i_{SC}) and open-circuit voltage (v_{OC}) are not required, making the proposal less invasive than other solutions. The proposed method is validated using the P–V curve of one PV module. Experimental work demonstrates the speed in the calculation of v_{MPP} and the feasibility of the proposed solution. In addition, this MPPT proposal requires only the typical and available measurements, namely, PV voltage and current. Consequently, the proposed method could be implemented in most PV applications.

Keywords: PV module; mathematical model; MPP reference generator; maximum power point trackers

MSC: 00-02; 00A05; 00A06; 00A69

1. Introduction

Nowadays, PV technology is widely used in several applications such as portable devices, home applications and large-scale projects. However, in order to obtain the maximum benefit of PV modules, the maximum power must be obtained. With this in mind, some operating conditions have been identified as the main challenges to be solved [1]. For example, sudden irradiance changes [2], temperature variations and partial shading conditions [3,4]. It should be noted that if the condition changes, then the maximum power point also changes and should be recalculated. These problems have been addressed in many contributions with different approaches.

In the literature, the well-known conventional tracking algorithms, such as Perturb and Observe (P&O), Incremental Conductance (InC), Fractional Open Circuit Voltage (FOCV), Fractional Short Circuit Current (FSCC) and others, are very popular due to their ease of implementation, high rate of success and low computational requirements. However, P&O and InC are prone to showing oscillatory behavior near the maximum power point (MPP) and therefore they can provide a low performance. As for FOCV and FSCC, they have poor accuracy as their main drawback. There are several reviews that show the advantages and disadvantages of the classical MPPT algorithms [1,2,5]; in such contributions it is shown that these algorithms can still be useful under certain conditions, such as uniform solar irradiation. Moreover, these basic solutions became the basis for more elaborated proposa that present important improvements in P&O [6,7], FOCV [8] and InC [9] techniques.

In this sense, another approach for solving the MPPT issue can be seen in the u of model-based non analytical approaches. In these cases, it is possible to find solutio based on fuzzy logic [10], neural networks [4], amongst others [2,11]. Such solutio have shown improvements over time and provided more efficient solutions at the cost complexity. Over time, hybrid schemes provide even better response to the challenges PV applications, as can be seen in reviews and in several contributions [3,4,7].

In general, it is possible to say that the better the performance, the more the complexi as can be observed in the reviews [1,2]. In fact, if high efficiency is demanded then hi computation cost and a high number of sensors is a usual requirement. For examp a highly efficient solution would require the use of voltage, current, temperature ar irradiance sensors.

In order to deal with the disadvantages of elaborated solutions, the improveme of basic approaches is very attractive [12]. Basically, the main drawbacks of Fraction Open Circuit Voltage (FOCV) or Fractional Short Circuit Current (FSCC) are the lack accuracy at the calculation of the reference voltage at MPP (v_{ref}) and the periodical conne tion/disconnection of the PV module [13]. The main characteristics of FOCV and FSCC a well presented in several papers in comparison with other methods [1,2,13]. It should noted that conventional algorithms are still useful under uniform solar irradiance. For t case of non-uniform solar irradiance, the conventional techniques have been mixed wi others in order to cope with the accuracy problem with good results [13].

Another interesting approach is the curve-fitting (CF) method [13–18]. This approa consists of the proposal of an equation that allows the calculation of v_{ref}. This meth makes use of manufacturing information or/and real time measurements in order provide a solution [14]. The main disadvantage of this approach is the large amount of da and knowledge of physical parameters [13,17,18]. Under this approach, some proposa even require irradiance (G) and temperature (T) measurements. Furthermore, this approa may require the measurement of v_{OC} and i_{SC}, which can be a disadvantage. Anoth important drawback is that ageing is not considered and for this reason some physic parameters may change (such as v_{OC}, i_{SC}, R_S, etc), producing errors in the calculated val of v_{ref}. However, the main goal of the curve-fitting method is to find an equation th describes, as precisely as possible, the I–V and/or P–V curves. In this sense, the propose equation, so far, provides an explicit solution for the v_{ref} value.

This paper proposes an alternative method for the Fractional Open Circuit Volta MPPT technique. The proposal allows us to obtain a precise value for v_{ref} based on so minimal measurements that should be updated periodically or when a change occu In this way, the main events that change the value of v_{ref} can be faced properly, such a temperature changes, sudden irradiation variations, ageing, etc. The proposed equati requires some minimal measurements that should be updated when a change occurs; th low data requirement is the main drawback of the CF method. This proposal requires thr points near the MPP and in exchange, it provides a precise value of v_{ref}, dealing with t inaccuracy problem presented using the basic approaches of FOCV and FSCC. In the lo term, the proposed method can also deal with the ageing problem due to its periodic natu Besides, the proposed method does not require measurements of v_{OC} and i_{SC}. Therefo PV module disconnections are not necessary. Moreover, the proposed method relies on t use of some basic algebraic equations that require very few computing resources; this important from a numerical analysis perspective [19].

2. Proposed Method

2.1. The MPPT Problem Formulation

A PV module produces electricity in the form of a current or a voltage. Under normal operation and with specific climate conditions, temperature, solar illumination, et the electrical behavior of a PV module can be illustrated, as with Figure 1. In this figur the relationship between current and available power versus voltage is shown. However, changes in temperature and solar irradiation can change the IV and PV curves and create a new MPP; these phenomena are widely reported in the available literature.

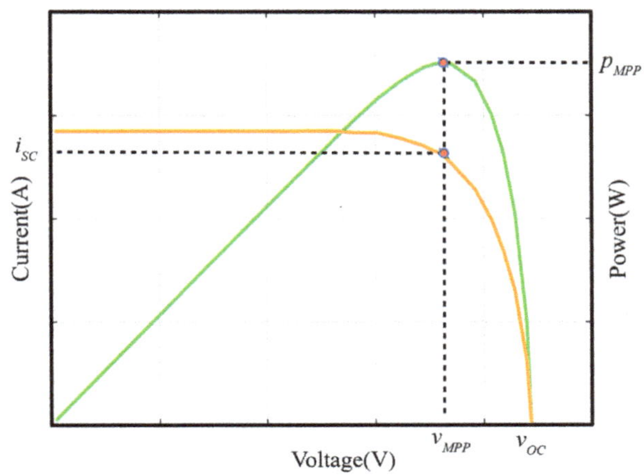

Figure 1. Photovoltaic I–V and P–V graphs.

Normally, a complete PV solar tracking system, as is depicted in Figure 2, allows the obtainment of maximum power by using a proper MPPT strategy and a suitable controller. Notice that the main contribution of this work is a method for the v_{ref} generation. In the available literature, a lot of proposals can be found about new control systems that effectively follow the provided value for v_{ref}; however, a proper generation of v_{ref} is a requirement as a previous stage before the controller design. In the following, this paper proposes a method for the calculation of the MPP voltage value (v_{MPP})—see Figure 1. Then, the calculated v_{MPP} voltage value will be used as the v_{ref} voltage by the controller—see Figure 2.

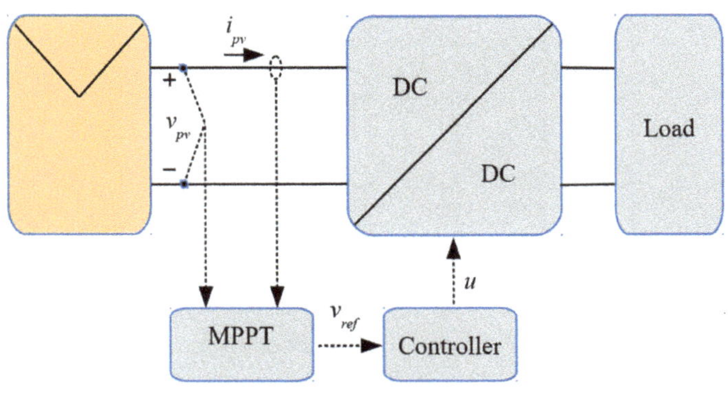

Figure 2. Typical solution for photovoltaic energy harvesting.

Despite the fact that equations for PV modules are well-known, it is not possible to obtain an equation for the v_{MPP} voltage because the main equation has a transcendental nature, making v_{MPP} calculation a challenging task. The original equations that describe the relationship between current and voltage are studied in several papers and exhibit a dependence of several factors which may be difficult to obtain [17]. Several proposals of analytical solutions for the MPP problem can be found in [14,17], and they provide equations for the calculation of a proper v_{MPP} voltage. However, the proposed equations require the measurement of i_{SC} and/or v_{OC}, photovoltaic current (i_{PV}) and voltage (v_{PV}), amongst other information. It should be noted that the measurement of i_{SC} and/or v_{O} gives the system a hard time because with the generation of two undesired situation short circuit and open circuit, both situations stop the power flow from the PV modules t the load.

2.2. Proposed Analitycal Solution

In the literature, it is a well-known fact that the mathematical relationship betwee voltage and current is expressed using a transcendental equation. Hence, it is not possib to find an explicit equation for the MPP voltage value (v_{MPP}). This paper proposes th using of the mathematical relationship between PV power and voltage in order to describ a small vicinity of the actual PV module operation.

The proposed solution begins with the identification of the zone of interest in the P– graph, as in Figure 3.

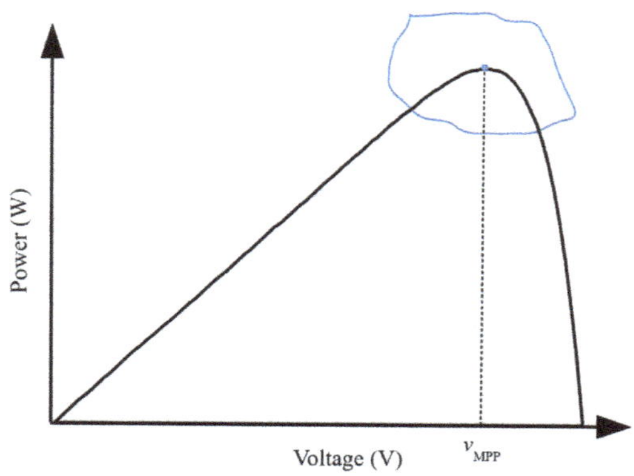

Figure 3. P–V curve illustrating the zone of interest.

Under this proposal, it is stated that the most important zone in the P–V curve co responds to the shaded area because it contains the MPP. It should be clear that, even i we have the complete curve information, the most important region is where the MP is contained. In order to have a definition for the zone of interest, the boundaries stab lished via the fractional method are used for this paper and represented with the followin equations [12,13]:

$$v_{MPP} = k * v_{OC} \quad (1$$

with

$$0.7 < k < 0.9 \quad (2$$

In this regard, the present proposal pretends to make an improved fractional metho by considering the established boundaries in the conventional fractional method, (1) an (2), and the general equation for a circle suggested as the main model of this proposa In comparison with other solutions of the same nature, this proposal does not pretend t generate an equation for the entire PV module behavior, but only for the zone of interes It should be noted that, under this approach, only the v_{OC} voltage is required for definin the initial boundaries.

The proposed model (3) belongs to the circle family and has the following structure:

$$(x - v_c)^2 + (y - p_c)^2 = r^2 \quad (3$$

where the variables x and y are v_{PV} and PV power (p_{PV}), respectively. The point (v_c, p_c represents the coordinates of the center of the circle and r the corresponding radius. Th proposed equation emulates the P–V curve behavior in the zone of interest. By takin the first derivative in (3) with respect to x, it is possible to demonstrate that the MPP corresponds with $x = v_c$, then the MPP problem is reduced to the calculation of the coordinates (v_c, p_c), where $v_c = v_{MPP}$.

In order to find the coordinates, the following procedure is proposed:

First, three arbitrary points on the P–V curve, $Q_1(v_1, p_1)$, $Q_2(v_2, p_2)$ and $Q_3(v_3, p_3)$, must be selected so that these points belong to the zone of interest, as shown in Figure 4.

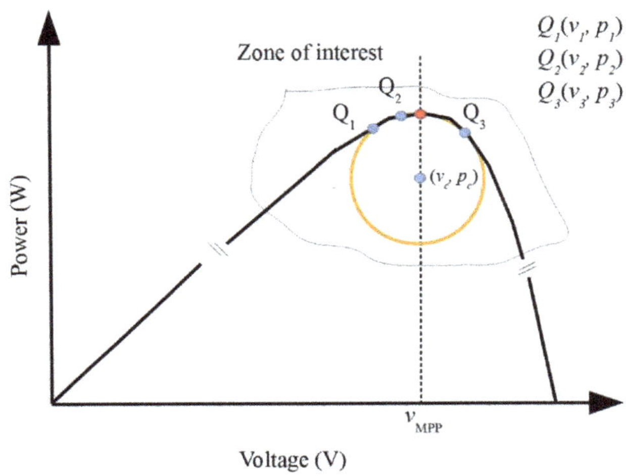

Figure 4. P–V curve with three arbitrary points.

Next, we need to find the equations for two perpendicular lines (l_1 and l_2) to the segments $\overrightarrow{Q_1Q_2}$ and $\overrightarrow{Q_2Q_3}$, which can be visualized in Figure 5. It is worth noting that the intersection of l_1 and l_2 will define the v_{MPP}. To define the equations for l_1 and l_2, we need to find the middle point of the segments $\overrightarrow{Q_1Q_2}$ and $\overrightarrow{Q_2Q_3}$, which are represented by the coordinates (x_1, y_1) and (x_2, y_2). Also, we need to find the slopes (m_{11} and m_{22}) of l_1 and l_2.

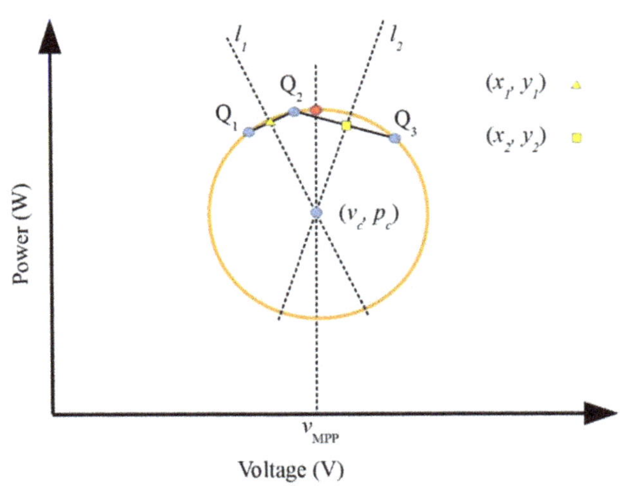

Figure 5. Proposed model within the zone of interest.

For this, the coordinates (x_1, y_1) and (x_2, y_2) can be calculated departing from the three arbitrary points, as described in (4) and (5):

$$x_1 = \frac{v_1 + v_2}{2}, \qquad y_1 = \frac{p_1 + p_2}{2}, \tag{4}$$

$$x_2 = \frac{v_2 + v_3}{2}, \qquad y_2 = \frac{p_2 + p_3}{2}. \tag{5}$$

While the slopes are calculated with (6) and (7):

$$m_{11} = -\frac{1}{m_1} = -\left(\frac{v_2 - v_1}{p_2 - p_1}\right), \tag{6}$$

$$m_{22} = -\frac{1}{m_2} = -\left(\frac{v_3 - v_2}{p_3 - p_2}\right). \tag{7}$$

Figure 5 contains the main elements of the proposed circle modeling.

In this way, the equations for l_1 and l_2 are given using:

$$y = m_{11}(x - x_1) + y_1 \tag{8}$$

$$y = m_{22}(x - x_2) + y_2 \tag{9}$$

Finally, the value of v_{MPP} is calculated with the intersection of l_1 and l_2 and is represented by Equation (10).

$$v_c = v_{MPP} = \frac{m_{22}x_2 - m_{11}x_1 + y_1 - y_2}{m_{22} - m_{11}} \tag{10}$$

In addition, we can find the values for p_c and r. However, these values are not relevant for the calculation of v_{MPP}.

It is worth mentioning that some operating conditions must be avoided when using Equation (10), which are summarized in the following:

(a) $p_1 \neq p_2$ and $p_2 \neq p_3$,
(b) $m_{22} \neq m_{11}$,

i.e., the PV power in the three different points, Q1, Q2 and Q3, must be different so that zero division can be avoided in Equations (6) and (7). Also, the slopes m_{11} and m_{22} must be different to avoid the indetermination of (10). Note that the conditions given in (a) and (b) can be guaranteed through the MPPT algorithm implementation in a digital processor.

It should be noted that all the previous elements of analytic geometry are widely known in the available literature. However, this proposal provides a solution to the MPPT problem with minimum measurements and computing requirements.

3. Experimental Results

The experimental validation of the present MPPT proposal is presented in this section. For this, two scenarios have been considered, which are detailed in the following subsections:

(a) Case 1: Offline test using I–V and P–V curves.
(b) Case 2: Online test under a closed-loop control operation.

In case one, the characteristics (I–V and P–V) curves of a PV module were obtained by using a variable resistance connected in the PV module terminals, as described in Figure Hence, three operating points were selected as shown in Table 1. Then, by applying the proposed method in this paper, the voltage at the maximum power point (v_{MPP}) was calculated departing from (10). For case two, a closed loop control operation was implemented in a digital platform, as illustrated in Figure 2.

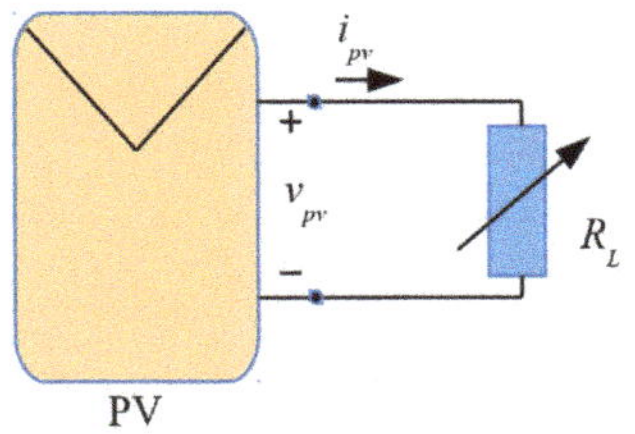

Figure 6. Test Bench for Solartec S72MC-175.

Table 1. Numerical evaluation with Solartec S72MC-175.

Parameter	Value	Equation
$Q_1(v_1, p_1)$	$Q_1(35.51, 174.52)$	-
$Q_2(v_2, p_2)$	$Q_2(36.57, 174.90)$	-
$Q_3(v_3, p_3)$	$Q_3(37.10, 174.44)$	-
(x_1, y_1)	(36.04, 174.71)	(4)
(x_2, y_2)	(36.835, 174.67)	(5)
m_{11}	−2.7894	(6)
m_{22}	1.11521	(7)
v_{MPP}	36.28 V	(10)

3.1. Case 1: Offline Test Using I-V and P-V Curves

In this case, the curves of the Solartec S72MC-175 PV module were obtained by using the simulation test bench depicted in Figure 6. The proposed test bench was implemented in Psim software and the functional model was used for the PV module.

The functional model in Psim requires V_{OC}, I_{SC}, maximum power voltage (V_M) and maximum power current (I_M). The parameter values included in the datasheet are $V_{OC} = 44.40V$, $I_{SC} = 5.30A$, $V_M = 36.30V$ and $I_M = 4.82A$. The I–V and P–V curves were obtained by varying the load resistance value (R_L). These measurements are presented in Appendix A. Note that the exact value for v_{MPP} corresponds with $V_M = 36.30V$. With this information, the effectiveness of the proposed method can be evaluated.

Once the I–V and P–V curves were available, three points were selected to feed the proposed algorithm. As a result, it was possible to calculate the v_{MPP} voltage using Equations (4)–(7) and (10). The numerical evaluation of the proposed method is summarized in Table 1.

In addition, Figure 7 shows all the elements considered in the proposed method.

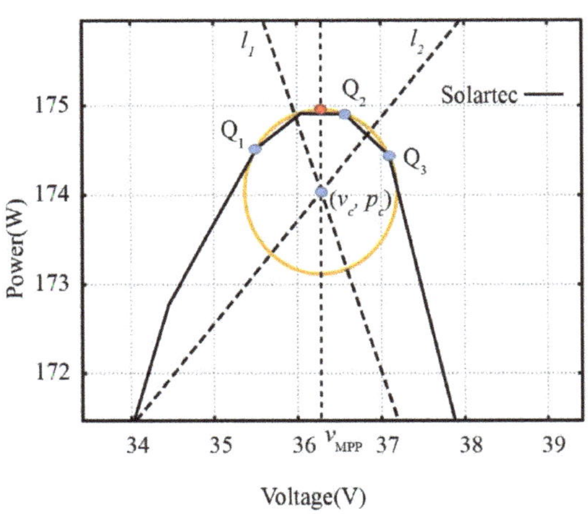

Figure 7. v_{MPP} with Solartec S72MC-175.

Finally, the error between the exact and calculated voltage values is presented in Table 2. In the table is also included a comparison with the error produced using the typica fractional method (see fourth column).

Table 2. Calculated error for v_{MPP} voltage.

Exact v_{MPP}	Calculated v_{MPP}	Fractional Method v_{MPP}	Proposed Method Error (%)	Fractional Method Error (%)
36.30 V	36.28 V	Between 31.08 V to 39.96 V	0.05%	14.3% (worst case)

Where the error value can be calculated with:

$$Error(\%) = \frac{|v_{exact} - v_{calculated}|}{v_{exact}} \times 100 \quad (11)$$

It should be noted that this section serves just for validation of the proposed methoc Using the offline information provided by the I–V and P–V curves, it was possible to calcu late the v_{MPP} value with minimal error. However, in real life applications, the calculation c v_{MPP} must be carried out in real time. This situation is addressed in the following sectio

3.2. *Case 2: Online Test Using Closed Loop Control*

In order to validate the proposed method, an experimental test bench was buil considering the elements showed in Figure 2. A Chroma programmable DC Power Suppl model 62050H-600S with Solar Array Emulation capabilities was used as the power sourc and connected to a dc-dc Boost converter with a rated capability of 350 W. Table 3 contain the main parameters of the power converter. A series connection of six batteries was usec as load with 12 volts in each battery. The proposed MPPT algorithm and the closed loo controller were implemented in a DS1104 dSpace digital board with a sampling frequenc of 70 kHz. The PWM technique, for the dc–dc converter, was implemented with analog circuits at a frequency of 10 kHz.

Table 3. Parameters of the dc–dc boost converter.

Parameter	Value
Mosfet	IRFP250N
Diode	STTH30R04W
L	1.5 mH
C_{in}	30 μ F
C_{out}	680 μ F

Figure 8 shows the simplified diagram of the experimental platform including dc–d boost converter details. The complete experimental platform is shown in Figure 9 anc illustrates the mentioned elements.

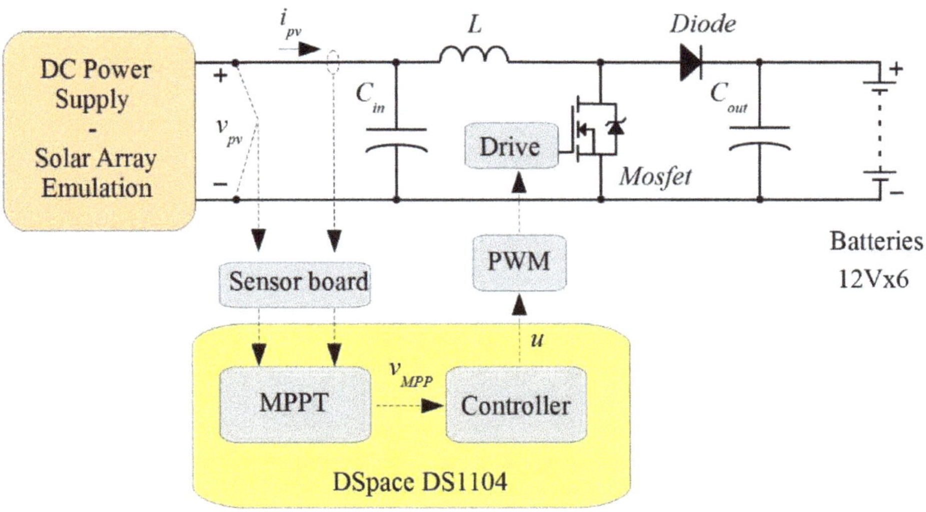

Figure 8. Simplified diagram.

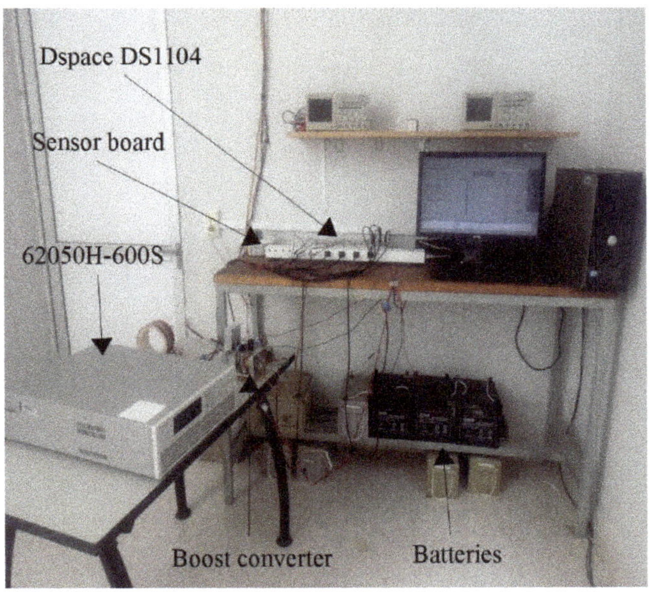

Figure 9. Experimental test bench.

It should be noted that the control strategy is based on a previous contribution of the authors and is based on a high-performance input–output linearization controller; details about the controller can be found in [20].

First, three arbitrary points of the PV curve were required as the input information of the proposed algorithm. This can be accomplished with an induced change in the setpoint reference of the closed-loop control. Figure 10a shows two setpoint changes (upper side) and their corresponding PV power (lower side). Its corresponding PV curve is illustrated in Figure 10b. The obtained results are summarized in Table 4 for a solar irradiance of 500 W/m^2.

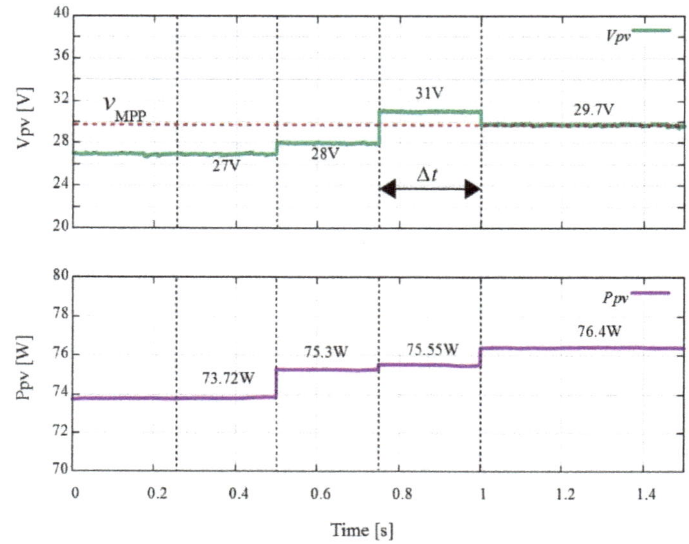

(a)

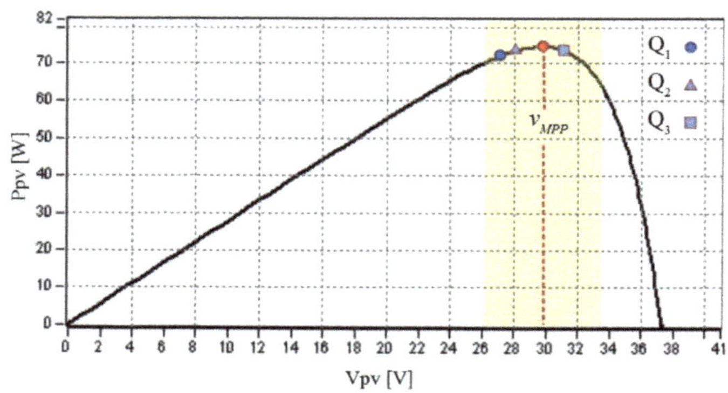

(b)

Figure 10. Three arbitrary points of the PV curve with 500 W/m^2. (**a**) Experimental generation of three points with setpoint changes and (**b**) location of the three points in the PV curve.

Table 4. Experimental selected points under an irradiance of 500 W/m^2.

PV Module	$Q_1(v_1,p_1)$	$Q_2(v_2,p_2)$	$Q_3(v_3,p_3)$
Solar Array Emulator	(27.0 V, 73.72 W)	(28.0 V, 75.30 W)	(31.0 V, 75.55 W)
Calculated v_{MPP}		29.69 V	
Exact v_{MPP}		29.73 V	
Error %		0.13%	

In order to prove that the proposed methodology can be applied under differen conditions, another set of points were generated with a solar irradiance of 1000 W/m Figure 11a shows the setpoint changes (upper side) and their corresponding PV powe (lower side). Its corresponding PV curve is illustrated in Figure 11b. The obtained resul are summarized in Table 5.

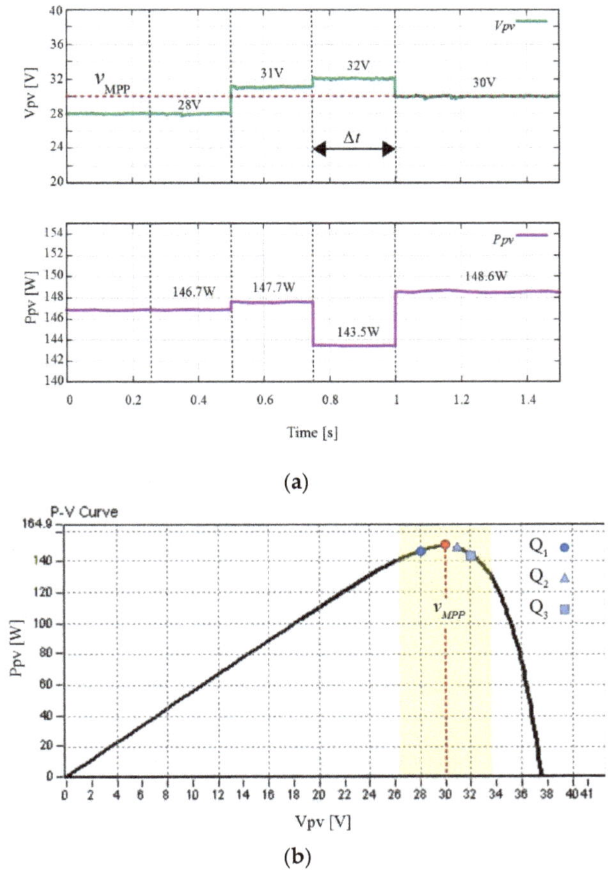

Figure 11. Three arbitrary points on the PV curve with 1000 W/m^2. (**a**) Experimental generation of three points with setpoint changes and (**b**) location of the three points on the PV curve.

Table 5. Experimental selected points under an irradiance of 1000 W/m^2.

PV Module	$Q_1(v_1,p_1)$	$Q_2(v_2,p_2)$	$Q_3(v_3,p_3)$
Solar Array Emulator	(28.00, 146.70)	(31.00, 147.70)	(32.00, 143.50)
Calculated v_{MPP}		30.14 V	
Exact v_{MPP}		30.00 V	
Error %		0.46%	

Notice that in Tables 4 and 5, the v_{MPP} value was calculated with Equation (10); such an equation involves purely algebraic operations, thus making a very fast algorithm. Another benefit is the absence of a v_{oc} and i_{sc} measurement, making this proposal less invasive than other solutions. This is important because the measurement of v_{oc} or i_{sc} produce a temporal stop of power flow from the PV module to the load. In addition, the arbitrary selected points of the P–V curve were selected according to (1) and (2).

It should be observed that the time between setpoint changes (Δt) in Figures 10 and 11 can be reduced. The employed Δt was selected only for the validation of the proposed algorithm and to clearly show the proposed method: $\Delta t \cong 0.25s$.

In real time applications Δt will be limited by the settling time (t_s) of the closed-loop controller. This situation is observed in Figure 12, where after the calculation of the new

v_{MPP} value and its corresponding actualization, a set point change occurs and t_s is required in order to obtain a new measurement under steady-state conditions.

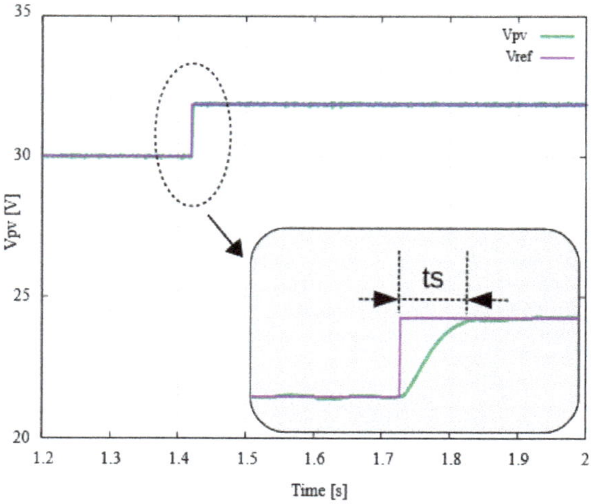

Figure 12. Transient time response (t_S) under closed-loop control.

Considering the transient response under a closed-loop controller, the following criteria of Δt selection is proposed.

$$\Delta t > 2 * t_s \quad (12)$$

In the present experiments, the closed-loop controller allows us to establish t_s with the following relationship:

$$t_s = 7 * \left(\frac{1}{f_{sw}}\right), \quad (13)$$

where f_{sw} stands for the PWM switching frequency on the dc–dc boost converter. More details about the high-performance closed-loop controller can be found in [20].

Based on the previous relationships, the minimum required time to perturb v_{ref} for taking a new measurement will be

$$\Delta t_{(min)} = 14 * \left(\frac{1}{f_{sw}}\right). \quad (14)$$

In the experiments, and with $f_{sw} = 10$kHz, we have $\Delta t_{(min)} = 1.4$ms as the time required to take a new measurement, which is illustrated in Figure 12, where $t_s = 0.7$ms.

Finally, the flowchart for the proposed method is included in Appendix B. The flowchart shows that the measurement of v_{OC} is only needed at the beginning of the operation. As a starting point, it is suggested that $v_{MPP} = 0.8 * v_{oc}$ and then Q_1, Q_2 and Q_3 can be measured. Next, by using the proposed method, a new v_{MPP} can be calculated. Then, the update of the voltage reference is applied by making $v_{ref} = v_{MPP}$. Note that the flowchart includes a delay time (t_D); this time is required between each iteration and is a user-defined parameter.

4. Discussion

In this paper, an improved Fractional Open Circuit Voltage (FOCV) MPPT method was presented, which requires only three points of the P–V curve of PV modules. Here, an analytical equation has been proposed in this paper by using the classical circumference equation, thereby allowing the calculation of the voltage at the maximum power point (v_{MPP}). This proposal has been validated through numerical and experimental tests by considering a closed-loop operation in the power converter. Furthermore, there is no need for i_{SC} measurement; also, v_{OC} would be required only at the beginning of the day, making it ideal for online applications. Currently, sudden irradiance changes produce changes in i_{SC} and have minimal impact in the v_{OC} value. For this reason, the proposed method can be employed to deal with this phenomenon. In contrast, temperature changes produce a direct impact on the v_{OC} voltage value, just as reported in the literature. For this reason, the proposed method flowchart includes an initial measurement of v_{OC}. However, after the initial measurement of v_{OC}, it is no longer required for the calculation of v_{MPP}. Indeed, the proposed method produces minimal interference between the PV module and the load by avoiding the measurement of i_{SC} and reducing the number of times v_{OC} is measured. Additionally, this proposal copes with the main disadvantage of analytical approaches that require a huge amount of data from manufacturer datasheets. As a final characteristic, the proposed algorithm has very few computational requirements.

Finally, note that the proposed method includes elements of the CF approach. In addition, the present proposal is considered as an improved FOCV method because it uses the same boundaries. However, in comparison with the regular FOCV, the proposed method can be used to calculate a precise value of v_{MPP}.

Author Contributions: Conceptualization and methodology, E.B.-B.; validation, E.B.-B. and D.R.E.-T.; formal analysis, E.B.-B. and D.R.E.-T.; resources, J.A.P.-S.; writing—original draft preparation, E.B.-B.; writing—review and editing, E.B.-B., D.R.E.-T., J.A.P.-S. and H.A.Á.-M.; project administration, Á.A.V.-H. and I.C.-M. All authors have read and agreed to the published version of the manuscript.

Funding: This research received no external funding.

Data Availability Statement: Not applicable.

Conflicts of Interest: The authors declare no conflict of interest.

Appendix A

Table A1. PV module data.

Solartec S72MC-175		
V	I	P
0.53	5.3	2.809
15.90	5.29	84.26
18.55	5.29	98.30
21.20	5.29	112.30
23.85	5.29	126.21
26.50	5.27	139.87
28.00	5.26	147.32
30.21	5.22	157.75
32.33	5.14	166.18
34.45	5.01	172.77
35.51	4.91	174.52
36.04	4.85	174.91
36.57	4.78	174.90
37.10	4.70	174.44
39.22	4.24	166.37
40.80	3.68	150.27
41.87	3.14	131.61
42.93	2.37	102.15
43.99	1.13	50.00
44.40	0	0

Appendix B

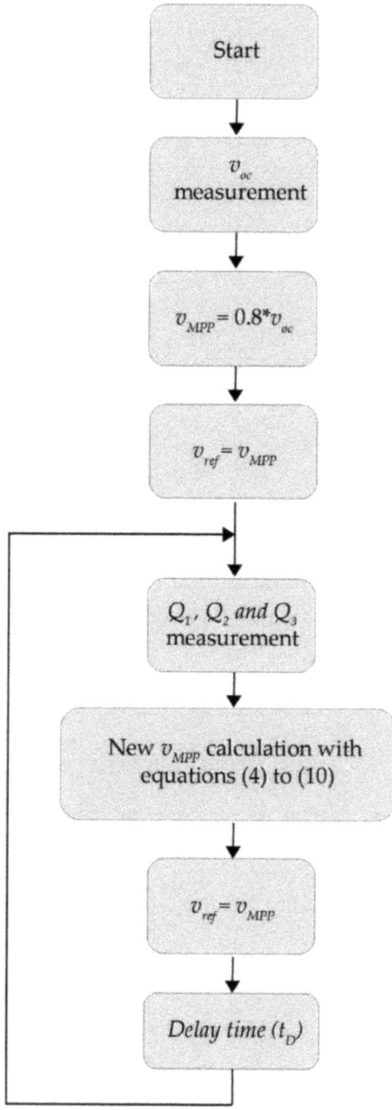

Figure A1. Suggested flowchart for the proposed method.

References

1. Bollipo, R.B.; Mikkili, S.; Bonthagorla, P.K. Hybrid, optimization, intelligent and classical PV MPPT techniques: Review. *CSEE Power Energy Syst.* **2020**, *7*, 9–33. [CrossRef]
2. Hanzaei, S.H.; Gorji, S.A.; Ektesabi, M. A Scheme-Based Review of MPPT Techniques With Respect to Input Variables Includin Solar Irradiance and PV Arrays' Temperature. *IEEE Access* **2020**, *8*, 182229–182239. [CrossRef]
3. Zhang, W.; Zhou, G.; Ni, H.; Sun, Y. A Modified Hybrid Maximum Power Point Tracking Method for Photovoltaic Arrays Und Partially Shading Condition. *IEEE Access* **2019**, *7*, 160091–160100. [CrossRef]
4. El-Helw, H.M.; Magdy, A.; Marei, M.I. A Hybrid Maximum Power Point Tracking Technique for Partially Shaded Photovolta Arrays. *IEEE Access* **2017**, *5*, 11900–11908. [CrossRef]

5. Troudi, F.; Jouini, H.; Mami, A.; Ben Khedher, N.; Aich, W.; Boudjemline, A.; Boujelbene, M. Comparative Assessment between Five Control Techniques to Optimize the Maximum Power Point Tracking Procedure for PV Systems. *Mathematics* **2022**, *10*, 1080. [CrossRef]
6. Restrepo, C.; Yanẽz-Monsalvez, N.; González-Castaño, C.; Kouro, S.; Rodriguez, J. A Fast Converging Hybrid MPPT Algorithm Based on ABC and P&O Techniques for a Partially Shaded PV System. *Mathematics* **2021**, *9*, 2228. [CrossRef]
7. Wang, S.-C.; Pai, H.-Y.; Chen, G.-J.; Liu, Y.-H. A Fast and Efficient Maximum Power Tracking Combining Simplified State Estimation With Adaptive Perturb and Observe. *IEEE Access* **2020**, *8*, 155319–155328. [CrossRef]
8. Bi, Z.; Ma, J.; Man, K.L.; Smith, J.S.; Yue, Y.; Wen, H. An Enhanced 0.8 V_{OC}-Model-Based Global Maximum Power Point Tracking Method for Photovoltaic Systems. *IEEE Trans. Ind. Appl.* **2020**, *56*, 6825–6834. [CrossRef]
9. Alsumiri, M. Residual Incremental Conductance Based Nonparametric MPPT Control for Solar Photovoltaic Energy Conversion System. *IEEE Access* **2019**, *7*, 87901–87906. [CrossRef]
10. Farah, L.; Hussain, A.; Kerrouche, A.; Ieracitano, C.; Ahmad, J.; Mahmud, M. A Highly-Efficient Fuzzy-Based Controller With High Reduction Inputs and Membership Functions for a Grid-Connected Photovoltaic System. *IEEE Access* **2020**, *8*, 163225–163237. [CrossRef]
11. Obukhov, S.; Ibrahim, A.; Zaki Diab, A.A.; Al-Sumaiti, A.S.; Aboelsaud, R. Optimal Performance of Dynamic Particle Swarm Optimization Based Maximum Power Trackers for Stand-Alone PV System Under Partial Shading Conditions. *IEEE Access* **2020**, *8*, 20770–20785. [CrossRef]
12. Baimel, D.; Tapuchi, S.; Levron, Y.; Belikov, J. Improved Fractional Open Circuit Voltage MPPT Methods for PV Systems. *Electronics* **2019**, *8*, 321. [CrossRef]
13. Ali, A.; Almutairi, K.; Padmanaban, S.; Tirth, V.; Algarni, S.; Irshad, K.; Islam, S.; Zahir, M.H.; Shafiullah, M.; Malik, M.Z. Investigation of MPPT Techniques Under Uniform and Non-Uniform Solar Irradiation Condition–A Retrospection. *IEEE Access* **2020**, *8*, 127368–127392. [CrossRef]
14. Pindado, S.; Cubas, J.; Roibás-Millán, E.; Bugallo-Siegel, F.; Sorribes-Palmer, F. Assessment of Explicit Models for Different Photovoltaic Technologies. *Energies* **2018**, *11*, 1353. [CrossRef]
15. Ostadrahimi, A.; Mahmoud, Y. Novel Spline-MPPT Technique for Photovoltaic Systems Under Uniform Irradiance and Partial Shading Conditions. *IEEE Trans. Sustain. Energy* **2021**, *12*, 524–532. [CrossRef]
16. González-Castaño, C.; Restrepo, C.; Revelo-Fuelagán, J.; Lorente-Leyva, L.L.; Peluffo-Ordóñez, D.H. A Fast-Tracking Hybrid MPPT Based on Surface-Based Polynomial Fitting and P&O Methods for Solar PV under Partial Shaded Conditions. *Mathematics* **2021**, *9*, 2732. [CrossRef]
17. Álvarez, J.M.; Alfonso-Corcuera, D.; Roibás-Millán, E.; Cubas, J.; Cubero-Estalrrich, J.; Gonzalez-Estrada, A.; Jado-Puente, R.; Sanabria-Pinzón, M.; Pindado, S. Analytical Modeling of Current-Voltage Photovoltaic Performance: An Easy Approach to Solar Panel Behavior. *Appl. Sci.* **2021**, *11*, 4250. [CrossRef]
18. Pindado, S.; Cubas, J. Simple mathematical approach to solar cell/panel behavior based on datasheet information. *Renew. Energy* **2017**, *103*, 729–738. [CrossRef]
19. Andrean, V.; Chang, P.; Lian, K. A Review and New Problems Discovery of Four Simple Decentralized Maximum Power Point Tracking Algorithms—Perturb and Observe, Incremental Conductance, Golden Section Search, and Newton's Quadratic Interpolation. *Energies* **2018**, *11*, 2966. [CrossRef]
20. Espinoza, D.; Barcenas, E.; Campos-Delgado, D.; De Angelo, C. Voltage-Oriented Input-Output Linearization Controller as Maximum Power Point Tracking Technique for Photovoltaic Systems. *IEEE Trans. Ind. Electron.* **2014**, *62*, 3499–3507. [CrossRef]

mathematics

Article

Blockchain-Driven Real-Time Incentive Approach for Energy Management System

Aparna Kumari [1], Riya Kakkar [1], Rajesh Gupta [1,*], Smita Agrawal [1], Sudeep Tanwar [1,*], Fayez Alqahtani [2], Amr Tolba [3], Maria Simona Raboaca [4,5,*] and Daniela Lucia Manea [6]

1 Department of Computer Science and Engineering, Institute of Technology, Nirma University, Ahmedabad 382481, Gujarat, India
2 Software Engineering Department, College of Computer and Information Sciences, King Saud University, Riyadh 12372, Saudi Arabia
3 Computer Science Department, Community College, King Saud University, Riyadh 11437, Saudi Arabia
4 Doctoral School, University Politehnica of Bucharest, Splaiul Independentei Street No. 313, 060042 Bucharest, Romania
5 National Research and Development Institute for Cryogenic and Isotopic Technologies—ICSI Rm. Vâlcea, Uzinei Street, No. 4, 240050 Râmnicu Vâlcea, Romania
6 Faculty of Civil Engineering, Technical University of Cluj-Napoca, Constantin Daicoviciu Street, No. 15, 400020 Cluj-Napoca, Romania
* Correspondence: rajesh.gupta@nirmauni.ac.in (R.G.); sudeep.tanwar@nirmauni.ac.in (S.T.); simona.raboaca@icsi.ro (M.S.R.)

Abstract: In the current era, the skyrocketing demand for energy necessitates a powerful mechanisr to mitigate the supply–demand gap in intelligent energy infrastructure, i.e., the smart grid. To handl this issue, an intelligent and secure energy management system (EMS) could benefit end-consumer participating in the Demand–Response (DR) program. Therefore, in this paper, we proposed real-time and secure incentive-based EMS for smart grid, i.e., *RI-EMS* approach using Reinforcemen Learning (RL) and blockchain technology. In the *RI-EMS* approach, we proposed a novel rewar mechanism for better convergence of the RL-based model using a Q-learning approach based on th greedy policy that guides the RL-agent for faster convergence. Then, the proposed *RI-EMS* approac designed a real-time incentive mechanism to minimize energy consumption in peak hours and reduc end-consumers' energy bills to provide incentives to the end-consumers. Experimental results shov that the proposed *RI-EMS* approach induces end-consumer participation and increases custome profitabilities compared to existing approaches considering the different performance evaluatio metrics such as energy consumption for end-consumers, energy consumption reduction, and total cos comparison to end-consumers. Furthermore, blockchain-based results are simulated and analyze with the help of deployed smart contracts in a Remix Integrated Development Environment (IDE with the parameters such as transaction efficiency and data storage cost.

Keywords: residential energy management; reinforcement learning; Q-learning; smart grid; blockchai technology; smart contracts; energy infrastructure

MSC: 68M25

Citation: Kumari, A.; Kakkar, R.; Gupta, R.; Agarwal, S.; Tanwar, S.; Alqahtani, F.; Tolba, A.; Raboaca, M.S.; Manea, D.L. Blockchain-Driven Real-Time Incentive Approach for Energy Management System. *Mathematics* **2023**, *11*, 928. https://doi.org/10.3390/math11040928

Academic Editors: Atanda Raji, Khaled M. Abo-Al-Ez and Alessandro Niccolai

Received: 16 January 2023
Revised: 3 February 2023
Accepted: 9 February 2023
Published: 12 February 2023

1. Introduction

The proliferation of energy demand necessitates the effective production and distribu tion of energy in modern grid infrastructure, i.e., a smart grid with an automated contro facility for an energy management system (EMS). Energy management (EM) can be realize using three ways, i.e., energy efficiency, strategic load growth, and Demand–Respons (DR). Strategic load and energy efficiency involve long-term planning problems and do no consider real-time planning, whereas DR is a mechanism that controls load in real time [1 There are several ways to implement load control in real time, such as direct load contro time-based techniques, and many more. In direct load control, an electric utility company (EUC) can switch off/on the end-consumer electric appliances and provide incentives to the end-consumer as per the agreement. Furthermore, the total energy consumption associated with the end-consumer in the time-based techniques remains the same. Only the time to consume energy is changed considering the varying price signal forwarded by the EUC to the consumer. It also changes the shape of the energy load curve with the help of minimization of peak-to-average ratio (PAR) and reduces the energy bill of the end-consumer [2,3]. Moreover, DR can be characterized [4] into price-based DR and incentive-based DR, both of which have been comprehensively investigated in smart grid systems [5,6].

The usage of the DR mechanism can be defined with the time of use (TOU), critical peak pricing (CPP), and real-time pricing (RTP) mechanisms. As per the available literature survey, it is found that end-consumers are comfortable with the TOU mechanism, and system complexity is also less. Still, it suffers from a rebound peak problem, which is not the scenario in RTP. In recent decades, different DR strategies have been presented that aim to control residential houses or commercial buildings [7,8]. For example, Sun et al. [9] study the importance of heating, ventilation, and air conditioning, i.e., (HVAC) with distributions and physical parameters. Next, Zhang et al. [10] presented a service pricing-based load balancing approach for residential end-consumers. In [11], a secure and effective real-time scheduling mechanism is proposed for residential DR. Next, Ruzbahani et al. [12] presented an optimal incentive-based DR program for smart homes. Most studies focused on the deterministic rule, abstract model, or mathematical approach that suffers from various issues. For example, optimality cannot be obtained by the deterministic rules in the dynamic energy systems that can cause financial losses. It is heavily dependent on the operator's skill and suffers from scalability issues (for example, game-theoretic or MILP optimization) due to the involvement of a high number of binary variables.

To tackle the aforementioned issues associated with optimality and scalability, one of the noteworthy solutions is Artificial Intelligence (AI), which has proved its effectiveness toward optimal decision making utilizing Deep Learning (DL) and Reinforcement Learning (RL). Several RL approaches, such as Q-learning, Deep Q-Network, etc., have been incorporated by researchers worldwide to mitigate the decision-making problems [13–15]. Then, Zheng et al. [16] focused on the behavioral coupling of end-consumers by incorporating an incentive-based integrated DR approach for multiple energy carriers. In [17], a priority double deep Q-learning approach is presented to improve residential EMS. Most of the existing techniques used in Q-learning are modeled as Markov Decision Processes (MDPs). Still, it has not been exploited fully with real-time incentives and the accessibility of data for all stakeholders such as smart grid, end-consumers, and EUC. Other challenges, such as confidentiality, security, and privacy, must also be considered for efficient and trustable EM. Therefore, blockchain technology is the only solution to handle all these challenges mentioned above.

Blockchain is a secure, immutable, distributed ledger technique (DLT) that contains a chain of data blocks to mitigate security and trust issues such as single-point-of-failure, anonymity, and data manipulation [18]. It has been adopted to monitor EMS securely in the smart grid environment efficiently. For example, the authors of [19] formulated a Stackelberg game approach for achieving optimal energy pricing for efficient energy trading. In [18], blockchain achievability is presented in a smart grid system. Next, Jindal et al. [1] projected a blockchain-based system, i.e., GUARDIAN, to ensure the security of DR. Later, researchers adopted the decentralized blockchain technology in EMS for residential research areas as well [20]. As a result, existing blockchain-based approaches have several limitations, such as data storage cost (relatively high), high energy consumption, low transaction efficiency, and the requirement of high bandwidth to access data in real time [21–23]. Table 1 shows the comparative analysis of energy management systems with the proposed approach, which highlights how the proposed approach surpasses the research gaps such as the reliability, data storage cost, and transaction efficiency of related research work with the help of blockchain and interplanetary file system (IPFS)-based framework. Motivate by the above-mentioned gap, this paper proposed the *RI-EMS* approach: a Real-tin Incentive-based Energy Management System using RL (i.e., Q-learning) and blockchai The proposed *RI-EMS* approach stores energy data transactions utilizing an off-chain da storage platform, i.e., IPFS, that improves the scalability, data storage cost, reliability, an throughput of the EM.

Table 1. Comparative analysis of various state-of-the-art EMSs with the proposed.

Author	Year	Objective	Pricing Mechanism	Pros	Cons
Zhang et al. [10]	2020	Presented a load dispatch energy storage method for residential area	Iteration algorithm	Reduced operation cost, convergent	Need to consider energy trading for dynamic energy loads, privacy issues
Kumari et al. [11]	2020	Implemented the smart contract to ensure secure energy trading for smart grid	No mechanism	High scalability, reduced storage cost, and low latency	Should focus on optimal pricing, efficiency, and energy consumption
Zheng et al. [16]	2020	Presented a DR model to obtain the incentives for multiple energy carriers	Incentive-based approach	Improved accuracy, reduced dissatisfaction cost	Reduced energy consumption and transaction efficiency is not focused
Mathew et al. [17]	2021	Proposed a DR learning model for an efficient residential EM	DR-based greedy policy	Optimized peak cost and peak load	Need to implement with larger state space for optimal incentive
Li et al. [19]	2018	Discussed a secure energy-trading system for the Industrial Internet of Things using consortium blockchain	Stackelberg game	Optimized price, secure against double-spending and adversary attacks	No discussion on energy consumption reduction and cost
Hupez et al. [24]	2022	Formulated a game-theoretical approach for efficient energy scheduling in residential communities	Non-cooperative game theory	Optimized incentive and fair	No discussion on energy consumption, data storage cost, and transaction efficiency
Bruno et al. [25]	2022	Presented a residential demand response management for optimal load scheduling	Genetic algorithm	Reduced energy cost and electricity bill	Reliability, data storage cost, and energy consumption need to be considered
The proposed approach	2022	Proposed a real-time incentive approach for EMS using blockchain	Q-learning	Optimal price, incentive, high efficiency, and reliability	-

1.1. Research Contributions

The following are the research contributions of this paper.

- This paper proposes an *RI-EMS* approach for DR based on Q-learning to prioritize the experience of an agent and for faster convergence of DR using an epsilon greedy policy.
- A novel real-time incentive mechanism is proposed using a smart contract for the end-consumer to motivate them to participate in DR due to the appropriate and optimal incentives obtained for each participant in the EM.
- The proposed *RI-EMS* approach is evaluated compared to the conventional approaches in terms of consumer participation, energy consumption reduction, transaction efficiency, and data storage cost.

1.2. Organization of the Paper

The rest of the paper is organized as follows. First, Section 2 highlights the system model and problem formulation of the proposed *RI-EMS* approach, and Section 3 discusses the proposed *RI-EMS* approach in detail. Next, Section 4 presents the performance evaluation of the *RI-EMS* approach. Finally, the paper is concluded with future work in Section 5.

2. System Model and Problem Formulation

This section presents the system model and problem formulation of the proposed *RI-EMS* approach.

2.1. System Model

The proposed *RI-EMS* approach (as shown in Figure 1) involves the utilization of a smart grid platform to optimize and preserve energy consumption for consumers with the incorporated blockchain network. Now, energy consumption associated with the consumers σ_i participating in the energy management scheme can be defined based on the different types of energy load (E_l) in the particular locality, i.e., residential or commercial. For that, initially, energy consumption data are considered to optimize the incentive for consumers as residential σ^r or commercial σ^c. Based on the residential or commercial consumers, we can contemplate the energy consumption to optimize it further using the smart grid.

Therefore, if we consider the case of residential consumers, then energy consumption is affected by various energy loads, i.e., thermal (θ_{σ^r}), time-shiftable (τ_{σ^r}), power-shiftable (μ_{σ^r}), and other non-controllable ($N^c_{\sigma^r}$) energy loads. Meanwhile, we assume the energy consumption associated with the commercial consumers to be affected by the controllable (β_{σ^c}) and non-controllable (δ_{σ^c}) energy loads. Next, we need to reduce the evaluated energy consumption of residential and commercial consumers with the help of a smart grid. Thus, consumers with higher energy consumption can be given an incentive, which will also encourage them to save more energy in real time [18]. It seems difficult for consumers to save energy due to the involvement of high energy consumption. To achieve the optimum energy consumption, we have formulated an incentive mechanism for the residential and commercial consumers by applying a Q-learning approach in which ϵ greedy policy is applied to attain reduced energy consumption and convergence for a better response from consumers. TOU-based EMS is introduced in a smart grid to incentivize consumers based on their energy usage. Furthermore, to ensure the fair incentive mechanism in the scheme, we have employed a blockchain network enabled with IPFS to store the data in a distributed and immutable manner to add data transactions in the blockchain network further (with the help of a smart contract) [26]. However, before storing the energy consumption data in IPFS, data should be legitimate and authenticated by the introduced validation authority (VA). Once data are authorized by VA, they can be made available for storage in IPFS based on smart contract execution. So, real-time data accessibility and storage can be performed over a blockchain network in a real-time incentive-based energy management scheme usin the smart grid.

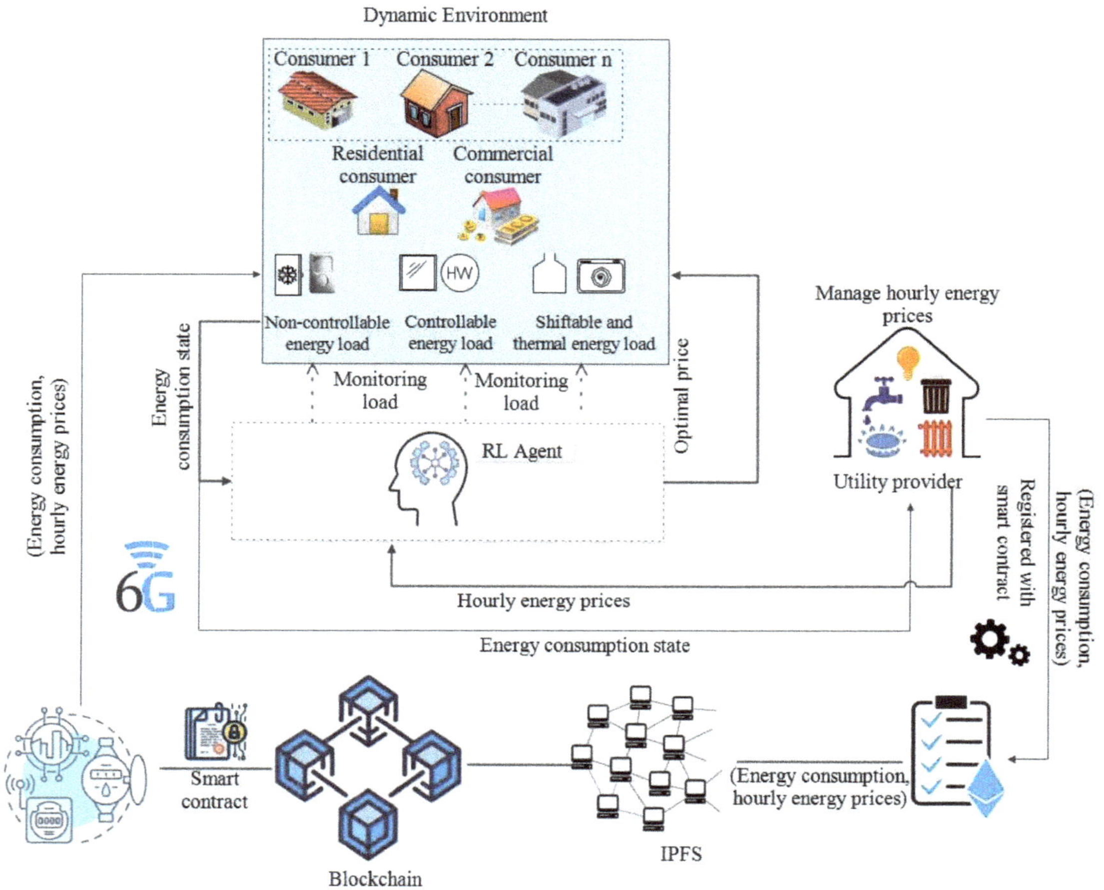

Figure 1. System model.

2.2. Problem Formulation

The proposed *RI-EMS* approach is a real-time incentive-based energy management sys tem in which s number of consumers $\{\sigma_1, \sigma_2, \sigma_{s'}, \ldots\ldots, \sigma_s\}$ are categorized into residentia σ^r and commercial consumers (σ^c) based on their energy consumption. However, to enabl the classification of consumers, energy consumption data can be defined based on th consumers, i.e., residential or commercial. Next, energy consumption associated with th residential consumer is determined by considering the energy loads, i.e., $\{\theta_{\sigma^r}, \tau_{\sigma^r}, \mu_{\sigma^r}, N^c_{\sigma^r}$ of various appliances. Similarly, the energy consumption corresponding to the commercia consumers also depends on the types of energy loads of appliances, which is assumed to b controllable γ_{σ^c} and non-controllable $N^c_{\sigma^r}$. Therefore, we can mention the various energ loads $(E_l^{\sigma^r}, E_l^{\sigma^c})$ which affect the energy consumption of the residential and commercia consumer. The aforementioned association is represented as follows:

$$E_l^{\sigma^r} = \begin{cases} \theta, & \text{if thermal load} \\ \tau, & \text{if time-shiftable load} \\ \mu, & \text{if power-shiftable load} \\ N^c, & \text{if non-controllable load} \end{cases} \quad (1$$

$$E_l^{\sigma^c} = \begin{cases} \beta, & \text{if controllable load} \\ \delta, & \text{if non-controllable load} \end{cases} \quad (2)$$

Thus, we have discussed the variables affecting the energy loads of residential and commercial consumers. Then, based on the several energy loads, we can focus on the energy consumption corresponding to the residential and commercial consumers. Firstly, we have to evaluate the energy consumption associated with the consumers $\epsilon_{\sigma_r}, \epsilon_{\sigma_c}$ based on the classification, i.e., residential and commercial. Therefore, the energy consumption of residential and commercial consumers can be calculated considering the energy demand ε of various energy loads at a time interval ξ, which can be mentioned as follows:

$$\varepsilon_{\sigma_r}^{\xi} = \varepsilon_{\theta_{\sigma_r}}^{\xi} + \varepsilon_{\tau_{\sigma_r}}^{\xi} + \varepsilon_{\mu_{\sigma_r}}^{\xi} + \varepsilon_{N_{\sigma_r}^{c}}^{\xi} \quad (3)$$

$$\varepsilon_{\sigma_c}^{\xi} = \varepsilon_{\beta_{\sigma_c}}^{\xi} + \varepsilon_{\delta_{\sigma_c}}^{\xi} \quad (4)$$

Next, we have to determine the reduction in energy consumption of residential and commercial consumers to achieve the maximum incentive for optimal energy management using the smart grid. However, the incentive obtained by the residential and commercial consumers depends on the reduction in energy consumption. Thus, the reduction in energy consumption of residential consumers can be calculated as follows:

$$\varepsilon_{\sigma_r^{Min}}^{\xi} = \left| \varepsilon_{\sigma_r^a}^{\xi} - \varepsilon_{\sigma_r^o}^{\xi} \right|^2 \quad (5)$$

where ε_a^{ξ} and ε_o^{ξ} denote the actual and objective energy consumption at a time interval ξ. Furthermore, objective energy consumption is decided based on previous energy usage in energy management. The actual energy consumption of residential consumers can be evaluated based on the types of energy loads along with their energy usage, which is mentioned as follows:

$$\varepsilon_{\sigma_r^a}^{\xi} = N_{\tau}^{\xi}\varepsilon_k + N_{\mu}^{\xi}\varepsilon_l + N_{N^c}^{\xi}\varepsilon_m - N_s^{\xi}\varepsilon_n \quad (6)$$

Similarly, the reduction in energy consumption of commercial consumers can be determined as follows:

$$\varepsilon_{\sigma_c^{Min}}^{\xi} = \left| \varepsilon_{\sigma_c^a}^{\xi} - \varepsilon_{\sigma_c^o}^{\xi} \right|^2 \quad (7)$$

Here, the actual energy consumption of commercial consumers can be calculated based on the types of energy load, i.e., controllable and non-controllable. As we have not considered the shiftable types of energy load for commercial consumers (not included in the scope of this research work), the calculation of actual energy consumption of commercial consumers with the number of controllable and non-controllable energy loads can be represented as follows:

$$\varepsilon_{\sigma_c^a}^{\xi} = N_{\beta}^{\xi}\varepsilon_k + N_{\delta}^{\xi} \quad (8)$$

Thus, we have considered the N and P number of energy loads for residential and commercial consumers to reduce energy consumption, which further leads to efficient and optimal energy management using a smart grid. In the proposed system, the main criteria of the smart grid are to provide an incentive to the consumers based on the reduction in energy consumption σ_r^{Min} and σ_c^{Min}. Therefore, we have deduced the objective function $C^O, C^{O'}$ to optimize the real-time incentive for consumers $\sigma_{r,c}$ with the help of a smart grid at a time interval (in hours) $\xi \in \{1, 2, \ldots\ldots, 24\}$, which can be mentioned as follows:

$$Optimize\ (C^O, C^{O'}) = \sum_{\xi=1}^{24} \sum_{(N,P)=1}^{4,2} \varepsilon_{\sigma_{r,c}^{Min},(N,P)}^{\xi} * \rho^s \quad (9)$$

where N and P number of energy loads affecting the energy consumption of residential and commercial consumers are considered for optimizing the incentive price based on the obtained reduced energy consumption $\sigma_{r,c}^{Min}$ with the help of smart grid ρ^s. Moreov the energy consumption data of residential and commercial consumers are stored in IPI immutable data storage after being made legitimate by VA. After being authenticated b VA, a smart contract runs to check the energy data's validity that must be stored in th IPFS. As a distributed and secure ledger, the blockchain network stores data transactio with improved cost-efficiency with the help of integrated IPFS protocol. Furthermore, th proposed incentive mechanism ensures the optimal energy management of the consume with the help of a smart grid. The real-time incentive mechanism works on the princip of an optimal Q-value and action-value function determined to achieve the reward fo consumers at a particular state in a dynamic environment. Furthermore, the employed 5 network helps to provide the incentive to consumers with high efficiency, availability, an reliability in the blockchain-based energy management system. A 5G wireless networ along with its low latency, high availability, and high data rate (DR) features, is bein considered in the blockchain and IPFS-based energy management scheme.

3. The Proposed Approach

Figure 2 shows the proposed *RI-EMS* approach, i.e., a blockchain-based real-tim incentive energy management scheme, which is divided into a 3-layered architectu consisting of an energy layer, incentive layer, and blockchain layer. These layers ar explained in detail to provide an overview of the proposed approach, which is represente as follows:

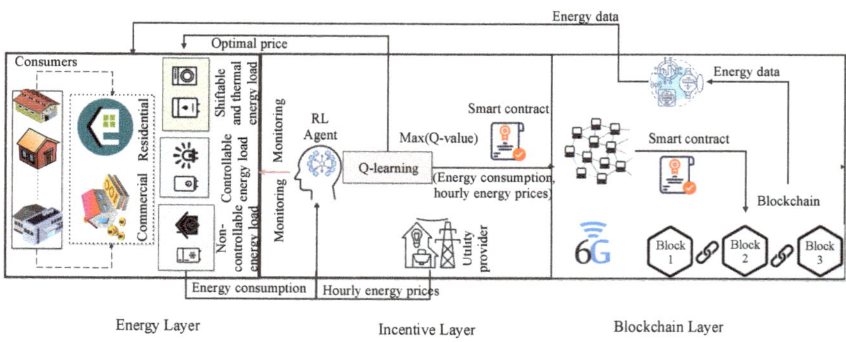

Figure 2. *RI-EMS*: The proposed approach [27,28] .

3.1. Energy Layer

The proposed scheme initiates with the energy layer, which involves collecting en ergy consumption data of residential and commercial consumers to deduce the maximu incentive for using various energy loads. The *RI-EMS* utilizes the Q-learning-based R approach to attain the real-time incentive by extracting the Q-value with the help of th Q-table. Consumers' energy consumption is affected by the usage of various appliance associated with the energy loads, i.e., thermal, time-shiftable, power-shiftable, and othe non-controllable loads for residential consumers and controllable and non-controllable en ergy loads for commercial consumers. The proposed scheme mainly focuses on optimizin the energy consumption associated with residential and commercial consumers using th smart grid. The energy consumption of consumers, along with their corresponding energ loads, can be represented as follows:

$$\epsilon_{\sigma_r} \rightarrow \{\theta_{\sigma^r}, \tau_{\sigma^r}, \mu_{\sigma^r}, N^c_{\sigma^r}\} \quad (10)$$

$$\sigma^c = \{\beta_{\sigma^c}, \delta_{\sigma^c}\} \quad (11)$$

Moreover, the energy consumption of residential and commercial consumers fluctuate based on the usage of appliances of various energy loads. Therefore, we have formulate the incentive mechanism for consumers based on the reduction in energy consumption explained in the incentive layer. Figure 3a shows the flowchart for the energy layer that mainly indicates the energy consumption associated with several energy loads, which the RL agent handles.

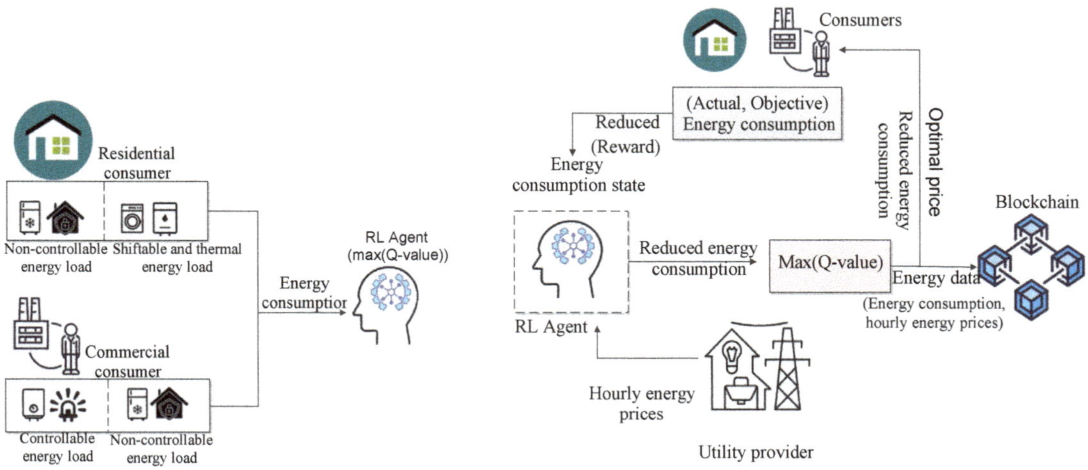

(**a**) Energy layer (**b**) Incentive layer

Figure 3. Flowchart for the proposed *RI-EMS* approach.

3.2. Incentive Layer—Reinforcement Learning Approach

The incentive layer serves as a middle layer between the energy and blockchain layer to provide real-time incentives to the residential and commercial consumers based on the optimized dynamic energy consumption $\varepsilon^{\xi}_{\sigma_r^{Min}}$ and $\varepsilon^{\xi}_{\sigma_c^{Min}}$ calculated using actual and objective energy consumption. We have applied the RL approach to obtain the minimized cost for residential and commercial consumers based on dynamic energy consumption. Furthermore, Figure 3b depicts that the RL method comprises multiple agents, i.e., residential and commercial consumers, whose main aim is to choose an action that yields the minimized cost in a dynamic environment. To implement the RL method, we need to consider three elements, i.e., state, action, and cost.

Assume S denotes the state set which represents the state of the RL agent, i.e., residential and commercial consumers $(s_{\sigma_r,\xi}, s_{\sigma_c,\xi})$ at a time interval ξ. Action is defined by A, which signifies the action of consumers $(a_{\sigma_r,\xi}, a_{\sigma_c,\xi}) \in (\varepsilon^{\xi}_{\sigma_r^{Min}}, \varepsilon^{\xi}_{\sigma_c^{Min}})$ to the dynamic environment to obtain the maximized cost. For example, how residential and commercial consumers act in a dynamic environment based on the dynamic energy consumption can decide their optimized cost C^{O}_{ξ} and $C^{O'}_{\xi}$. After obtaining the optimized cost, the dynamic environment can be forwarded to the next state $(s_{\sigma_r,\xi+1}, s_{\sigma_c,\xi+1})$. However, we have already discussed the associated energy loads of residential and commercial consumers and how they influence energy consumption. We have evaluated the reduced energy consumption of the consumers with the help of actual and objective energy consumption. Based on the calculated reduced energy consumption, we have deduced an objective function specifying the cost for residential and commercial consumers varying with the reduced energy consumption, and the smart grid ensures the low pricing for consumers.

Therefore, firstly, we can define the optimal policy Δ for agents, i.e., residential an commercial consumers, to optimize the objective cost evaluated using reduced energ consumption. Thus, the action-value function $Q_{\Delta_{\sigma_r}}$ for residential consumers considerin the state, action, and policy can be represented as follows:

$$Q_{\Delta}(s_{\sigma_r,\xi}, a_{\sigma_r,\xi}) = \sum_{j=\xi+1}^{T} \omega^{j-\xi-1} C^{O}_{j-1} | s_{\sigma_r,\xi}, a_{\sigma_r,\xi} \quad (12)$$

$$\forall s_{\sigma_r,\xi} \in S, a_{\sigma_r,\xi} \in A \quad (13)$$

where ω denotes the discount factor associated with the residential consumers. Similarl we can calculate the action-value function $Q_{\Delta_{\sigma_c}}$ for commercial consumers with the hel of objective cost $C^{O'}$. Moreover, the optimality of the action value for residential an commercial consumers is represented by $Q^*_{\Delta}(s_{\sigma_r,\xi}, a_{\sigma_r,\xi})$ and $Q^*_{\Delta'}(s_{\sigma_c,\xi}, a_{\sigma_c,\xi})$.

Furthermore, agents, i.e., $(s_{\sigma_r,\xi}, s_{\sigma_c,\xi})$ should take action $(a_{\sigma_r,\xi}, a_{\sigma_c,\xi})$ to maximize th reward or incentive $\eta(s_{\sigma_r,\xi}, a_{\sigma_r,\xi})$ using the Q-learning approach based on the policy at particular state $(s_{\sigma_r,\xi}, s_{\sigma_c,\xi})$. The Q-learning approach works on the principle of Q-valu $\Omega(s_{\sigma_r,\xi}, a_{\sigma_r,\xi})$ by preparing the Q-table containing the action $(a_{\sigma_r,\xi}, a_{\sigma_c,\xi})$ and state $(s_{\sigma_r,\xi}, s_{\sigma_c,\xi}$ As a result, the flow of the incentive layer with Q-value is considered an important aspect t obtain the optimal price, which is further forwarded to the consumers based on the reduce energy consumption calculated using actual and objective energy consumption. Therefor the calculation of the Q-value for residential consumers is represented as follows:

$$\Omega(s_{\sigma_r,\xi}, a_{\sigma_r,\xi}) \leftarrow \Omega(s_{\sigma_r,\xi}, a_{\sigma_r,\xi}) + \beta(\eta_{\xi+1}(s_{\sigma_r,\xi}, a_{\sigma_r,\xi}) + \omega \, max \, \Omega(s_{\sigma_r,\xi+1}, a_{\sigma_r,\xi}) - \Omega(s_{\sigma_r,\xi}, a_{\sigma_r,\xi})) \quad (14)$$

where β represents the learning rate which lies in the range of [0, 1] and ω is considered a the discount factor associated with the action–value pair calculated to maximize the cos objective function of residential consumers. Similarly, we can calculate the optimization c incentive or reward $\eta'(s_{\sigma_c,\xi}, a_{\sigma_c,\xi})$ for commercial consumers based on the Q-value, whic is expressed as follows:

$$\Omega(s_{\sigma_c,\xi}, a_{\sigma_c,\xi}) \leftarrow \Omega(s_{\sigma_c,\xi}, a_{\sigma_c,\xi}) + \beta(\eta'_{\xi+1}(s_{\sigma_c,\xi}, a_{\sigma_c,\xi}) + \kappa \, max \, \Omega(s_{\sigma_c,\xi+1}, a_{\sigma_c,\xi}) - \Omega(s_{\sigma_c,\xi}, a_{\sigma_c,\xi})) \quad (15)$$

Furthermore, Algorithm 1 shows how the Q-learning approach can be used to dete mine the optimization of the Q-value for residential and commercial consumers with th help of optimal policy in terms of time complexity of O(e) (which represents the number c episodes to compute the optimization of Q-value), which is expressed as follows:

Therefore, we have applied the Q-learning approach to maximize the incentiv $\eta(s_{\sigma_r,\xi}, a_{\sigma_r,\xi})$ and $\eta'(s_{\sigma_c,\xi}, a_{\sigma_c,\xi})$ for residential and commercial consumers based on th dynamic energy consumption that is considered as the action $(\varepsilon^{\xi}_{\sigma_r Min}, \varepsilon^{\xi}_{\sigma_c Min})$ taken by th consumers for each state of $s_{\sigma_r,\xi}, s_{\sigma_c,\xi}$ at a time interval ξ. After obtaining the incentiv mechanism for consumers using the Q-learning approach, the secure storage of reduce energy consumption has been explained in the blockchain layer, which focuses on real-tim incentive energy storage with the help of the introduced IPFS.

$$\{\delta, \delta'\} = argmax(\Omega(s_{\sigma_r}), \Omega(s_{\sigma_c}), \xi) \quad (16)$$

Algorithm 1 Incentive for Consumers using Q-learning

Input: $s_{\sigma_r,\xi}, s_{\sigma_c,\xi}, a_{\sigma_r,\xi}, a_{\sigma_r,\xi}, Q_\Delta, Q_{\Delta'}, \xi$
Output: Optimized incentive

procedure INCENTIVE_CONSUMER($s_{\sigma_r,\xi}, s_{\sigma_c,\xi}, \xi$)
  if $\sigma \in \sigma_r$ **then**
    for ξ time interval < 0 **do**
      $AssignQ - value \rightarrow 0$
      for E **do**pisode e
        Calculate action value for residential consumer
        Assign State $s_{\sigma_r,\xi}$
        $Q_\Delta = \sum_{j=\xi+1}^{T} \omega^{j-\xi-1} C_{j-1}^{O} | s_{\sigma_r,\xi}, a_{\sigma_r,\xi}$
        Compute incentive $\eta(s_{\sigma_r,\xi}, a_{\sigma_r,\xi})$
        Transit to new state $s_{\sigma_r,\xi+1}$
        Compute optimization of Q-value
        $\delta = argmax(\Omega(s_{\sigma_r,\xi}, a_{\sigma_r,\xi}))$
      end for
    end for
  else
    for ξ time interval < 0 **do**
      $AssignQ - value \rightarrow 0$
      for Episode e **do**
        Calculate action value for commercial consumer
        Assign State $s_{\sigma_c,\xi}$
        $Q_{\Delta'} = \sum_{j=\xi+1}^{T} \omega^{j-\xi-1} C_{j-1}^{O'} | s_{\sigma_c,\xi}, a_{\sigma_c,\xi}$
        Compute incentive $\kappa(s_{\sigma_c,\xi}, a_{\sigma_c,\xi})$
        Transit to new state $s_{\sigma_c,\xi+1}$
        Compute optimization of Q-value
        $\delta' = argmax(\Omega(s_{\sigma_c,\xi}, a_{\sigma_c,\xi}))$
      end for
    end for
  end if
end procedure

3.3. Blockchain Layer

Ethereum blockchain, as a secure and decentralized platform, is introduced to ensure secure and real-time incentive energy management for consumers implemented with a value-based Q-learning algorithm. To accomplish real-time data accessibility and energy consumption data stored securely in the blockchain, IPFS as an immutable peer-to-peer protocol is employed in the system to improve the scalability and reliability of the communication between multiple agents in the dynamic environment [29]. Initially, VA as an authorizing entity is considered to confirm the identity of consumers participating in energy management. To legitimize the authorization of data storage in IPFS, an intelligent contract run as a self-executable code to check the authenticity of energy consumption data. If it becomes authenticated for data storage, then IPFS as a cost-efficient protocol allocates hash keys ϕ_{σ_r} and ψ_{σ_c} to residential and commercial consumers. Next, the consumers containing the hash keys ϕ_{σ_r} and ψ_{σ_c} provided by the IPFS can use them as data access and storage keys to perform the transactions of real-time energy management in the blockchain network.

Algorithm 2 depicts how energy data can be stored securely with the help of a blockchain network considering the time complexity, i.e., O(s), and O(s') associated with the number of residential and commercial consumers request for data storage. Furthermore, the security of energy management transactions of the consumers needs to be ensured in the blockchain network. For that, we have utilized the pair of keys, i.e., public key and private key of consumers $Pc_{k_{\sigma_r}}, Pe_{k_{\sigma_r}}$ and $Pc_{k_{\sigma_c}}, Pe_{k_{\sigma_c}}$ using asymmetric public key cryptograpl to preserve the energy management of consumers in the dynamic environment, which denoted by D_E:

Algorithm 2 Blockchain-based algorithm for secure energy data storage

Input: $\sigma^r, \sigma^c, IPFS^{hk}, VA$
Output: Add energy data transactions to the blockchain

procedure ENERGY_DATA($\phi_{\sigma_r}, \phi_{\sigma_c}, \sigma^r, \sigma^c$)
  if $\sigma \in \sigma^r$ **then**
    for $x = 1, 2, \ldots, s$ **do**
      $IPFS^{hk} \leftarrow data\_requests(\sigma^r)$
      $\sigma^r \xleftarrow{authorize} VA$
      Execute smart contract
      if $\sigma^r \in$ authorized **then**
        $\sigma^r \xleftarrow{\phi_{\sigma_r}} IPFS^{hk}$
        $blockchain \leftarrow Add\_data\ (\sigma^r)$
        Secure data storage in the blockchain
      else
        Invalid consumer
      end if
    end for
  else if $\sigma \in \sigma^c$ **then**
    for $y = 1, 2, \ldots, s'$ **do**
      $IPFS^{hk} \leftarrow data\_requests(\sigma^c)$
      $\sigma^c \xleftarrow{authorize} VA$
      Execute smart contract
      if $\sigma^c \in$ authorized **then**
        $\sigma^r \xleftarrow{\psi_{\sigma_c}} IPFS^{hk}$
        $blockchain \leftarrow Add\_data\ (\sigma^r)$
        Secure data storage in the blockchain
      else
        Invalid consumer
      end if
    end for
  end if
end procedure

$$D^h((\sigma_r, \sigma_c), D_E) = ((\phi_{\sigma_r}, \psi_{\sigma_c}), D_E) \quad (1$$

$$\varphi^{Pc_k(\sigma_r, \sigma_c)}(Ds_d^{Pe_k^{\alpha r}})(D^h((\sigma_r, \sigma_c), D_E) = D^h((\sigma_r, \sigma_c), D_E) \quad (1.$$

where D^h signifies the hash digest of the energy transactions of consumers σ_r and σ_c in tl dynamic environment D_E. Ds_d represents the digital signature of consumers associate with their private key $\{Pe_{k_{\sigma_r}}, Pe_{k_{\sigma_c}}\}$. Furthermore, Figure 4 shows the basic working the blockchain layer in which energy data optimized from the incentive layer are store securely in the blockchain network through the IPFS intermediary protocol. Then, tl smart grid operator manages the energy data that can be forwarded to consumers based c their reduced energy consumption.

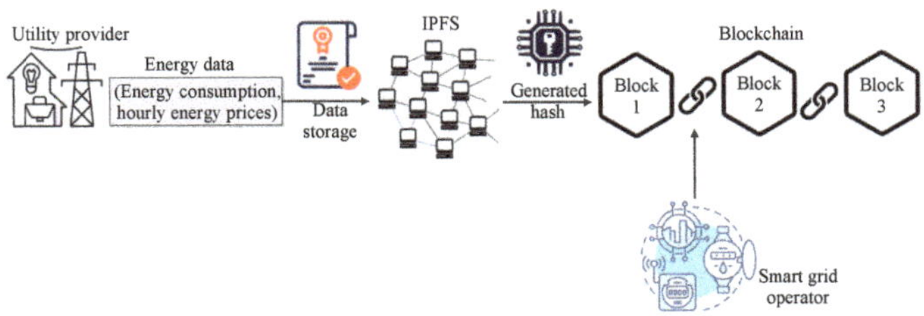

Figure 4. Blockchain layer.

4. Performance Evaluation

This section gives an overview of the performance evaluation of the proposed *RI-EMS* approach. The proposed *RI-EMS* approach is implemented with python high-level programming language on the Windows operating system with the configuration of Intel(R) Core(TM) CPU of 2.60 GHz and 8 GB RAM to maximize the incentive for consumers based on the Q-learning approach considering the performance metrics such as energy consumption for end-consumer, energy consumption reduction, and total cost comparison. Furthermore, blockchain-based results are evaluated and analyzed by deploying the smart contracts in Remix IDE with the help of parameters such as transaction efficiency and data storage cost.

4.1. Dataset Description

The performance evaluation of the proposed *RI-EMS* approach is conducted using the standard dataset, i.e., Open Energy Information (openEI) [30]. It contains energy consumption data for residential houses and commercial buildings as well. Then, the pre-processing of the energy data is performed with a sci-kit-learn library to tackle noise, Not-a-Number (NaN), missing values, duplicate values, etc. Next, the critical load data (such as AC and other appliances) is obtained from Pecanstreet [31]. Then, hourly energy prices are considered from PJM Data Miner as 2nd August 2022 [32]. Finally, Table 2 shows the several simulation parameters considered for implementing and predicting the results for the proposed *RI-EMS* approach.

Table 2. Simulation settings.

Particular	Values
ξ	1 h
Peak hour	5 PM to 12 PM
Mid-peak	8 AM to 5 PM
Off-peak	12 AM to 8 AM
$\delta_{\mathfrak{C}}$	0.01
ϕ	0.001
β	{0,1}

4.2. Energy Consumption Reduction and Comparative Analysis

Figure 5a highlights the energy consumption of the end-consumers by considering the distinguished non-controllable and controllable energy loads. In the proposed approach, the energy consumption of commercial consumers has been calculated with the help of controllable and non-controllable energy demand. Furthermore, the reduction in energy consumption is determined using the respective consumer's actual and objective energy consumption. The graph depicts the acquired energy consumption of the proposed approach in a time interval (hours) of [0, 25]. It can be observed from the graph that a controllable energy load yields higher energy consumption than non-controllable energy consumption, which leads to the increased incentive of consumers in the case of a non-controllable energy load.

Figure 5b presents the energy consumption reduction due to the incentive mechanis used in the proposed approach. Here, energy demand is marked in orange color, an consumption of energy by the consumer is marked in green. The dotted line represents th hourly energy prices. This graph depicts the consumption reduction in peak hours an the increase in consumption in non-peak hours; for example, in the morning (1 AM to AM), consumption is high, and during peak hours, consumption is reduced to receive mo incentives from the consumer. Furthermore, the proposed *RI-EMS* approach is compare with the baseline approach such as Gurobi optimizer [33] by having the same simulatio parameters setting. Figure 6 depicts the costs comparison, which comprises total energ consumption reduction and discomfort costs to the end-consumer. It is evident from th graph that the proposed *RI-EMS* approach learns through a trial and error mechanism an performs well with the increasing number of episodes compared to the baseline model.

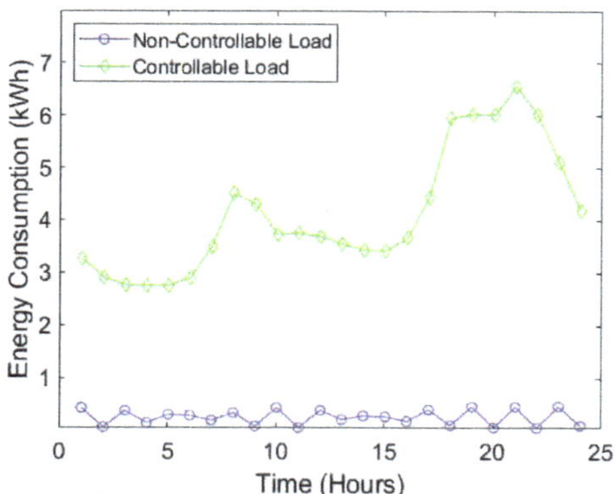

(**a**) Energy consumption for end-consumer

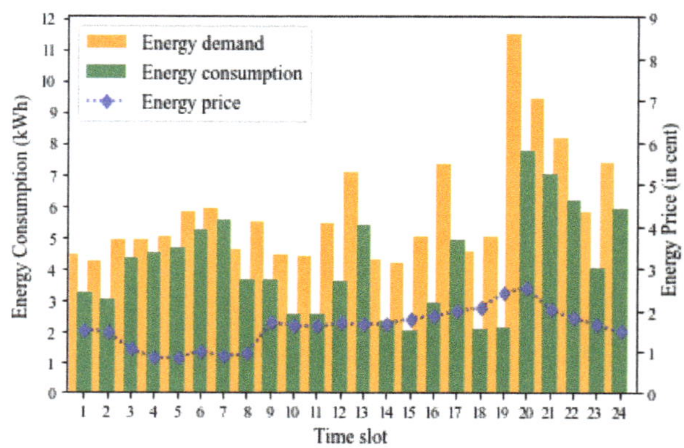

(**b**) Energy consumption reduction

Figure 5. Comparative analysis: (**a**) Energy consumption by a particular end-consumer, (**b**) Energ consumption reduction with the proposed *RI-EMS* approach.

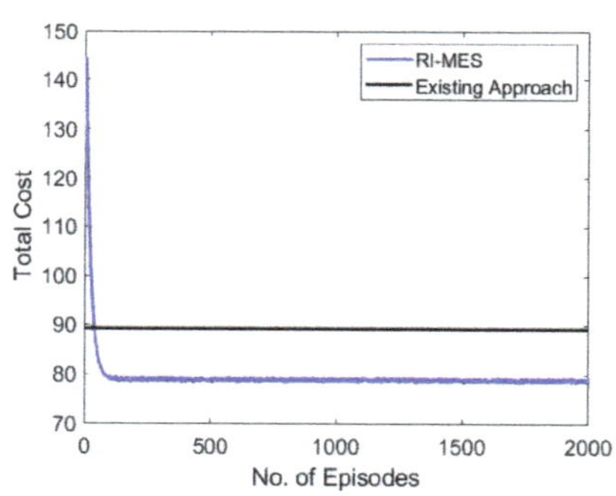

Figure 6. Comparison of *RI-EMS* with existing approach.

4.3. Transaction Efficiency

Figure 7a shows the transaction efficiency comparison considering two scenarios i which one scenario is to perform data transactions in the proposed *RI-EMS* approac with IPFS. Another scenario focuses on performing the data transactions in the propose approach with blockchain. It can be perceived from the scenarios that transaction efficienc seems to lie at the same level when fewer data transactions are performed between multipl agents. However, with the exponential increase in the number of data transactions, th proposed *RI-EMS* approach with IPFS exhibits quite improved transactions efficienc compared with the proposed scheme with blockchain. This is because IPFS works o generating a hash, which must be assigned to the consumers for secure and cost-efficient data storage.

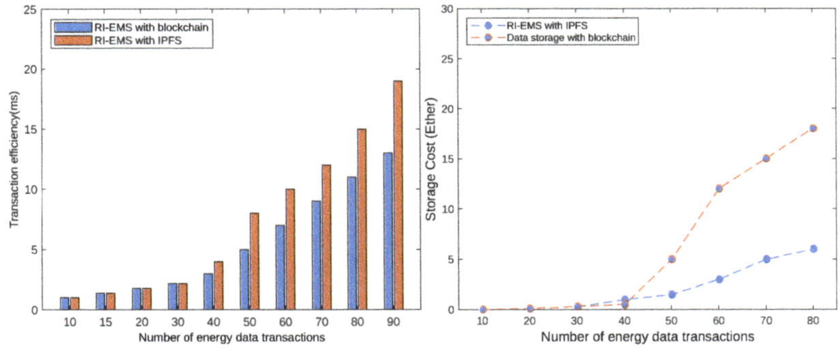

(**a**) Transaction efficiency comparison (**b**) Storage cost analysis

Figure 7. Comparative analysis: (**a**) Transaction efficiency comparison for the proposed *RI-EMS* approach, (**b**) Storage cost analysis for proposed *RI-EMS* approach.

4.4. Data Storage Cost

In this subsection, we have focused on the data storage cost of the *RI-EMS* approach to ensure cost-efficient energy management for consumers. Therefore, we have focused on the data storage cost of the Ethereum blockchain network, which is a decentralized and secure platform. Initially, we highlight an important metric, i.e., gas price for a single word, which is denoted by Gp_w. Furthermore, the gas price Gp^K for 1 KB of energy consumption data correlates with Gp^w in which Gp^K can be calculated as $20 * 10^3$ *Gas* and Gp^w can be written in the form of expression $(2^{10}/256) * (20 * 10^3)$ *Gas*. Furthermore, data storage cost C^W associated with W number of words in a blockchain can be computed with the parameters gas price and Ethereum price (gs_{bc}, ET_{bc}). Therefore, considering ether value (Ev) as 10^9, data storage cost can be expressed in the form $(w * G)/Ev$ to calculate the cost in USD as $(gs_{bc} * C^W) * ET_{bc}$ [34].

Storage Cost Analysis

The aforementioned computation for data storage cost proves that using blockchain as a data storage platform incurs huge costs, which can demotivate consumers from utilizing the energy of appliances associated with various energy loads. As a result, Figure 7b shows the data storage cost analysis of the proposed *RI-EMS* approach considering the data storage as blockchain and IPFS. Finally, the graph exhibits relatively low storage cost when using IPFS as data storage with the proposed *RI-EMS* approach. Moreover, when fewer consumers are involved in the energy data transactions, then the data storage cost for both platforms lies at the exact alignment. Still, with the exponential surge in the number of energy data transactions, the requirement of data storage cost for the proposed *RI-EMS* approach with IPFS is relatively lower than the blockchain data storage. The main reason for the cost-efficient behavior of IPFS is that it stores consumers' energy data by generating a hash, which requires a lower cost than the blockchain (stores a whole block of data).

5. Conclusions

The growth of smart homes has increased the research on EMS across the globe. So, in this paper, we presented an incentive-based EMS for smart grid, *RI-EMS* integrated with blockchain technology in real time. We have adopted the DR-based Q-learning approach to optimize the incentive for residential and commercial consumers based on the calculated reduction in energy consumption. We have categorized consumers based on the several energy loads to obtain insights into energy consumption. Moreover, we have formulated a real-time incentive mechanism based on the action-value function and Q-value applied using the Q-learning approach implemented in the python programming langua to obtain the reward or incentive for consumers. The consumer incentive mechanis has been optimized based on the ϵ greedy policy to guide multiple agents for bett convergence. Finally, the performance of the proposed *RI-EMS* approach is simulat against important metrics, i.e., consumer participation, energy consumption reduction, a total cost comparison to end-consumers. Next, blockchain-based results are implement by deploying the smart contracts in Remix IDE in terms of transaction efficiency and da storage cost.

In the future, we will implement a DL model with the Q-learning approach to obta the optimum energy consumption for consumers managed by the multiple agents. DL a Q-learning approaches can improve the incentive for consumers monitored by multip agents. Furthermore, we can consider a real-time and dynamic scenario to implement t blockchain-based technology for efficient and optimal EM in smart homes.

Author Contributions: Conceptualization: S.T., S.A., M.S.R., F.A. and R.G.; writing—original dr preparation: R.K., A.K., R.G. and S.A.; methodology: S.T., M.S.R., F.A., A.T. and A.K.; writing—revi and editing: R.K., R.G., S.T., S.A., D.L.M. and A.T.; Investigation: R.K., F.A., A.K., D.L.M. and S. Visualization: S.T., M.S.R., S.A., A.K., A.T. and R.G. All authors have read and agreed to the publish version of the manuscript.

Funding: This work was funded by the Researchers Supporting Project number (RSP2023R509) Ki Saud University, Riyadh, Saudi Arabia and was partially supported by UEFISCDI Romania and M through BEIA projects AutoDecS, SOLID-B5G, AISTOR, Hydro3D, EREMI, FinSESCo, CREATE a by European Union's Horizon Europe research and innovation program under grant agreement N 101081061 (PLENTY-LIFE). This work is supported by Ministry of Research, Innovation, Digitizati from Romania by the National Plan of R & D, Project PN 19 11, Subprogram 1.1. Institution performance-Projects to finance excellence in RDI, Contract No. 19PFE/30.12.2021 and a grant the National Center for Hydrogen and Fuel Cells (CNHPC)—Installations and Special Objectives National Interest (IOSIN).

Institutional Review Board Statement: Not applicable.

Informed Consent Statement: Not applicable.

Data Availability Statement: No data is associated with this research work.

Acknowledgments: This work was funded by the Researchers Supporting Project numb (RSP2023R509) King Saud University, Riyadh, Saudi Arabia and was partially supported by UEF CDI Romania and MCI through BEIA projects AutoDecS, SOLID-B5G, AISTOR, Hydro3D, EREM FinSESCo, CREATE and by European Union's Horizon Europe research and innovation program u der grant agreement No. 101081061 (PLENTY-LIFE). This work is supported by Ministry of Researc Innovation, Digitization from Romania by the National Plan of R & D, Project PN 19 11, Subprogra 1.1. Institutional performance-Projects to finance excellence in RDI, Contract No. 19PFE/30.12.20 and a grant of the National Center for Hydrogen and Fuel Cells (CNHPC)-Installations and Speci Objectives of National Interest (IOSIN).

Conflicts of Interest: The authors declare no conflict of interest.

Abbreviations

Acronym	Definition
AI	Artificial intelligence
CPP	Critical peak pricing
DLT	Distributed ledger technique
DR	Data rate
DL	Deep learning
DR	Demand response

EMS	Energy management system
EUC	Electric utility company
EM	Energy management
IPFS	Inteplanetary file system
IDE	Integrated development environment
MDP	Markov decision process
NaN	Not-a-number
PAR	Peak-to-average ratio
RTP	Real-time pricing
RL	Reinforcement learning
TOU	Time of use
VA	Validation authority

References

1. Jindal, A.; Aujla, G.S.S.; Kumar, N.; Villari, M. GUARDIAN: Blockchain-based Secure Demand Response Management in Smart Grid System. *IEEE Trans. Serv. Comput.* **2019**, *13*, 613–624. [CrossRef]
2. Jindal, A.; Singh, M.; Kumar, N. Consumption-Aware Data Analytical Demand Response Scheme for Peak Load Reduction in Smart Grid. *IEEE Trans. Ind. Electron.* **2018**, *65*, 8993–9004. [CrossRef]
3. Asef, P.; Taheri, R.; Shojafar, M.; Mporas, I.; Tafazolli, R. SIEMS: A Secure Intelligent Energy Management System for Industrial IoT Applications. *IEEE Trans. Ind. Inform.* **2023**, *19*, 1039–1050. [CrossRef]
4. Paterakis, N.G.; Erdinç, O.; Catalao, J.P. An overview of Demand Response: Key-elements and international experience. *Renew. Sustain. Energy Rev.* **2017**, *69*, 871–891. [CrossRef]
5. Kumari, A.; Vekaria, D.; Gupta, R.; Tanwar, S. Redills: Deep Learning-Based Secure Data Analytic Framework for Smart Grid Systems. In Proceedings of the 2020 IEEE International Conference on Communications Workshops (ICC Workshops), Dublin, Ireland, 7–11 June 2020; pp. 1–6. [CrossRef]
6. Miao, H.; Chen, G.; Zhao, Z.; Zhang, F. Evolutionary Aggregation Approach for Multihop Energy Metering in Smart Grid for Residential Energy Management. *IEEE Trans. Ind. Inform.* **2021**, *17*, 1058–1068. [CrossRef]
7. Basnet, S.M.; Aburub, H.; Jewell, W. Residential demand response program: Predictive analytics, virtual storage model and its optimization. *J. Energy Storage* **2019**, *23*, 183–194. [CrossRef]
8. Chen, T.; Bu, S.; Liu, X.; Kang, J.; Yu, F.R.; Han, Z. Peer-to-Peer Energy Trading and Energy Conversion in Interconnected Multi-Energy Microgrids Using Multi-Agent Deep Reinforcement Learning. *IEEE Trans. Smart Grid* **2022**, *13*, 715–727. [CrossRef]
9. Sun, Y.; Elizondo, M.; Lu, S.; Fuller, J.C. The impact of uncertain physical parameters on HVAC demand response. *IEEE Trans. Smart Grid* **2014**, *5*, 916–923. [CrossRef]
10. Zhang, W.; Wei, W.; Chen, L.; Zheng, B.; Mei, S. Service pricing and load dispatch of residential shared energy storage unit. *Energy* **2020**, *202*, 117543. [CrossRef]
11. Kumari, A.; Tanwar, S. A Reinforcement Learning-based Secure Demand Response Scheme for Smart Grid System. *IEEE Internet Things J.* **2021**, *9*, 2180–2191. [CrossRef]
12. Ruzbahani, H.M.; Karimipour, H. Optimal incentive-based demand response management of smart households. In Proceedings of the 2018 IEEE/IAS 54th Industrial and Commercial Power Systems Technical Conference (I & CPS), Niagara Falls, ON, Canada, 7–10 May 2018; pp. 1–7. [CrossRef]
13. Lu, R.; Hong, S.H. Incentive-based demand response for smart grid with reinforcement learning and deep neural network. *Appl. Energy* **2019**, *236*, 937–949. [CrossRef]
14. Ma, R.; Yi, Z.; Xiang, Y.; Shi, D.; Xu, C.; Wu, H. A Blockchain-Enabled Demand Management and Control Framework Driven by Deep Reinforcement Learning. *IEEE Trans. Ind. Electron.* **2023**, *70*, 430–440. [CrossRef]
15. Lu, R.; Jiang, Z.; Wu, H.; Ding, Y.; Wang, D.; Zhang, H.T. Reward Shaping-Based Actor-Critic Deep Reinforcement Learning for Residential Energy Management. *IEEE Trans. Ind. Inform.* **2022**, 1–12. [CrossRef]
16. Zheng, S.; Sun, Y.; Li, B.; Qi, B.; Shi, K.; Li, Y.; Tu, X. Incentive-Based Integrated Demand Response for Multiple Energy Carriers Considering Behavioral Coupling Effect of Consumers. *IEEE Trans. Smart Grid* **2020**, *11*, 3231–3245. [CrossRef]
17. Mathew, A.; Jolly, M.J.; Mathew, J. Improved residential energy management system using priority double deep Q-learning. *Sustain. Cities Soc.* **2021**, *69*, 102812. [CrossRef]
18. Kumari, A.; Gupta, R.; Tanwar, S.; Tyagi, S.; Kumar, N. When blockchain meets smart grid: Secure energy trading in demand response management. *IEEE Netw.* **2020**, *34*, 299–305. [CrossRef]
19. Li, Z.; Kang, J.; Yu, R.; Ye, D.; Deng, Q.; Zhang, Y. Consortium Blockchain for Secure Energy Trading in Industrial Internet of Things. *IEEE Trans. Ind. Inform.* **2018**, *14*, 3690–3700. [CrossRef]
20. Kumari, A.; Shukla, A.; Gupta, R.; Tanwar, S.; Tyagi, S.; Kumar, N. ET-DeaL: A P2P Smart Contract-based Secure Energy Trading Scheme for Smart Grid Systems. In Proceedings of the IEEE INFOCOM 2020-IEEE Conference on Computer Communications Workshops (INFOCOM WKSHPS), Toronto, ON, Canada, 6–9 July 2020; IEEE: Piscataway, NJ, USA, 2020; pp. 1051–1056.
21. Zhang, L.; Cheng, L.; Alsokhiry, F.; Mohamed, M.A. A Novel Stochastic Blockchain-Based Energy Management in Smart Cities Using V2S and V2G. *IEEE Trans. Intell. Transp. Syst.* **2022**, *20*, 915–922. [CrossRef]

22. AlSkaif, T.; Crespo-Vazquez, J.L.; Sekuloski, M.; van Leeuwen, G.; Catalão, J.P.S. Blockchain-Based Fully Peer-to-Peer Energ Trading Strategies for Residential Energy Systems. *IEEE Trans. Ind. Inform.* **2022**, *18*, 231–241. [CrossRef]
23. Singh, R.; Tanwar, S.; Sharma, T.P. Utilization of blockchain for mitigating the distributed denial of service attacks. *Secur. Pr* **2020**, *3*, e96. [CrossRef]
24. Hupez, M.; Toubeau, J.F.; Atzeni, I.; Grève, Z.D.; Vallée, F. Pricing Electricity in Residential Communities Using Game-Theoretic Billings. *IEEE Trans. Smart Grid* 2022, *early access*. [CrossRef]
25. Mota, B.; Faria, P.; Vale, Z. Residential load shifting in demand response events for bill reduction using a genetic algorithn *Energy* **2022**, *260*, 124978. [CrossRef]
26. Kumari, A.; Tanwar, S. A secure data analytics scheme for multimedia communication in a decentralized smart grid. *Multime Tools Appl.* **2022**, *81*, 34797–34822. [CrossRef]
27. Wen, L.; Zhou, K.; Li, J.; Wang, S. Modified deep learning and reinforcement learning for an incentive-based demand respons model. *Energy* **2020**, *205*, 118019. [CrossRef]
28. Salazar, E.J.; Jurado, M.; Samper, M.E. Reinforcement Learning-Based Pricing and Incentive Strategy for Demand Response i Smart Grids. *Energies* **2023**, *16*, 1466. [CrossRef]
29. Gupta, R.; Reebadiya, D.; Tanwar, S.; Kumar, N.; Guizani, M. When Blockchain Meets Edge Intelligence: Trusted and Securit Solutions for Consumers. *IEEE Netw.* **2021**, *35*, 272–278. [CrossRef]
30. OpenEI. Open Energy Information: Smart Meters Data from Houses. Available online: https://openei.org/datasets/files/961 pub (accessed on 29 July 2022).
31. Pecan Street Dataport. Available online: https://www.pecanstreet.org/dataport/ (accessed on 18 July 2021).
32. pjm Data Miner.Available online: https://www.pjm.com/markets-and-operati\ons/etools/data-miner-2.aspx (accessed on 1 January 2021).
33. Gurobi Optimization. Available online: http://www.gurobi.com (accessed on 29 July 2022).
34. REMIX: The Native IDE for Web3 Development. Available online: https://remix.ethereum.org/ (accessed on 28 December 2022

mathematics MDPI

Article

Optimal Integration of Battery Systems in Grid-Connected Networks for Reducing Energy Losses and CO_2 Emissions

Luis Fernando Grisales-Noreña [1,*], Oscar Danilo Montoya [2,3] and Alberto-Jesus Perea-Moreno [4,*]

1 Department of Electrical Engineering, Faculty of Engineering, Universidad de Talca, Curicó 3340000, Chile
2 Grupo de Compatibilidad e Interferencia Electromagnética (GCEM), Facultad de Ingeniería, Universidad Distrital Francisco José de Caldas, Bogotá 110231, Colombia
3 Laboratorio Inteligente de Energía, Universidad Tecnológica de Bolívar, Cartagena 131001, Colombia
4 Departamento de Física Aplicada, Radiología y Medicina Física, Universidad de Córdoba, Campus de Rabanales, 14071 Córdoba, Spain
* Correspondence: luis.grisales@utalca.cl (L.F.G.-N.); aperea@uco.es (A.J.P.-M.)

Abstract: This work addressed the problem regarding the optimal integration of battery systems (BS) in grid-connected networks (GCNs) with the purpose of reducing energy losses and CO_2 emissions, for which it formulates a mathematical model that considers the constraints associated with the operation of GCNs in a distributed generation environment that includes BS and variable power generation related to photovoltaic (PV) distributed generation (DG) and demand. As solution strategies, three different master–slave methodologies are employed that are based on sequential programming methods, with the aim to avoid the implementation of commercial software. In the master stage, to solve the problem regarding the location and the type of batteries to be used, parallel-discrete versions of the Montecarlo method (PMC), a genetic algorithm (PDGA), and the search crow algorithm (PDSCA) are employed. In the slave stage, the particle swarm optimization algortihm (PSO) is employed to solve the problem pertaining to the operation of the batteries, using a matrix hourly power flow to assess the impact of each possible solution proposed by the master–slave methodologies on the objective functions and constraints. As a test scenario, a GCN based on the 33-bus test systems is used, which considers the generation, power demand, and CO_2 emissions behavior of the city of Medellín (Colombia). Each algorithm is executed 1000 times, with the aim to evaluate the effectiveness of each solution in terms of its quality, standard deviation, and processing times. The simulation results obtained in this work demostrate that PMC/PSO is the master–slave methodology with the best performance in terms of solution quality, repeatability, and processing time.

Keywords: grid connected network; optimization algorithm; master-slave strategy; parallel processing; photovoltaic generation; battery systems; energy loss; environmental emissions

MSC: 65K05; 90C26; 90C27

Citation: Grisales-Noreña, L.F.; Montoya, O.D.; Perea-Moreno, A.-J. Optimal Integration of Battery Systems in Grid-Connected Networks for Reducing Energy Losses and CO_2 Emissions. *Mathematics* **2023**, *11*, 1604. https://doi.org/10.3390/math11071604

Academic Editor: Jinfeng Liu

Received: 17 February 2023
Revised: 18 March 2023
Accepted: 24 March 2023
Published: 26 March 2023

1. Introduction

1.1. General Context

In the last decades, the continuous growth of the global population and the wide implementation of electrical devices have generated an increased energy demand, which is generally supplied with fossil fuels. This increases CO_2 emissions and the energy losses associated with energy transport, directly affecting the quality of electrical services and the life conditions of GCN users [1]. With the aim to mitigate this issue, in the last years, grid operators and researchers have been given the task of integrating environmentally friendly distributed energy resources, which involves the smart integration and operation of renewable energy resources, battery systems, capacitor banks, and static compensators, among others [2,3]. BS have been the most widely installed and studied distributed energy resources in recent years, as they allow managing the energy of the grid, mitigating the variability of renewable energy resources, and improving the economic, technic and environmental conditions of the network (i.e., the reduction of energy purchasing cos energy losses, and CO_2 emissions, as well as voltage profile improvements, among ot ers) [4]. In turn, GCNs are the most widely studied electrical grids [5], as they are the mo developed networks around the world and the ones with the most technical problem (i.e., energy losses, voltage profile limit violations, and line overloadability, among ot ers) [6]. Furthermore, since they are located in cities and towns, fossil fuel-based generatic contributes to CO_2 emissions in their corresponding regions, affecting the health of the inhabitants. Based on the above, the authors of this work focus on the problem regardi the selection, location, and operation of BS in GCNs, with the aim to reduce energy loss and CO_2 emissions.

1.2. State of the Art

Aiming to take advantage of the different benefits associated with the integration BS in electrical GCNs, many researchers have proposed different strategies for the corre integration of BS in this kind of grid [7,8]. A large portion of these works has been oriente towards developing mathematical models that describe all of the technical and operatin constraints involved in the operation of electrical networks within an environment distributed energy resources (DERs) [9,10]. These mathematical models formulate th objective functions or goals to be achieved through the optimal integration and operatio of BS in GCNs, with two the most studied being the reduction of energy production cos and energy losses [11,12]. In [13,14], a literature review on the objective functions use in the optimal integration of BS in electrical systems was carried out, finding that a larg number of works focus on economic and technical indicators. Therefore, objective functio related to environmental indicators are a topic that requires exploration. By analyzing th different mathematical formulations reported in the literature to represent the proble regarding the optimal operation of BS in GCNs [13], it was possible to notice that the s of constraints that make up the problem under study must include active and reactiv constraints, the power limitations of conventional and renewable generation resource the power limitations and state of charge of the batteries, and the current line and voltag profile limits that represent the operating constraints of the GCN. It is necessary to includ all of these constraints into the mathematical formulation, with the aim to reap the benefi associated with the studied objective functions, while also including the variations in pow demand and renewable generation that are involved in the real behavior of a GCN.

To solve the problem regarding the optimal integration of BS in GCNs while conside ing technical and environmental aspects, many works have been reported in the specialize literature. An example of this is the work reported in [15], whose authors proposed solution methodology for sizing and operating distributed generators and energy storag systems in an electrical system located in India, with the aim to improve its technical cond tions. In this work, the authors considered the variation in power generation and deman as well as all of the constraints that represent the analyzed electrical grid. However, thi work considers a mono-nodal electrical system that ignores the complications related to th transport lines. Furthermore, the authors do not use comparison methods nor analyze th processing times required by the proposed solution. The work by [16] used the artifici electric field algorithm to size and operate distributed generators and BS in AC electric networks, considering the reduction of CO_2 emissions and the improvement of the technic aspects of the grid as objective functions. In this work, the authors compare the best an average solutions obtained by means of the proposed methodology against those of othe works reported in the literature. However, they use a mono-nodal grid as a test system an do not analyze the reported processing times and standard deviation values. In additio this work does not analyze the effectiveness of the studied solution methods regarding pr cessing times and repeatability. The authors of [17] also locate and size BS in a mono-nod grid. In this work, they describe the mathematical formulation for the integration of B and other energy resources that compose the electrical grid, (i.e., distributed generator and loads, among others). This study considers economical and technical indicators as objective functions, comparing the results obtained with those of other works reported in the literature. The main problem with these works is associated with the fact that multinodal GCNs are more widely used in real life, which implies several constraints related to voltage profiles and line currents in their mathematical formulations, thus increasing the complexity of the problem. However, the analyzed works offer important information on the variable power generation and demand of electrical grids, as well as data related to the implementation of smart optimization methods in the problem regarding the optimal operation and location of BS in GCNs.

Regarding the integration and operation of BS while considering multi-nodal grids, different works based on specialized software and sequential programming methods have been proposed in the last years [8], with the latter being the most used for solving the problem regarding the integration of BS in GCN, as this kind of solution methodologies avoid the use of specialized software, which causes an increase in complexity and implementation costs [15]. Furthermore, exact optimization methods such as convex optimization and specialized software are not commonly used, since these solution methodologies are often stuck in local optima. This occurs due to the nonlinearities generated by the discrete variables that represent the selection of the bus where the battery will be located, as well as the kind of battery to be used. Based on the above, a large number of solution methodologies have been reported in the literature which use master–slave strategies based on sequential programming methods. Here, the master stage is entrusted with solving the location and selection problem of the BS to be installed in the GCN, while the slave stage deals with the power operation schedule of the BS, allowing to obtain the best impact on the objective function. These optimization methods use discrete (master stage) and continuous variables (slave stage) and complex solution space exploration processes to find the optimal power configuration for locating and operating batteries in GCNs.

Most of works reported in the literature focus on a single objective of the two analyzed in this research, namely, reducing the energy losses [13]. An example of this is the work presented in [18], where the authors propose a methodology based on genetic algorithms to solve the problem concerning the optimal integration of BS in GCNs. The authors compare their results to those of other methods reported in the literature in terms of the best solution. However, this work does not include or analyze the standard deviation and processing times required by the solutions methods. The authors of [19] use the coalition formation algorithm for solving the problem regarding the optimal integration of BS in GCN. Their results demonstrate the effectiveness of the proposed solution methodology. However, they do not analyze the processing times and the standard deviation values of the solution strategies under study. A methodology based on the coyote optimization algorithm is proposed in [20] with the purpose of reducing the energy losses in an electrical network by integrating BS. To assess the performance of this approach, different works in the literature are used for the sake of comparison. The authors of this work do not evaluate the repeatability and processing times of the solution methods employed. Another proposal for reducing energy losses is made in [21], where the authors use a sensitivity indicator to locate and operate BS in an electrical grid. These methods get stuck in local optima, as heuristic methods cannot escape from bad solution regions. The authors of [22] propose a methodology based on a discrete version of a continuous method for solving the problem regarding the integration of BS in GCNs, aiming to improve the operating conditions of the electrical system. This work highlights the advantage of using modified conventional continuous optimization methods (discrete versions) in order to solve problems involving discrete and binary variables. In the same way, different authors have recently addressed the problem regarding the selection and location of BS in GCNs [23–25].

The reduction of environmental impacts, however, is still a topic in development, with a small number of works reported in the literature. The work reported in [26] aimed to reduce CO_2 emissions and energy power losses by using a non-linear mathematical formulation, which was solved with the GAMS software. Nevertheless, the results obtained were not compared with those of other works reported in the literature. By using specialize software, the authors of [27] obtained the optimal location and operation scheme of BS i an electrical grid, solving a second-order cone programming model with the MATLAB CV tool. This work used different test systems, but comparison methods were not considere In [28], a mixed-integer linear programming model was proposed for representing th problem pertaining to the location and operation of BS in electrical networks, in order reduce investments and CO_2 emissions from fossil fuel-based power generation. This wo considered different kinds of batteries and load curves, with the aim to analyze the effect different devices on the operation of an electrical system. Note that the works reported i the literature for reducing CO_2 emissions with the operation of BS use specialized softwa for solving the proposed mathematical models. This is explained by the fact that th mathematical models used are still being built and validated. Therefore, there is a need t propose mathematical models that guarantee the correct operation of the grid when thi environmental indicator is used, as well as solutions based on sequential programmin methods that avoid the use of specialized software.

By analyzing the state of the art, it was possible to notice that it is currently necessar to propose optimization methodologies that address the problem regarding the optima location and operation of BS in GCNs with the aim to reduce energy losses and CO_2 emi sions. Furthermore, these new methodologies must guarantee the best results in terms c technical and environmental indices, with the aim to obtain resilient strategies that conside the needs of the GCN and the community while avoiding the implementation of specialize software, which increases the costs and complexity of the solution methodologies [29 Furthermore, these methodologies must be compared against other approaches, with th aim to identify the solution methodology with the best performance in terms of solutio quality, repeatability, and processing times.

1.3. Scope and Main Contributions

Recognizing the advantage of discretizing continuous optimization methods for solv ing problems with binary and discrete variables, as well as the current needs to solv the optimal integration problem of BS in GCNs for reducing the energy losses and CO emissions, the authors of this paper propose a complete mathematical formulation c the problem regarding the selection, location, and operation of BS in GCN for reducin energy losses and CO_2 emissions. All this is conducted while including all constraint related to the electrical network (global power balance and line current and voltage profil limits), conventional and distributed generators (power limits), and BS (discharging an charging power limits and state of charge limits). Furthermore, with the aim to use th high-performance methodologies reported in the literature for solving electrical problem similar to that studied herein, three discrete versions of some optimization methods wer used in the master stage. The first of these is the parallel Montecarlo algorithm (PMC) [30 which employs a random search process to find the solution with the best performanc while taking advantage of parallel processing, using all workers in the computer to reduc processing times. Moreover, following the suggestions made in the literature, this pape generated two parallel-discrete versions of two continuous optimization methods: th genetic (PDGA) and crow search (PDCSA) algorithms. For the slave stage, the particl optimization algorithm proposed in [31] was adapted. This algorithm was developed fo operating batteries in direct current (DC) grids, with no application or validation in alte nating current (AC) networks. By combining the three optimization methods proposed i master stage and PSO, it was possible to obtain three new master–slave methodologies fo solving the problem under study, namely, PMC/PSO, PDGA/PSO, and PDCSA/PSO (here inafter called PMC, PDGA, and PDCSA for the sake of simplicity). In addition, with th aim to evaluate the objective function and constraints related to each solution offered b these strategies, aiming for the shortest processing times and the best convergence, thi study used the hourly power flow matrix based on successive approximations, as propose in [23], which allows considering variations in distributed generation and power demanc

As a test scenario, an adapted version of the 33-bus test system was used, which represents the technical and environmental conditions of the city of Medellín (Colombia), while considering the operation of three photovoltaic (PV) generators with maximum power point tracking, which is the traditional way to operate renewable generation technology in this country. Finally, to evaluate the performance of the proposed solution methodologies, 1000 executions of each one were carried out, with the aim to evaluate the minimum and average solution, as well as the standard deviation and average processing times. This analysis allowed selecting the optimization methodology with the best performance for solving the problem regarding the optimal integration and operation of BS in GCNs.

The main contributions of this paper regarding academic and industrial applications are described below:

1.3.1. Academic Contributions

- A mathematical model for the optimal integration of BS in GCN whose objective function is the reduction of energy losses and CO_2 emissions, observing all of the constraints that represent the operation of a GCN in an environment of variable distributed generation and power demand.
- A discrete codification for the problem regarding the location and selection of BS.
- A continuous codification for the problem regarding the operation of the batteries located in the GCN.
- Three new master–slave strategies (PMC, PDGA, and PDCSA) for solving the problem regarding the optimal integration of BS in GCNs.
- The identification of PMC as the master–slave strategy with the best performance in terms of solution quality and its repeatability and processing times for solving the problem under study. This optimization methodology could be used in future works for the sake of comparison, with the aim to obtain methodologies with a better performance.

1.3.2. Industrial Applications

- A mathematical formulation that allows the grid operators to quantify energy losses and CO_2 emissions before and after considering the integration of BS in GCNs.
- An effective and fast optimization method based on sequential programming, which allows determining the location and operation scheme of multiple batteries within the grid, with the purpose of reducing the energy losses and CO_2 emissions while observing all operating constraints.

1.4. Paper Organization

The remainder of this paper is organized as follows. Section 2 describes the mathematical formulation of the problem regarding the optimal selection, location, and operation of BS in GCNs. Section 3 presents the proposed master–slave methodologies. Section 4 describes the GCN used as a test scenario and explains the PV generation and demand curves, as well as the technical and operating parameters of the electrical systems located in Medellín. Finally, Sections 5 and 6, respectively, present the simulation results obtained by the proposed methodologies, as well as the conclusions and future works derived from this research.

2. Mathematical Formulation

In this mathematical formulation, two objective functions are employed which aim for reducing the energy losses and CO_2 emissions in GCNs. Furthermore, this section describes all of the constraints related to the technical limitations of the devices that make up the grid, as well as the operation limits associated with voltage profiles and line currents.

$$FO_1 = \min Eloss = \min\left(\sum_{h \in \Omega_{\mathcal{H}}} \sum_{i \in \Omega_{\mathcal{N}}} \sum_{j \in \Omega_{\mathcal{N}}} Y_{ij} v_{i,h} v_{j,h} \cos\left(\theta_{i,h} - \theta_{j,h} - \varphi_{ij}\right) \Delta t \right) \quad (1)$$

The first objective function used in this paper corresponds to the reduction of th energy losses related to energy transport in the GCN, which is presented in Equation (Here, $\Omega_{\mathcal{H}}$ and $\Omega_{\mathcal{N}}$ represent the periods of time contained in the horizon under stu and the total of nodes that make up the GCN, respectively. Furthermore, Y_{ij} and φ_{ij} a the magnitude and angle of the admittance of the line that interconnects nodes i and respectively. $v_{j,h}$ and $v_{i,h}$ are the voltage profile magnitudes of buses i and j, while $\theta_{i,h}$ ar $\theta_{j,h}$ are their angles, respectively. Finally, Δt is associated with the duration of each peri of time (1 h for this work).

As it could be appreciated in Equation (1), this mathematical formulation does n implicitly include the variables associated with the power supplied by the BS. Howev the effect of the location and operation of these devices is considered in the nodal volta profiles included in the equation.

$$FO_2 = \min CO_2\, emissions = \min\left(\sum_{h \in \Omega_{\mathcal{H}}} \sum_{i \in \Omega_N} X_i^{cg} P_{i,h}^{cg} CE_i^{cg} \Delta t + X_i^{gd} P_i^{gd} C_h^{gd} CE_i^{gd} \Delta t \right) \quad (2)$$

Equation (2) describes the mathematical formulation proposed to represent the secor objective function, i.e., the reduction of CO_2 emissions generated by the power suppli by the conventional and distributed generators located in the grid. In this equation, X and X_i^{gd} are the binary variables, which take a value of 1 if a conventional or distribut generator is located at bus i, respectively; otherwise, they take a value of 0. $P_{i,h}^{cg}$ and $P_{i,h}^{gd}$ a the power supplied by the conventional and distributed generators at bus i in the hour CE_i^{cg} and CE_i^{gd} are the emissions factors for the two generation technologies considere in this work. C_h^{gd} is a variable that represents the behavior of the distributed generat installed at bus i in the hour h. This factor is in p.u. and changes every hour as a functio of the technology used and the potential of the renewable energy in the region whe the DG is located. In this objective function, the variables associated with the problem integrating BS in the GCN are implicit.

The BS optimal integration problem is composed of multiple technical and operatir constraints, which apply for all buses in the GCN and the period considered in the tir horizon analyzed.

$$X_i^{cg} p_{i,h}^{cg} + X_i^{cg} C_h^{gd} p_{i,h}^{gd} \pm X_i^B P_{i,h}^B - p_{i,h}^d = v_{i,h} \sum_{j \in \Omega_{\mathcal{N}}} Y_{ij} v_{j,h} \cos\left(\theta_{i,h} - \theta_{j,h} - \varphi_{ij}\right) \quad (3)$$

The first constraint is associated with the the active power balance in the electric network. In this equation, X_i^B is a binary variable that takes the value of 1 or 0 if a batte is located or not at bus i, respectively, while $p_{i,h}^B$ is the active power supplied or demande by the BS located at bus i in the hour h.

$$q_{i,h}^{cg} - Q_{i,h}^d = v_{i,h} \sum_{j \in \Omega_{\mathcal{N}}} Y_{ij} v_{i,h} \sin\left(\theta_{i,h} - \theta_{j,h} - \varphi_{ij}\right) \quad (4)$$

Equation (4) ensures the reactive power balance in the grid. In this equation, $q_{i,h}^{cg}$ an $Q_{i,h}^d$ are, respectively, the reactive power generated and demanded by the convention generators and loads located at bus i in the hour h. By analyzing this equation, it can b noted that this work does not consider the injection of reactive power by the distributed generator and batteries located in the GCN.

$$P_i^{cg,\min} \leq p_{i,h}^{cg} \leq P_i^{cg,\max} \quad (5)$$

The maximum ($P_i^{cg,\min}$) and minimum ($P_i^{cg,\max}$) power to be supplied by the conventional generator located at bus i are modeled in Equation (5).

$$Q_i^{cg,\min} \leq q_{i,h}^{cg} \leq Q_i^{cg,\max} \quad (6)$$

The reactive power limits associated with the conventional generators are presented in (6), where $Q_i^{cg,\min}$ and $Q_i^{cg,\max}$ denote the minimum and maximum reactive power to be supplied by these generators, respectively.

$$P_i^{gd,\min} \leq p_i^{gd} \leq P_i^{gd,\max} \quad (7)$$

Equation (7) represents the power limits of the distributed generator located at bus i in the hour h. In this equation, $P_i^{gd,\min}$ and $P_i^{gd,\max}$ denote the minimum and maximum power, respectively, which are a function of the technology and renewable potential in the region where the generator is located.

$$P_{B,i}^{charg_{max}} \leq p_{i,h}^{B} \leq P_{B,i}^{disch_{max}} \quad (8)$$

$$P_{B,i}^{disch_{max}} = \frac{C_i^B}{td_i^B} \quad (9)$$

$$P_{B,i}^{charg_{max}} = -\frac{C_i^B}{tc_i^B} \quad (10)$$

The power in the batteries of the electrical system is controlled by Equation (8). The maximum charge and discharge powers are limited by $P_{B,i}^{charg_{max}}$ and $P_{B,i}^{disch_{max}}$. To calculate these values, Equation (9) and (10) are used, where C_i^B is the nominal power capacity of the BS located at bus i, while tc_i and tc_i are the charge and discharge times, respectively, required by the battery type, which is related to the BS technology.

$$SOC_{i,h}^{B} = SOC_{i,h-1}^{B} - \phi_i^B P_{i,h}^{B} \Delta t \quad (11)$$

Equation (11) allows calculating the state of charge at the hour h of the battery located at bus i ($SOC_{i,h}^{B}$). This equation requires the state of charge of the previous hour ($h-1$), the charging coefficient of the battery located at bus i (ϕ_i^B), the power supplied or stored by the same battery at the hour h ($P_{i,h}^{B}$), and its time of charge or discharge, (Δt). To calculate ϕ_i^B, Equation (12) is calculated, which is expressed in terms of the previously described parameters. On the other hand, Equations (13) and (14) define the initial (SOC_i^0) and final (SOC_i^f) state of charge of the battery located at bus i.

$$\phi_i^B = \frac{1}{td_i^B P_{B,i}^{discharg_{max}}} = \frac{1}{tc_i^B P_{B,i}^{charg_{max}}}, \{\forall i \in \Omega_B, \forall h \in \Omega_{\mathcal{H}}\} \quad (12)$$

$$SOC_{i,h=0}^{B} = SOC_i^0, \ \{\forall i \in \Omega_B\} \quad (13)$$

$$SOC_{i,h=24}^{B} = SOC_i^f, \ \{\forall i \in \Omega_B\} \quad (14)$$

Finally, with the aim to integrate the operating constraints of the electrical distribution system, the mathematical formulation includes Equations (15) and (16), which ensure that the voltage profiles and the current that flows through the lines are within the technical limits set by the electrical operator and the manufacturer. In these equations, $V_i^{\min}$ an $V_i^{\min}$ correspond to the minimum and maximum nodal voltage at bus i, respectively, whi $I_{ij,h}$ and $I_{ij}^{\max}$ are the current flowing through the line that interconnects nodes i and respectively, whose maximum level is set during the design of the electrical network.

$$V_i^{\min} \leq v_{i,h} \leq V_i^{\max}, \{ \ \forall i \in \Omega_{\mathcal{N}}, \quad \forall h \in \Omega_{\mathcal{H}} \ \} \quad (15)$$

$$I_{ij,h} \leq I_{ij}^{\max} \{ \ \forall ij \in \Omega_{\mathcal{N}}, \quad \forall h \in \Omega_{\mathcal{H}} \ \} \quad (16)$$

3. Proposed Solution Methodologies

3.1. Master–Slave Methodology and Codifications Used

To solve the problem regarding the optimal selection, location, and operation of BS i GCNs, this paper proposes the master–slave methodology illustrated in Figure 1.

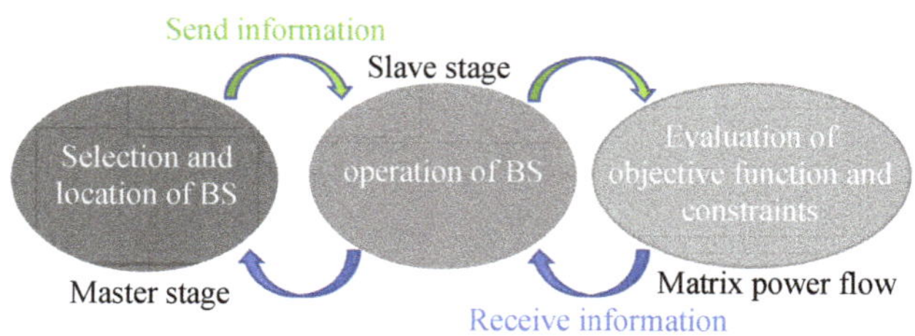

Figure 1. Proposed master–slave methodology.

The master stage is entrusted with the selection and location of the batteries, a discret problem that requires identifying the buses and battery types to be installed in the gri For its codification, a vector of size 1 × 6 was used, where the number of columns corre sponds to the three locations and types of batteries be to installed (Section 4). Figure 2 i presented as an example, where three A-, B-, and C-type batteries were located at buses 3 12, and 3, respectively.

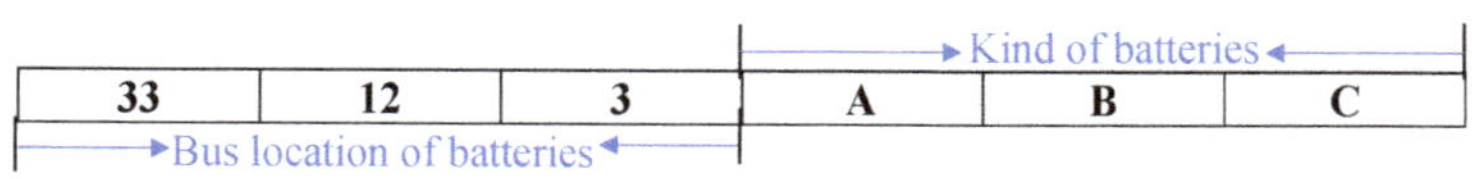

Figure 2. Codification proposed for selecting and locating batteries in the GCN.

The slave stage is responsible for finding a battery power dispatch that allows fo the maximum possible reduction of the objective function, using the location and type c batteries provided by the master stage. To this effect, the codification proposed in Figure is employed. This codification includes a vector of size 1 × 72, in whose columns ar 24 variables associated with the states of charge of each battery in the different periods c the time horizon analyzed.

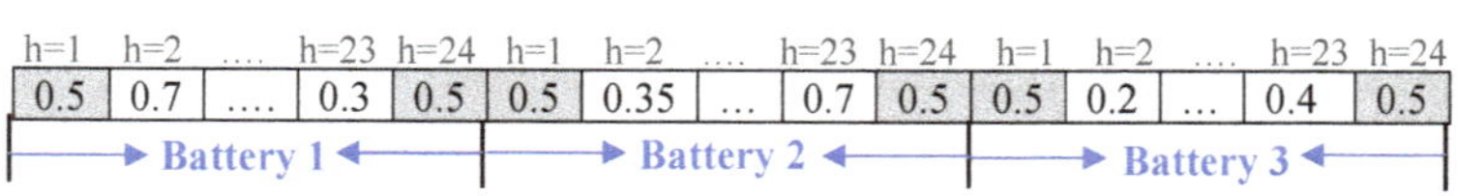

Figure 3. Codification used to find the operation scheme of the batteries selected and located by th master stage.

Due to the nature of electrical systems, in order to evaluate the impact on energy losse and CO_2 emissions, as well as the constraints that make up the problem, it is necessary t determine the power flow for the different periods, with the aim to analyze the effect of th power generated and demanded by the loads, PV generators, and batteries installed in th

GCN. After evaluating each period of time and obtaining the values associated with the objective function and constraints, this information is summarized with the aim to evaluate the effect of the batteries on the grid during an average day of operation. A power flow evaluation within a multi-hour scenario is known as an hourly power flow. In this work, a matrix hourly power flow based on successive approximations (MHPF) was selected, given the excellent results reported in [23]. Algorithm 1 describes this method:

Algorithm 1: Algorithm proposed for the matrix hourly power flow based on successive approximations

```
Data: Load the electrical system data;
Load the BS information provided by the master–slave strategy: types of BS,
  location, and operation;
Load V_dh^t (with t = 0), ϵ, and t_max data;
for t = 0 : t_max do
    Evaluate the MHPF using Equation (17);
    if max(|V_dh^{t+1} − V_dh^t|) ≤ ϵ then
        Solution achieved;
        Result: V_dh = V_dh^{t+1}.
        break;
    else
        V_dh^t = V_dh^{t+1};
Summarize the objective functions obtained in each period of time;
Penalize the objective function if a constraint is violated;
Return the objective function to the slave stage;
```

In the first step, the MHPF loads the electrical system data described in Section 4. Then, the information provided by the master–slave strategy is loaded (i.e., regarding the selection, location, and operation of the batteries during an average day. Next, these data on the BS are integrated into the hourly power flow, and the voltage profiles for all buses in the 24 h of operation are loaded as 1 < 0 by using $\mathbb{V}_{dh}^{t+1}$, which is a matrix of size $|d|x|\mathcal{H}|$, with $|d|$ being the number of demand buses and $|\mathcal{H}|$ the entirety of the time period analyzed (24). Furthermore, to control the iterative process, a maximum number of iterations (t_{max}) for the MHPF of 10,000 and a convergence error (ϵ) of $1e^{-10}$ were set. These values allow for a fast convergence and were heuristically selected.

$$\mathbb{V}_{dh}^{t+1} = -\mathbf{Y}_{dd}^{-1}\left[\left(ones \oslash \mathbb{V}_{dh}^{t,*}\right) \circ \left(\mathbb{S}_{dh} - \mathbb{S}_{pvh}\right)^{*} + \mathbf{Y}_{ds}\mathbb{V}_{sh}\right] \tag{17}$$

After setting the initial parameters of the algorithm, the iterative process to solve the hourly power flow begins. In each iteration, the hourly power flow is simultaneously evaluated in all periods of time by using Equation (17). This is made possible by the fact that this power flow method uses the Hadamard product ($\circ$) and division ($\oslash$). In this equation, $\mathbb{V}_{dh}^{t+1}$ and $\mathbb{V}_{dh}^{t}$ represent the demand bus voltages in all periods considered in the current and previous iteration. These matrices are of size $|d|x|\mathcal{H}|$, where $|d|$ is the number of demand buses and $|\mathcal{H}|$ the total of periods in the time horizon analyzed. Furthermore, in this equation, *ones* is a matrix of ones, and $\mathbb{S}_{dh}$ and $\mathbb{S}_{pvh}$ correspond to a matrix composed of the power demand and the PV power generated in all periods. These matrices have the same size as $\mathbb{V}_{dh}^{t}$. Finally, in this equation, $\mathbf{Y}_{dd}$ and $\mathbf{Y}_{ds}$ denote the components of the admittance matrix generated between the demand (d) and slack buses (s), with $\mathbb{V}_{sh}$ being the voltage in the slack buses, which are composed of (1 < 0) at all times (i.e., the nominal voltage of the GCN). In each iteration of the MHPF, this equation is evaluated by using the hourly voltage profiles of the current and previous iteration.

Then, in order to evaluate the hourly power flow via Equation (17), the MHPF evaluates the stopping criterion (convergence error). To this effect, the current and previous voltage profiles are compared. If ϵ is achieved, the iterative process ends; otherwise, it co tinues. When ϵ or t_{max} is achieved, the objective function values obtained for the differe periods are summarized, obtaining the objective function related to the whole operati day. Furthermore, the constraints associated with each period of time are analyzed. If a of them is violated, a high value is added to the objective function, with the aim to disca the solution from the master–slave strategy. This strategy allows obtaining a solution good quality that satisfies all of the constraints involved in the studied problem. Final the MHPF returns the objective function to the slave stage, with the aim to continue wi the iterative process of the master–slave strategy proposed in this work.

3.2. Master Stage

This work used three discrete optimization methods for the master stage: a parall version of the Montecarlo method (PMC) and two parallel-discrete versions of tradition continuous optimization methods, i.e., the genetic and crow search algorithms. The sele tion of these methods was based on the excellent results reported in the literature wi regard to the solution of similar electrical engineering problems [18,29,32,33]. This subse tion outlines the iterative process of each of these algorithms. For a whole description each algorithm, please refer to the cited papers.

3.2.1. Parallel Montecarlo Method (PMC)

The PMC is a random optimization method that evaluates multiple randomly pr posed scenarios, thus allowing for a good-quality solution in a previously defined numb of iterations. In each iteration, the PMC generates a population that contains differe individuals, each of which represents a possible solution to the problem. After evaluati the objective function of each individual and confirming that it satisfies all constrain the individual with the best solution (incumbent) is included in an elite list. This, in ord to select the best solution from this list at the end of the iterative process. The rando exploration of the algorithm allows covering the solution space in reduced processing tim with low standard deviation values. The original PMC was proposed in [30]. Howev with the aim to obtain the best performance, this paper used a PSO to tune the PM parameters, as per the suggestions made by [29], obtaining 10 as the maximum number iterations ($iter_{max}$). In each iteration, a population of 8 individuals was used, as this is th maximum number of workers in the workstation used. This limitation is explained by th fact that, in parallel processing, a population size higher than the number of workers is n recommended, as this does not ensure a single parallel process. This limitation applies all solution algorithms used in this work. The iterative process of the PMC is presented Algorithm 2 and described below.

Algorithm 2: Algorithm proposed for the PMC

```
Data: Read PMC parameters
for t = 1 : iter_max do
    Randomly generate the first population;
    Evaluate the objective function of the population by using the slave stage
      (parallel process);
    Include the best solution of the population in the elite list;
Select the best solution of the elite list as the solution to the problem;
Print the solution;
```

The PMC employs an iterative process that generates a random population by usir the discrete codification proposed in Figure 2. After that, using the location and batte type proposed by each individual, the slave stage is used to evaluate the objective functic This stage uses PSO and the MHPF to find the power schedule (supply or storage) of th batteries in the GCN. This also includes the aim to obtain the minimum possible objectiv function (energy losses or CO_2 emissions), satisfying the set of constraints that compo the problem under study. Therefore, each individual of the population must be processed by the slave stage, which implies long processing times. This is addressed by using parallel processing, with the aim to evaluate multiple individuals at the same time. After evaluating the objective function values of the population, the best solution is included in the elite list, a process that is repeated iteration to iteration until the maximum number of iterations is reached. When this occurs, the best solution of the elite list is selected as the solution to the problem. This solution contains the location, selection, and operation of the batteries for a day which yields the lowest objective function value.

3.2.2. Parallel-Discrete Genetic Algorithm (PDGA)

The parallel-discrete version of the genetic algorithm performs the selection, recombination, and mutation steps of traditional genetic algorithms (GA). However, this discrete version of the GA uses the codification proposed in Figure 2 and adapts the recombination and mutation steps to work with discrete variables, but the nature of the process is the same [23]. Algorithm 3 describes the iterative process of the PDGA.

Algorithm 3: Iterative process of the PDGA

```
Data: Read PDGA parameters
for t = 1 : iter_max do
    if iter == 1 then
        Randomly generate the first population;
        Evaluate the objective function of the population by using the slave stage
          (parallel process);
        Select the best solution as the incumbent;
    else
        Update the population by performing selection, recombination,
          and mutation;
        Evaluate the objective function of the population by using the slave stage
          (parallel process);
        Update the best solution;
        if Has the stopping criterion been met? then
            End the iterative process and select the incumbent as the solution to
              the problem;
            Break;
        else
            Continue;
```

The PDGA starts by reading all of its parameters. In the particular case of this work, a population size of 8 individuals was used, as well as a recombination of 1 point and the mutation of 1 individual, and, as a stopping criterion, an $iter_{max}$ of 1000 was employed. The tuning process of this algorithm was carried out with the same PSO used for PMC. This algorithm generates the first population by using the codification in Figure 2 and a random process. Subsequently, the objective function of each individual is evaluated by using the slave stage and the MHPF. With this information, the individual with the lowest objective function value is selected as the incumbent of the problem.

From the second iteration until the iterative process ends, the PDGA updates the population via selection, recombination, and mutation. Then, the objective function and constraints of each individual are calculated. Based on this information, the incumbent of the problem is updated (the best solution). After that, the stopping criterion ($iter_{max}$) is evaluated. If it is met, the iterative process ends, and the incumbent is selected as the solution to the problem; otherwise, the algorithm continues. It is important to highlight that, as in the PMC, the incumbent contains the location, types, and operation scheme of the batteries that allow obtaining the lowest possible objective functions.

3.2.3. Parallel-Discrete Crow Search Algorithm (PDCSA)

The parallel-discrete version of the crow search algorithm uses the hunting strategies of crows and takes advantage of parallel processing to evaluate the objective function in each iteration of the algorithm. The iterative process of the PDCSA is presented in Algorithm 4 and described below.

Algorithm 4: Iterative process of the PDCSA

Data: Read PDCSA parameters
for $t = 1 : iter_{max}$ **do**
  if *iter* $==$ 1 **then**
    Randomly generate the first population of crows;
    Evaluate the objective function of the population by using the slave stage (parallel process);
    Store all individuals in the population (current position of the crows);
    Select the crow with the best solution as the incumbent;
  else
    Update the population by using the information of the incumbent, the population, and random values;
    Evaluate the objective function of the population by using the slave stage (parallel process);
    Store all individuals in the population (current position of the crows);
    Update the incumbent;
    if *Has the stopping criterion been met?* **then**
      End the iterative process and print the incumbent as the solution;
      Break;
    else
      Continue;

The conventional CSA works with a population of crows that, iteration to iteration, take the decision to follow the leader or go their own way [34,35]. In the first iteration, the PDCSA reads the parameters and randomly generates the initial population by using the discrete codification proposed in Figure 2. Then, as with the PMC and the PDGA, this optimization algorithm evaluates the objective function of the population via the slave stage and a parallel process. Subsequently, the information of all individuals is stored and the crow with the best solution is selected as the incumbent, i.e., the leader.

From the second iteration until the iterative process ends, the position of all crows is updated. In other words, the information of the population is renewed. To this effect, each individual decides to follow the leader or take a different path. This is made possible by using a random value. In this case, if the random value is higher than 0.5, the crow follows the leader; otherwise, its position is updated by using random values between the maximum and minimum ones allowed (number of buses and battery types). After updating the position of the crows, these values are stored and the incumbent is updated. Subsequently, it is verified whether the maximum number of iterations has been met. If this is true, the iterative process ends, and the incumbent is printed as the solution; otherwise, the algorithm continues.

As with the other solution methodologies employed in the master stage, the PDCSA was tuned via the PSO suggested in [29]. Thus, a population size of 8 individuals and a maximum number of iterations of 1000 were obtained.

3.3. *Slave Stage*

For solving the problem regarding the optimal operation of BS in GCN, this paper uses the PSO proposed in [31], given the excellent results reported by the authors. Furthermore, as the Montercalo method and the genetic and crow search algorithms have been traditionally used the in the literature for solving continuous problems, as is the case of the one studied herein, these optimization methods were validated in this research, with a low performance in comparison with PSO. Additionally, to obtain these results, it is necessary to present a lot of information that does not contribute to the state of the art. For this reason, only PSO was used in the slave stage, which is presented in Algorithm 5 and described below.

Algorithm 5: Iterative process proposed for the PSO used in the slave stage

```
Data: Read PSO parameters
for t = 1 : iter_max do
    if iter == 1 then
        Randomly generate the position of the particles (initial population);
        Evaluate the objective function of all particles by using the MHPF;
        Select the initial position of the particles as the best solution and store all of
          the objective functions obtained;
        Select the particle with the best objective function value and its position as
          the incumbent;
    else
        Update the position of the swarm by using the information of the
          particles and the incumbent;
        Evaluate the objective function of all particles by using the MHPF;
        Update the best particle position and its objective function;
        Select the particle with the best objective function and its position as the
          incumbent;
        if Has the stopping criterion been met? then
            End the iterative process and print the incumbent as the solution;
            Break;
        else
            Continue;
```

The PSO used to solve the problem under study starts by reading the optimizer parameters. Using the same methodology as the master stage, the following values were obtained: a population of 60 individuals, a maximum number of iterations of 971, a cognitive constant of 1.5922, a social constant of 2, and an initial and final inertia of 0.0022 and 0.0477, respectively. Note that the PSO does not consider the population size as the maximum number of workers; it does not employ parallel processing.

In order to read the initial parameters, the PSO generates the initial population by randomly spreading the particles throughout the solution space. This step is carried out by using the codification presented in Figure 3, which assigns the state of charge for each period of operation for the three batteries located by the master stage. After that, the objective functions of all particles are evaluated by using the MHPF and observing all constraints. If any constraint is violated, the objective function is penalized with a value of 100,000. This value was heuristically obtained for both objective functions under study. With the values of the objective functions, the first iteration selects the initial position of the particles as the best solution and stores all of these values as the best ones found by each particle. Furthermore, it selects the particle with the best objective function as the incumbent, storing its location and objective function value.

From the second iteration until the end of the iterative process, the location of the particle swarms is updated by using the information on the best particle position and the incumbent. Then, the objective function of the swarms is calculated by analyzing the constraints in order to penalize any solution that violates the technical and operating limits. With the objective function values, the best position and objective function of the particles are updated, as well as the incumbent of the problem. Subsequently, it is verified whether the stopping criterion has been met. If this is true, the iterative process ends, and the incumbent returns to the master stage with the power dispatch of the BS located in t GCN; otherwise, the iterative process continues until the maximum number of iterations reached, sending the information of the last incumbent to the master stage.

4. Test Scenarios and Considerations

Figure 4 presents the electrical diagram of the GCN used in this work.

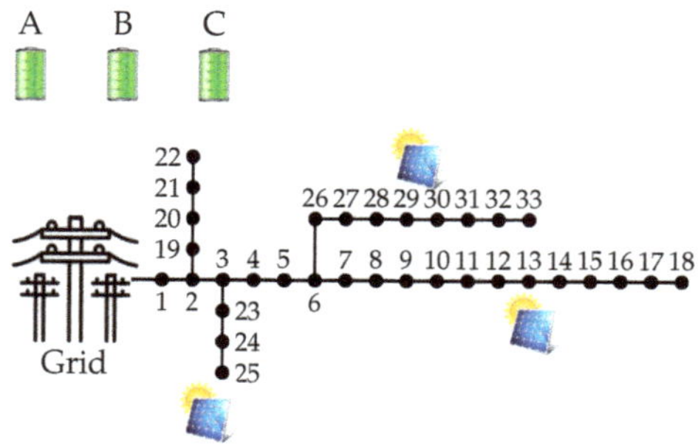

Figure 4. Electrical diagram of the GCN.

Table 1 describes the electrical parameters of the test system regarding line and pow demand capacity. This table presents, from left to right, the line number l, the sending b i, the receiving bus j, the resistance and reactance of the line that interconnects buses i ar j, the active and nominal power demanded at bus j, and the maximum current allowe by each line considered. Furthermore, for the voltage profile limits, this paper follows tl Colombian electrical regulations for electrical distribution networks, which stipulates bus voltage variation of $+/-$ 10% of the main generator's nominal voltage [36]. In th particular case, the nominal and base voltage is 12.66 kV, with a base power of 100 kW.

In this figure, it can be noted that this work proposes a modified version of tl 33-bus test system, which is highly used in the literature to validate planing and operatic strategies in electrical distribution networks [37–39]. The test scenario employed conside the power energy solar production, power demand, and CO_2 emissions from convention generators (electrical grid) of the city of Medellín (Colombia), as well as PV-DGs and thr different kinds of lithium-ion batteries (types A, B, and C), with different power capaciti and charge and discharge times [27]. Lithium-ion batteries are a type of rechargeab battery which uses the reversible reduction of lithium ions to store energy. They are high used in the literature because they have a higher energy density, a higher efficiency, and longer useful life. Traditional lead acid batteries allow 1500 life-cycles, while lithium batte technology offers a duration of up to 2500 [40].

In the test system used, the PV-DGs were located at buses 13, 25, and 36, with nomin power capacities of 1.125, 1.320, and 0.999 MW, respectively [29]. The behavior of the sol energy production and power demand of Medellín was taken from [23]. In the particul case of PV generation, the temperature and solar radiation data reported by NASA [4 were used, as well as technical data on the polycrystalline PV panels, in order obtain a cur that represents the average behavior of the solar production in an average operation da (see Figure 5a). Furthermore, this study obtained a power demand curve that represen the average behavior of the users in Medellín by using power demand data reported by tl local operator, Empresas Públicas de Medellín, [42] (see Figure 5b). All of the data used elaborate these curves correspond to 2019. This year was selected with the aim to elimina the effect of the COVID-19 pandemic on power consumption.

Table 1. Technical parameters of the 33-node test system (urban network).

Line l	Node i	Node j	R_{ij} (Ω)	X_{ij} (Ω)	P_j (kW)	Q_j (kVAr)	I_{ij}^{max} (A)
1	1	2	0.0922	0.0477	100	60	385
2	2	3	0.4930	0.2511	90	40	355
3	3	4	0.3660	0.1864	120	80	240
4	4	5	0.3811	0.1941	60	30	240
5	5	6	0.8190	0.7070	60	20	240
6	6	7	0.1872	0.6188	200	100	110
7	7	8	1.7114	1.2351	200	100	85
8	8	9	1.0300	0.7400	60	20	70
9	9	10	1.0400	0.7400	60	20	70
10	10	11	0.1966	0.0650	45	30	55
11	11	12	0.3744	0.1238	60	35	55
12	12	13	1.4680	1.1550	60	35	55
13	13	14	0.5416	0.7129	120	80	40
14	14	15	0.5910	0.5260	60	10	25
15	15	16	0.7463	0.5450	60	20	20
16	16	17	1.2890	1.7210	60	20	20
17	17	18	0.7320	0.5740	90	40	20
18	2	19	0.1640	0.1565	90	40	40
19	19	20	1.5042	1.3554	90	40	25
20	20	21	0.4095	0.4784	90	40	20
21	21	22	0.7089	0.9373	90	40	20
22	3	23	0.4512	0.3083	90	50	85
23	23	24	0.8980	0.7091	420	200	85
24	24	25	0.8960	0.7011	420	200	40
25	6	26	0.2030	0.1034	60	25	125
26	26	27	0.2842	0.1447	60	25	110
27	27	28	1.0590	0.9337	60	20	110
28	28	29	0.8042	0.7006	120	70	110
29	29	30	0.5075	0.2585	200	600	95
30	30	31	0.9744	0.9630	150	70	55
31	31	32	0.3105	0.3619	210	100	30
32	32	33	0.3410	0.5302	60	40	20

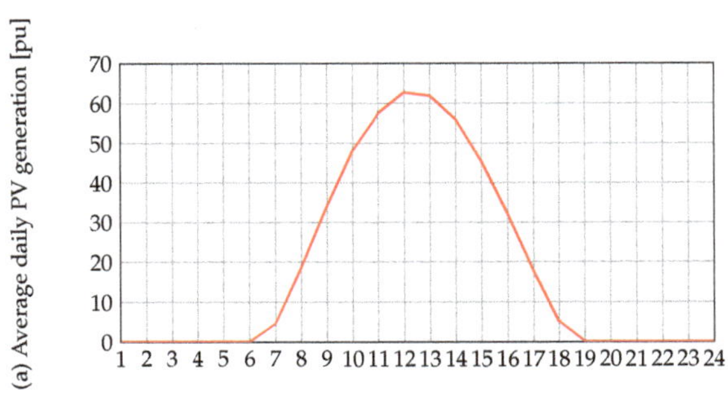

Figure 5. *Cont.*

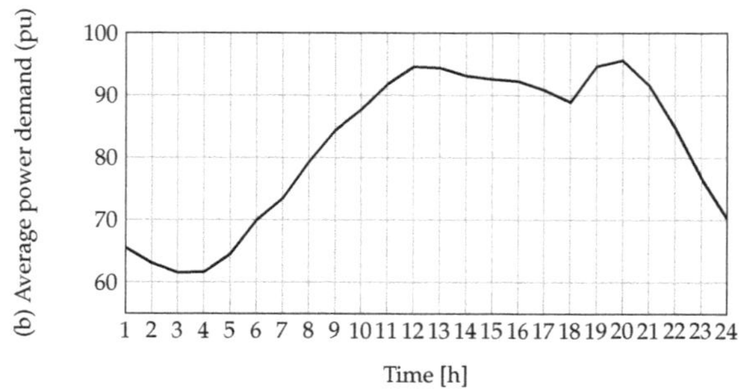

Figure 5. Average daily PV power generation power demand in the city of Medellín (Colombia).

This work considered three kinds of lithium-ion batteries, denoted with types A, and C. Table 2 describes the technical parameters of the BS employed. It presents, from le to right, the type of BSS, the nominal capacity in kWh, and the charge and discharge tim in hours. With these values and the equations presented in Section 2 of this manuscript, is possible to obtain all parameters for the operation of the batteries [18]. As the maximu and minimum SOC for these batteries, the following limits were set for the lithium-io batteries [17]: 0.1 (10%) and 0.9 (90%), respectively. Finally, to obtain the best performanc out of the BS, an initial and final state of charge of 0.5 (50%) was used, following th suggestions made in [31].

Table 2. Parameters of the batteries.

Type	Capacity (kWh)	Charge Time (h)	Discharge Time (h)
A	1000	4	4
B	1500	4	4
C	2000	5	5

Finally, in order to calculate the CO_2 emissions associated with the generators locate in the grid, this work considered 0.1644 kg of CO_2 per kWh as the emissions factor for th conventional generators, as well as a value of 0 kg of CO_2 per kWh for the PV-DGs, as thi kind of generator does not emit greenhouse gases or release carbon-based pollutants whe producing energy [43]. The authors of this paper acknowledge the environmental impac of constructing PV modules, just as well as the fact that this technology does not affec environmental conditions when used for generating energy.

5. Simulation Results

This section presents the simulation results obtained after evaluating the master–slav methodologies proposed for solving the problem regarding the optimal integration of B in GCNs with the aim to reduce energy losses and CO_2 emissions. All simulations wer carried out in the Matlab 2023 software, on a Dell Workstation with an Intel(R) Xeon(R E5-1660 v3 3.0 GHz processor, 16 GB DDR4 RAM, and a 480 GB 2.5″ solid state hard driv with 8 workers running on Windows 11 Pro. All simulations were executed 1000 time in order to evaluate performance in terms of the average solution and processing time as well as regarding the standard deviation.

Table 3 presents the minimum solutions (i.e., the highest reduction in the objectiv function) and the average reductions achieved by the three different master–slave metho ologies, as well as the standard deviation and the average processing times.

Table 3. Simulation results obtained by the proposed master–slave methodologies

	Minimum Solution		Average Solution	
Method	Eloss (kWh)	Emissions (Ton CO_2)	Eloss (kWh)	Emissions (Ton CO_2)
PMC	2350.8270	9864.5471	2358.9454	9866.72854
PDGA	2336.0684	9862.8580	2347.9000	9865.1420
PDCA	2354.5459	9864.7264	2367.0639	9867.1920
	Standard deviation (%)		**Processing time (s)**	
Method	Eloss	Emissions	Eloss	Emissions
PMC	0.2391	0.0112	75.0580	75.4806
PDGA	0.3829	0.0147	7477.4304	7338.0373
PDCA	0.4592	0.0143	6792.9205	7440.9693

To analyze the impact of the master–slave strategies on the GNC, the energy losses and CO_2 emissions were analyzed without considering the BS installed in the grid. Thus, the base scenario involved variable power demand and the PV distributed generators operating in maximum power point tracking (MPPT) mode (Figure 5). This scenario obtained values of 2484.5746 kWh for energy losses and 9887.4082 kg of CO_2 (9.88 Ton) for the CO_2 emissions.

Figure 6a compares these values against those of the PMC, PDGA, and PDCSA. This figure presents the minimum and average reductions obtained by the solution methodologies for both objective functions with regard to the base case. Note that all solution methods reduce the objective functions. In the particular case of *Eloss*, a minimum reduction of 130.0287 kWh was obtained, while the average reduction was 117.5107 kWh (5.2334% and 4.72961%, respectively). These reductions are significant for the GCN; in order to highlight their importance, note that they imply a reduction of 42.8914 MWh for a year of operation.

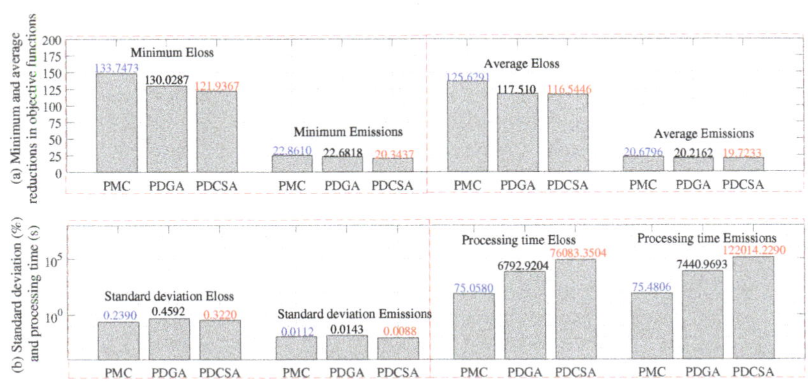

Figure 6. Reductions obtained by the proposed master–slave methodologies regarding the base case: (**a**) minimum and average reductions of the objective function and (**b**) standard deviation (percentage) and processing times.

The obtained emissions reductions are presented in Figure 6a. In the particular case of the minimum emissions, the master–slave strategies obtained an average a value of 22.6818 kg of CO_2. The average reduction in this environmental index (after 100 executions) was 20.2162 kg of CO_2. With respect to the base case, these values correspond to reductions of 0.2294% and 0.2044%, respectively. As in the case of the *Eloss*, considering a year of operation, the optimization methods would achieve a total reduction of 7.37 Ton of CO_2

on average, thus demonstrating the environmental importance and effectiveness of t integration, selection, and smart operation of BS in GCNs.

Finally, Figure 6b presents the standard deviation and the processing times requir by the solution methodologies. In terms of the former, average values of 0.3220% a 0.0112% were obtained for *Eloss* and *Emissions*, respectively. These values demonstrate t repeatability of the methodologies under study. In terms of the processing times, avera times of 6792.9204 (*Eloss*) and 7440.969347 s (*Emissions*) were obtained. These processi times are short given the complexity of the problem and its large solution space, and the values show the importance of the matrix hourly power flow used for calculating t objective functions and constraints in all evaluated scenarios.

Figure 6 highlights, in blue and red, the methods with the best and worst performan respectively. By analyzing this figure, it is possible to appreciate that, in all indicato the PMC achieved the best results, which makes it the best solution method among tho analyzed in this study. Figure 7 illustrates the improvements obtained by the PMC wh compared to the other solution methods.

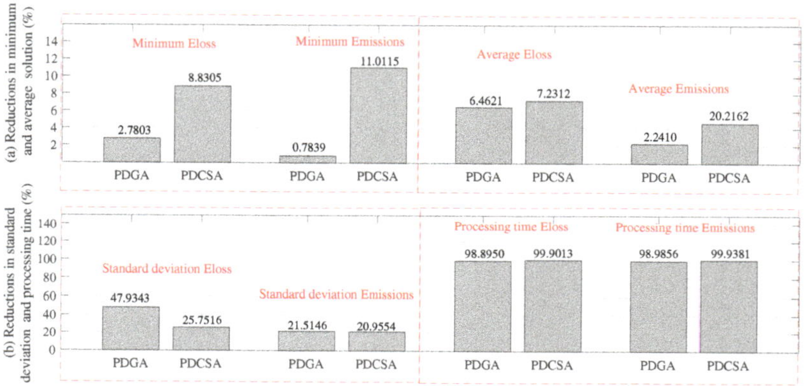

Figure 7. Percent reductions obtained by the PMC with regard to the other comparison metho (**a**) in the minimum and average objective function values, (**b**) in the standard deviation and proce ing times.

Figure 7a presents the reductions obtained by the PMC with regard to the minimu objective function values, i.e., 5.8054% and 5.8977% when compared to the other soluti methodologies. Furthermore, the PMC achieved reductions of 6.8467% and 3.4328% the average *Eloss* and *Emissions*, respectively. By analyzing the standard deviation, it possible to calculate average reductions of 36.8430% and 21.2350% in *Eloss* and *Emissio* Finally, the PMC is the fastest solution method, as its processing times for calculating t *Eloss* and *Emissions* were reduced by 99.3982% and 99.4618%, respectively, thus demo strating the superiority of the PMC with respect to the PDGA and the PDCSA.

Finally, with the purpose of demonstrating that the PMC satisfies all technical ar operating limits set for the GCN located in Medellín, Figures 8–10 are presented. It important to highlight that all master–slave strategies in this paper satisfy the technic and operating constraints associated with a GCN in an environment of DERs. Howev this article only describes and analyzes the technical and operating behavior of the PM as explaining the performance of the other methods would require a lot of unnecessa information. Furthermore, in future works, comparisons should only be made with th most efficient method, which is the PMC.

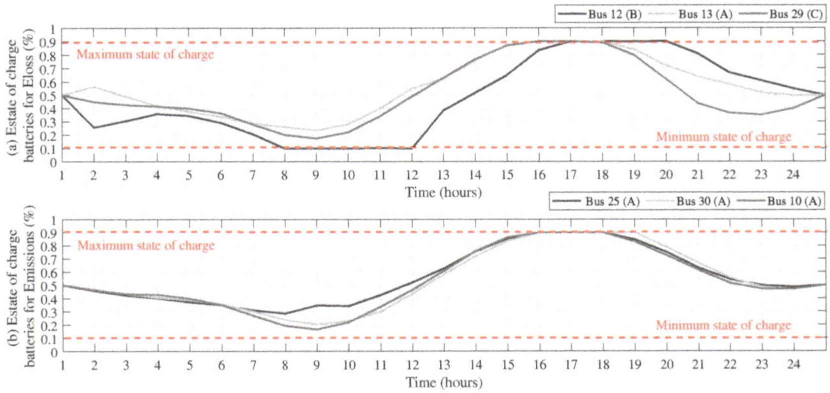

Figure 8. State of charge set by the PMC: reductions in (**a**) energy losses and (**b**) CO_2 emissions.

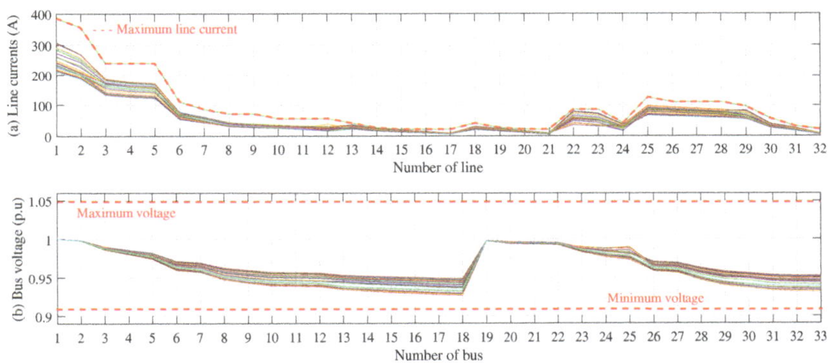

Figure 9. Values obtained by the PMC regarding energy loss reductions: (**a**) line current (A), (**b**) bus voltage (p.u.).

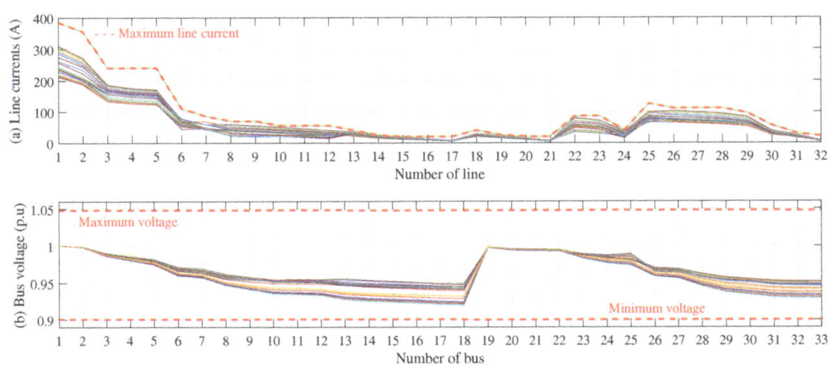

Figure 10. Values obtained by the PMC regarding CO_2 emissions reductions: (**a**) line current (A), (**b**) bus voltage (p.u.).

Figure 8 describes the dynamics of the state of charge of the three BS integrated into the GCN, considering an average day of power demand and PV generation in Medellín. It is important to highlight that, following the suggestions made in [31] for obtaining the best performance of the batteries, all BS start and finish with 0.5 (50%) of the SOC. Figure 8a illustrates that, for *Eloss*, BS were installed at buses 12, 13, and 29 (types B, A, and C, respectively). Note that the behavior of the power supplied is similar for all three batterie with low dynamics in the first hours, complete charging before hour 17—when the pow demand of the system increases—and maximum demand on hour 20. The battery installe at bus 12 supplies energy until it achieves the final state of charge (50%), the battery at bu 13 achieves the final SOC on hour 24, and the one at bus 29 supplies energy until hour 2 starting its charging process until the last hour of the horizon, when it achieves the fin SOC. Note that, according to this figure, all BS satisfy their maximum and minimum SO of lithium-ion batteries: 0.1 (10%) and 0.9 (90%), respectively.

The operation of the BS regarding the reduction of CO_2 emissions is illustrated i Figure 8b. In this case, the BS were located at buses 25, 30, and 10 (all of them type A). Th batteries follow the same dynamics: they start at 50% SOC, discharging all batteries unt hour 9. They start the charging process from this hour until hour 16, and they discharg until hour 24, achieving the final SOC (50%). The batteries satisfy the state-of-charge limi at all times.

Finally, Figures 9 and 10 present the line currents and voltage profiles for the differe operation hours. By analyzing the current limits, it is possible to note that the maximu line current limits are satisfied at all times. In the same way, all voltage profiles are withi the voltage limits set for the GCN (+/− 10% of the nominal voltage: 1 p.u.).

The above demonstrates that the solution obtained by the PMC with regard to th objective functions satisfies all operating and technical constraints of the mathematic model for the problem studied herein.

6. Conclusions and Future Work

This work formulated the problem regarding the optimal integration and operatio of BS in GCN in order to reduce energy losses and CO_2 emissions. As solution method three different master–slave methodologies were proposed. In the master stage, the PM PDGA, and PDCSA were employed for selecting and locating three different BS types in GCN. Furthermore, the slave stage used PSO for the operation of the batteries, as well as matrix hourly power flow to calculate the objective functions and evaluate the technic and operating constraints involved in the mathematical formulation. Finally, with th aim to identify the solution methodology with the best performance, each algorithm wa executed 1000 times, analyzing the best and average solutions, the standard deviatio and the processing times. The 33-bus test system was used for validation, which wa adapted to represent the power demand and PV power generation of the city of Medellí (Colombia). This city constitutes an excellent test scenario, given its high energy losse and CO_2 emissions levels, as well as its excellent conditions for PV generation (this kind c renewable energy is widely used in the city). In this paper, the PV-DGs were considere to operate in maximum power point tracking mode, with the aim to make the best out c this resource.

All methods achieved excellent results in terms of solution quality and processin times. The master–slave strategies obtained average reductions of 4.72% and 0.20% regar ing energy losses and CO_2 emissions for an average operation day, respectively. Thes reductions are equivalent to 42.89 MWh and 7.37 Ton of CO_2 in a year of operation. Thes values are significant for the operation of grid-connected electrical distribution system as they imply commercial and environmental benefits. In addition, the proposed solutio methodologies reported a low standard deviation, with average values of 0.3220% an 0.01124% for energy losses and CO_2 emissions, respectively. Moreover, in a problem a demanding as the integration of BS in GCN, the implementation of a matrix hourly powe flow based on successive approximations allowed reducing the processing times by abou 68%, with values of 6792.92 and 7440.969347 s regarding energy losses and CO_2 emission With this information, it can be concluded that these strategies allow solving the problen regarding the selection, location, and operation of multiple BS in a GCN in about 2 h which allows electrical operators to evaluate multiple generation and demand scenario as well as different electrical systems, in reduced times, which is very important for public bidding processes.

The above demonstrates that all master–slave strategies are suitable for solving the problem under study. However, the PMC was the best solution methodology in terms of solution quality, repeatability, and processing time, for which it obtained average reductions of 5.13%, 29.03%, and 99.42%, respectively.

The main limitation associated with the proposed methodology corresponds to the implementation of single-objective optimization algorithms, which is why a multi-objective analysis is not possible. However, the proposed methodologies obtained the best results regarding the reduction of energy losses and CO_2 emissions.

Future work could consider the implementation of new optimization methods that allow improving the results reported in this paper. Furthermore, it is possible to include variations in the power supplied by the PV generators, with the aim to achieve the best solution quality. This, while allowing for the relocation of PV generators in the GCN. In addition, other kinds of distributed energy resources could be included, such as capacitors and reactive static compensators, among others, with the aim to increase the reductions in energy losses and CO_2 emissions. Finally, the mathematical formulation could include economical indicators with regard to the cost of the BS, by using multi-objective functions that consider the improvement of technical, economical, and environmental indicators.

Author Contributions: Conceptualization, methodology, software, and writing (review and editing), L.F.G.-N., O.D.M. and A.-J.P.-M. All authors read and agreed to the published version of the manuscript.

Funding: This research received no external funding.

Acknowledgments: This research was supported by University of Talca (Chile), Universidad Distrital Francisco José de Caldas (Colombia), and Universidad de Córdoba (España).

Conflicts of Interest: The authors declare no conflict of interest.

Abbreviations

BS	Battery systems.
GCN	Grid-connected network.
PV	Photovoltaic.
DG	Distributed generator.
PMC	Parallel-discrete version of the Montecarlo method.
PDGA	Parallel-discrete version of the genetic algorithm.
PDSCA	Parallel-discrete version of the search crow algorihm.
PSO	Particle swarm optimization algorithm.
CO_2	Carbon dioxide.
GAMS	General Algebraic Modeling System.

eferences

Kumar, C.M.S.; Singh, S.; Gupta, M.K.; Nimdeo, Y.M.; Raushan, R.; Deorankar, A.V.; Kumar, T.A.; Rout, P.K.; Chanotiya, C.; Pakhale, V.D.; et al. Solar energy: A promising renewable source for meeting energy demand in Indian agriculture applications. *Sustain. Energy Technol. Assessments* **2023**, *55*, 102905. [CrossRef]

Ahmadipour, M.; Othman, M.M.; Salam, Z.; Alrifaey, M.; Ridha, H.M.; Veerasamy, V. Optimal load shedding scheme using grasshopper optimization algorithm for islanded power system with distributed energy resources. *Ain Shams Eng. J.* **2023**, *14*, 101835. [CrossRef]

Henrique, L.F.; Silva, W.N.; Silva, C.C.; Dias, B.H.; Oliveira, L.W.; de Almeida, M.C. Optimal siting and sizing of distributed energy resources in a Smart Campus. *Electr. Power Syst. Res.* **2023**, *217*, 109095. [CrossRef]

Gómez, J.C.; Morcos, M.M. Impact of EV battery chargers on the power quality of distribution systems. *IEEE Trans. Power Deliv.* **2003**, *18*, 975–981. [CrossRef]

Qian, K.; Zhou, C.; Allan, M.; Yuan, Y. Modeling of load demand due to EV battery charging in distribution systems. *IEEE Trans. Power Syst.* **2010**, *26*, 802–810. [CrossRef]

6. Resch, M.; Buehler, J.; Klausen, M.; Sumper, A. Impact of operation strategies of large scale battery systems on distribution g planning in Germany. *Renew. Sustain. Energy Rev.* **2017**, *74*, 1042–1063. [CrossRef]
7. Zichen, W.; Changqing, D. A comprehensive review on thermal management systems for power lithium-ion batteries. *Ren Sustain. Energy Rev.* **2021**, *139*, 110685. [CrossRef]
8. Twaha, S.; Ramli, M.A. A review of optimization approaches for hybrid distributed energy generation systems: Off-grid a grid-connected systems. *Sustain. Cities Soc.* **2018**, *41*, 320–331. [CrossRef]
9. Ramadesigan, V.; Northrop, P.W.; De, S.; Santhanagopalan, S.; Braatz, R.D.; Subramanian, V.R. Modeling and simulation lithium-ion batteries from a systems engineering perspective. *J. Electrochem. Soc.* **2012**, *159*, R31. [CrossRef]
10. Allison, J. Robust multi-objective control of hybrid renewable microgeneration systems with energy storage. *Appl. Therm. E* **2017**, *114*, 1498–1506. [CrossRef]
11. Yang, Y.; Bremner, S.; Menictas, C.; Kay, M. Battery energy storage system size determination in renewable energy systems: review. *Renew. Sustain. Energy Rev.* **2018**, *91*, 109–125. [CrossRef]
12. Sharma, P.; Naidu, R.C. Optimization techniques for grid-connected pv with retired ev batteries in centralized charging stati with challenges and future possibilities: A review. *Ain Shams Eng. J.* **2022**, 101985. [CrossRef]
13. Grisales-Noreña, L.F.; Restrepo-Cuestas, B.J.; Cortés-Caicedo, B.; Montano, J.; Rosales-Muñoz, A.A.; Rivera, M. Optimal Locati and Sizing of Distributed Generators and Energy Storage Systems in Microgrids: A Review. *Energies* **2023**, *16*, 106. [CrossRef
14. Khezri, R.; Mahmoudi, A.; Aki, H. Optimal planning of solar photovoltaic and battery storage systems for grid-connect residential sector: Review, challenges and new perspectives. *Renew. Sustain. Energy Rev.* **2022**, *153*, 111763. [CrossRef]
15. Kumar, N.M.; Chopra, S.S.; Chand, A.A.; Elavarasan, R.M.; Shafiullah, G. Hybrid renewable energy microgrid for a resident community: A techno-economic and environmental perspective in the context of the SDG7. *Sustainability* **2020**, *12*, 39 [CrossRef]
16. Naderipour, A.; Kamyab, H.; Klemeš, J.J.; Ebrahimi, R.; Chelliapan, S.; Nowdeh, S.A.; Abdullah, A.; Marzbali, M.H. Optim design of hybrid grid-connected photovoltaic/wind/battery sustainable energy system improving reliability, cost and emissi *Energy* **2022**, *257*, 124679. [CrossRef]
17. Wang, P.; Wang, W.; Xu, D. Optimal sizing of distributed generations in DC microgrids with comprehensive consideration system operation modes and operation targets. *IEEE Access* **2018**, *6*, 31129–31140. [CrossRef]
18. Grisales-Noreña, L.; Montoya, O.D.; Gil-Gonzalez, W. Integration of energy storage systems in AC distribution networks: Optim location, selecting, and operation approach based on genetic algorithms. *J. Energy Storage* **2019**, *25*, 100891. [CrossRef]
19. Wei, C.; Fadlullah, Z.M.; Kato, N.; Stojmenovic, I. On optimally reducing power loss in micro-grids with power storage devic *IEEE J. Sel. Areas Commun.* **2014**, *32*, 1361–1370. [CrossRef]
20. Yuan, Z.; Wang, W.; Wang, H.; Yildizbasi, A. A new methodology for optimal location and sizing of battery energy storage syste in distribution networks for loss reduction. *J. Energy Storage* **2020**, *29*, 101368. [CrossRef]
21. Karanki, S.B.; Xu, D.; Venkatesh, B.; Singh, B.N. Optimal location of battery energy storage systems in power distribution netwo for integrating renewable energy sources. In Proceedings of the 2013 IEEE Energy Conversion Congress and Exposition, Denv CO, USA, 15–19 September 2013; pp. 4553–4558.
22. Mohamed, S.; Shaaban, M.F.; Ismail, M.; Serpedin, E.; Qaraqe, K.A. An efficient planning algorithm for hybrid remote microgri *IEEE Trans. Sustain. Energy* **2018**, *10*, 257–267. [CrossRef]
23. Grisales-Noreña, L.F.; Rosales-Muñoz, A.A.; Cortés-Caicedo, B.; Montoya, O.D.; Andrade, F. Optimal Operation of PV Sources DC Grids for Improving Technical, Economical, and Environmental Conditions by Using Vortex Search Algorithm and a Matr Hourly Power Flow. *Mathematics* **2023**, *11*, 93. [CrossRef]
24. Revankar, S.R.; Kalkhambkar, V.N. Grid integration of battery swapping station: A review. *J. Energy Storage* **2021**, *41*, 1029 [CrossRef]
25. Zhan, W.; Wang, Z.; Zhang, L.; Liu, P.; Cui, D.; Dorrell, D.G. A review of siting, sizing, optimal scheduling, and cost-bene analysis for battery swapping stations. *Energy* **2022**, 124723. [CrossRef]
26. Molina-Martin, F.; Montoya, O.D.; Grisales-Noreña, L.F.; Hernández, J.C.; Ramírez-Vanegas, C.A. Simultaneous minimization energy losses and greenhouse gas emissions in AC distribution networks using BESS. *Electronics* **2021**, *10*, 1002. [CrossRef]
27. Gil-González, W.; Montoya, O.D.; Grisales-Noreña, L.F.; Escobar-Mejía, A. Optimal Economic–Environmental Operation of BE in AC Distribution Systems: A Convex Multi-Objective Formulation. *Computation* **2021**, *9*, 137. [CrossRef]
28. Terlouw, T.; AlSkaif, T.; Bauer, C.; van Sark, W. Multi-objective optimization of energy arbitrage in community energy stora systems using different battery technologies. *Appl. Energy* **2019**, *239*, 356–372. [CrossRef]
29. Grisales-Noreña, L.F.; Gonzalez Montoya, D.; Ramos-Paja, C.A. Optimal sizing and location of distributed generators based PBIL and PSO techniques. *Energies* **2018**, *11*, 1018. [CrossRef]
30. Martinez, J.A.; Guerra, G. A parallel Monte Carlo method for optimum allocation of distributed generation. *IEEE Trans. Pow Syst.* **2014**, *29*, 2926–2933. [CrossRef]
31. Grisales-Noreña, L.F.; Montoya, O.D.; Ramos-Paja, C.A. An energy management system for optimal operation of BSS in D distributed generation environments based on a parallel PSO algorithm. *J. Energy Storage* **2020**, *29*, 101488. [CrossRef]
32. Abdelaziz, A.Y.; Fathy, A. A novel approach based on crow search algorithm for optimal selection of conductor size in radi distribution networks. *Eng. Sci. Technol. Int. J.* **2017**, *20*, 391–402. [CrossRef]

3. da Cruz Souza, J.; Soares, S.F.; de Paula, L.C.M.; Coelho, C.J.; de Araújo, M.C.U.; da Silva, E.C. Bat algorithm for variable selection in multivariate classification modeling using linear discriminant analysis. *Microchem. J.* **2023**, *187*, 108382. [CrossRef]
4. Askarzadeh, A. A novel metaheuristic method for solving constrained engineering optimization problems: Crow search algorithm. *Comput. Struct.* **2016**, *169*, 1–12. [CrossRef]
5. Grisales-Noreña, L.F.; Cortés-Caicedo, B.; Alcalá, G.; Montoya, O.D. Applying the Crow Search Algorithm for the Optimal Integration of PV Generation Units in DC Networks. *Mathematics* **2023**, *11*, 387. [CrossRef]
6. Instituto Colombiano de Normas Técnicas y Certificación (ICONTEC). TENSIONES Y FRECUENCIA NOMINALES EN SISTEMAS DE ENERGÍA ELÉCTRICA EN REDES DE SERVICIO PÚBLICO NTC1340. *Bogotá DC* **2004**.
7. Noreña, L.F.G.; Rivera, O.D.G.; Toro, J.A.O.; Paja, C.A.R.; Cabal, M.A.R. Metaheuristic optimization methods for optimal power flow analysis in DC distribution networks. *Trans. Energy Syst. Eng. Appl.* **2020**, *1*, 13–31. [CrossRef]
8. Dolatabadi, S.H.; Ghorbanian, M.; Siano, P.; Hatziargyriou, N.D. An enhanced IEEE 33 bus benchmark test system for distribution system studies. *IEEE Trans. Power Syst.* **2020**, *36*, 2565–2572. [CrossRef]
9. Shawon, S.M.R.H.; Liang, X.; Janbakhsh, M. Optimal Placement of Distributed Generation Units for Microgrid Planning in Distribution Networks. *IEEE Trans. Ind. Appl.* **2023**. [CrossRef]
0. May, G.J.; Davidson, A.; Monahov, B. Lead batteries for utility energy storage: A review. *J. Energy Storage* **2018**, *15*, 145–157. [CrossRef]
1. NASA. NASA Prediction Of Worldwide Energy Resources, Washington, D.C., United States. Available online: https://power.larc.nasa.gov/ (accessed on 21 January 2023).
2. XM SA ESP. Sinergox Database, Colombia. Available online: https://sinergox.xm.com.co/Paginas/Home.aspx (accessed on 21 January 2023).
3. Tawalbeh, M.; Al-Othman, A.; Kafiah, F.; Abdelsalam, E.; Almomani, F.; Alkasrawi, M. Environmental impacts of solar photovoltaic systems: A critical review of recent progress and future outlook. *Sci. Total Environ.* **2021**, *759*, 143528. [CrossRef]

mathematics

MDPI

Article

Coordinated Economic Operation of Hydrothermal Units with HVDC Link Based on Lagrange Multipliers

Ali Ahmad [1,2], Syed Abdul Rahman Kashif [2], Arslan Ashraf [1], Muhammad Majid Gulzar [3,4], Mohammed Alqahtani [5] and Muhammad Khalid [4,6,7,*]

1. Department of Electrical Engineering, University of Central Punjab, Lahore 54000, Pakistan
2. Department of Electrical Engineering, University of Engineering and Technology, Lahore 54890, Pakistan
3. Department of Control & Instrumentation Engineering, King Fahd University of Petroleum & Minerals, Dhahran 31261, Saudi Arabia
4. Interdisciplinary Research Center for Renewable Energy and Power Systems (IRC-REPS), King Fahd University of Petroleum & Minerals, Dhahran 31261, Saudi Arabia
5. Department of Industrial Engineering, King Khalid University, Abha 62529, Saudi Arabia
6. Electrical Engineering Department, King Fahd University of Petroleum & Minerals, Dhahran 31261, Saudi Arabia
7. SDAIA-KFUPM Joint Research Center for Artificial Intelligence, King Fahd University of Petroleum & Minerals, Dhahran 31261, Saudi Arabia

* Correspondence: mkhalid@kfupm.edu.sa

Abstract: Coordinated operation of hydrothermal scheduling with HVDC links considering netwo constraints becomes a vital issue due to their remote location and recent induction in the existin power system. The nonlinear and complex nature of the problem introduces many variables an constraints which results in a heavy computational burden. A widespread approach for handlin these complexities is to reformulate the problem by several linearization methods. In this paper, a L grange multipliers-based method is proposed for the solution of hydrothermal economic schedulin including HVDC link. This method solves equality constraint optimization problems. The linear pr gramming approach is embedded with the Lagrange method to consider both equality and inequali constraints. The proposed technique has been used on piecewise linear variables and constrain of the system considering generation, water volume, and line power flow limits. The formulate method efficiently minimizes the operational cost of thermal units and maximizes the utilization hydro units while meeting all generation, water volume, and the HVDC link constraints. The metho was successfully implemented in two scenarios of a case study. In the first scenario, hydrotherm scheduling was performed on the typical network without an HVDC line limit and equal nodal pric were found with minimal thermal generation cost of $278,822.3. In the second scenario, the propose method optimally dispatches units to meet the HVDC line limit and minimizes thermal generatio cost to $279,025.4 while satisfying hydro, thermal, and other operating constraints. Both scenarios a implemented for a 24 h period. The results have been presented to illustrate the performance of th proposed method.

Keywords: linear programming; economic dispatch; hydrothermal scheduling; HVDC link; Lagrang multipliers; optimal power flow

MSC: 49-11

Citation: Ahmad, A.; Kashif, S.A.R.; Ashraf, A.; Gulzar, M.M.; Alqahtani, M.; Khalid, M. Coordinated Economic Operation of Hydrothermal Units with HVDC Link Based on Lagrange Multipliers. *Mathematics* **2023**, *11*, 1610. https://doi.org/10.3390/math11071610

Academic Editors: Atanda Raji and Khaled M. Abo-Al-Ez

Received: 25 February 2023
Revised: 15 March 2023
Accepted: 22 March 2023
Published: 27 March 2023

1. Introduction

Recently, an increase in energy demand and fossil fuel prices has raised the cos of energy generated by thermal power plants. The world has reduced the use of costl and inefficient thermal plants by inducting renewable energy resources and hydel powe generation [1,2]. The energy prices can also be minimized by optimal scheduling o resources [3,4]. Conventionally, power is transmitted to loads through AC transmissio lines. However for long distance, the HVDC link becomes more efficient and cost effective as compared to the conventional mode of transmission [5]. Therefore, the coordinated economical operation of hydrothermal units with HVDC links has emerged as an interesting research area in today's world of increasing energy stress.

Different researchers have discussed different aspects of hydrothermal scheduling considering HVDC systems. Hydrothermal scheduling is a complex and non-linear constrained optimization problem. Inclusion of HVDC links increases the complexity of the problem that can be addressed by good computational tools. Hydrothermal scheduling problems were successfully solved using conventional methods such as Lagrange multipliers method (LMM) [6], Newton–Raphson method [7,8], gradient search algorithm [9], mixed integer programming [10], and dynamic programming [11]. A number of heuristic and metaheuristic algorithms are used to solve hydrothermal scheduling problems [12–14]. A few recently used algorithms in hydrothermal scheduling operation are rigid cuckoo search algorithm [15], firefly and accelerated particle swarm optimization [16], Lagrangian relaxation [13], grasshopper optimization algorithm (GOA) [17], and artificial bee colony algorithm [18]. However, heuristic algorithms cannot deal effectively with premature convergence problems. Additionally, when dealing with the large number of variables in optimization, the computation time of heuristic algorithms increases drastically. Each metaheuristic technique has some weaknesses and strengths to find near the optimal solutions for hydrothermal scheduling [19]. Meanwhile, mathematical programming methods are computationally fast and provide stable solution each time [20]. In unit commitment and economic dispatch cases, the linear programming (LP) method gives better results than the genetic algorithm (GA) technique [21]. The literature shows that hydrothermal scheduling has been effectively performed by the Lagrange multipliers method [6,22].

The induction of HVDC links in the existing power system requires an optimal power flow (OPF) study for realistic hydrothermal scheduling. Formulation and implementation of HVDC systems for OPF has been performed in reference [23]. Combined economic operation of point-to-point VSC-HVDC and AC grids is also performed based on Lagrange multipliers efficiently [24]. Joint operation of two area HVDC links has been performed in reference [20] to improve the operational economy and efficiency. Optimal scheduling of fixed head hydrothermal scheduling considering wind uncertainty is conducted in [25]. The literature survey signifies that hydrothermal scheduling was performed using various optimization techniques under various operating conditions to minimize the fuel price of thermal units.

A careful review of the above-mentioned excellent research shows that the existing methods of the economical scheduling of hydrothermal power plants only consider the price minimization of thermal units subject to meeting the load, losses, and water discharge constraints. These calculations ignore the network details and result in snubbing effects of transmission branch loading and bus voltages [9]. Therefore, economical hydrothermal dispatching does have an important effect on line flows, and under HVDC line constraints, these effects need to be taken into account.

The main contributions of this research are:

- Formulation of complex hydrothermal scheduling problem.
- Modelling of AC grids to add network constraints using DC optimal power flow (DCOPF) in the existing scheduling problem.
- Induction of HVDC link with line flows limitation constraints in hydrothermal problem.
- Linearization of quadratic cost curves of thermal generators to deal with inequality constraints.
- Implementation of linear programming-based Lagrange multipliers methods on a case study to check the robustness of the proposed method.

In this paper, a novel hydrothermal scheduling problem with an HVDC link is formulated to meet the load demand and network constraints. The formulation is general and allows to find out the economic impact of HVDC link power usage on hydrothermal-based AC system. An algorithm is developed to solve such a diverse problem using lin programming-based Lagrange multipliers method as it is fast and less prone to converger issues and more deterministic in nature as compared to existing popular metaheuris algorithms. To the best knowledge of the authors, such a coordinated economic operati of hydrothermal units with HVDC links keeping in view the power network constrair has not been discussed before.

The rest of the paper is arranged as follows. Section 2 presents the detailed formulati of the concerned problem and proposes the solution methodology. Section 3 explains t proposed research methodology for a case study of a power system. Section 4 demonstra the results of the case study under two different scenarios to validate the effectiveness the proposed methodology. Section 5 concludes the discussion and presents some poir for future work.

2. Problem Formulation

The coordinated economic operation of hydrothermal units with HVDC link aims minimize the operating cost of thermal units while maximizing the utilization of availal reservoir water volume, fulfilling the HVDC line flow limits and satisfying the load pow balance. Figure 1 shows the generalized network diagram considered for this scenario.

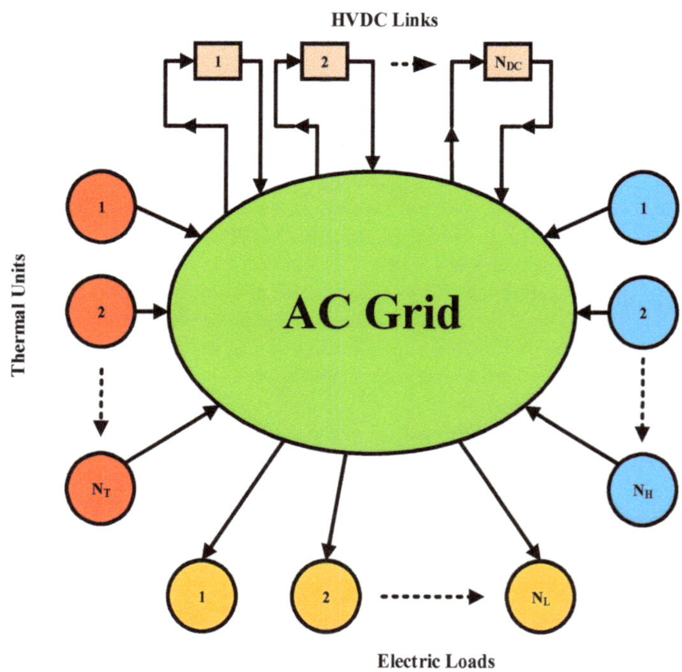

Figure 1. Generalized network diagram.

Figure 1 shows an electric network in which a number of hydel and thermal pow plants are embedded. The inward arrows show that the power is delivered to the netwo The loads are attached to the network. Here, outward arrows show the flow of power fro the network to the loads. The HVDC blocks take power from certain buses (represent by outward arrows) and deliver it to other buses of the network (represented by inwa arrows). The idea is to propose a Lagrange function for such a network considering all t generation, load, and network constraints. The following steps are being followed.

2.1. Hydrothermal Problem Formulation

The objective of hydrothermal is to utilize a given volume of water such that to minimize the production cost of N_T generating units subject to constraints of transmission lines, generators, and HVDC line limits. Therefore, the problem is formulated in general as $minimize\sum_{j=1}^{J_{max}}\sum_{t=1}^{N_T}\left[F_t\left(P_{tj}\right)\right]$ subject to constraints $p(x) = 0$ and $q(x) \leq 0$, where $p(x)$ and $q(x)$ are equality and inequality constraints, respectively. Equation $p(x)$ represents the network DC optimal power flow (DCOPF)-based load power balances, while Equation $q(x)$ denotes water volume discharge and minimum and maximum limits of hydro and thermal power generators. The cost of thermal generating units is approximated as a quadratic cost rate given in (1). Similarly, the water flow from hydro power plants in j time intervals is approximated using (2). The expressions (1) and (2) are used for the incremental cost of thermal power and fictitious cost of water, respectively, that must be paid to satisfy power balance, water volume, HVDC line limit, and generation constraints.

$$\sum_{j=1}^{J_{max}}\sum_{t=1}^{N_T}\left[F_t\left(P_{tj}\right)\right] = \sum_{j=1}^{J_{max}}\sum_{t=1}^{N_T}\left[a_t\left(P_{tj}\right)^2 + b_t P_{tj} + c_t\right] \tag{1}$$

$$\sum_{j=1}^{J_{max}} n_j \sum_{h=1}^{N_H} q_h\left(P_{hj}\right) = \sum_{j=1}^{J_{max}} n_j \sum_{h=1}^{N_H}\left[x_h\left(P_{hj}\right)^2 + y_h P_{hj} + z_h\right] = q_{TOT} \tag{2}$$

2.2. HVDC Line Flow Problem Formulation

A point-to-point connected HVDC system consists of two converter stations, namely rectifier and inverter stations, as shown in Figure 2. Both converter stations are connected to AC systems on 'r' and 'i' nodes through filters and tap-changing transformers. Inverter station maintains DC bus voltage within limits and rectifier station controls the active power flow at specified value P_{flow} [24]. The power transfer in the system can be in either direction. In the previous research, the investigations have been carried out to optimize the controller gains to control the active power transfer in HVDC system [26]. HVDC problem formulation consists of the following steps.

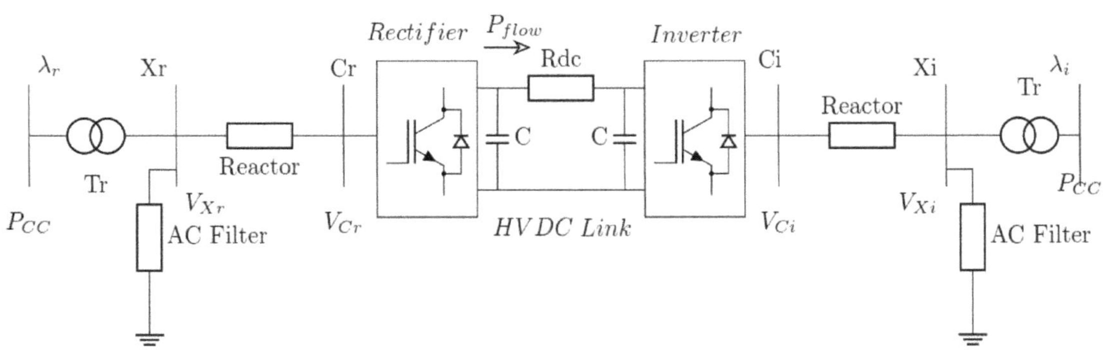

Figure 2. HVDC system.

Modelling of the point-to-point HVDC grids has been extensively performed to couple the VSC station with an AC network as an ideal voltage source either through an impedance or by phase shifting transformer for power flow analysis [27–31]. Similarly, intensive literature exists in the domain of economic dispatch for renewables ([32–36] and references therein). It is vital for economic dispatch applications to formulate the power flow equations of HVDC link from inverter to rectifier and vice versa. The DC power flow expression fro inverter and rectifier, respectively, are given in (3).

$$P_{dc_i} = \left(V_i^2 - V_i V_r\right) G_{dc}, P_{dc_r} = \left(V_r^2 - V_i V_r\right) G_{dc} \tag{3}$$

where V_i and V_r are inverter and rectifier voltages, respectively, and $G_{dc} = \frac{1}{R_{dc}}$.

To formulate the combined equations for HVDC power flow and AC grids, some bas assumptions are made on AC grid: (a) neglect conduction losses, i.e., $G_i = G_r = 0$, (b) s $V_i = V_r = 1\ p.u$, and (c) due to very small angular difference, set $\sin(\theta_r - \theta_i) \cong (\theta_r - \theta_i)$ Hence, (6) takes the form for nodal power balances in various nodes of HVDC grid an defines the equality constraints for HVDC link. The resultant expression becomes as give in (4):

$$P_{ri} = (B_r - B_i)(\theta_r - \theta_i) \tag{4}$$

where B_r,B_i and θ_r,θ_i represent susceptance and phase angle on rectifier and inverter sid respectively. Susceptance is the inverse of reactance offered by AC filter, reactors, an transformer on either side of the converter.

2.3. AC Network Problem Formulation

The nodal power balance is formulated based on the operating point of the AC syster The power balance relation in (5) exhibits equality constraints $p(x)$ as a function of nod power generation, demand, and calculated power in (6).

$$\Delta P_k = P_{gk} - P_{dk} - P_k^{cal} = 0 \tag{5}$$

$$P_K^{cal} = G_{kk} V_k^2 + \sum_{m \epsilon k} V_k V_m [G_{km} \cos(\theta_k - \theta_m) + B_{km} \sin(\theta_k - \theta_m)] \tag{6}$$

where V_k, V_m and θ_k, θ_m are the voltage magnitudes and phase angles of the transmissio lines linking buses 'k' and 'm', respectively. Moreover, G and B are conductance an susceptance of lines connecting the respective busses, respectively.

Under steady state condition of AC grid, $V_k = V_m = 1\ p.u$ and with negligible pow loss $G_{kk} = G_{km} = 0$, the expression (6) reduces to $P_K^{cal} = \sum_{m \epsilon k} B_{km}(\theta_k - \theta_m)$. Then, th expression can be generalized for lossless AC grids having α nodes in (7).

$$\begin{bmatrix} P_1 \\ \vdots \\ P_\alpha \end{bmatrix} = \begin{bmatrix} B_{11} & \cdots & B_{1\alpha} \\ \vdots & \ddots & \vdots \\ B_{\alpha 1} & \cdots & B_{\alpha\alpha} \end{bmatrix} \begin{bmatrix} \theta_1 \\ \vdots \\ \theta_\alpha \end{bmatrix} \Longrightarrow [P] = [B][\theta] \tag{7}$$

where P is nodal power, B is susceptance, and θ is the nodal phase angle.

The mostly used HVDC links occur either in embedded or decoupled form wit the AC grids [24]. These models are used for the investigation of optimal power flow i HVDC links connected to AC systems. The embedded HVDC link model deduces that bot sending and receiving end converter stations consider same phase angles to AC grid. Th expressions (4) and (7) provide the relationship between two AC grids connected HVD link for sending and receiving end power flow. The mathematical expressions (1), (2), (4 and (7) are added in the Lagrange function to augment DCOPF-based AC grid details an HVDC flow limit in hydrothermal scheduling. The resultant expression is given in (8).

2.4. Overall Problem Formulation

The objective of the research is to minimize the thermal generation cost subject to me the load balance, generator limit, HVDC line flow, and water consumption constraint Based on the objective and constraints, the Lagrange function is stated as given in (8);

$$\mathcal{L} = \sum_{j=1}^{J_{max}} \left[\left(n_j \sum_{t=1}^{N_b} F_t(P_{tj}) \right) + \lambda_j^{N_b} \left([B_x][\theta]_j - \left(P_{thj} - P_{dj} \right) \right) + \lambda_j^{N_b+1} \left((P_{ri})_j - P_{HVDC_{set}} \right) + \gamma \left(n_j \sum_{h=1}^{N_H} q_h(P_{hj}) - q_{TOT} \right) + \mu \left[g\left(P_{tj}^{min}, P_{tj}^{max}, P_{hj}^{min}, P_{hj}^{max} \right) \right] \right] \quad (8)$$

where n_j is the scheduling interval in hours, N_b is the number of buses, F_t is the thermal cost function which needs to be minimized subjected to power balance constraint $\left([B_x][\theta]_j - \left(P_{thj} - P_{dj} \right) \right)$ and water storage constraint $\left(n_j q_h \left(P_{hj} \right) = q_{TOT} \right)$, $\lambda_j^{N_b}$ is Lagrange multiplier which shows nodal marginal price of bus number N_b in j^{th} time interval in $\frac{\$}{MWh}$, P_{dj} is the load in each time interval, P_{thj} is the sum of power generated by thermal and hydro units in each time interval, $(P_{ri})_j$ is the actual power flow on HVDC link from rectifier to inverter, $P_{HVDC_{set}}$ is the line flow limit of the HVDC link, P_{tj}^{min} and P_{tj}^{max} are minimum and maximum limits of thermal generators, P_{hj}^{min} and P_{hj}^{max} are minimum and maximum limits of hydro generators, γ is Lagrange multiplier which shows fictious cost of water, H is number of hydro units, and q_{TOT} is the total water volume available for hydro power generation.

2.5. Constraints

The constraints of hydro and thermal power plants along with load balance and transmission limits are explained.

- Load balance constraints

$$P_{thj} = P_{dj} + [B_x][\theta]_j \quad (9)$$

where P_{thj}, P_{dj}, and $[B_x][\theta]_j$ are the total thermal and hydropower, load demands, and transmission line flows in j^{th} time intervals, respectively.

- Thermal plant generation limit

$$P_{tj}^{min} \le P_{tj} \le P_{tj}^{max} \quad (j = 1, 2 \ldots, 24\ \text{h}) \quad (10)$$

where P_{tj}^{min} and P_{tj}^{max} are the minimum and maximum output of the N_T thermal power plants in j^{th} time intervals, respectively.

- Hydro plant generation limit

$$P_{hj}^{min} \le P_{hj} \le P_{hj}^{max} \quad (j = 1, 2 \ldots, 24\ \text{h}) \quad (11)$$

where P_{Hj}^{min} and P_{Hj}^{max} are the minimum and maximum output of the N_H hydro power plants in j^{th} time intervals, respectively.

- Water volume limit

$$n_j \sum_{h=1}^{N_H} q_h\left(P_{hj}\right) = q_{TOT} \quad (j = 1, 2 \ldots, 24\ \text{h}) \quad (12)$$

where q_{TOT} is the total water volume available for power generation and $q_h\left(P_{hj}\right)$ is the water flow rate at the output power of P_{hj} hydro unit in n_j number of hours. Constant head is assumed to consume total available water volume for power generation in the jth time intervals.

- HVDC line limitation

$$P_{ri_j}^{min} \le P_{ri_j} \le P_{ri_j}^{max} \quad (j = 1, 2 \ldots, 24\ \text{h}) \quad (13)$$

This means power flow of lines is calculated by DC load flow to check the capac limitations of HVDC link. In real scheduling, HVDC line limitations are obeyed as specifi electric quantity trade limitations lines between regional distributors of electricity.

The proposed approach is a novel alternative to address fundamental problems, it is structured to augment DC power flow in hydrothermal scheduling. Moreover, it general to harmonize the number of HVDC links.

As already discussed, the quadratic cost expression, given in (1) of thermal pow plants, does not include the power system network details, transmission line paramete transmission line congestion constraints, HVDC links, etc. [37,38]. The expressions and (2) only consider equality constraints of load balance and water volume. Wh considering the network constraints without line limits in an economic dispatch, t gradient method is used to solve the Lagrange equation [9]. Moreover, the Lagran function considering all network constraints, hydrothermal, HVDC links, the generato extreme limits, and line limitations can be optimized by linear programming (LP). Hen linear programming (LP) technique is used to solve combined hydrothermal scheduli with DCOPF considering generators' inequality and HVDC line limit constraints. As p LP, the quadratic cost expression of thermal power plants does not satisfy the requiremen Therefore, a piecewise linear approximation of the cost function given in (1) is perform The segment wise slopes of piecewise linear approximation are developed using (14).

$$S_a = \frac{F_{a+1} - F_a}{P_{a+1} - P_a} \quad \textit{for } a = 1, 2, 3, \ldots\ldots\ldots\ldots N_s \text{ (Number of slope segments)} \qquad (1$$

In (14), the number of slope segments (N_s) depends on the minimum and maximu limits of thermal power generators and step size defined by the user.

Previously, (7) gave important information about the nodal power injection consid ing the network admittances without real part and help in finding the angles. However, does not provide the line power flows. Therefore, the proposed approach considers t line flows, network constraints, generation limits, etc., formally in (15).

$$\left.\begin{array}{c} Minimize\ F(P_t) = F\left(P_t^{min}\right) + \sum_{a=1}^{N_s} \sum_{t=1}^{N_T} S_{at} P_{at} \\ Subject\ to\ P = B\theta \\ P_B = (D \times E) \times \theta \\ -P_B^{max} \leq P_B \leq P_B^{max} \\ q_{TOT} = \sum_{h=1}^{N_H} q_h(P_h) \\ 0 \leq P_{gk} \leq P_{gk}^{max}, \forall k \in \{generator_{buses}\} \end{array}\right\} \qquad (1$$

where N_s is the number of slope segments, N_T is the number of thermal generators, $\boldsymbol{P}$ the nodal power injections at all nodes, θ is nodal angle, P_B is line flow for AC and HVD system, D and E are node-arc matrices, and $-P_B^{max}$ and P_B^{max} are line flow limits. The q_{T} is the total water volume available for generation, N_H is the number of hydro units, a q_h is the water discharge rate to generate hydro power P_h in an interval. Moreover, (15) similar to (8) with constraints mentioned as Lagrange multipliers.

3. Research Methodology

The proposed linear programming based on Lagrange multipliers research method ogy shown in Figure 3 has been successfully implemented on a case study power syste network shown in Figure 4 by carrying out the following steps.

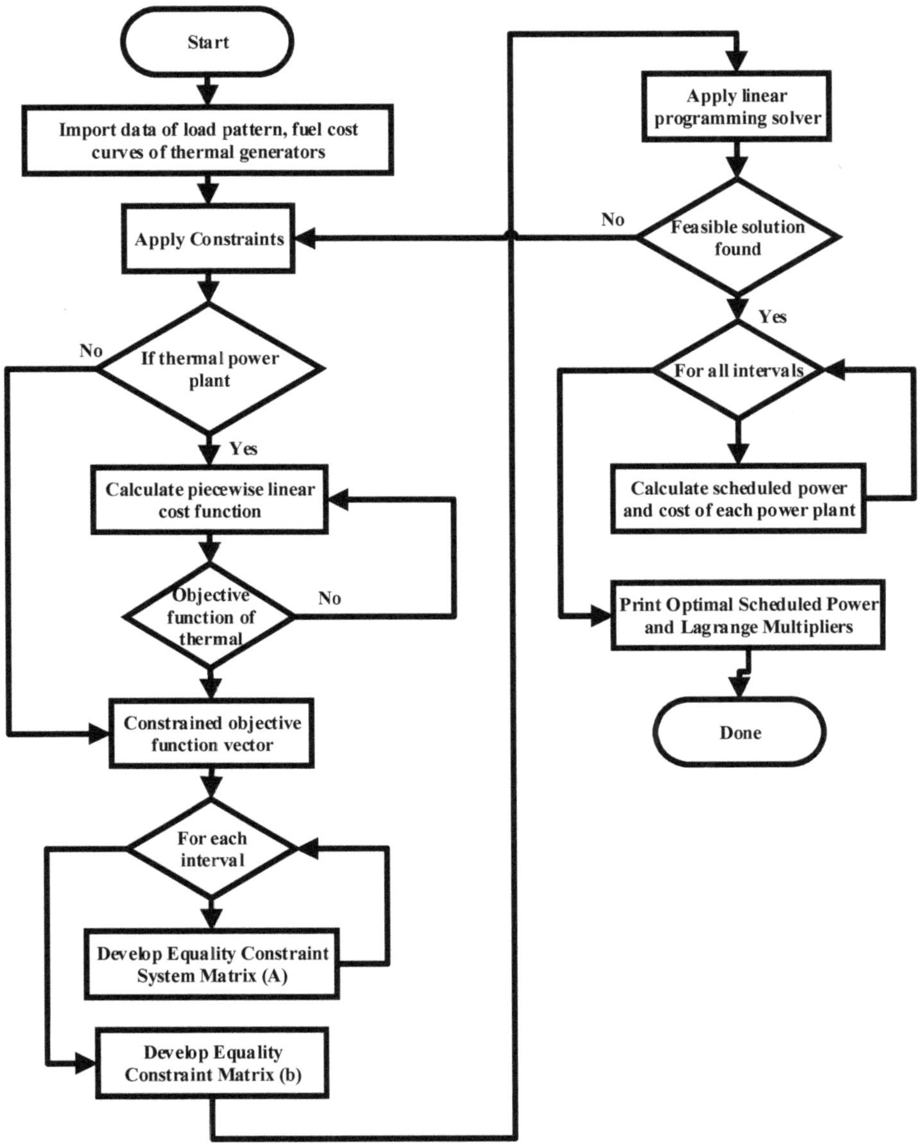

Figure 3. Algorithm flow chart.

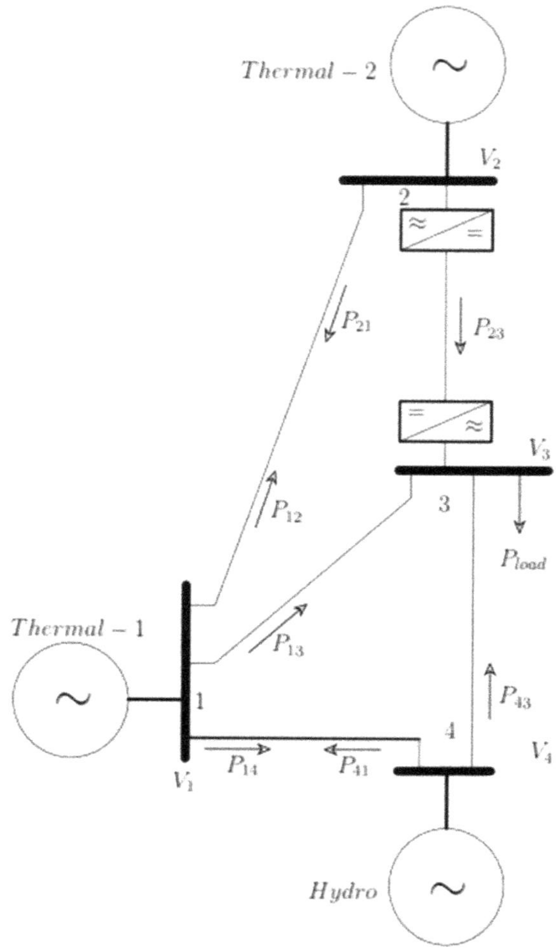

Figure 4. Four (4) bus AC system with HVDC link.

3.1. Objective Function

The objective function is given in (16);

$$f(x) = Ax = \begin{bmatrix} s_1 & s_2 & 0 & 0 & 0 & 0 & 0 & 0 & 0 & 0 & 0 & 0 & 0 \end{bmatrix} \begin{bmatrix} P_{t1} & P_{t2} & P_h & P_{B1} & P_{B2} & P_{B3} & P_{B4} & P_{B5} & \theta_1 & \theta_2 & \theta_3 & \theta_4 & W \end{bmatrix}^T \quad (16)$$

In (16), the dimensions of 's_1' and 's_2' (in the row) depend on the number of slope based on step size, while P_t^{min} and P_t^{max} depend on the limits of the thermal power statio

3.2. Output Vector

The output vector 'x' for the given case study network and constraints is define in (17).

$$\begin{aligned} x &= \begin{bmatrix} P_t & P_h & P_B & \theta & W \end{bmatrix}^T \\ &= \begin{bmatrix} P_{t1} & P_{t2} & P_h & P_{B1} & P_{B2} & P_{B3} & P_{B4} & P_{B5} & \theta_1 & \theta_2 & \theta_3 & \theta_4 & W \end{bmatrix}^T \end{aligned} \quad (17)$$

where P_t is the incremental generation vector for two thermal generators connected to bus- and bus-2, P_h is hydropower generation, P_B is the line power flows vector for five branche

θ is the vector of angles for all four buses in radian, and W is the water volume variable. All these variables are added as equality constraints given in (15) to a single matrix relation. For single matrix relation, one kind of equality constraint is given as injected powers and another kind of equality constraint is given as line power flows as in (15), resulting in (18) and (19).

$$-P_B + (D \times E) \times \theta = 0 \quad (18)$$

$$-P + B\theta = 0 \quad (19)$$

3.3. Equality Constraints

The equality constraints are given in (18) and (19). Matrix (Ax = b) form of equality constraints requires dimensions. The dimensions of equality constraints based on the case study are:

- Number of columns: As the vector x has 13 $\times$ 1, matrix A should have 13 columns to multiply x.
- Number of rows: As there are five branches, four buses, and one water volume constraint, (18) and (19) will have to contribute a total of ten (10) rows to matrix A.

Therefore, the dimension of matrix A will be 10 $\times$ 13. Let us start from line power flow as in (18). The D-matrix is given in (20). The node-arc matrix **E**, is given as (21):

$$D = \begin{bmatrix} B_1 & 0 & 0 & 0 & 0 \\ 0 & B_2 & 0 & 0 & 0 \\ 0 & 0 & B_3 & 0 & 0 \\ 0 & 0 & 0 & B_4 & 0 \\ 0 & 0 & 0 & 0 & B_5 \end{bmatrix} \quad (20)$$

$$E = \begin{bmatrix} 1 & 0 & 0 & -1 \\ 1 & -1 & 0 & 0 \\ 0 & 1 & -1 & 0 \\ 0 & 0 & -1 & 1 \\ 1 & 0 & -1 & 0 \end{bmatrix} \quad (21)$$

The product $D \times E$ required to find line power flows is given as (22):

$$D \times E = \begin{bmatrix} B_1 & 0 & 0 & 0 & 0 \\ 0 & B_2 & 0 & 0 & 0 \\ 0 & 0 & B_3 & 0 & 0 \\ 0 & 0 & 0 & B_4 & 0 \\ 0 & 0 & 0 & 0 & B_5 \end{bmatrix} \begin{bmatrix} 1 & 0 & 0 & -1 \\ 1 & -1 & 0 & 0 \\ 0 & 1 & -1 & 0 \\ 0 & 0 & -1 & 1 \\ 1 & 0 & -1 & 0 \end{bmatrix} = \begin{bmatrix} B_1 & 0 & 0 & -B_1 \\ B_2 & -B_2 & 0 & 0 \\ 0 & B_3 & -B_3 & 0 \\ 0 & 0 & -B_4 & B_4 \\ B_5 & 0 & -B_5 & 0 \end{bmatrix} \quad (22)$$

The elements of (22) will be placed in the upper right corner of matrix A from columns 9 to 12. The first eight columns of the top five rows will be multiplied to generated powers (P_{t1}, P_{t2}, P_h) and line power flows ($P_{B1}, P_{B2}, P_{B3}, P_{B4}, P_{B5}$) variables. As generation variables are not used within the line power flow equations, the first three (3) columns of the top five rows will be zeros. The columns (4–8) will be zeros, except a single element in each row will be -1 to obtain corresponding line power flows.

DC power flow equations corresponding to (19) are written in matrix A. The augmented DC power flow matrix is given as (23):

$$B = \begin{bmatrix} B_{11} & -B_{12} & -B_{13} & -B_{14} \\ -B_{21} & B_{22} & -B_{23} & -B_{24} \\ -B_{31} & -B_{32} & B_{33} & -B_{34} \\ -B_{41} & -B_{42} & -B_{43} & B_{44} \end{bmatrix} \quad (23)$$

Elements of (23) will be placed at the lower right of matrix A from rows 6 to 9 a columns 9 to 12. The first three (3) columns of matrix A are reserved for generation variabl The expression (19) requires negative injections for all buses and injected power is P_{th} – However, load variables are not added in matrix A, which will be placed on the right si of the expression Ax = b in matrix b. Additionally, generated power from generation pla will be placed with a negative sign in the rows starting from 6 to 9 and the first three columns with respect to the respective bus in matrix A. Hydro power plant and wa volume constraints are added to the last (10th) row and column 3rd and 13th of the mat A, respectively. The resultant expression Ax = b in matrix form is given in (24):

$$Ax = b => \begin{bmatrix} 0 & 0 & 0 & -1 & 0 & 0 & 0 & 0 & B_1 & 0 & 0 & -B_1 & 0 \\ 0 & 0 & 0 & 0 & -1 & 0 & 0 & 0 & B_2 & -B_2 & 0 & 0 & 0 \\ 0 & 0 & 0 & 0 & 0 & -1 & 0 & 0 & 0 & B_3 & -B_3 & 0 & 0 \\ 0 & 0 & 0 & 0 & 0 & 0 & -1 & 0 & 0 & 0 & -B_4 & B_4 & 0 \\ 0 & 0 & 0 & 0 & 0 & 0 & 0 & -1 & B_5 & 0 & -B_5 & 0 & 0 \\ -1 & 0 & 0 & 0 & 0 & 0 & 0 & 0 & B_{11} & -B_{12} & -B_{13} & -B_{14} & 0 \\ 0 & -1 & 0 & 0 & 0 & 0 & 0 & 0 & -B_{21} & B_{22} & -B_{23} & -B_{24} & 0 \\ 0 & 0 & 0 & 0 & 0 & 0 & 0 & 0 & -B_{31} & -B_{32} & -B_{33} & -B_{34} & 0 \\ 0 & 0 & -1 & 0 & 0 & 0 & 0 & 0 & -B_{41} & -B_{42} & -B_{43} & B_{44} & 0 \\ 0 & 0 & q(2)*n & 0 & 0 & 0 & 0 & 0 & 0 & 0 & 0 & 0 & 1 \end{bmatrix} \begin{bmatrix} P_{t1} \\ P_{t2} \\ P_h \\ P_{B1} \\ P_{B2} \\ P_{B3} \\ P_{B4} \\ P_{B5} \\ \theta_1 \\ \theta_2 \\ \theta_3 \\ \theta_4 \\ W \end{bmatrix} = \begin{bmatrix} 0 \\ 0 \\ 0 \\ 0 \\ 0 \\ P_{t1}^{min} \\ P_{t2}^{min} \\ -P_{load} \\ P_h^{min} \\ W_{start} + inflow - q(1)*n \end{bmatrix} \tag{24}$$

3.4. Inequality Constraints

These constraints are simple and are given in (25):

$$\begin{bmatrix} 0 \\ 0 \\ 0 \\ -P_{B1}^{max} \\ -P_{B2}^{max} \\ -P_{B3}^{max} \\ -P_{B4}^{max} \\ -P_{B5}^{max} \\ -\pi \\ -\pi \\ -\pi \\ -\pi \\ W_{end} \end{bmatrix} \leq \begin{bmatrix} P_{t1} \\ P_{t2} \\ P_h \\ P_{B1} \\ P_{B2} \\ P_{B3} \\ P_{B4} \\ P_{B5} \\ \theta_1 \\ \theta_2 \\ \theta_3 \\ \theta_4 \\ W \end{bmatrix} \leq \begin{bmatrix} step \\ step \\ P_h^{max} \\ P_{B1}^{max} \\ P_{B2}^{max} \\ P_{B3}^{max} \\ P_{B4}^{max} \\ P_{B5}^{max} \\ \pi \\ \pi \\ \pi \\ \pi \\ W_{start} + \max(w) \end{bmatrix} \tag{25}$$

In (25), the dimensions of P_{t1} and P_{t2} depend on the number of slope segments bas on P_{t1}^{min} and P_{t1}^{max} values of the thermal power generating stations. Hence, the row-wi size of columns 1 and 2 will depend on the number of slope segments. The expressio (24) and (25) are applicable to single load interval. For 24 h or daily load intervals, the equations are modified as in (26):

$$Ax = b => \begin{bmatrix} A_{11} & 0 & 0 & \ldots\ldots\ldots & \ldots & 0 \\ C_{11} & A_{22} & 0 & \ldots\ldots\ldots & \ldots & 0 \\ 0 & C_{22} & A_{33} & \ldots\ldots\ldots & \ldots & 0 \\ \vdots & \vdots & \vdots & \vdots\vdots\vdots\vdots\vdots & \vdots & \vdots \\ 0 & 0 & 0 & \ldots\ldots\ldots & A_{(i-1)(j-1)} & 0 \\ 0 & 0 & 0 & \ldots\ldots\ldots & C_{(i-1)(j-1)} & A_{ij} \end{bmatrix} \begin{bmatrix} x_{11} \\ x_{21} \\ x_{31} \\ \vdots \\ \ldots \\ x_{i1} \end{bmatrix} = \begin{bmatrix} b_{11} \\ b_{21} \\ b_{31} \\ \vdots \\ \ldots \\ b_{i1} \end{bmatrix} \tag{26}$$

where A_{ij} and $C_{(i-1)(j-1)}$ matrices have dimensions 10×13, having ten rows and thirteen columns for each load interval; $i = j$ is the number of load intervals $(1, \cdots\cdots 24\text{ h})$. The elements of matrix A are shown in (24) for a single load interval. The elements of matrix C are given as in (27):

$$C = \begin{bmatrix} Zeros(9 \times 12) & Zeros(9 \times 1) \\ Zeros(1 \times 12) & -1 \end{bmatrix} \quad (27)$$

Equation (27) is only used as leftover water volume in one interval to be used in the next load interval in constraint expression (24). Inequality constraints for a single load period given in (25) are modified for 24 h load intervals accordingly. The algorithm flow chart of the proposed approach is shown in Figure 3 to implement (8) using (15) with constraints from (9) to (13) and realistic equations from (16) to (27) for all load periods in MATLAB Software.

The detailed step-by-step implementation procedure of the proposed approach in MATLAB Software is given in Table 1.

Table 1. Step-by-step implementation of proposed procedure.

Step-by-Step Implementation of Proposed Procedure for Coordinated/Optimal Economic Operation of Hydrothermal Units with HVDC Link Based on Lagrange Multipliers	
1	Consider a power system network having 'N_b' number of buses and 'N_M' number of branches (shown in Figure 4).
2	Calculate the susceptance of each line following nodal power injection using (5), (6), and (7) of AC network and (4) for HVDC network or directly follow (10) which is common for both hydrothermal-based AC and HVDC power system.
3	Find slopes of cost functions of all thermal generators using (14) and define an objective function based on (16).
4	Formulate the DCOPF coordinated economic dispatch problem using (8) and (15)
5	Find a single matrix for all equality constraints, load balance, line power flows, and nodal power injections using (9), (18), and (19), respectively. This can be executed using node-arc incidence matrix product ($D \times E$) using (22).
6	Embed hydropower plant variable and water volume constraint using (11) and (12). Then, develop a standard form of the matrix ($Ax = b$) using (24) for linear programming (LP).
7	Find parameters for b-matrix (of $Ax = b$) using load and generation buses of power system network (shown in Figure 4) and objection function given in (15).
8	Define inequality constraints of the power system network under study using (10), (11), and (13).
9	Apply the standard linear programming (LP) method using MATLAB software.
10	Check constraints. If all constraints are satisfied, then procedure is done. Otherwise, go to step 3.
11	Print the optimal operating schedule and nodal price of each bus.

4. Results and Discussion

A case study is designed to verify the effectiveness of the proposed methodology. Four bus AC systems with two AC grids are interconnected by point-to-point HVDC transmission link, shown in Figure 4. It is ensured that the modelled HVDC link agrees with (4). The cost curves of the thermal power plant and flow rate characteristics of the hydro power plant are shown in Tables 2 and 3, respectively.

Table 2. Thermal plants characteristics.

Unit	$a_t\left[\frac{\$}{\text{MW}^2\text{h}}\right]$	$b_t\left[\frac{\$}{\text{MWh}}\right]$	$c_t\left[\frac{\$}{\text{h}}\right]$	P_t^{min}(MW)	P_t^{max}(MW)
Thermal-1	0.0033	10.8	1200	25	875
Thermal-2	0.003	12.6	1710	40	600

Table 3. Hydro plant characteristics.

Unit	$x_h\left[\frac{\text{AF}}{\text{MW}^2\text{h}}\right]$	$y_h\left[\frac{\text{AF}}{\text{MWh}}\right]$	$z_h\left[\frac{\text{AF}}{\text{h}}\right]$	P_h^{min}(MW)	P_h^{max}(MW)	q_{TOT}(AF)
Hydro	0	4.9	50	0	500	30,000

The power system network has five transmission lines connected to four differe buses. The transmission line connected between bus-2 and bus-3 is considered an HVD link. This line has point-to-point converter stations connected to the respective buses. is assumed that all the AC lines have equal admittance with the assumption of zero re parts. Thermal unit-1 connected to bus-1 can generate a minimum power of 25 MW and maximum power of 875 MW. However, thermal unit-2 connected to bus-2 can produce minimum of 40 MW and a maximum of 600 MW. The hydro plant connected to bus-4 ca produce a maximum of 500 MW output. The initial volume of water is 15,000 Acre-feet (A The allowed inflow is 1250 AF per hour. For the 24 h period, the total water available f power generation is 30,000 AF. The required ending capacity of the reservoir is 15,000 A

Two scenarios are implemented in Figure 4 network: (i) DCOPF including hydrothe mal scheduling without line trading limitation on the HVDC line and (ii) DCOPF includin hydrothermal scheduling with line trading limitation ($P_{HVDC} = 400$ MW) on the HVD line. Only HVDC line has direction power flow constrain, whereas all the other lines hav flexibility in the flow of power on either side.

The load is connected to bus-3 and the daily load on the four-bus system is shown Figure 5.

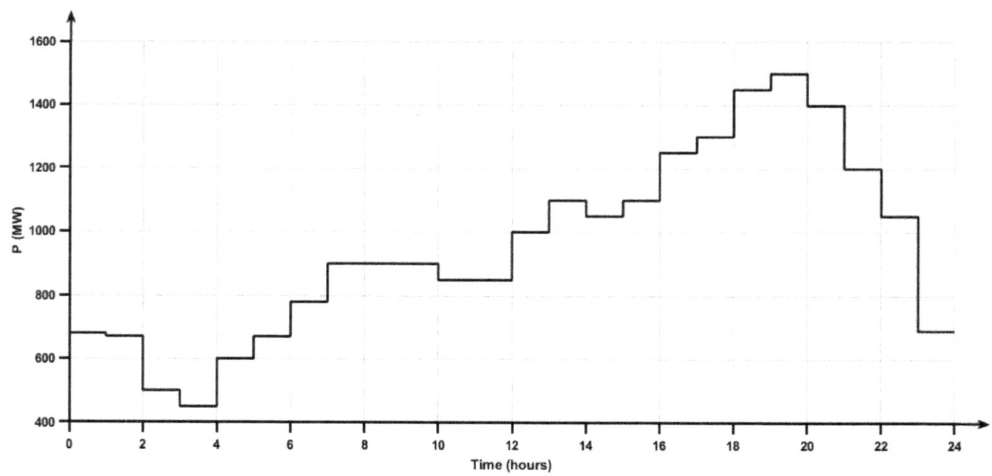

Figure 5. Load pattern on network.

Figure 6 shows the power contribution (bar charts) of each generating station to me the load demand. The power shared by the hydro unit to load is based on the availabili of water for power generation. The power shared by both thermal units to load demand optimized based on their cost rate characteristics shown in Table 2. Therefore, the cost thermal power units is the minimum for typical load demand in a specific time period.

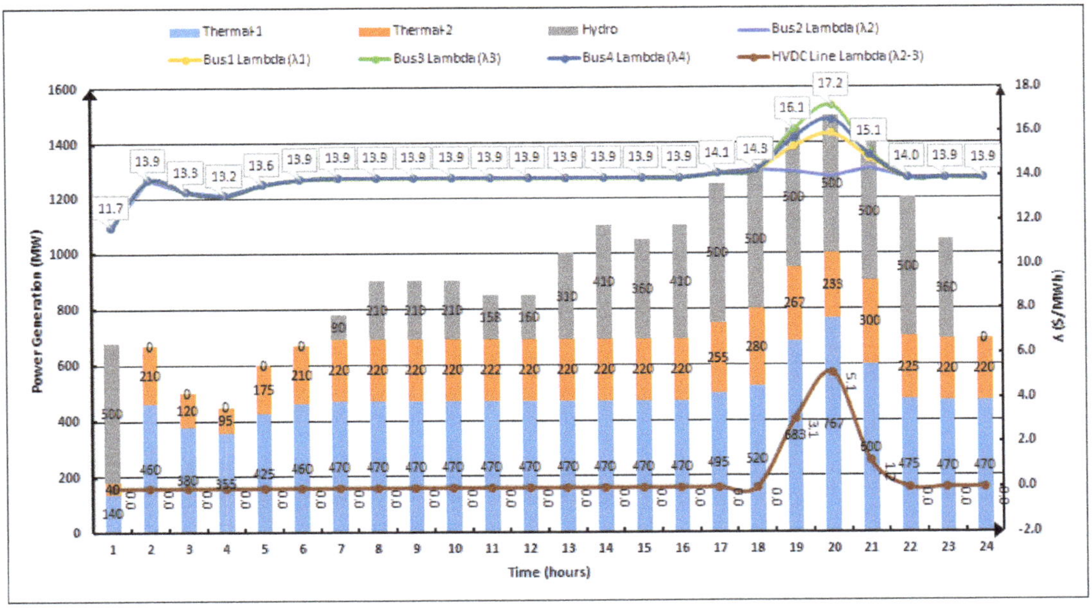

Figure 6. Hydrothermal scheduled power, bus price, and line limit price.

Due to combined investigations of economic hydrothermal scheduling and DCOPF at the same time, the system lambdas (λ) represent the bus location marginal price (LMP) or nodal price and line congestion price instead of the generator's incremental cost. Figure 6 shows both lambdas (λ) under two different scenarios.

4.1. Scenario-1: Infinite HVDC Line Capacity

When the power system network shown in Figure 4 is operating under normal conditions, there is no capacity limit on the HVDC transmission line. Then, the lambda (λ) of each bus is constant for a specific load period, as shown in Figure 6. In this figure, for the first load interval, the load demand is 680 MW, and the nodal price is fixed (11.7 $/MWh) for each bus. In the first interval, the hydro plant contribution is maximum. During the second load interval, load demand is 670 MW, and the nodal price is fixed (13.9 $/MWh) for each bus with zero power contribution by the hydro plant. Similarly, for load periods 1 to 18th and 22nd to 24th, nodal prices of all buses are fixed when there is an infinite line capacity limit. The resultant HVDC line limit lambda (λ2-3) is zero for these load intervals due to the infinite line capacity limit; the network considered it as a single bus. The respective power flow in each line branch and each interval is shown in Figure 7. The optimal power generation of each generating station to meet the load demand in each interval is shown in Figure 8.

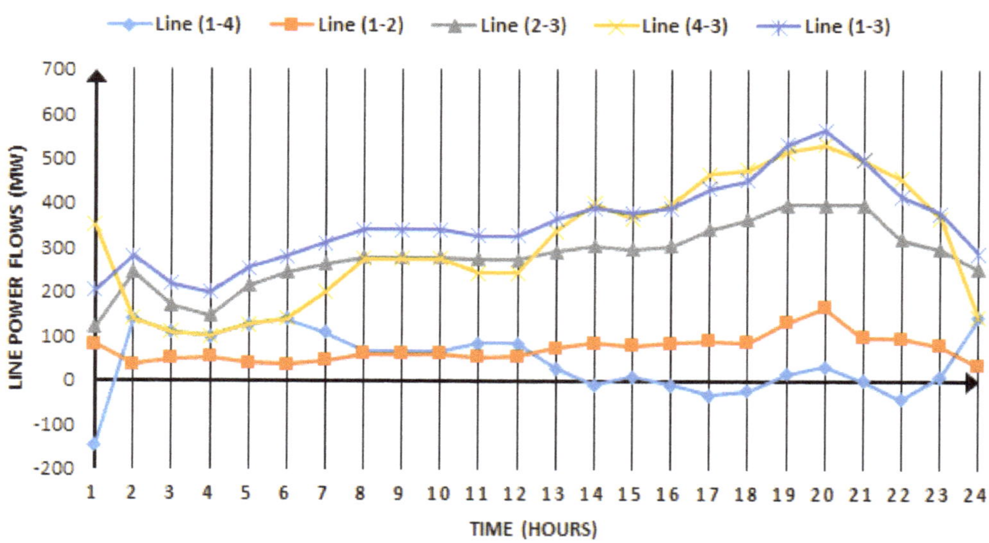

Figure 7. Line power flow during each load interval.

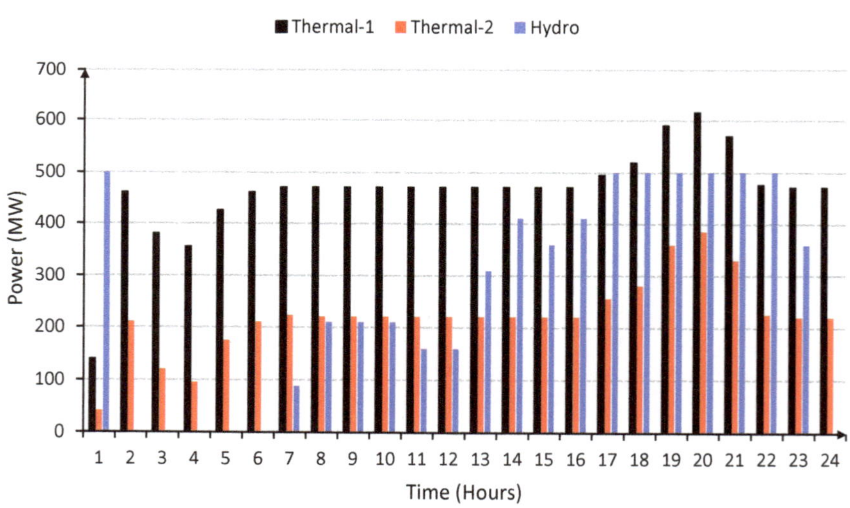

Figure 8. Generators optimal power during each load interval in infinite line capacity.

4.2. Scenario-2: Limited HVDC Line Capacity

In this case, line trading limitation $P_{Line2-3} = P_{HVDC} = 400$ MW is added on the HVD link shown in Figure 4. The load power demand during time intervals 19, 20, and 21 1450 MW, 1500 MW, and 1400 MW, respectively, as shown in Figure 5. During these tim intervals, the power shared by a hydro unit is 500 MW, thermal-1 is 683 MW, 767 MW, an 600 MW, and thermal-1 is 267 MW, 233 MW, and 300 MW, as given in Figure 6. Due HVDC line limitations, the power is diverted to other transmission lines of the networ Hence, the network does not consider a single bus which results in different buses' nod lambdas (λ) price and line (λ) price. This makes the nodal price at bus-1 = λ_1 = 15.4 \$/MW bus-2 = λ_2 = 14.2 \$/MWh, bus-3 = λ_3 = 16.1 \$/MWh, and bus-4 = λ_4 = 15.7 \$/MWh, shown in Figure 7. The line limit price in three different time intervals increased fro 0 \$/MWh to 3.1 \$/MWh and 5.1 \$/MWh due to line congestion heating and losses. Th line limit price is shown in Figure 7. Moreover, the optimal power generation of eac generating station to meet the load demand in each interval with limited line capacity is shown in Figure 9.

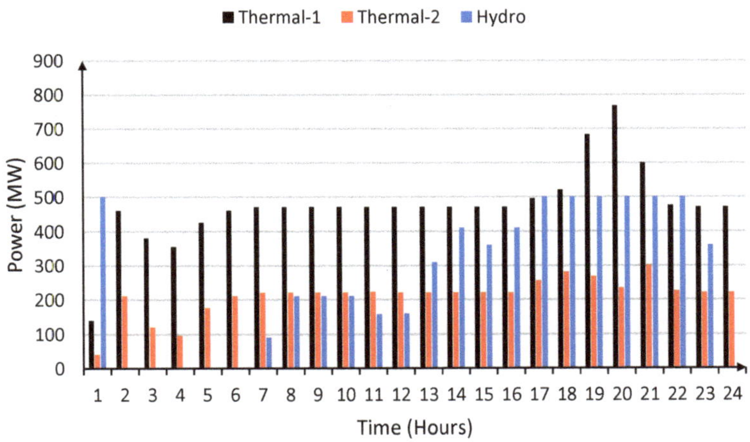

Figure 9. Generators optimal power during each load interval in limited line capacity.

The results described in both scenarios depict that all the buses' nodal prices are the same for specific time interval loads when there are no line limitations. Additionally, the line price is zero. Meanwhile, when there are line capacity limitations, the bus nodal price increases from 14.3 \$/MWh to 17.2 \$/MWh and the line price increases from 0 \$/MWh to 5.1 \$/MWh. This results in total cost of power generation by the thermal-1 unit in 24 h period being \$170,921.4 and the thermal-2 unit being \$108,104 with P_{HVDC} = 400 MW limit. However, thermal-1 unit in 24 h period costs \$166,739.8 and the thermal-2 unit costs \$112,082.5 without HVDC line limitation. The proposed approach optimally scheduled the outputs of thermal generating stations under both scenarios to minimize the operating cost. It can be concluded that limited capacity HVDC links may change the optimal operating points of all generating stations throughout the load intervals. Hence, the electricity market will affect generation companies (GENCOs) and transmission system operators (TSO). Therefore, the proposed approach will be helpful in complex hydrothermal scheduling including embedded HVDC lines in existing AC networks for GENCOs and TSO to find their power generation price, bus nodal price, and line limit price.

The effectiveness of the proposed approach has been compared using interior-point and dual-simplex method. The results are presented in Table 4.

Table 4. Comparison of different techniques.

Parameters	Proposed Method	Interior-Point Method	Dual-Simplex Method
No. of iterations	6	10	292
Computational time (sec)	1.05	1.58	2.93

The interior-point method solves formulated problems in 10 iterations while dual-simplex uses 292 iterations without changing any other results. However, the proposed method takes only six iterations to solve the problem.

5. Conclusions

This research article has investigated the optimal operating point of hydrothermal power plants on AC networks with the limited capacity of an HVDC line under the different operating constraints. Modelling of AC power system including VSC-HVDC link has been described for hydrothermal scheduling. The method is computationally efficient as considers linear model of AC network with HVDC transmission link including DCOI Nodal bus power injection matrix is modified to embed line flow constraint in DCOI The quadratic cost curves of thermal generators are linearized segment wise to inclu the minimum and maximum generation limits. The hydro power plant is operated maximum power (P_h = 500 MW) to obtain maximum efficiency. HVDC transmission li flow was successfully limited at P_{HVDC} = 400 MW. The Lagrange multipliers method used for optimal operation of hydrothermal plants on power networks. The propos formulation is applied to two scenarios of a case study. In the first scenario of the ca study, the total thermal generation cost comes out to $278,822.3. In the second scenar of the case study, the total cost of thermal generation is $279,025.4. The difference in co in both scenarios is minimum. It is observed in both scenarios, with the change in loa that this algorithm optimally selects the thermal generator to redispatch to meet the lo demand and other line constraints with minimum cost.

The proposed approach can be extended to perform hydrothermal scheduling c seasonal load changes and deregulated electricity market in the future.

Author Contributions: Conceptualization, A.A. (Ali Ahmed); data curation, S.A.R.K.; formal ana sis, A.A. (Arslan Ashraf) and M.M.G.; funding acquisition, M.A. and M.K.; investigation, S.A.R. methodology, M.M.G.; project administration, M.K.; resources, A.A. (Ali Ahmed), A.A. (Arsl Ashraf), M.A. and M.K.; software, A.A. (Ali Ahmed) and A.A. (Arslan Ashraf); supervision, M. validation, A.A. (Arslan Ashraf) and M.M.G.; visualization, M.A.; writing—original draft, A.A. (Ahmed); writing—review and editing, S.A.R.K. and M.K. All authors have read and agreed to t published version of the manuscript.

Funding: The authors would like to acknowledge the support received from Saudi Data and Authority (SDAIA) and King Fahd University of Petroleum and Minerals (KFUPM) under SDAI KFUPM Joint Research Center for Artificial Intelligence Grant No. JRC-AI-RFP-08, Dhahran 312 KSA. Additionally, the authors would like to thank King Khalid University for its financial suppo to this study.

Data Availability Statement: Not applicable.

Conflicts of Interest: The authors declare no conflict of interest.

Nomenclature

P_{hj}	Output power (MW) of hth hydro unit in j^{th} period
P_{tj}	Output power (MW) of tth thermal unit in j^{th} period
$F_t\left(P_{tj}\right)$	Fuel cost rate ($/hour) for t^{th} unit in j^{th} period
$q_h(P_{hj})$	Water flow rate (Acre-feet/hour) for h^{th} unit in j^{th} period
N_T	Number of thermal power plants
N_H	Number of hydro power plants
N_b	Number of buses
N_M	Number of lines (branches)
$P_{thj} = P_{tj} + P_{hj}$	Total output power (MW) of tth thermal and h^{th} hydro unit in j^{th} period
J_{max}	Maximum number of periods
n_j	Number of hours in j^{th} period
a_t, b_t, c_t	Cost coefficients of tth thermal unit
x_h, y_h, z_h	Water flow rate coefficients of hth hydro unit
q_{TOT}	Total water volume available for power generation
$\mathcal{L}$	Lagrange function
λ, γ, μ	Lagrange multipliers
P_{flow}	HVDC line power flow limit
R_{dc}	Resistance of HVDC line

λ_r	Locational marginal Price (LMP) of rectifier bus
λ_i	Locational marginal Price (LMP) of inverter bus
PCC	Point of common coupling (PCC)
Tr	Coupling transformer
V_{X_r}, V_{C_r}	Voltage at bus X_r and C_r on rectifier side
V_{X_i}, V_{C_i}	Voltage at bus X_i and C_i on inverter side
C	DC link capacitor
ΔP_k	k^{th} bus nodal power balance
P_{gk}	Power generation on k^{th} bus
P_{dk}	Power demand on k^{th} bus
P_k^{cal}	Calculated nodal power on k^{th} bus
G_{km}	Conductance of line connecting bus node k and m
B_{km}	Susceptance of line connecting bus node k and m
θ	Nodal phase angle
Subscript k, m	Indicate the nodal bus
Subscript i, r	Indicate the inverter and rectifier, respectively
V	Bus voltage magnitude
$p.u$	Per unit quantity
DCOPF	Direct current optimal power flow
HVDC	High voltage direct current
N_s	Number of segments of quadratic cost function
S_a	Slope of quadratic cost function
T	Transpose of matrix
W	Water volume
P_B	Line power flow

eferences

Kazantsev, Y.V.; Glazyrin, G.V.; Khalyasmaa, A.I.; Shayk, S.M.; Kuparev, M.A. Advanced Algorithms in Automatic Generation Control of Hydroelectric Power Plants. *Mathematics* **2022**, *10*, 4809. [CrossRef]

Sibtain, D.; Gulzar, M.M.; Murtaza, A.F.; Murawwat, S.; Iqbal, M.; Rasool, I.; Hayat, A.; Arif, A. Variable structure model predictive controller based gain scheduling for frequency regulation in renewable based power system. *Int. J. Numer. Model. Electron. Netw. Devices Fields* **2022**, *35*, e2989. [CrossRef]

Tan, K.; Tian, Y.; Xu, F.; Li, C. Research on Multi-Objective Optimal Scheduling for Power Battery Reverse Supply Chain. *Mathematics* **2023**, *11*, 901. [CrossRef]

Grisales-Noreña, L.F.; Cortés-Caicedo, B.; Alcalá, G.; Montoya, O.D. Applying the Crow Search Algorithm for the Optimal Integration of PV Generation Units in DC Networks. *Mathematics* **2023**, *11*, 387. [CrossRef]

Gul, M.; Tai, N.; Huang, W.; Nadeem, M.H.; Ahmad, M.; Yu, M. Technical and Economic Assessment of VSC-HVDC Transmission Model: A Case Study of South-Western Region in Pakistan. *Electronics* **2019**, *8*, 1305. [CrossRef]

Bento, P.; Pina, F.; Mariano, S.; Calado, M.D.R. Short-term Hydro-thermal Coordination By Lagrangian Relaxation: A New Algorithm for the Solution of the Dual Problem. *KnE Eng.* **2020**, 728–742. [CrossRef]

Gulzar, M.M.; Murawwat, S.; Sibtain, D.; Shahid, K.; Javed, I.; Gui, Y. Modified Cascaded Controller Design Constructed on Fractional Operator 'β'to Mitigate Frequency Fluctuations for Sustainable Operation of Power Systems. *Energies* **2022**, *15*, 7814. [CrossRef]

Saini, O.; Chauhan, A. Optimal Operation of Short-Term Variable-Head Hydrothermal Generation Scheduling Using the Differential Evolution Method, Newton-Raphson Method and Heuristic Search Method. In Proceedings of the 1st International Conference on Recent Innovation in Electrical, Electronics and Communication System, RIEECS, Bhubaneswar, India, 27–28 July 2018.

Wood, A.J. *Power Generation, Operation, and Control*, 3rd ed.; John Wiley & Sons: Hoboken, NJ, USA, 2014.

Jian, J.; Pan, S.; Yang, L. Solution for short-term hydrothermal scheduling with a logarithmic size mixed-integer linear programming formulation. *Energy* **2019**, *171*, 770–784. [CrossRef]

Gjerden, K.S.; Helseth, A.; Mo, B.; Warland, G. Hydrothermal scheduling in Norway using stochastic dual dynamic programming; a large-scale case study. In Proceedings of the 2015 IEEE Eindhoven PowerTech, Eindhoven, The Netherlands, 29 June–2 July 2015; pp. 1–6. [CrossRef]

Fakhar, M.S.; Liaquat, S.; Kashif, S.A.R.; Rasool, A.; Khizer, M.; Iqbal, M.A.; Baig, M.A.; Padmanaban, S. Conventional and Metaheuristic Optimization Algorithms for Solving Short Term Hydrothermal Scheduling Problem: A Review. *IEEE Access* **2021**, *9*, 25993–26025. [CrossRef]

Iqbal, M.; Gulzar, M.M. Master-slave design for frequency regulation in hybrid power system under complex environment. *IET Renew. Power Gener.* **2022**, *16*, 3041–3057. [CrossRef]

14. Gulzar, M.M. Maximum Power Point Tracking of a Grid Connected PV Based Fuel Cell System Using Optimal Control Technic *Sustainability* **2023**, *15*, 3980. [CrossRef]
15. Zheyuan, C.; Hammid, A.; Kareem, A.; Jiang, M.; Mohammed, M.; Kumar, N. A Rigid Cuckoo Search Algorithm for Solvi Short-Term Hydrothermal Scheduling Problem. *Sustainability* **2021**, *13*, 4277. [CrossRef]
16. Liaquat, S.; Fakhar, M.S.; Kashif, S.A.R.; Rasool, A.; Saleem, O.; Padmanaban, S. Performance Analysis of APSO and Fire Algorithm for Short Term Optimal Scheduling of Multi-Generation Hybrid Energy System. *IEEE Access* **2020**, *8*, 177549–1775 [CrossRef]
17. Zeng, X.; Hammid, A.T.; Kumar, N.M.; Subramaniam, U.; Almakhles, D.J. A grasshopper optimization algorithm for optin short-term hydrothermal scheduling. *Energy Rep.* **2021**, *7*, 314–323. [CrossRef]
18. Hassan, T.U.; Alquthami, T.; Butt, S.E.; Tahir, M.F.; Mehmood, K. Short-term optimal scheduling of hydro-thermal power pla using artificial bee colony algorithm. *Energy Rep.* **2020**, *6*, 984–992. [CrossRef]
19. Azad, A.S.; Rahaman, S.A.; Watada, J.; Vasant, P.; Vintaned, J.A.G. Optimization of the hydropower energy generation usi Meta-Heuristic approaches: A review. *Energy Rep.* **2020**, *6*, 2230–2248. [CrossRef]
20. Zhao, J.; Zhang, Y.; Liu, Z.; Hu, W.; Su, D. Distributed multi-objective day-ahead generation and HVDC transmission jo scheduling for two-area HVDC-linked power grids. *Int. J. Electr. Power Energy Syst.* **2021**, *134*, 107445. [CrossRef]
21. Nemati, M.; Braun, M.; Tenbohlen, S. Optimization of unit commitment and economic dispatch in microgrids based on gene algorithm and mixed integer linear programming. *Appl. Energy* **2018**, *210*, 944–963. [CrossRef]
22. Bento, P.M.R.; Mariano, S.J.P.S.; Calado, M.R.A.; Ferreira, L.A.F.M. A Novel Lagrangian Multiplier Update Algorithm Short-Term Hydro-Thermal Coordination. *Energies* **2020**, *13*, 6621. [CrossRef]
23. Ergun, H.; Dave, J.; Van Hertem, D.; Geth, F. Optimal Power Flow for AC–DC Grids: Formulation, Convex Relaxation, Line Approximation, and Implementation. *IEEE Trans. Power Syst.* **2019**, *34*, 2980–2990. [CrossRef]
24. Castro, L.M.; González-Cabrera, N.; Guillen, D.; Tovar-Hernández, J.; Gutiérrez-Alcaraz, G. Efficient method for the optin economic operation problem in point-to-point VSC-HVDC connected AC grids based on Lagrange multipliers. *Electr. Power Sy Res.* **2020**, *187*, 106493. [CrossRef]
25. Nguyen, T.T.; Pham, L.H.; Mohammadi, F.; Kien, L.C. Optimal Scheduling of Large-Scale Wind-Hydro-Thermal Systems w Fixed-Head Short-Term Model. *Appl. Sci.* **2020**, *10*, 2964. [CrossRef]
26. Ahmad, A.; Kashif, S.A.R.; Nasir, A.; Rasool, A.; Liaquat, S.; Padmanaban, S.; Mihet-Popa, L. Controller Parameters Optimizati for Multi-Terminal DC Power System Using Ant Colony Optimization. *IEEE Access* **2021**, *9*, 59910–59919. [CrossRef]
27. Gulzar, M.M.; Sibtain, D.; Ahmad, A.; Javed, I.; Murawwat, S.; Rasool, I.; Hayat, A. An Efficient Design of Adaptive Moc Predictive Controller for Load Frequency Control in Hybrid Power System. *Int. Trans. Electr. Energy Syst.* **2022**, *2022*, 78942 [CrossRef]
28. Baradar, M.; Ghandhari, M. A Multi-Option Uni Fi Ed Power Flow Approach for Hybrid AC/DC Grids Incorporating. *IE Trans. Power Syst.* **2013**, *28*, 2376–2383. [CrossRef]
29. Gonzalez-Torres, J.C.; Damm, G.; Costan, V.; Benchaib, A.; Lamnabhi-Lagarrigue, F. A Novel Distributed Supplementary Cont of Multi-Terminal VSC-HVDC Grids for Rotor Angle Stability Enhancement of AC/DC Systems. *IEEE Trans. Power Syst.* **2020**, 623–634. [CrossRef]
30. Sibtain, D.; Gulzar, M.M.; Shahid, K.; Javed, I.; Murawwat, S.; Hussain, M.M. Stability Analysis and Design of Variable Step-Si P&O Algorithm Based on Fuzzy Robust Tracking of MPPT for Standalone/Grid Connected Power System. *Sustainability* **20** *14*, 8986. [CrossRef]
31. Cao, J.; Du, W.; Wang, H.F.; Member, S. An Improved Corrective Security Constrained OPF for Meshed AC/DC Grids wi multi-terminal VSC-HVDC. *IEEE Trans. Power Syst.* **2015**, *31*, 485–495. [CrossRef]
32. Al-Sakkaf, S.; Kassas, M.; Khalid, M.; Abido, M.A. An Energy Management System for Residential Autonomous DC Microgr Using Optimized Fuzzy Logic Controller Considering Economic Dispatch. *Energies* **2019**, *12*, 1457. [CrossRef]
33. Salman, U.; Khan, K.; Alismail, F.; Khalid, M. Techno-Economic Assessment and Operational Planning of Wind-Battery Distribut Renewable Generation System. *Sustainability* **2021**, *13*, 6776. [CrossRef]
34. Alhumaid, Y.; Khan, K.; Alismail, F.; Khalid, M. Multi-Input Nonlinear Programming Based Deterministic Optimizati Framework for Evaluating Microgrids with Optimal Renewable-Storage Energy Mix. *Sustainability* **2021**, *13*, 5878. [CrossRef]
35. Khalid, M. Wind Power Economic Dispatch—Impact of Radial Basis Functional Networks and Battery Energy Storage. *IE Access* **2019**, *7*, 36819–36832. [CrossRef]
36. Khalid, M.; Ahmadi, A.; Savkin, A.V.; Agelidis, V.G. Minimizing the energy cost for microgrids integrated with renewable ener resources and conventional generation using controlled battery energy storage. *Renew. Energy* **2016**, *97*, 646–655. [CrossRef]
37. Gulzar, M.M.; Iqbal, A.; Sibtain, D.; Khalid, M. An Innovative Converterless Solar PV Control Strategy for a Grid Connect Hybrid PV/Wind/Fuel-Cell System Coupled with Battery Energy Storage. *IEEE Access* **2023**. [CrossRef]
38. Gulzar, M.M.; Sibtain, D.; Khalid, M. Cascaded Fractional Model Predictive Controller for Load Frequency Control in Multiar Hybrid Renewable Energy System with Uncertainties. *Int. J. Energy Res.* **2023**. [CrossRef]

mathematics

MDPI

rticle

Reinforcement Learning-Enabled Electric Vehicle Load Forecasting for Grid Energy Management

I. Zulfiqar [1,2], Nahar F. Alshammari [3] and M. B. Rasheed [4,*]

1 Department of Telecommunication Systems, Bahauddin Zakariya University, Multan 60000, Pakistan
2 Department of Electrical Engineering, University of Engineering and Technology, Lahore 54890, Pakistan
3 Department of Electrical Engineering, Faculty of Engineering, Jouf University, Sakaka 72388, Saudi Arabia
4 Escuela Politécnica Superior, Universidad de Alcalá, ISG, 28805 Alcalá de Henares, Spain
* Correspondence: muhammad.rasheed@uah.es

Abstract: Electric vehicles are anticipated to be essential components of future energy systems, as they possess the capability to assimilate surplus energy generated by renewable sources. With the increasing popularity of plug-in hybrid electric vehicles (PHEVs), conventional internal combustion engine (ICE)-based vehicles are expected to be gradually phased out, thereby decreasing greenhouse gases and reliance on foreign oil. Intensive research and development efforts across the globe are currently concentrated on developing effective PHEV charging solutions that can efficiently cater to the charging needs of PHEVs, while simultaneously minimizing their detrimental effects on the power infrastructure. Efficient PHEV charging strategies and technologies are necessary to overcome the obstacles presented. Forecasting PHEV charging loads provides a solution by enabling energy delivery to power systems based on anticipated future loads. We have developed a novel approach, utilizing machine learning methods, for accurately forecasting PHEV charging loads at charging stations across three phases of powering (smart, non-cooperative, and cooperative). The proposed Q-learning method outperforms conventional AI techniques, such as recurrent neural and artificial neural networks, in accurately forecasting PHEV loads for various charging scenarios. The findings indicate that the Q-learning method effectively predicts PHEV loads in three scenarios: smart, non-cooperative, and cooperative. Compared to the ANN and RNN models, the forecast precision of the QL model is higher by 31.2% and 40.7%, respectively. The Keras open-source set was utilized to simulate three different approaches and evaluate the efficacy and worth of the suggested Q-learning technique.

Keywords: Q-learning; electric vehicles; artificial neural network; plug-in hybrid electric vehicles

MSC: 68T07

itation: Zulfiqar, M.; Alshammari, .F.; Rasheed, M.B. Reinforcement earning-Enabled Electric Vehicle oad Forecasting for Grid Energy anagement. *Mathematics* **2023**, *11*, 80. https://doi.org/10.3390/ ath11071680

cademic Editors: Atanda Raji and haled M. Abo-Al-Ez

eceived: 2 March 2023
evised: 24 March 2023
ccepted: 28 March 2023
ublished: 31 March 2023

1. Introduction

Electric vehicles (EVs) have been mainly integrated into power grids because of their significant benefit over traditional combustion engine vehicles, which includes bolstered energy self-sufficiency and diminished levels of carbon emissions [1]. There have been numerous creative works on EV recharge infrastructure [2,3]. The surging demand for EVs is introducing intricacy to the energy grid system, leading to difficulties in efficient management. One of the significant apprehensions for executives in this world is accurately projecting the charging capacity of automobiles, to determine their energy consumption requirements [4,5]. Recently, plug-in hybrid electric vehicle (PHEV) energy control methods have used reinforcement learning (RL) [6]. The incorporation of forecasting methods was also used in [7] to explore the merging of PHEVs into microgrids. Furthermore, Ref. [8] considered blockchain technology for trading energy. Additionally, Ref. [9] proposed PHEV energy management with data-driven techniques to estimate battery discharge depth.

The forecasting of EV charging station load has drawn a lot of interest in the literatu For instance, in [10], the load demand for Ev was predicted using online ride-hailir services. To model the load demands of PHEVs and EVs, the authors examined t stochastic load demand of EVs [11]. A statistical utilization strategy has been examined forecast the PHEVs load profile. The energy shift at midday and night has been predict using the inverse load flow and a localized charging time-shift procedure [12]. For examp Refs. [13,14] used charging stations with battery storage systems to supplement ener or capacity or to delay demand. Intelligent PHEV charging demand forecasting in [1 aims to keep a constant daily network load profile while ensuring that each PHEVs lo requirement is fulfilled in [16]. Ensemble learning predicts load consumption of cooperati PHEVs using innovative data mining to forecast charging sites [17].

1.1. Literature Survey

The battery charging of PHEVs is primarily done at night because everyday vehic travel considerably lowers the charging time. Due to the high number of these vehicl early night charging of vehicles damages power infrastructure traits like overloading ar fast ramping of energy sources. Increased EV adoption may also lead to more seve grid features like feeder overcrowding, unfavourable peak demand, increased power lo decreased load factor, harmonic distortion, phase unbalance, etc. The charging of PHE during high demand has these negative impacts [18]. Smart recharge for electric vehicl is the remedy to the problems mentioned above [19]. EVSC can effectively handle E charging, particularly at night, to satisfy the network's technical restrictions. According data, vehicles are typically only driven 4 to 5 percent of the time, with the remainder of th day being spent in parking lots or basements [20]. The extra storage space provided by th batteries of electric vehicles can be used to create a large-scale energy storage system [21

PHEV owners have expressed high satisfaction with the cooperative charging EVs, and grid characteristics are considered when deciding the number of EVs to b charged per time interval [22]. In addition to considering the use of renewable energ production in a power network, the EVSC process should also consider PHEV owners an technological limitations on renewable-based power [23]. PHEVs have the potential contribute significantly to helping keep or enhance the normal functioning of a power gri PHEV charging via EVSC can also provide extra services like frequency regulation [2 The schematic diagram of EVSC is shown in Figure 1. Both centralized and decentralize types of charging—in which electric vehicles are refuelled in intelligent houses, at home by fleet operators, as well as parking lots occupied by EVs parked overnight—can be use to refill the batteries of electric cars. EVSCs flexible charging and auxiliary services ca reduce the energy cost of plug-in cars by as much as 60% when compared to convention vehicles [25].

A company that operates a fleet of electric vehicles must be able to forecast its loa in order to optimize the number of EVs, their charging rates and available charging sit Additionally, more energy-efficient use of existing resources can be enabled by predictin the future capacity needs [26–28]. The load profile of EVs in particular geographic regio results from the habits people have while driving and contributes to the overall deman for electricity [29,30]. A method has been developed to forecast electric vehicle trav patterns and arrival/departure times. In addition, an ensemble learning-based forecastin method can predict the number of charging stations required in any given area [31,3 The proposed method was developed by combining three different types of learnin algorithms: Linear regression (to find the weighting for each primary learner), recurre neural networks, and long short-term memory. The ability of three neural network mode to predict the load profile in an EVS has been examined in other research [33]. The Radi Basis Function technique has produced better results in load forecasting than the tradition method, but it also had a higher error rate and processing cost. In [34], six distinct dee learning techniques have been examined from the EV usage prediction perspective. Th methods included canonical LSTM, ANN, RNN, gated recurrent units, bi-directional LSTM forecasting algorithms, and stacked auto-encoders.

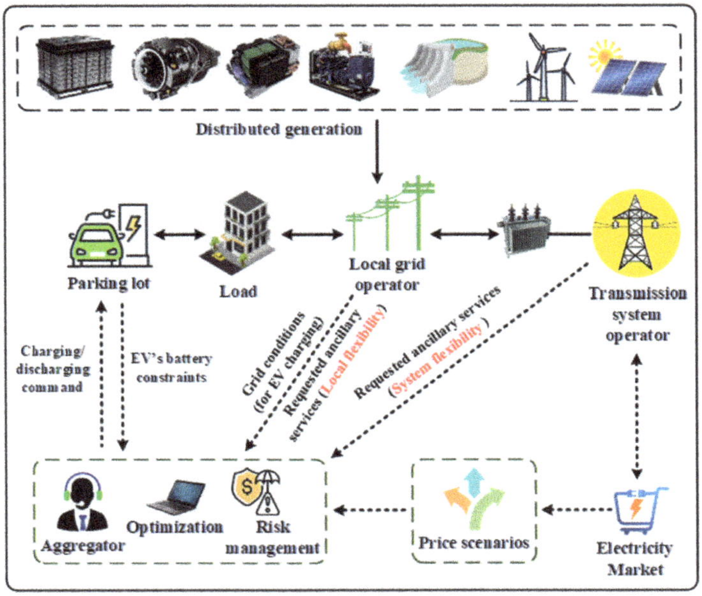

Figure 1. A diagrammatic representation illustrating the charging process of plug-in hybrid electric vehicles within power distribution networks.

1.2. Motivation and Problem Statement

Deep learning (DL) techniques are increasingly being used to improve the prediction performance of highly nonlinear systems to reduce the computational burden on a design and allow the real-time application of complex controllers [34]. The literature achieves the aforementioned goals by utilizing a variety of machine learning (ML) methods [35]. Artificial neural networks (ANN) were mainly used in [36] for datasets with no time dependency between the accessible datasets. Time-varying data sets are commonly processed using recurrent neural networks (RNNs).RNNs and CNNs are two popular AI techniques used to gather data for text and media applications, respectively [12]. SARSA and Q-learning are two additional reinforcement learning techniques that facilitate real-time learning of intricate problems. The action-reward system underlies these techniques [32–37]. Consequently, the RL technique may impose a different reward for each activity that moves an agent from one condition to another. This paper foresees PHEV loads using one of the RL techniques, the Q-learning (QL) approach. Out of all possible actions and states, the QL method can identify the best ones. As a result, by using the outputs of other ML techniques, like ANN, LSTM, RNN, GRU, and CNN, as initial inputs, the efficacy of the QL technique can be improved [38].

Despite the variety of methods used to predict PHEV load demand [11–13], more research still needs to be focussed on an overarching tactic that encompasses all PHEV load demand scenarios, including smart, cooperative, and non-cooperative situations. In this study, the authors have explored the potential of an RL-based QL approach in forecasting PHEV load under diverse scenarios. This research evaluated three commonly used artificial intelligence techniques–RNN, ANN, and QL–to determine their efficacy in prediction. The investigation indicates that implementation of the QL technique, which leverages the initial outcomes of the ANN and RNN methodologies, yields enhanced predictive precision.

1.3. Real Contribution

The leading contributions and innovations of this article are as follows:

- A comprehensive investigation of load forecasting challenges for PHEV chargi remains imperative through implementing RL, a potent tool for combining multi ML models [6]. To address this matter, a new approach based on QL for forecasti load at EV charging stations has been introduced in this publication. QL has be frequently employed in problems with distinct states and actions. Our proposed model is suitable for the control task that demands ongoing response to the prevaili circumstances. The system comprises neural networks and deduces the different for the state evolution of unknown epistemic uncertainty. This solution prese an opportunity to enhance the operational efficiency of PHEV charging, while a serving as a mechanism of reinforcement for energy dispatching within power gri
- The recommended RL methodology for determining the optimal framework for PH load forecasting encompasses smart, cooperative, and non-cooperative scenarios. T developed QL approach exhibits superior efficiency, precision, and flexibility in PH load estimation when compared to traditional ANN and RNN models. Furtherm incorporating modifications such as adjusting the epoch, hidden layer, and no quantities can significantly augment the accuracy of PHEV charging load predictio as evidenced by empirical analyses.

1.4. Paper Organization

This article is organized as follows. Section 2 presents the technical background the three models utilized in this paper. Section 3 describes the charging behaviour the PHEVS. Section 4 presents the proposed framework for the PHEVs load forecasti Sections 5 and 6 show the evaluation criteria and QL-model forecasting performance w different network depths. The implementation of ANN, RNN and the proposed mo and simulation results considering the test cases are described in Section 7. Sensitiv analysis, validation of the proposed model, QL in terms of speed, flexibility and accur and discussion are presented in Sections 8, 9, 10 and 11, respectively. Finally, Section concludes the article.

2. Technical Background

The three models—ANN, RNN, and QL—used in this paper are briefly introduced this part.

2.1. ANN

Figure 2a shows the schematic unit of ANN [23]. The Artificial Neural Network (AN comprises three distinctive layers i.e., input layer (X), hidden layer (H), and output lay (Y) that operate using multiple nodes. As shown in Figure 2a, a weight (ω_{ij}) is used every line of the ANN technique in between each pair of successive layers (ith previo layer, and jth present layer). The Equation (1) can be used to determine each output nod

$$NO_j^p = \sigma_j^p\left(\sum_{i=0}^{n_{p-1}} (NO_i^{p-1}\omega_{ij}) + b_j^p\right)$$

In the context of neural networks, the jth output of the pth layer is represented by N whereas NO_j^{p-1} represents the jth output of the $p-1$th layer. Similarly, b_j^p and σ_j^p resp tively refer to the bias and activation function of the jth node in the pth layer. Furtherm n_{p-1} denotes the number of nodes in the $p-1$th layer.

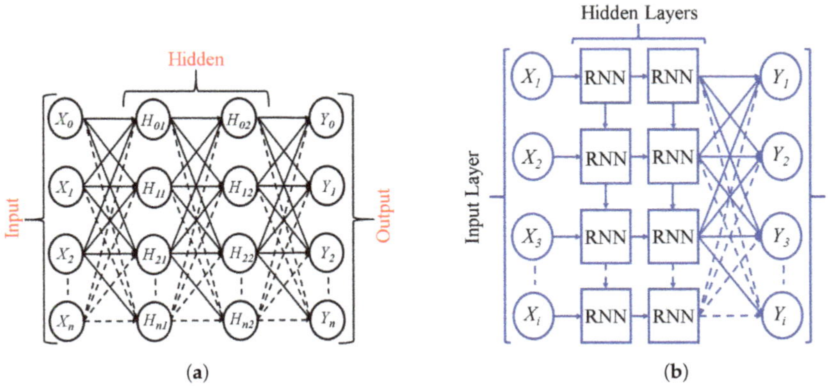

Figure 2. ANN and RNN Blocks Techniques. (**a**) ANN Block. (**b**) RNN Block.

2.2. RNN

The RNN technique's block layout is depicted in Figure 2b. The time-series data set's future is predicted using RNNs [24]. It can be applied to various chores, including predicting weather patterns, green energy trends, stock market trends, and other activities. The line the RNN method uses between each hidden layer's node is the primary distinction between it and the ANN technique. The output of the RNN unit is shown in Equation (2):

$$\begin{cases} r_t = b + Ks_{t-1} + Wx_t \\ s_t = f(r_t) \\ q_t = g(d + Us_t) \end{cases} . \quad (2)$$

At time t, the state of the hidden layers is denoted by s_t, while at time $t-1$, it is indicated by s_{t-1}. Similarly, the input and output of the RNN unit at time t are represented by r_t and q_t, correspondingly. K, W and U are the weight matrices of s_{t-1}, x_t, and q_t respectively. Both f and g are the activation functions (softmax, tanh, etc.) of memory and output of the RNN unit, while b and d are the biases.

2.3. QL

The QL method, which uses the Markov decision process to select the most appropriate action from all possible options, is one of the model-free RL techniques [25,26]. QL falls under the category of off-policy methods. When choosing the best future action, the QL technique selects the initial state, reward, current state, and accessible activities in a very similar way to the SARSA. The primary equation, which is used by the QL technique to choose, presented in (3):

$$\begin{cases} Q^{\text{new}}(c_t, m_t) \leftarrow Q(c_t, m_t) + \beta(R_t \\ \quad + d_f \max_m Q(c_{t+1}, m_t) - Q(c_t, m_t)) \end{cases} \quad (3)$$

where c_t and m_t are the current states and current actions of the QL technique. $Q(c_t, m_t)$ is the Q-value updated Q-value of c_t and m_t can be denoted as $Q^{\text{new}}(c_t, m_t)$. The QL method has a learning rate β that satisfies the condition $0 \leq \beta \leq 1$, while the discount factor d_f is between 0 and 1 ($0 \leq d_f \leq 1$). The optimal future value estimation is given by $\max_m(Q(c_{t1}, m_t))$. The proposed QL reward for the future action and state is R_t. The initial conditions for the Q-value in this proposed technique are zero. The optimal solution for PHEV loads is found using the QL method. The proposed QL technique updates the Q-value depending on c_t and m_t after each iteration. The best course of action for the subsequent steps is determined using Equation (4):

$$m_t = arg \max_m(Q(c_{t+1}, m_t)). \quad (4)$$

3. Charging Behavior of PHEVs

The charging strategy, charging time, amount of charged PHEVs, battery capacity, battery state of charge (B_{soc}), and start time are all unknown variables that affect PHEV load behaviour. The three charging cases for PHEVs—smart, cooperative, and non-cooperative—are explored in this article.

3.1. Non-Cooperative PHEV Charging

The simplest charging approach for PHEVs is the non-cooperative strategy, which involves plugging them into charging stations. PHEVs are believed to commute daily from their homes in the morning to their homes in the evening. In this instance, around 5:30 p.m. when most people get home, most PHEVs are hooked up and begin charging. As a result, this strategy considers a limited range uniformly distributed probability density function (PDF) with a charging start time of 5:30 p.m. Such a PDFs mathematical form is provided in Equation (5) [25]:

$$y(\tau_x) = \frac{1}{u-v}, \; v = 18, \; u = 19, \; u \leq \tau_x \leq v \quad (5)$$

where u and v are constant values, τ_x is the charging start time of PHEV.

3.2. Cooperative PHEV Charging

Stockholders in the cooperative charging plan typically link their cars to chargers during off-peak hours to prevent confrontation with the sunset peak hours, when power costs are higher. Consequently, to save money on their electricity prices, people defer charging until after 9:30 p.m. The cooperative charging strategy presented in Equation (6) is then modelled using the provided pdf:

$$y(\tau_x) = \frac{1}{u-v}, \; v = 21, \; u = 24, \; u \leq \tau_x \leq v \quad (6)$$

3.3. Smart PHEV Charging

A smart charging plan schedules charging when energy is most reasonably priced, demand is at its lowest point, or excess capacity exists. All smart charging techniques adhere to the fundamental principle that a vehicle should only be charged when doing so benefits both the owner and the utility. A typical pdf, as shown in Equation (7), illustrates the difficulty of choosing the charging start time using various smart charging options [24]:

$$y(\tau_x) = \frac{1}{\beta\sqrt{2\pi}} exp^{\left(-\frac{1}{2}\left(\frac{\tau_{start}-\alpha}{\beta}\right)^2\right)}, \quad \alpha = 1, \quad \beta = 3. \quad (7)$$

where α is the mean of values and β is the covariance.

Once the PHEV is connected to the home charger, the battery begins to charge. The battery's leftover B_{soc} is calculated using the vehicle's daily mileage. The ratio of available energy to maximal stored energy is known as a battery's B_{soc}. The daily mileage of a car is said to follow a log-normal PDF presented in Equation (8):

$$y(z) = \frac{1}{z\beta\sqrt{2\pi}} exp^{\frac{-(\ln(z)-\alpha)^2}{2\beta^2}}, \quad z > 0. \quad (8)$$

The B_{soc} at the time of plug-in is then calculated using the vehicle's driven mile and all-electric range (AE_{range}) as follows in Equation (9):

$$B_{soc} = \begin{cases} 0, & z > AE_{range} \\ \frac{AE_{range}-z}{AE_{range}} \times 100\%, & z \leq AE_{range}. \end{cases} \tag{9}$$

A multitude of PHEVs exist, including the PHEV-20, PHEV-30, PHEV-40, and PHEV-60, each characterized by a unique AE_{range} denoted by a numerical subscript. The PHEV-20 has been selected as a primary example in the investigation of market potential over time [29], but consideration may be extended to additional PHEV models given the unrestricted nature of the methodology. Equation (10) is used to determine the charging time of a PHEV [29].

$$\tau_d = \frac{B_{cap} \times (1 - B_{soc}) \times D_{depth}}{\eta \times T}. \tag{10}$$

where B_{cap}, D_{depth}, η and T and the PHEV battery capacity (in kWh), depth of discharge of PHEV battery, charging efficiency, and PHEV charger rate (kW), respectively. The charging types for PHEVs are outlined in the Table 1, and are contingent on the charging levels provided by the charger [20]. Ethics dictate that the first two charging stages are suitable for PHEVs that are charged domestically. As the charging stations with the third stage are specifically constructed for public transportation, this write-up excludes their inclusion. As per Table 2, PHEVs are segregated into four groups, with each group possessing characteristic market and share traits. In this study, the market share of PHEVs can be represented with a discrete distribution. Each PHEV group is arbitrarily selected from the market shares shown in Equations (11) and (12) using a normal distribution.

$$\alpha_{B_{cap}} = \frac{\text{Min } B_{cap} + \text{Max } B_{cap}}{2} \tag{11}$$

$$\beta_{B_{cap}} = \frac{\text{Min } B_{cap} - \text{Max } B_{cap}}{4} \tag{12}$$

The proposed QL approach can compile training data to forecast PHEV charging demand using three simulated charging methods: Smart, non-cooperative, and cooperative, based on the (5) to (12). Min and Max represent the minimum and maximum values, respectively.

Table 1. Various Charger Types for PHEVs.

Charging Type	I/P Voltage	P_{max} (KW)
Level-I (AC)	120 V_{ac}	1.42
Level-II (AC)	208–240 V_{ac}	11.5
Level-III (AC)	208–240 V_{ac}	97
Level-III (DC)	208–600 V_{dc}	239

Table 2. Various Classes for PHEVs.

Class	Market Share	B_{cap} (Min–Max)
Mini Vehicle	0.2	8–12
Mid Size Vehicle	0.3	14–18
Economy Vehicle	0.3	10–14
Light Truck	0.3	19–23

4. Proposed QL-Based PHEV Charging Load Forecasting Framework

In this article, using the results of RNN and ANN, we developed a novel RL-bas model that optimally forecasts the load demands of smart, cooperative, and non-cooperat PHEVs.The proposed model has two major parts: (1) Data Preprocessing Pha (2) QL-based forecasting considering the optimal results of ANN and RNN as init inputs. Figure 3 shows the structure of the proposed reinforcement learning-based forec ing model.

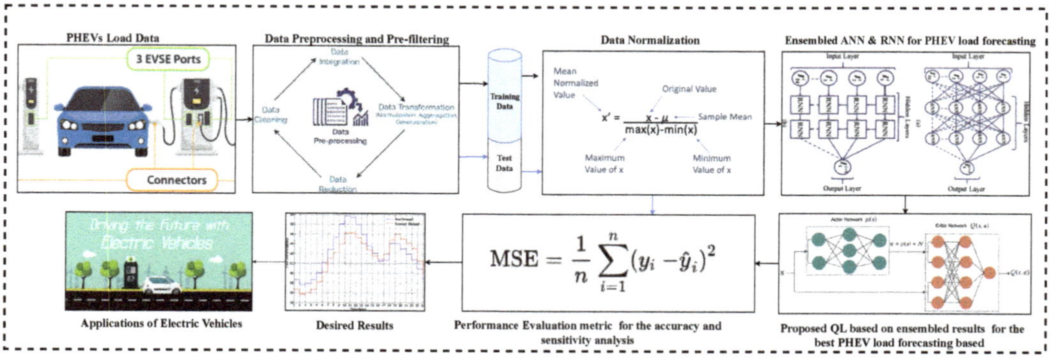

Figure 3. Proposed QL based PHEVs load forecasting Model.

4.1. Data Source

Caltech has made the ACN-Data [39] available to academics for research purposes; t dataset comprises EV charging events from various workplaces. More than 30,000 chargi events are organized in the dataset from two Californian workplace charging locations, Jet Propulsion Laboratory, and the Caltech campus. For this article, our test data consi of charging information collected from the charging station on the Caltech campus. T ACN (Adaptive Charging Network) is open to the public and equipped with a 50 kW rapid charger, frequently used by non-drivers. The dataset covers details of all transacti made within the architecture of ACN, with 54 EV supply equipment. The dataset's relat data categories are described in Table 3. The exactitude of load forecasting depends on d pre-processing because the initial data will contain anomalies and missing numbers.

Table 3. Data fields in ACN-Data.

Field	Description
Time of connection	The plugs in time of users.
Accomplished charging time	The time of last non-zero charging rate.
Time of disconnection	The unplugs time of users.
kWh supply	Supplied energy measurement.
Session ID	Unique identity for the session.
Station ID	Unique identity of the EV Supply Equipment

4.2. Preprocessing Module

The EV charging stack data were initially analyzed and refined for inquiry. To enab real-time charging control based on energy usage during charging periods, it is imperati to establish the average hourly charging capacity of the stations, taking into account t variability of data encountered due to its random nature. To achieve this, the dataset w pre-processed, whereby inaccurate data points were replaced with the average chargi load observed during a given time slot across different days. This step was essential ensure accurate forecasting without being impacted by erroneous data.

For practical reasons, the data pertaining to the charging load operation were bifurcated into training and test sets using a proportion of 0.7/0.3. The model was suitably trained with the training instances, while the final function evaluation utilized the test collection. Multiple neural networks were employed to learn from the charging load data; however, the efficiency of these networks was significantly influenced by the size of the data. As a resolution, the input data of the forecast model were altered through min-max normalisation, which constrained the information within a specific limit (0–1), as illustrated in Equation (13):

$$Y' = \frac{Y - y_{\min}}{y_{\max} - y_{\min}} \tag{13}$$

4.3. QL-Based Forecasting Module

The three primary learners, ANN, RNN, and QL, were trained using the training set by the training technique. Following the last training session, predictions were made using the training data and prepared elementary learners. Forecasts from the training data and the proper labels were then set as the features. During the testing process, predictions were generated using the trained base of learners on test data. The forecasts made by the primary learners were then used to test the learned weighted averaging model. After reversing normalization, the outcomes were obtained, and the test labels and marks used to calculate the prediction accuracy.

- When using the ANN technique to forecast the load on PHEVs, the input and output ANN units should be chosen appropriately. Due to the time series nature of PHEVS load data, the ANN unit utilized prior PHEV load data. The baseline 24-h PHEVs load data were helpful for more accurate one-hour-ahead load forecasting. The deployment of ANN and RNN for predicting one hour ahead is shown in Figure 4.
- The proposed QL method for the PHEVs load forecasting used the previous days' ANN and RNN forecasting results. In hopes of identifying the best day-ahead PHEV load forecasting, the proposed QL approach chose the best course of action based on the output of ANN and RNN. The proposed QL model's reward function is shown in Equation (14):

$$R_t(\gamma_j) = \frac{1}{\exp\left|(J_{Optimal_t} - \gamma_j(J_{ANN}/RNN_t))\right|} \tag{14}$$

  where γ_j is the random number of the jth action. The proposed approach chose one of two possible courses of action for each time horizon ($j = 0$ for ANN and $j = 1$ for RNN). The power forecast for the ANN or RNN method was (J_{ANN/RNN_t}). Additionally, $J_{Optimal_t}$ contains the PHEV data for the tth period. The ANN and RNN predicted results were used in the devised method to predict the PHEV future load and compare it with actual data. Thus, 24 optimal actions were chosen in the proposed QL approach for the day ahead horizon. Therefore t can be any integer between 1 and 24 for the day ahead horizon. $J_{Optimal_t}$, J_{ANN_t}, and J_{RNN_t} were the optimal, ANN, and RNN PHEV load of the tth day, respectively. The γ_j values between 0 and 2 were used to improve the ability of the QL technique to find the best search locations. When the action was 1, the RNN technique chose the reward function. Alternatively, the reward function was determined from the results of the ANN method.

QL is a learning system that is built on values. Value-based algorithms update the value function based on a calculation (particularly Bellman equation). The other policy-based version calculates the value function using the most recent policy improvement's greedy policy. QL is an ad hoc student. This implies that it learns the value of the best strategy independent of the agent's behaviours. On the other hand, an on-policy learner learns the value of the policy being carried out by the agent, including the exploration stages, and it will find an optimum strategy that considers the exploration implicit the approach.

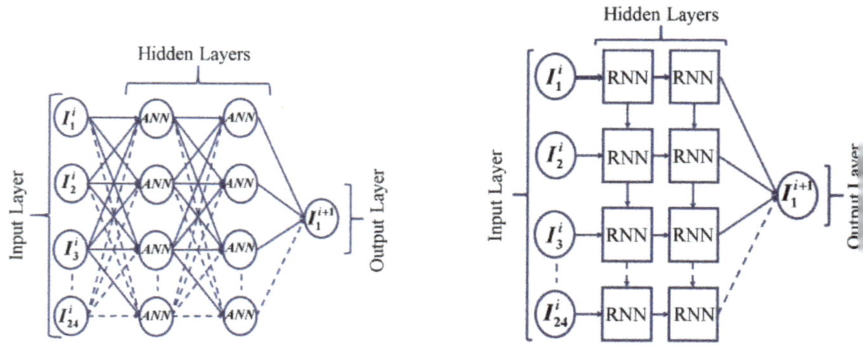

(a) (b)

Figure 4. Employing artificial neural network (ANN) and recurrent neural network (RNN) metho ologies to forecast plug-in hybrid electric vehicle (PHEV) demand. (**a**) The PHEV load forecasting ANN method. (**b**) The PHEV load forecasting by RNN method.

5. Evaluation Criteria

The predicting performance of the analyzed techniques was evaluated using the mea square error (MSE), a standard evaluation criterion. Considering P^{test} and $\hat{P}^{test}$ are th actual and expected loads of the EV charging station at time step t, respectively. The MS measured the errors between the real and the forecasting values, which can be formulate in Equation (15) as below:

$$MSE = \sqrt{\frac{\sum_{t=1}^{T}\left(P_t^{\text{test}} - \hat{P}_t^{\text{test}}\right)^2}{T}} \quad (1$$

6. QL-Model Forecasting Performance with Different Network Depths

We evaluated the predicting effectiveness in comparison to various network leve The results of the Mean Squared Error (MSE) and Mean Absolute Error (MAE) that we found when the layers were changed from two to six when using the prepared datas are shown in Figure 5. The proposed QL model worked best with five layers. As th network depth increased from two to five, the MSE and MAE values became small and the prediction performance improved. Properly boosting network depth and setting significantly enhanced the predicting performance. Due to increased parameter redundanc and reduced data variety, as a network's depth grows, it may overfit and suffer fro performance loss. The training period needed for each layer of QL is listed in Table 4. A layers increased, factors for training optimization increased, causing a longer required tim More layers were needed to improve forecast accuracy, resulting in time consumption.

Table 4. Training time for different layers of QL.

Layer of QL	CPU Time (s)
2	499.48
3	798.21
4	1401.98
5	2001.73
6	2312.19

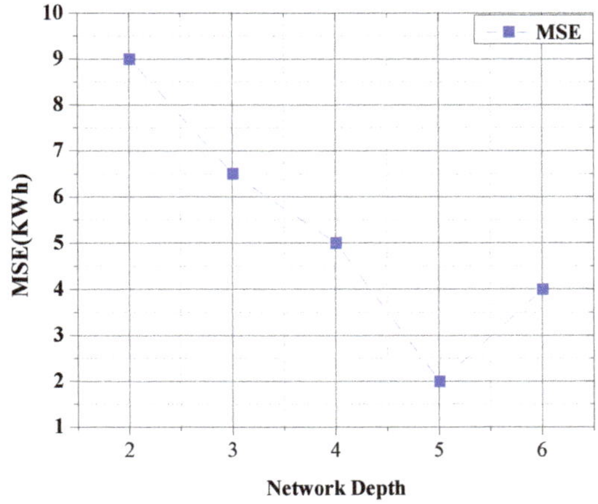

(a)

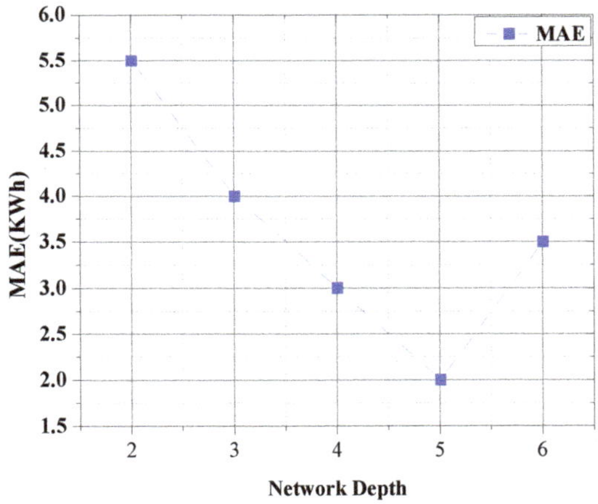

(b)

Figure 5. The metrics (**a**) MSE and (**b**) MAE for the QL model, adjusted for alterations in network depth. (**a**) MSE versus different network layers. (**b**) MAE versus different network layers.

6.1. Convergence of the QL

In Figure 6, the state $(c, m) = (0, 50)$ marks the convergence of the QL model. The optimal value function $V(0, 50) = 160.83$ is approached as $\min_{m \in \mathcal{A}} Q(0, 50, m)$ converges. The synchronized parallel Q-learning procedure involves offline learning of the QL model, in which batch updates of all state-action pairs (Q-functions) occur at each training instance. The step sizes utilized in this process are inversely proportional to the number of visits (updates) to each combination of state and action.

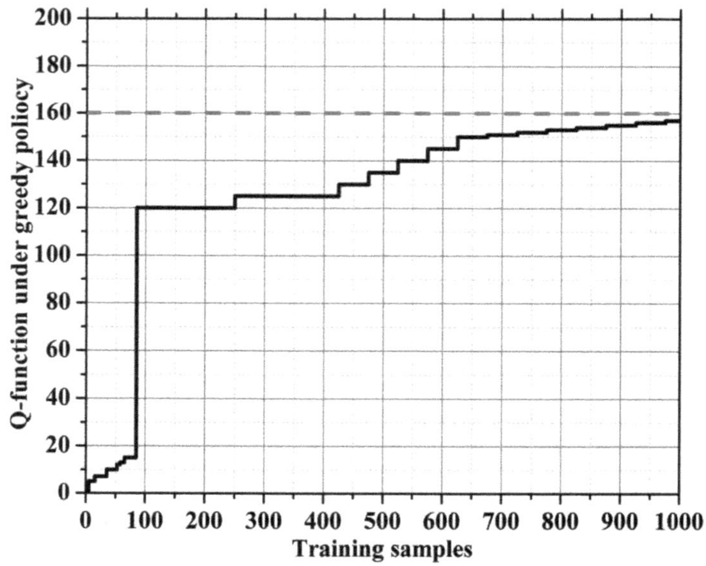

Figure 6. The convergence of the QL at state $(c, m) = (0, 50)$. The red dotted line signifies the id value, while the black dotted line denotes the point at which convergence occurs.

7. Test Cases Simulations and Results

This section compares three techniques (RNN, ANN, and QL) in various scenari to show the benefits and efficacy of the proposed ML techniques. The number of hidd layers used for RNN and ANN is shown in Table 5. The number 24 is the input for the M methods. (For example, the charging load for the following hour is forecast using the mo current 24-h data). This analysis estimated the devised PHEV load forecasting approa using the free and open-source Keras software [26]. Additionally, MATLAB was utilized build the original PHEV data for various charging techniques.

Table 5. Examining and projecting the charging demand of PHEVs utilizing ANN, RNN, and Q approaches (MSE, MAPE and Epoch) across various penetrations (30%, 60% and 90%).

Techniques	Charging Strategy	Penetration (%)	MSE (KW)	Epoch	MAPE(%)
ANN-1 (non-cop)	Non-cooperative	30	4.2	3000	4.3371
RNN-1 (non-cop)			9.3	1000	2.9189
QL-1 (non-cop)			0.79	10,000	2.4741
ANN-1 (Cop)	Cooperative	30	7.06	3000	4.4901
RNN-1 (Cop)			9.12	500	3.0129
QL-1 (Cop)			6.21	10,000	2.7210
ANN-1 (Smart)	Smart	30	6.23	3000	4.3210
RNN-1 (Smart)			6.38	500	2.7112
QL-1 (Smart)			5.30	10,000	2.4214
ANN-2 (non-cop)	Non-cooperative	60	1.67	3000	5.2121
RNN-2 (non-cop)			25.63	500	3.1489
QL-2 (non-cop)			1.30	10,000	2.9741

Table 5. *Cont.*

Techniques	Charging Strategy	Penetration (%)	MSE (KW)	Epoch	MAPE(%)
ANN-2 (Cop) RNN-2 (Cop) QL-2 (Cop)	Cooperative	60	9.54 9.67 9.37	3000 500 10,000	5.7371 4.2189 3.1451
ANN-2 (Smart) RNN-2 (Smart) QL-2 (Smart)	Smart	60	7.12 7.34 5.23	3000 500 10,000	5.9871 4.0189 2.9741
ANN-3 (non-cop) RNN-3 (non-cop) QL-3 (non-cop)	Non-cooperative	90	0.0031 0.0019 0.0019	1000 1000 1000	3.2171 2.7189 2.2741
ANN-3 (Cop) RNN-3 (Cop) QL-3 (Cop)	Cooperative	90	1.23 10.3 0.889	3000 500 10,000	3.3371 2.8189 2.1741
ANN-3 (Smart) RNN-3 (Smart) QL-3 (Smart)	Smart	90	7.11 7.45 4.45	3000 500 10,000	3.6371 2.7189 2.0741

7.1. Load Forecasting of Non-Cooperative PHEVs Charging

This article predicts non-cooperative PHEV load demands using ML approaches such as RNN, ANN, and the proposed QL techniques. A predicted load with a 30% PHEV adoption is shown in Figure 7a. The necessary information is given in Table 5, along with the epoch numbers and means square errors (MSEs) used for the different ML techniques in this work during the training phase. ANN and QL methods have lower MSEs but higher accuracies than the RNN strategy. The more significant epoch number used during training was the primary factor that makes the ANN technique perform better in terms of accuracy than the RNN method. The non-cooperative PHEV hourly load forecasting with 60% penetration is shown in Figure 7b. The charging of the PHEV load started at 5:30 p.m. and ended at 5:30 a.m. depicted in Figure 7. Additionally, the AI technique had a 60% penetration of PHEV load and an insignificant MSE (significant accuracy). Table 5 shows that the RNN method had a more significant mean squared error (MSE) and smaller epoch number than other techniques. The devised QL method predicted PHEV load using ANN or RNN outcomes with lower MSEs than the other two methods, which makes this approach more accurate. This illustration shows how the QL approach was more valuable than ANN and RNN for predicting PHEV loads.

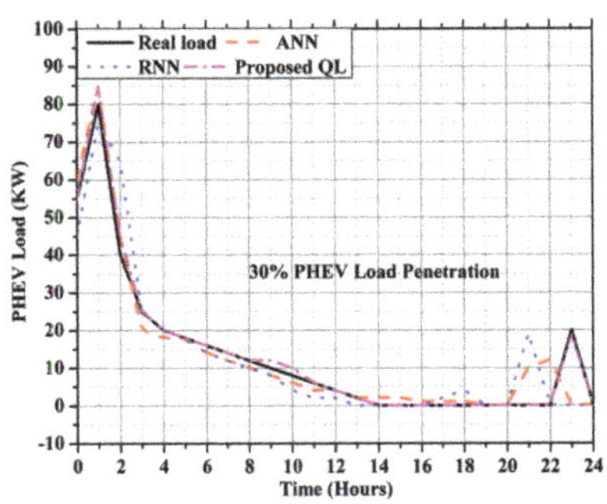

(**a**)

Figure 7. *Cont.*

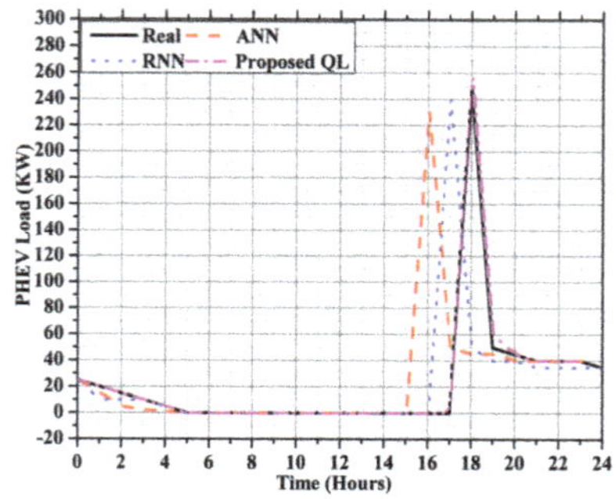

(**b**)

Figure 7. Non-cooperative charging PHEV load forecasting with a 30% and 60% penetratio
(**a**) Non-cooperative charging PHEV load forecasting with a 30% penetration. (**b**) Non-cooperati
charging PHEV load forecasting with a 60% penetration.

7.2. Load Forecasting of Cooperative PHEVs Charging

The predicted load consumption for cooperative PHEVs is given in this subsectio When PHEV penetration is 30%, Figure 8a shows the cooperative PHEVs hourly loa forecasting with 30% penetration using ANN, RNN, and QL algorithms. Training iteratio (1000 (ANN), 3000 (RNN), and 10,000 (QL)) were used. The cooperative PHEV chargir strategy lasted from 7:00 p.m. to 11:00 a.m. as shown in Figure 8. As shown in Table 5, th QL method predicted the cooperative PHEVs hourly charging load more precisely (wi a reduced MSE) than the ANN and RNN approaches. Additionally, the output of the Q method in Figure 8 demonstrates less deviation from the actual data.

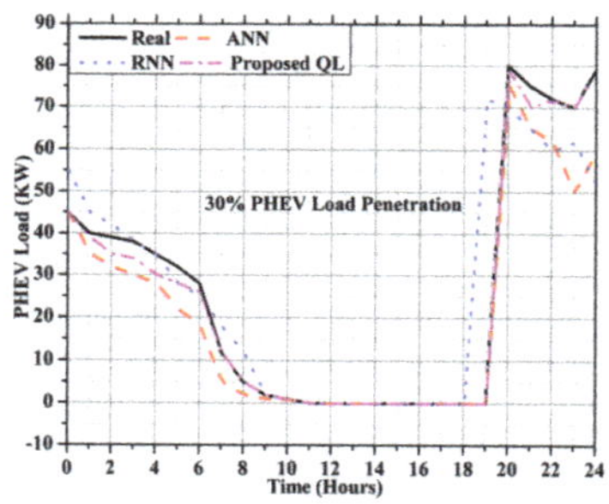

(**a**)

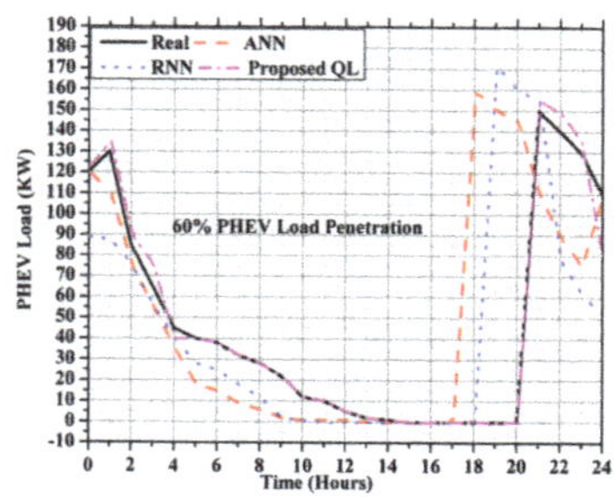

(**b**)

Figure 8. Cooperative charging PHEV load forecasting with a 30% and 60% penetrations. (**a**) Coope ative charging PHEV load forecasting with a 30% penetration. (**b**) Cooperative charging PHEV loa forecasting with a 60% penetration.

7.3. Load Forecasting of Smart PHEVs Charging

The 60% and 30% penetration levels of the smart PHEVs charging hourly load are forecasted in this subsection. The training iterations (1000 (ANN), 3000 (RNN), and 10,000 (QL)) are used. Figure 9 shows that all techniques were capable of forecasting PHEV loads with accuracy. However, because the RNN method employed fewer iterations than the ANN and QL techniques, its error in calculating the PHEV load was more important. The proposed QL approach has a reduced MSE (a better precision) than the ANN and RNN methods because it chose the optimal policy from the potential actions for each state (each hour). The proposed strategy's accuracy can be improved by applying the QL method to more extensive and diverse data sets.

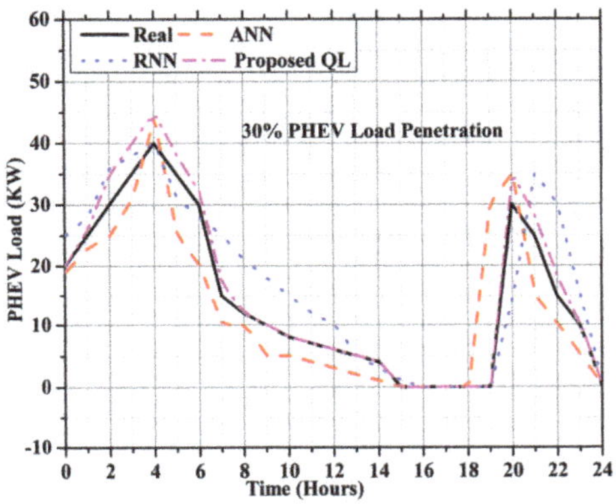

(**a**)

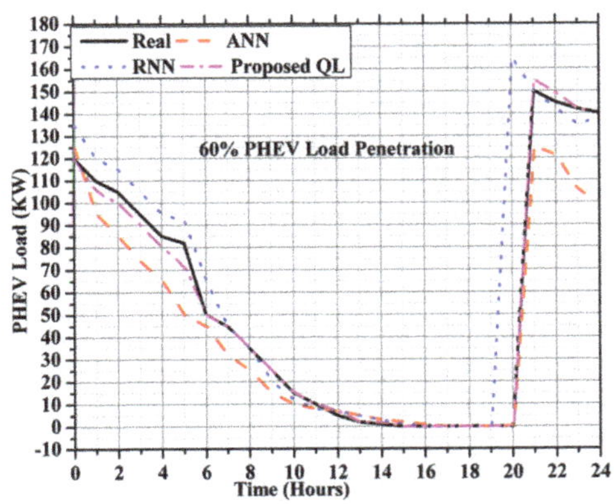

(**b**)

Figure 9. Smart charging PHEV load forecasting with a 30% and 60% penetrations. (**a**) Smart charging PHEV load forecasting with a 30% penetration. (**b**) Smart charging PHEV load forecasting with a 60% penetration.

8. Sensitivity Analysis of Three Charging Techniques

Figure 10 shows the sensitivity evaluations of three PHEV charging load strategies implemented with different hidden layers, node numbers, and several neurons in the proposed QL. Figure 10a shows how the proposed QL method's forecasting accuracy varied with different epoch spans when increased from three to four hidden layers, each having 200–400 nodes. Table 5 shows that when more nodes and hidden layers were used in the ML model, it accurately predicted PHEV loads for the cooperative charging method. Table 5 shows that the quality of a forecast was improved by increasing the number of nodes in each hidden layer. It can be seen from the Figure 10b that the QL technique for the non-cooperative PHEV load forecasting was unable to model the PHEV load filling consisten The more layers and nodes a QL has, the better it can handle large datasets. Figure 1 shows the sensitivity analysis of the QL method for smart PHEV load forecasting wi various nodes and hidden layers when penetration was 30%. Therefore, it can accurat predict PHEV load if the proposed QL method employed more nodes and hidden lay than other techniques—such as ANNs or RNNs.

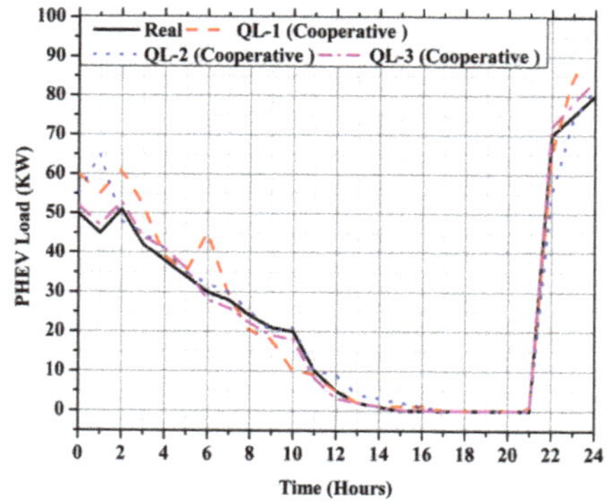

(**a**)

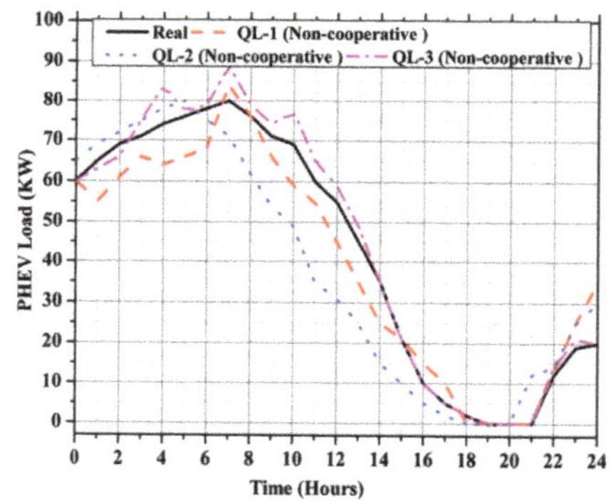

(**b**)

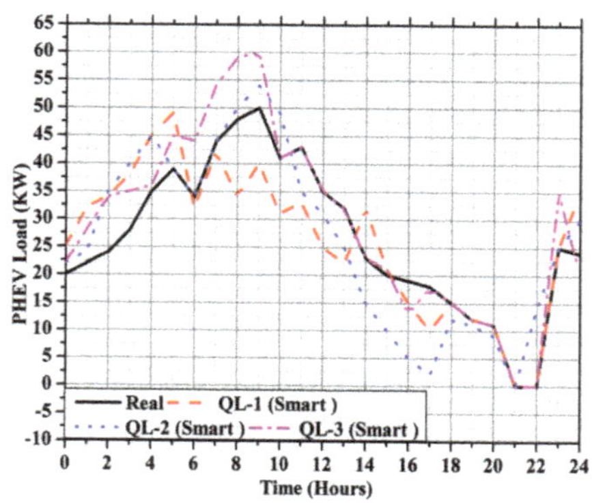

(**c**)

Figure 10. QL sensitivity analysis for the three PHEV charging (Cooperative, Non-cooperative a Smart) (kW). (**a**) QL sensitivity analysis for cooperative PHEV charging (kW). (**b**) QL sensitivi analysis for non-cooperative PHEV charging (kW). (**c**) QL sensitivity analysis for smart PHE charging (kW).

9. Validation of Proposed QL

To produce accurate forecasting results for comparison, the parameters and hyper-parameters of the QL approach should be developed appropriately. The validation test selected these values through trial and error. For instance, the MAPE evaluated QL efficiency, as seen in Figure 11. The QL with five hidden layers among ten guaranteed the best MAPE and the QL selected 41 hidden neurons among fifty to achieve the smallest MAPE.

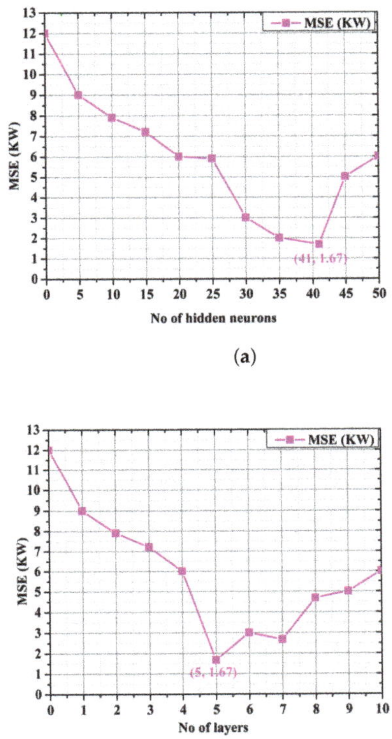

(b)

Figure 11. Demonstration of QLs validity with a various number of hidden neurons and layers. (**a**) Demonstration of QLs validity with various numbers of hidden neurons. (**b**) Demonstration of QLs validity with various numbers of hidden layers.

10. QL in Terms of Speed, Flexibility and Accuracy

The flexibility, speed, and accuracy of the three DL methods are discussed with their ability to predict PHEV charging loads.

10.1. Faster Speed

In this paper, we integrate two ML methods—ANN and RNN—to enhance the QL method's real-time ability to forecast PHEV load. Our three ML methods focus on training times but vary substantially across applications. When ANN and RNN have trained accurately with PHEV data, their networks can efficiently predict the input data to implement the QL technique in real-time. Therefore, utilizing ANN and RNN techniques to their full potential can significantly improve the QL technique's capability to carry out this process quickly.

10.2. Improved Accuracy

It is essential to examine all DL techniques currently being used; after evaluating the three ML techniques described in this article, the most efficient QL for forecasting PHEV charging loads was chosen. From the results of the simulations, predictions from M techniques may be inaccurate if there are substantial variations from the initial chargir points. Therefore, the QL approach, which gets its data from ANN and RNN, may precise forecast optimal PHEV charging. As Table 5 demonstrates, increasing the epoch numb can reduce forecasting errors.

10.3. Flexibility

The simulation results show that ML techniques can be used to model PHEVs l training charge methodologies. Any complex load-charging scenario can be variab predicted by increasing the training iterations for ML techniques. The estimate for PHI charging potential was correct as a consequence. Forecasting PHEV loads for cooperati non-cooperative, and smart charging can be elevated by combining nodes and hidd layers of ANN and RNN techniques. Moreover, the QL technique's capacity to choose t best course of action out of numerous choices can be improved by having more preci estimates of PHEV load charging.

11. Discussion

A significant study is being done to increase the accuracy of load forecasting PHEV charging sites, which can both guide the efficient dispatch of the power syste and advance the growth of EVs. This study proposes a QL method for forecasting demand at electric car charging sites. The data preprocessing unit handles the origir dataset, builds time series data, and adds characteristics to prediction data in our techniq We use the QL forecasting technique and approximate the posterior distribution usi variational inference. Results show the success of this method in predicting PHEV chargi station demand. Comparing the devised method to some forecast methods further indica its better performance. The designed model is around 38.2%, 41.5% lower in MSE a MAE, and 17.9% higher than the standard techniques, according to the findings of po forecasting. The devised strategy can be used to forecast load for PHEV charging static in the real world, as evidenced by its outstanding performance on various datasets.

12. Conclusions

This manuscript presents a novel QL approach for PHEV load forecasting employ ANN and RNN techniques. Various PHEV models, such as smart, cooperative, and n cooperative, are analyzed in this study. Our results indicate that the proposed QL meth achieved accurate predictions of PHEV load charging by leveraging data from the AN and RNN techniques. Importantly, our findings suggest that the QL method outperform the ANN and RNN techniques in predicting PHEV loads accurately, as substantiated the simulation outcomes. Table 5 shows that in the worst-case scenario for PHEV charg (smart charging), the QL method exhibited superior performance compared to traditic ANN and RNN techniques, surpassing them by over 60%. Furthermore, increasing number of iterations demonstrated that the ANN technique delivers more precise dictions of PHEV loads (MSE) than the RNN approach. The QL methodology un consideration demonstrates superior tracking capabilities for PHEV loads when compa to ANN and RNN techniques, exhibiting greater precision and adaptability in the proc Further enhancements in the form of adjustments to the hidden layer, epoch, and n numbers have been shown to significantly enhance the accuracy of PHEV charging l forecasts, as corroborated by the case studies.The forecasting methodology developed l demonstrates the potential for future implementation in expansive power networks contain complex PHEV loads. The performance of this probabilistic forecasting techni can be improved through ongoing optimization efforts. Reliable load projections deri from PHEV charging facilities will facilitate a structured approach to electric vehicle ch ing, and effectively reduce energy usage variations while optimizing distribution netw resource allocation. We are committed to further forecasting the load from various char stations, including those with AC, DC, and higher power capabilities. We plan to ext our suggested model to more PHEV charging data sets. Additionally, we intend to utilize this methodology in additional applications, including dispersed grid capacity forecasting and wind power forecasting, to expand its potential impact.

Author Contributions: Conceptualization, M.Z.; Methodology, M.B.R.; Software, M.Z.; Validation, M.Z., N.F.A. and M.B.R.; Investigation, M.B.R.; Writing—original draft, M.Z. and M.B.R.; Writing—review & editing, M.B.R.; Visualization, N.F.A.; Project administration, N.F.A.; Funding acquisition, N.F.A. All authors have read and agreed to the published version of the manuscript.

Funding: The authors extend their appreciation to the Deputyship for Research and Innovation, Ministry of Education in Saudi Arabia for funding this research work through project number 223202.

Data Availability Statement: The dataset utilized in our publication can be accessed at the following link: https://ev.caltech.edu/dataset, accessed on 2 March 2023.

Conflicts of Interest: The authors affirm that there are no conflicts of interest.

References

1. Taherzadeh, E.; Javadi, S.; Dabbaghjamanesh, M. New optimal power management strategy for series plug-in hybrid electric vehicles. *Int. J. Auto. Technol.* **2018**, *19*, 1061–1069. [CrossRef]
2. Moeini, A.; Zhao, H.; Wang, S. A current-reference-based selective harmonic current mitigation PWM technique to improve the performance of cascaded H-bridge multilevel active rectifiers. *IEEE Trans. Ind. Electron.* **2018**, *65*, 727–737. [CrossRef]
3. Moeini, A.; Wang, S. The state of charge balancing techniques for electrical vehicle charging stations with cascaded H-bridge multilevel converters. In Proceedings of the 2018 IEEE Applied Power Electronics Conference and Exposition (APEC), San Antonio, TX, USA, 4–8 March 2018; pp. 637–644.
4. Liu, Z.; Wen, F.; Ledwich, G. Optimal planning of electric-vehicle charging stations in distribution systems. *IEEE Trans. Power Deliv.* **2012**, *28*, 102–110. [CrossRef]
5. Zheng, Y.; Dong, Z.Y.; Xu, Y.; Meng, K.; Zhao, J.H.; Qiu, J. Electric vehicle battery charging/swap stations in distribution systems: Comparison study and optimal planning. *IEEE Trans. Power Syst.* **2013**, *29*, 221–229. [CrossRef]
6. Liu, T.; Hu, X.; Hu, W.; Zou, Y. A heuristic planning reinforcement learning-based energy management for power-split plug-in hybrid electric vehicles. *IEEE Trans. Ind. Informat.* **2019**, *15*, 6436–6445. [CrossRef]
7. Chen, C.; Duan, S. Optimal integration of plug-in hybrid electric vehicles in microgrids. *IEEE Trans. Ind. Informat.* **2014**, *10*, 1917–1926. [CrossRef]
8. Kang, J.; Yu, R.; Huang, X.; Maharjan, S.; Zhang, Y.; Hossain, E. Enabling localized peer-to-peer electricity trading among plug-in hybrid electric vehicles using consortium blockchains. *IEEE Trans. Ind. Informat.* **2017**, *13*, 3154–3164. [CrossRef]
9. Xie, S.; Qi, S.; Lang, K. A data-driven power management strategy for plug-in hybrid electric vehicles including optimal battery depth of discharging. *IEEE Trans. Ind. Informat.* **2019**, *16*, 3387–3396. [CrossRef]
10. Xing, Q.; Chen, Z.; Zhang, Z.; Huang, X.; Leng, Z.; Sun, K.; Chen, Y.; Wang, H. Charging demand forecasting model for electric vehicles based on online ride-hailing trip data. *IEEE Access* **2019**, *7*, 137390–137409. [CrossRef]
11. Alizadeh, M.; Scaglione, A.; Davies, J.; Kurani, K.S. A scalable stochastic model for the electricity demand of electric and plug-in hybrid vehicles. *IEEE Trans. Smart Grid.* **2014**, *5*, 848–860. [CrossRef]
12. Koyanagi, F.; Uriu, Y. A strategy of load leveling by charging and discharging time control of electric vehicles. *IEEE Trans. Power Syst.* **1998**, *13*, 1179–1184. [CrossRef]
13. Ortega-Vazquez, M.A.; Bouffard, F.; Silva, V. Electric vehicle aggregator/system operator coordination for charging scheduling and services procurement. *IEEE Trans. Power Syst.* **2013**, *28*, 1806–1815. [CrossRef]
14. Dai, Q.; Cai, T.; Duan, S.; Zhao, F. Stochastic modeling and forecasting of load demand for electric bus battery-swap station. *IEEE Trans. Power Deliv.* **2014**, *29*, 1909–1917. [CrossRef]
15. Kisacikoglu, M.C.; Erden, F.; Erdogan, N. Distributed control of PEV charging based on energy demand forecast. *IEEE Trans. Ind. Inform.* **2018**, *14*, 332–341. [CrossRef]
16. Gomez-Quiles, C.; Asencio-Cortes, G.; Gastalver-Rubio, A.; Martinez-Alvarez, F.; Troncoso, A.; Manresa, J.; Riquelme, J.C.; Riquelme-Santos, J.M. A novel ensemble method for electric vehicle power consumption forecasting: Application to the Spanish system. *IEEE Access* **2019**, *7*, 120840–120856. [CrossRef]
17. Xydas, E.S.; Marmaras, C.E.; Cipcigan, L.M.; Hassan, A.S.; Jenkins, N. Electric vehicle load forecasting using data mining methods. In Proceedings of the IET Hybrid and Electric Vehicles Conference 2013 (HEVC 2013), London, UK, 6–7 November 2013; pp. 1–6.
18. Nguyen, V.L.; Tran-Quoc, T.; Bacha, S.; Nguyen, B. Charging strategies to minimize the peak load for an electric vehicle fleet. In Proceedings of the IECON 2014-40th Annual Conference of the IEEE Industrial Electronics Society, Dallas, TX, USA, 29 October–1 November 2014; pp. 3522–3528. [CrossRef]
19. Mal, S.; Chattopadhyay, A.; Yang, A.; Gadh, R. Electric vehicle smart charging and vehicle-to-grid operation. *Int. J. Parallel Emergent Distrib. Syst.* **2013**, *28*, 249–265. [CrossRef]

20. Yilmaz, M.; Krein, P.T. Review of benefits and challenges of vehicle-to-grid technology. In Proceedings of the 2012 IEEE Energy Conversion Congress and Exposition (ECCE), Raleigh, NC, USA, 15–20 September 2012; pp. 3082–3089. [CrossRef]
21. [CrossRef] Oshnoei , A.; Kheradmandi, M.; Muyeen, S.M. Robust control scheme for distributed battery energy storage system in load frequency control. *IEEE Trans. Power Syst.* **2020**, *35*, 4781–4791. [CrossRef]
22. Ahmadi, A.; Tavakoli, A.; Jamborsalamati, P.; Rezaei, N.; Miveh, M.R.; Gandoman, F.H.; Heidari, A.; Nezhad, A.E. Power quality improvement in smart grids using electric vehicles: A review. *IET Electr. Syst. Transp.* **2019**, *9*, 53–64. [CrossRef]
23. Weis, A.; Jaramillo, P.; Michalek, J. Estimating the potential of controlled plug-in hybrid electric vehicle charging to reduce operational and capacity expansion costs for electric power systems with high wind penetration. *Appl. Energy* **2014**, *115*, 190–204. [CrossRef]
24. Zhang, J.; Yan, J.; Liu, Y.; Zhang, H.; Lv, G. Daily electric vehicle charging load profiles considering demographics of vehicle users. *Appl. Energy* **2020**, *274*, 115063. [CrossRef]
25. Zhang, J.; Wang, Z.; Liu, P.; Zhang, Z. Energy consumption analysis and prediction of electric vehicles based on real-world driving data. *Appl. Energy* **2020**, *275*, 115408. energy.2020.115408. [CrossRef]
26. Mackenzie, J.; Roddick, J.F.; Zito, R. An evaluation of HTM and LSTM for short-term arterial traffic flow prediction. *IEEE Trans. Intell. Transp. Syst.* **2019**, *20*, 1847–1857. [CrossRef]
27. Alquthami, T.; Zulfiqar, M.; Kamran, M.; Milyani, A.H.; Rasheed, M.B. A performance comparison of machine learning algorithms for load forecasting in smart grid. *IEEE Access* **2022**, *10*, 48419–48433. [CrossRef]
28. Zulfiqar, M.; Kamran, M.; Rasheed, M.B.; Alquthami, T.; Milyani, A.H. A hybrid framework for short term load forecasting with a navel feature engineering and adaptive grasshopper optimization in smart grid. *Appl. Energy* **2023**, *338*, 120829. [CrossRef]
29. Li, Q.; Wang, F.; Wang, J.; Li, W. LSTM-based SQL injection detection method for intelligent transportation system. *IEEE Trans. Veh. Technol.* **2019**, *68*, 4182–4191. [CrossRef]
30. Schmidhuber, J. Deep learning in neural networks: An overview. *Neural Netw.* **2015**, *61*, 85–117. [CrossRef]
31. Rostami, M.-A.; Kavousi-Fard, A.; Niknam, T. Expected cost minimization of smart grids with plug-in hybrid electric vehicles using optimal distribution feeder reconfiguration. *IEEE Trans. Ind. Informat.* **2015**, *11*, 388–397. [CrossRef]
32. Qian, K.; Zhou, C.; Allan, M.; Yuan, Y. Modeling of load demand due to EV battery charging in distribution systems. *IEEE Trans. Power Syst.* **2010**, *26*, 802–810. [CrossRef]
33. Li, G.; Zhang, X.-P. Modeling of plug-in hybrid electric vehicle charging demand in probabilistic power flow calculations. *IEEE Trans. Smart Grid* **2012**, *3*, 492–499. [CrossRef]
34. Zurada, J.M. *Introduction to Artificial Neural Systems*; West Publishing Company: St. Paul, MN, USA, 1992; Volume 8.
35. Hecht-Nielsen, R. Theory of the backpropagation neural network. In *Neural Networks for Perception*; Elsevier: Amsterdam, The Netherlands, 1992; pp. 65–93.
36. Rummery, G.A.; Niranjan, M. *On-Line Q-Learning Using Connectionist Systems*; Department of Engineering, University of Cambridge: Cambridge, UK, 1994 ; Volume 37.
37. Chollet, F. Keras. 2015. Available online: https://keras.io (accessed on 22 February 2023).
38. Liu, D.; Zeng, P.; Cui, S.; Song, C. Deep Reinforcement Learning for Charging Scheduling of Electric Vehicles Considering Distribution Network Voltage Stability. *Sensors* **2023**, *23*, 1618. [CrossRef]
39. Lee, Z.J.; Li, T.X.; Low, S.; Assoc Comp, M. ACN-Data: Analysis and Applications of an Open EV Charging Dataset. In Proceedings of the E-Energy'19: Proceedings of the 10th ACM International Conference on Future Energy Systems, Phoenix, AZ, USA, 25–28 June 2019; pp. 139–149.

mathematics

MDPI

Article

Strategic Electricity Production Planning of Turkey via Mixed Integer Programming Based on Time Series Forecasting

Gökay Yörük [1], Ugur Bac [2], Fatma Yerlikaya-Özkurt [2] and Kamil Demirberk Ünlü [2,*]

1 Graduate School of Natural and Applied Sciences, Atilim University, 06830 Ankara, Turkey
2 Department of Industrial Engineering, Atilim University, 06830 Ankara, Turkey
* Correspondence: demirberk.unlu@atilim.edu.tr

Abstract: This study examines Turkey's energy planning in terms of strategic planning, energy policy, electricity production planning, technology selection, and environmental policies. A mixed integer optimization model is proposed for strategic electricity planning in Turkey. A set of energy resources is considered simultaneously in this research, and in addition to cost minimization, different strategic level policies, such as CO_2 emission reduction policies, energy resource import/export restriction policies, and renewable energy promotion policies, are also considered. To forecast electricity demand over the planning horizon, a variety of forecasting techniques, including regression methods, exponential smoothing, Winter's method, and Autoregressive Integrated Moving Average methods, are used, and the best method is chosen using various error measures. The optimization model constructed for Turkey's Strategic Electricity Planning is obtained for two different planning intervals. The findings indicate that the use of renewable energy generation options, such as solar, wind, and hydroelectric alternatives, will increase significantly, while the use of fossil fuels in energy generation will decrease sharply. The findings of this study suggest a gradual increase in investments in renewable energy-based electricity production strategies are required to eventually replace fossil fuel alternatives. This change not only reduces investment, operation, and maintenance costs, but also reduces emissions in the long term.

Keywords: time series forecasting; strategy planning; electricity production; integer programming

MSC: 37M10; 90C05

Citation: Yörük, G.; Bac, U.; Yerlikaya-Özkurt, F.; Ünlü, K.D. Strategic Electricity Production Planning of Turkey via Mixed Integer Programming Based on Time Series Forecasting. *Mathematics* **2023**, *11*, 1865. https://doi.org/10.3390/math11081865

Academic Editors: Atanda Raji, Khaled M. Abo-Al-Ez and Aleksandr Rakhmangulov

Received: 14 February 2023
Revised: 7 April 2023
Accepted: 12 April 2023
Published: 14 April 2023

1. Introduction

Strategic energy planning is the process of coming up with long-term energy policies that will affect the future of energy systems in a region or the whole country. Due to globalization, fast population growth, and countries' efforts to become more industrialized, the demand for energy and natural resources has grown a lot. The demand for energy services is expected to grow by 1.3% per year until the year 2040 [1]. The main sources of energy are hydraulic, nuclear power, and thermal. Renewable energy sources can also be thought of as alternatives to traditional energy sources, such as wind, sunlight, geothermal heat, waterpower, and biomass [2].

A lot of greenhouse gases are made by the main energy sources, which causes global warming. On the other hand, the main problems with using renewable energy resources are the high initial investment costs, the unknown operational risks, and the need to choose different locations for facilities [2]. Due to these problems, energy resources are not used as well as they could be, thus, optimal planning is very important and can help ensure sustainability and protect the natural balance.

In this study, a strategic level energy planning model is proposed for Turkey. This model can be used to figure out different ways to produce electricity during the planning horizon, considering strategic goals, resource limits, available demand, emission goals and limits, and other factors. To forecast electricity demand over the planning horizon, a variety of forecasting techniques, including regression methods, exponential smoothing Winter's method, and Autoregressive Integrated Moving Average (ARIMA) methods, are used, and the best method is chosen using various error measures. Thus, this ensures that the information needed in optimization modeling is accurately predicted.

In the proposed optimization model different types of alternative energy sources are considered, such as fossil, renewable, and nuclear. The cost to build, run, maintain, and fuel each type of power plant is different. The levelized cost concept is used so that all costs can be measured in the same way. This concept is defined as "the average cost over the lifetime of the electricity generation plan per MWh of electricity generated" [3]. Optimization methods are needed to choose the best portfolio of ways to produce electricity that minimizes costs while meeting operational constraints and long-term goals. Mathematical modeling is used to find the best way to reduce the total levelized cost of all power plants that are running during the planning period.

All possible energy resources, such as fossil fuels, renewable energy resources, nuclear energy, and so on, are considered in the production of electricity. As a result, in terms of "classifying energy problems based on energy type," our problem is an electricity planning problem. Furthermore, we consider energy resources, such as solid fuels, oil/gas, renewable energy sources, and nuclear energy sources. The research is a general energy planning problem in this regard. The research problem includes a set of alternative energy policies, such as reducing CO_2 emissions, using fewer fossil fuels in electricity generation, and utilizing more renewable energy resources. In terms of application, it falls into the category of "energy policy analysis." In addition, the proposed model performs strategic level energy planning over the planning horizon while meeting annual total electricity demand throughout the year. In this regard, the problem in this study can also be considered as an "Energy Power Planning" issue.

In the rest of this study, Section 2 gives a detailed review of the related literature. Section 3 goes over the methods used to figure out how much electricity costs and how much energy it uses, and the prices and methods used to predict how much electricity will be used. In Section 4, the results are analyzed based on a real-world application. Finally, in Section 5, the study is summed up and future research ideas are given.

2. Literature Review

Multiple objectives are handled concurrently in energy planning issues, making it a strong application area for operations research. As a result, the number of studies in energy planning is expanding in the literature [4]. In the literature, problems, such as energy efficiency improvement, energy decision-making, energy investment and planning, energy plant selection, selection of the most suitable energy alternative, energy resource sharing, and energy source reliability, are studied in terms of energy optimization applications [4]. Several researchers focus on decision-making challenges in the literature. The most commonly used multi-criteria approaches in the literature are the Analytic Hierarchy Process (AHP), Preference Ranking Organization Method for Enrichment Evaluation (PROMETHEE), Elimination and Choice Translating Reality (ELECTRE), and Technique for Order of Preference by Similarity to Ideal Solution (TOPSIS) methods [5–13].

Since most energy resources are used in response to climatic circumstances, optimization approaches are necessary for the design, planning, and control phases of energy management. Challenges in the energy sector are complicated, unclear, and involve several associated parties. As a result, the choices are constrained by many restrictions. As there are so many choice factors and parameters, they are technically complicated to solve. It is seen that a variety of studies are available in the literature [14–28].

Moreover, there are several evaluations on energy planning issues. These works examine and categorize the literature on energy planning and suggest some future research areas. For example, ref. [29] examined energy supply models to assess investment options and expansions. It was assumed that demand quantities and input costs were known, and the model determined the investment choice with the lowest cost over time. Energy type (fossil, nuclear, single hydro, or hydro), energy transport options, investment, and replacement schedules, and optimal mode of system operation are all factors to consider. Ref. [30] examined African countries' electrical planning studies that used both qualitative and quantitative methodologies. Ref. [31] studied Turkey's generation plans up to 2023 and assessed the viability of the 2023 Vision. The capacity objectives were examined, and projections were made using a semi-empirical electrical demand model. Other recent studies on Turkish electricity markets are [32–34].

On the other hand, there are also many studies conducted on electricity demand forecasting. Indeed, electricity demand forecasting is classified into three categories: short-term, mid-term, and long-term forecasting. Short-term predictions range from one hour to one week, mid-term forecasts range from one week to a year, and long-term forecasts span more than a year [35,36]. Demand forecasting is a prominent study topic since it occurs in practically every system that involves production and customers. In the literature, several techniques and models have been established for electricity demand forecasting, such as Holte Winters exponential smoothing approach, multivariate adaptive regression splines, ARIMA, and support vector regression [37–43]. Another classification of demand forecasting is based on the degree of mathematical analysis involved in the forecasting process. These approaches are classified as quantitative and qualitative. Qualitative approaches include the Delphi method and curve fitting. Regression, machine learning by [44], smoothing approaches by [45], deep learning by [46,47], and the Box-Jenkins methodology by [48], on the other hand, are examples of quantitative methods.

To summarize, electricity planning is a prominent research field in the literature, and several works examine various elements of energy planning. Our study varies from the previous research for many reasons. First, we address the energy planning problem for Turkey over several time horizons. Some investigations have been undertaken in Turkey, such as [49,50]. These studies also look at energy planning in Turkey, but only in the past, therefore they do not address the present situation.

Furthermore, we evaluate the government's most recent strategic aims under current strategic plans in our research. As a result, we throw light on the near future. This analysis takes into consideration not just Turkey's present installed energy capacity, but also projected energy investments and closures. In other words, in addition to present capacity, prospective power plants must be opened or shuttered within the timeframes specified. First, demand is forecasted for several planning horizons (e.g., 10 and 20 years) utilizing a set of forecasting approaches in this work. The best forecasting technique is chosen using several error measurements, and the prediction provided by the chosen approach is incorporated into the mathematical model.

3. Methodology

Framework developed for the strategic planning of electricity production in Turkey is summarized as shown in Figure 1. Forecasted data is the main requirement of strategic plans. To generate this data there are different time series forecasting methods available in the literature. All of these methods require past-time data to be used in the forecasting of the future. Our framework begins with the gathering of the energy demands in the past and these data were used to evaluate different time series forecasting methods under different performance metrics, as explained in Section 4. Following the evaluation phase, the selected forecasting method is used to forecast energy demand in the years 2021–2040 and this data is used in the mixed integer programming model formulated to plan the number and capacities of energy plants according to their types. Different types of time series forecasting methodologies have been utilized and the one which has the best performance metrics is used to get the future demand of electricity loads. The forecasted electricity loads are used as the input of the model. The formulated mathematical model has been coded in General Algebraic Modeling System (GAMS) software which includes built-in solvers to find the optimal solution for different types of mathematical models. Runs are made in GAMS for two different time intervals and three additional scenarios.

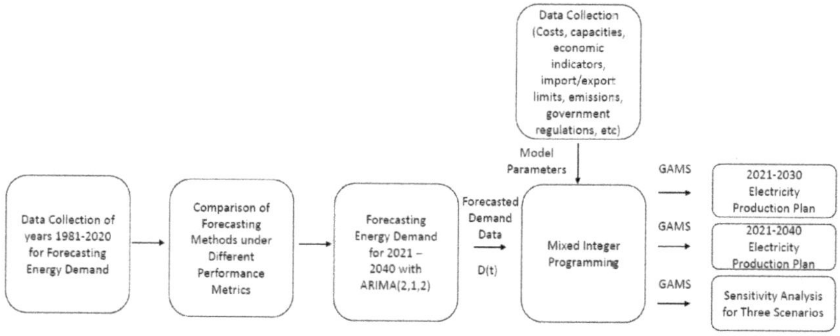

Figure 1. Framework of the Proposed Methodology.

3.1. Time Series Forecasting

Determining the most appropriate energy resources to satisfy the yearly energy demand requires consideration of alternative electricity generation options, such as fossil fuels, renewable energy, and nuclear energy, and selecting a subset of these energy generation technologies while taking government strategic goals and environmental issues into account. To assist this goal a methodology is developed to forecast yearly energy demands and by using these demand forecasts a mathematical optimization model is formulated to select the most appropriate energy resources to satisfy these demands. The developed methodology is applied to the Turkey case to identify investment decisions for the next twenty years. To accomplish this goal, demand data for Turkey from 1981 to 2020 is used to forecast demand between 2021 and 2040. The steps described by [51] were followed to determine the best forecasting model that fits the demand data.

Quantitative forecasting methods are used in this study to generate long-term forecasts of electricity demand. Regression Analysis (linear/exponential/beta growth and first, second, and third order polynomial equations), Double Exponential Smoothing, Winters' Method (linear, additive, and multiplicative models), and ARIMA for different autoregressive, differencing, and moving average parameters are among the forecasting methods considered. Regression analysis is a set of statistical methods for estimating the relationships between one or more independent variables and a dependent variable. The least squares method is used to approximate the model parameters, resulting in an improved model. For parameter estimation, the method minimizes the sum of squares.

$$Y_t = \beta_0 + \beta_1 t + e_t, \quad (1)$$

where t is the time, β_0 is the constant, β_1 is the average difference from one period to the next, and e_t is the error term. The regression model can be linear or nonlinear and different mathematical models can be used for modeling the input data, such as linear growth, exponential growth, beta growth, the polynomial of the first order, second order, third order, etc. In this study, different regression types are used for finding the best regression model that fits the demand data and third order polynomial model produced the best fits.

Double exponential smoothing uses a level and a trend component at each period. This method has two smoothing weights for updating the components in each period. Winters' Method smooths the data and provides forecasts for the short to medium term using Holt–Winters' exponential smoothing. When both seasonality and trend are present, this procedure can be beneficial. These two components can have a linear, multiplicative, or additive relationship. Winters' Method generates dynamic estimates for three variables: level, trend, and seasonal. When there is no seasonality in the data, the linear model is used, and this method is known as the Holt–Winters nonseasonal algorithm. When the seasonal pattern in the data does not change with the data size, an additive model is used. When seasonal patterns in data depend on data size, a multiplicative model is used. ARIMA model is a modified traditional technique that is used for modeling time series [40,41,51]. In this research, the forecasting accuracy measures are calculated for the comparison of different forecasting methods. Scale dependent measures are used commonly whose scale depends on the scale of the data. Root mean square errors (RMSE) and mean absolute errors (MAE) are the most commonly used scale dependent measures while percentage errors are scale independent, they are frequently used to compare forecasting performance across different data sets. Mean absolute percentage error (MAPE) and coefficient of determination score (R^2) are the most commonly used percentage error measures [51].

3.2. *Optimization Model*

In this section, the assumptions, sets and indices, parameters, decision variables, and mathematical model are presented.

3.2.1. Assumptions

- Even if construction is completed within the previous year, the new power plants are expected to be operational at the beginning of the following year;
- The availability factor determines the maximum working hours of power plants while taking maintenance and other resource requirements into account. Unplanned interruptions and plant failures are considered at the operational level;
- Due to the closing dates of the power plants are unknown due to the government's information privacy policy, it is assumed that the existing facilities will be operational without interruption until the end of the planning period;
- Future costs are calculated using average escalation rates that are determined for each cost component, and future cash flows are calculated using an average interest rate;
- There is no significant variation or dramatic change in economic indicators and demand patterns, and they continue to follow the long-term trend;
- The potential energy resources in Turkey will not change significantly over the planning horizon;
- Power plant basic data, efficiency, initial investment costs, and CO_2 emissions are assumed to be constant over time.

3.2.2. Set, Indices, Parameters, and Decision Variables

The set of indices used in the mathematical model is as follows:

I:	*Set of energy resources used for electricity production, indexed by i;* *I = {lignite, hard coal, imported coal, natural gas, uranium};*
J:	*Set of electrical generation power plant types, indexed by j;* *J = {Fluized Lignite, Elbistan Lignite, Hard Coal, Imported Coal, Natural Gas, Nuclear, Hydroelectric, Wind, solar, Geothermal};*
K:	*Set of power plant categories indexed by k;* *K = {Renewables (R), Fossil Fuels (F), Nuclear (N)};*
Jk:	*Set of power plants that are in resource category k;* *JR = {Hydroelectric, Wind, Solar, Geothermal};* *JF = {Fluized Lignite, Elbistan Lignite, Hard Coal, Imported Coal, Natural Gas};* *JN = {Nuclear};*
T:	*Set of years considered in the planning period, indexed by t;* *t = {2021,2023, ... ,T}.*

The parameters of the model are:

Cj,tinv:	*Capital investment cost of type j power plant at year t ($);*
Cj,tOM:	*Operation and maintenance cost of type j plant at year t;*
Cj,tfuel:	*Fuel cost of type j power plan at year t ($);*
Ej,t:	*Total energy generation of type j power plant at year t (MWh);*
Tj:	*The operational lifetime of type j power plant (year);*
Tjconst:	*Construction time of type j power plant (year);*
ICapj:	*Installed capacity of type j power plant (MW);*
β_j:	*Availability percentage of type j power plant (%);*
β_j^{hour} :	*Availability factor of type j power plant (h/year);*
LCj:	*Levelized cost of type j power plant ($/MWh);*
Ctimp:	*Unit import cost in year t ($/MWh);*
Ctexp:	*Export revenue in year t ($/MWh);*
explimit:	*Annual export limit (MWh);*
implimit:	*Annual import limit (MWh);*
AVLj:	*Number of type j power plants that are operational before the planning horizon;*
PLNjt:	*Number of type j power plants that are already planned to be opened before the planning horizon at year t*
Dt:	*Electric Demand in Year t (MWh);*
NJRopr:	*Maximum number of renewable power plants that can be in operation in a year (calculated based on resource potential);*
H_t^{num}:	*Maximum number of hydroelectric power plants that can be opened in year t (calculated based on construction capacity in Turkey);*
ε_j :	*The CO_2 emission factor of type j power plant (ton/MWh);*
ε_t^{limit} :	*Emission limit of CO_2 in year t (ton);*
Yt:	*Percentage of renewable power plant capacity in year t (%);*
M:	*A sufficiently large number;*
r:	*Interest rate (%);*
ef:	*Escalation rate for fuel type f (%);*
e_{OM} :	*Escalation rate for operation and maintenance costs (%).*

The escalation rate is the price increase for goods and services caused by a variety of factors, such as inflation, supply, and demand, engineering changes, or other similar causes. Using historical data, average escalation rates for fuel types, and operation and maintenance costs are estimated. Lastly, the decision variables are as follows:

xjt:	*Number of type j power plants opened in year t;*
wjt:	*Number of type j power plants closed in year t;*
Njt:	*Total number of type j power plants in year t;*
vj:	*Binary variable, 1 if the capacity of type j power plants is increased, 0 otherwise;*
yjt:	*The energy supply of type j power plant in year t (MWh);*
expt:	*Electric energy exported in year t (MWh);*
impt:	*Electric energy imported in year t (MWh);*
z:	*Total levelized cost of power plants.*

3.2.3. Mathematical Model

A mathematical optimization model is required to decide on the optimal combination of different power plants and capacities required from each type. For this purpose, a linear mixed integer programming model is formulated. As with all linear programming models, real-life objectives and constraints are represented as mathematical equations in the proposed model. Each equation formulated for this purpose is explained in detail below.

The objective function of the model seeks to minimize the total cost, which has three sub goals. The first one is to minimize the levelized cost of power plants operating within the planning horizon. The levelized cost is the average cost per MWh of electricity generated over the life of a power plant. The lifetime cost of a power plant is expressed in terms of generation cost in $/MWh. Investment costs, operation, and maintenance costs, and fuel costs are all included in the levelized cost. The second part of the objective function minimizes the total energy import costs, and finally, the last part of the objective function maximizes the total energy export revenues.

$$Minimize\ z = \sum_{t\epsilon T}\sum_{j\epsilon J}\beta_j^{hour} ICap_j N_{jt} LC_j + \sum_{t\epsilon T} C_t^{imp} imp_t - \sum_{t\epsilon T} C_t^{exp} exp_t \tag{2}$$

$$LC_j = \frac{Total\ Capital\ and\ Operation\ Costs\ of\ Power\ Plant\ j\ During\ Lifetime}{Net\ Electricity\ Generation\ of\ Power\ Plant\ j\ During\ Lifetime}$$

$$LC_j = \frac{C_{j,0}^{inv} + \sum_{t=1}^{T_j}\left[\left(C_{jt}^{inv} + C_{jt}^{OM} + C_{jt}^{fuel}\right)/(1+r)^t\right]}{\sum_{t=1}^{T_j}\left[E_{jt}\right]}$$

$$E_{jt} = \beta_j^{hour}\left(ICap_j\right)$$

$$C_{jt}^{OM} = C_{j0}^{OM}(1+e_{OM})^t$$

$$C_{jt}^{fuel} = C_{j0}^{fuel}\left(1+e_f\right)^t$$

Constraints (3) and (4) are the flow balance constraints, and they ensure that the sum of already existing type j power plants in year t before the planning period, type j power plants that have already planned to be opened or closed in year t, and new type j power plants opened or closed in year t equals to the total number of type j power plants in year t. Constraint (4) ensures that the number of type j power plants from the previous period $(t-1)$ is updated accordingly in the following years.

$$AVL_j + PLN_{jt} + x_{jt} - w_{jt} = N_{jt}\ \text{ for } \forall j \in J \text{ and } t = 2021 \tag{3}$$

$$N_{j,t-1} + PLN_{jt} + x_{jt} - w_{jt} = N_{jt} \text{ for } \forall j \in J \text{ and } t > 2021 \tag{4}$$

Constraints (5) and (6) control the opening and closing decisions for power plants. A power plant type can be either opened or closed within the planning horizon in a given year, but not both.

$$x_{jt} \le M\ v_j\ \text{ for } \forall j \in J\ \text{ and }\ t \in T \tag{5}$$

$$w_{jt} \le M\ \left(1 - v_j\right)\ \text{ for } \forall j \in J\ \text{ and }\ t \in T \tag{6}$$

Constraint (7) prevents a power plant from being operational before the construction time at the start of the planning horizon. For example, because nuclear power plants take seven years to build, it is not possible to open one during the first seven years of the planning horizon.

$$x_{jt} = 0\ \text{ for } \forall j \in J\ \text{ and }\ t \le T_j^{const} \tag{7}$$

Constraint (8) confirms that the total electricity generation capacity of type j power plants cannot exceed the total installed electricity generation capacity of type j power plants in any year during the planning horizon. The availability factor of a type j power plant, the installed capacity of a unit type j power plant, and the number of type j power plants in operation in year t are multiplied to calculate the electricity generation capacity of a type j power plant in year t.

$$y_{jt} \le \beta_j^{hour} ICap_j N_{jt}\ \text{ for } \forall j \in J\ \text{ and }\ t \in T \tag{8}$$

The total electrical energy generated by power plants plus total imports minus total exports should be greater than or equal to the forecasted demand at year t. This constraint is formulized as follows:

$$\sum_{j\in J}\left[y_{jt}+imp_t-exp_t\right]\geq D_t \text{ for } \forall t\in T \tag{9}$$

Constraints (10) and (11) confirm that total exports and imports in year t cannot exceed total export and import limits. The capacity of the transmission lines, which connect the importing and exporting countries, determines export and import limits. There are independent transmission lines between the countries in each direction (export and import); thus we have different limits for exports and imports.

$$exp_t\leq exp^{limit} \text{ for } \forall t\in T \tag{10}$$

$$imp_t\leq imp^{limit} \text{ for } \forall t\in T \tag{11}$$

Renewable power plants should at least generate a certain percentage of the total installed capacity. This percentage is determined by the government and stated in the government's strategic goals.

$$\sum_{j\in J_r} ICap_jN_{jt}\geq \gamma_t\sum_{j\in J} ICap_jN_{jt} \text{ for } \forall t\in T \tag{12}$$

Constraint (13) confirms that power plants' total CO_2 emissions should be less than or equal to the CO_2 emission limit. CO_2 emissions are proportional to the amount of electricity generated by power plants.

$$\sum_{j\in J}\varepsilon_j y_{jt}\leq \varepsilon_t^{limit} \text{ for } \forall t\in T \tag{13}$$

Constraint (14) ensures that nuclear power plants are not closed due to the government's strategic goals.

$$v_{NuclearPP}=1 \tag{14}$$

Due to construction capacity, Constraint (15) limits the number of hydroelectric power plants built each year.

$$x_{HydroelectricPP,t}\leq H_t^{num} \text{ for } \forall t\in T \tag{15}$$

Constraint (16) ensures that the total number of type j renewable energy plants does not exceed renewable capacity. Potentials for each type of renewable energy plant are defined for each country. For example, in the case of solar energy, the potential is determined by the angle of solar radiation, total sunbathing time, the total area suitable for solar farms, and so on. Wind power plant potentials are determined by wind speed, duration, and the total area reserved for wind farms. The flow rates of the rivers, available areas for power plants, and construction time and capacity limitations are all considered when determining hydroelectric potential. The potential of geothermal energy is determined by the amount of thermal water and its temperature.

$$N_{jt}\leq N_j^{opr} \text{ for } \forall j\in J_R \text{ and } t\in T \tag{16}$$

Finally, the sign restrictions of the model are as follows:

$$\begin{gathered} x_{jt}, w_{jt}, \text{ and } N_{jt}\geq 0 \text{ and } integer \forall j\in J \text{ and } t\in T\\ v_j\in\{0,1\}\forall j\in J\\ y_{jt}, exp_t, \text{ and } imp_t\geq 0\forall j\in J \text{ and } t\in T \end{gathered} \tag{17}$$

4. Time Series Analysis and Application of the Model

4.1. Time Series Analysis

For Turkey, demand data from 1981 to 2020 are used to forecast demand for the years 2021–2030 and 2021–2040 [52], respectively. As a result, 40 observations are gathered. The obtained data shows an increase over the years. First, the data is represented graphically to determine whether or not it is stationary. Several tests and analyses, including the Augmented Dickey–Fuller Test (ADF) based on the unit root process, were used to analyze the demand data. At a 5% significance level, the unit root test implies nonstationary. As [51] suggests we differentiate the time series and this time we reject the null hypothesis of ADF. Thus, the time series investigated is integrated as order 1 ($I(1)$). The first difference in the time series is used for the ARIMA model in this study.

The demand data from 1980 to 2020 is used to forecast demand between 2021–2030 and 2021–2040. First, various regression methods are used, including beta growth, exponential growth, and first, second, and third order polynomial equations. The third-order regression model produced the best fit of these methods. Different ARIMA models, on the other hand, are considered, and the best model is found to be ARIMA (2,1,2), which produces the best fit. Furthermore, double exponential smoothing, and Holt-additive, Winter's multiplicative, and linear models, are considered. Table 1 compares selected forecasting methods in terms of different performance measures. When the performance results of all statistical models in this table are examined, it is seen that especially R^2 and MAPE values show very good performance results. Since these selected models produce extremely good and sufficient results for the given data, the statistical modeling approach is preferred instead of learning-based models.

Table 1. Comparison of Forecasting Methods.

Performance Metrics	Nonlinear Regression (Third Order Polynomial)	Exponential (Double)	Holt–Winters (Additive)	Holt–Winters (Multiplicative)	Holt–Winters (Linear)	ARIMA (2,1,2)
RMSE	3984	4986	5600	5329	4771	3236
MAPE	2.160%	3.162%	3.541%	3.247%	2.844%	1.702%
MAE	2776	3689	4181	3842	3519	2322
R^2	99.82%	99.69%	99.62%	99.65%	99.72%	99.87%

Based on these analyses, the ARIMA (2,1,2) model outperformed the other forecasting methods across all performance metrics. As a result, the ARIMA (2,1,2) model can be used to forecast electricity demand in Turkey. Hence, ARIMA forecast results are used in our mathematical model.

4.2. Application of the Mathematical Model

In this section, we will look at Turkey's strategic energy production planning problem and apply the mathematical model defined in Section 3. The parameters' values are gathered from a variety of sources, including Turkey Electricity Transmission Company (TEA), the International Energy Agency (IEA), the Turkish Statistical Institute, and others.

The proposed model is studied in terms of constraints and goal function during model verification. The model is shown to perform correctly, and all imposed conditions are met as expected. The validation stage confirms that the mathematical model's outputs are appropriate when compared to real-world strategies and goals. The suggested model estimates power plant types and total installed capacities based on operational limitations and strategic goals. The model's outputs meet the strategic goals set by institutions, such as Energy Market Regulatory Authority's (EMRA), strategic plans. As a result, we find that the suggested model accurately captures the real system. The model is solved with GAMS optimization software for two distinct periods, namely 2021–2030 and 2021–2040.

The number of power plants in operation between 2020 and 2030 is shown in Table 2. In general, the number of fossil power plants is decreasing while the number of renewable power plants is increasing. Specifically, the coal power plants (Elbistan lignite, fluidized lignite, hard coal, and imported coal) are scheduled to close within the next ten years. Natural gas power plants, another type of fossil fuel power plant, are reduced in numbers (from 37 to 11), but not completely closed. On the other hand, the number of renewable options, such as wind, solar, and hydroelectric power plants is increasing. Geothermal power plants, the other renewable option, are scheduled to close within the planning horizon. Finally, in 2024, five preplanned nuclear power plant modules are put into service and used within the planning interval. In general, due to strategic goals, such as renewable share constraints and emission limits, the renewable share is increasing.

Table 2. Number of Power Plants (N_{jt}) for the 2020–2030 Period.

Power Plant (j)	2020	2021	2022	2023	2024	2025	2026	2027	2028	2029	2030
Elbistan Lignite	8	3	3	0	0	0	0	0	0	0	0
Fluized Lignite	48	0	0	0	0	0	1	0	0	0	0
Geothermal	54	54	54	33	4	3	2	1	1	0	0
Hard Coal	3	1	1	0	0	0	0	0	0	0	0
Hydroelectric	214	214	214	214	214	225	277	285	288	291	295
Imported Coal	18	0	0	0	0	0	0	0	0	0	0
Natural Gas	37	37	37	37	28	25	19	17	15	13	11
Nuclear	0	0	0	0	5	5	5	5	5	5	5
Solar	133	133	133	233	352	452	552	652	752	852	950
Wind	221	221	321	464	564	664	764	864	964	1064	1164

The total installed capacity of power plants follows a similar pattern to the number of power plants. Table 3 shows that the shares of wind, solar, hydroelectric, and nuclear power plants are increasing while the share of other power plants is decreasing. Along with strategic goals, the total share of fossil fueled power plants decreases significantly.

Table 3. Installed Capacities of Power Plants (MW) for the 2021–2030 Period.

Power Plant (j)	2021	2022	2023	2024	2025	2026	2027	2028	2029	2030
Elbistan Lignite	1080	1080	0	0	0	0	0	0	0	0
Fluidized Lignite	0	0	0	0	0	150	0	0	0	0
Geothermal	1620	1620	990	120	90	60	30	30	0	0
Hard Coal	300	300	0	0	0	0	0	0	0	0
Hydroelectric	28,676	28,676	28,676	28,676	30,150	37,118	38,190	38,592	38,994	39,530
Imported Coal	0	0	0	0	0	0	0	0	0	0
Natural Gas	25,900	25,900	25,900	19,600	17,500	13,300	11,900	10,500	9100	7700
Nuclear	0	0	0	5000	5000	5000	5000	5000	5000	5000
Solar	6650	6650	11,650	17,600	22,600	27,600	32,600	37,600	42,600	47,500
Wind	8840	12,840	18,560	22,560	26,560	30,560	34,560	38,560	42,560	46,560

During the planning horizon, the total supply of natural gas decreases significantly. Wind and solar, on the other hand, are becoming increasingly important. In addition, the hydroelectric contribution increases marginally. Finally, nuclear will make a consistent contribution beginning in 2024. Wind, hydroelectric, solar, natural gas, and nuclear power plant options are listed in decreasing order of contribution in 2030.

Finally, Figure 2 depicts the total emissions from all power plants. As can be seen, emissions decrease over the planning horizon after a slight increase. This is because the contribution of renewable resources is increasing while the contribution of fossil fuels is decreasing. As a result, emissions decrease over time in tandem with the strategic goals.

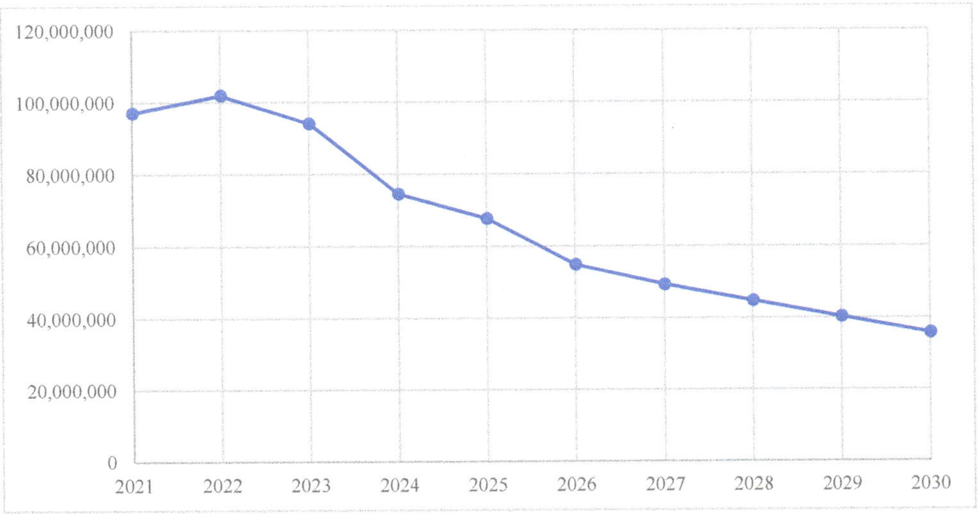

Figure 2. CO_2 Emissions between 2021–2030 (tons).

Table 4 displays the number of power plants that will be operational between 2020 and 2040. In general, the number of fossil power plants is decreasing, while the number of renewable power plants is increasing. Specifically, coal power plants (Elbistan lignite, fluidized lignite, hard coal, and imported coal) and natural gas power plants are scheduled to close within the next 20 years. On the other hand, the number of renewable options, such as wind, solar, and hydroelectric power plants, is increasing. Geothermal power plants, the other renewable option, are scheduled to close within the planning horizon. Additionally, between 2024 and 2032, five preplanned nuclear power plant modules are put into service and used within the planning interval. In general, due to strategic goals, such as renewable share constraints and emission limits, the renewable share is increasing.

Table 4. Number of Power Plants (N_{jt}) for the 2020–2040 Period.

Power Plant (*j*)	**2020**	**2021**	**2022**	**2023**	**2024**	**2025**	**2026**	**2027**	**2028**	**2029**	**2030**
Elbistan Lignite	8	3	3	0	0	0	0	0	0	0	0
Fluidized Lignite	48	0	0	0	0	0	1	0	0	0	0
Geothermal	54	54	54	33	4	3	3	0	0	0	0
Hard Coal	3	1	1	0	0	0	0	0	0	0	0
Hydroelectric	214	214	214	214	214	225	277	288	299	311	335
Imported Coal	18	0	0	0	0	0	0	0	0	0	0
Natural Gas	37	37	37	37	28	25	19	17	15	12	8
Nuclear	0	0	0	0	5	5	5	5	5	5	5
Solar	133	133	133	233	352	452	552	652	752	852	1052
Wind	221	221	321	464	563	663	762	856	940	1040	1188
Power Plant (*j*)	**2031**	**2032**	**2033**	**2034**	**2035**	**2036**	**2037**	**2038**	**2039**	**2040**	
Elbistan Lignite	0	0	0	0	0	0	0	0	0	0	
Fluized Lignite	0	0	0	0	0	0	0	0	0	0	
Geothermal	0	0	0	0	0	0	0	0	0	0	
Hard Coal	0	0	0	0	0	0	0	0	0	0	
Hydroelectric	347	357	367	378	390	402	415	428	442	456	
Imported Coal	0	0	0	0	0	0	0	0	0	0	
Natural Gas	0	0	0	0	0	0	0	0	0	0	
Nuclear	10	10	10	10	10	10	10	10	10	10	
Solar	1152	1252	1352	1452	1552	1652	1752	1852	1951	2050	
Wind	1198	1198	1200	1200	1200	1200	1200	1200	1200	1200	

The total installed capacity of power plants follows a similar pattern to the numbe[r] of power plants. Table 5 shows that the shares of wind, solar, hydroelectric, and nuclea[r] power plants are increasing, while the share of other power plants is dropping. Along wit[h] strategic goals, the total share of fossil fuel power plants drop dramatically.

Table 5. Installed Capacities of Power Plants (MW) for the 2021–2040 Period.

Power Plant (j)	2021	2022	2023	2024	2025	2026	2027	2028	2029	2030
Elbistan Lignite	1080	1080	0	0	0	0	0	0	0	0
Fluidized Lignite	0	0	0	0	0	150	0	0	0	0
Geothermal	1620	1620	990	120	90	90	0	0	0	0
Hard Coal	300	300	0	0	0	0	0	0	0	0
Hydroelectric	28,676	28,676	28,676	28,676	30,150	37,118	38,592	40,066	41,674	43,282
Imported Coal	0	0	0	0	0	0	0	0	0	0
Natural Gas	25,900	25,900	25,900	19,600	17,500	13,300	11,900	10,500	8400	7000
Nuclear	0	0	0	5000	5000	5000	5000	5000	5000	5000
Solar	6650	6650	11,650	17,600	22,600	27,600	32,600	37,600	42,600	47,600
Wind	8840	12,840	18,560	22,520	26,520	30,480	34,240	37,600	41,600	44,440
Power Plant (j)	**2031**	**2032**	**2033**	**2034**	**2035**	**2036**	**2037**	**2038**	**2039**	**2040**
Elbistan Lignite	0	0	0	0	0	0	0	0	0	0
Fluidized Lignite	0	0	0	0	0	0	0	0	0	0
Geothermal	0	0	0	0	0	0	0	0	0	0
Hard Coal	0	0	0	0	0	0	0	0	0	0
Hydroelectric	44,890	46,498	47,838	49,178	50,652	52,260	53,868	55,610	57,352	59,228
Imported Coal	0	0	0	0	0	0	0	0	0	0
Natural Gas	5600	0	0	0	0	0	0	0	0	0
Nuclear	5000	10,000	10,000	10,000	10,000	10,000	10,000	10,000	10,000	10,000
Solar	52,600	57,600	62,600	67,600	72,600	77,600	82,600	87,600	92,600	97,550
Wind	47,520	47,920	47,920	48,000	48,000	48,000	48,000	48,000	48,000	48,000

During the planning horizon, the overall supply of natural gas and other coal power plants is reduced to zero. Wind, solar, and hydroelectric solutions, on the other hand, contribute much more. Finally, as additional power plants are built, nuclear will make a consistent contribution from 2024 to 2032 and from 2032 to 2040. Solar, hydroelectric, wind, and nuclear power plant choices are listed in decreasing order of contribution in 2040.

Figure 3 shows the total emissions from all power stations. As can be observed, emissions drop across the planning horizon following a minor increase caused by increased demand. This is because the contribution of renewable resources is increasing while the contribution of fossil fuels is declining. As a result, emissions decrease over time in tandem with the strategic goals.

As the last step in the mathematical model sensitivity analysis and validations are conducted and we found no evidence of disruptions in the model. Additionally, scenario analyses are completed, and they are summarized in the following paragraphs.

The first scenario is the case with no preplanned plants. In contrast to the base scenario, it is envisaged that no preplanned power plants will be operational within the planning horizon. The goal here is to see the model's ideal selections considering the available power plants at the beginning of the planning horizon. For example, in the base scenario, it is envisaged that several nuclear power reactors will be operational in different years. In this scenario, the model determines the number of new power plants to be built, and it will be possible to see whether or not these preplanned power plants are chosen. Results regarding the installed capacities of power plants for the first scenario are visualized as shown in Figures 4 and 5 for 2021–2030 and 2021–2040 planning horizons, respectively. According to this scenario, wind and solar power plant capabilities do not differ much from the baseline scenario for the same period. Natural gas power plants are utilized instead of nuclear energy, which was used in the base scenario, hence the percentage of natural gas in 2030 is 6.7% greater than in the base scenario. In contrast, the share of hydroelectric power plants declined by 3.7% by 2030 when compared to base scenario. Wind and solar power plant capabilities do not differ much from the baseline scenario for the same period. Natural gas had a 0% share in the base scenario, whereas nuclear power plants had a 4.6% share. However, in this situation, nuclear power plants are not used, and natural gas plants are not completely shut down. Natural gas capacity will account for 6.2% of total capacity by 2040. Finally, the share of hydroelectric power plants falls by roughly 2%.

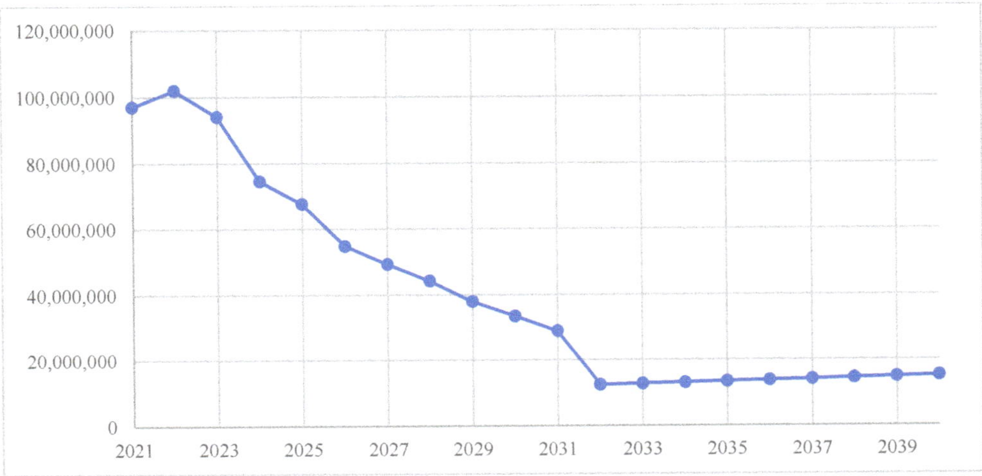

Figure 3. CO_2 emissions between 2021–2040 (tons).

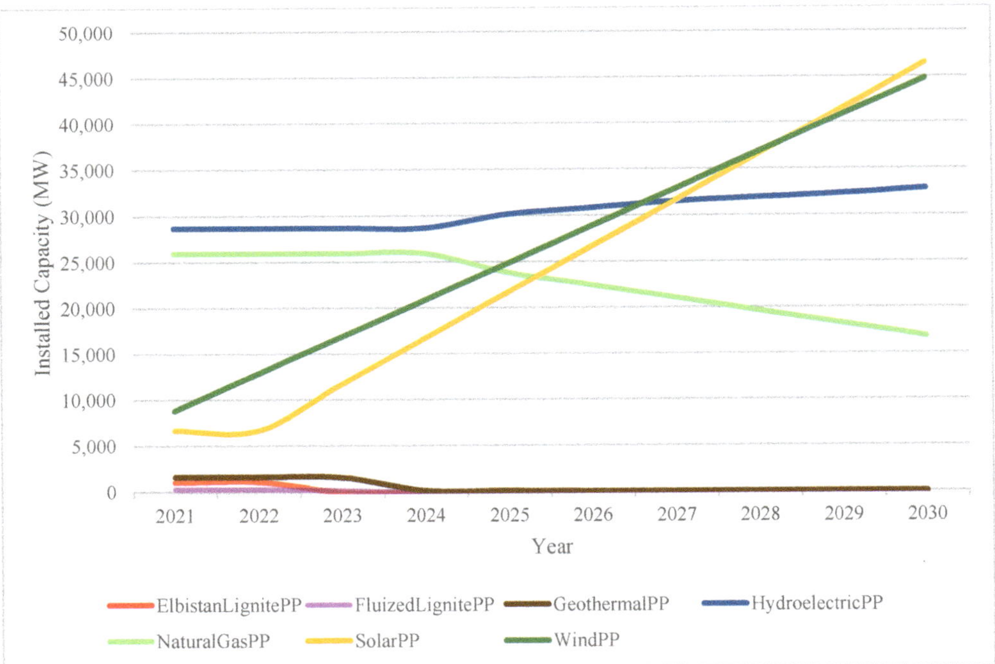

Figure 4. Installed capacities for the first scenario in the 2021–2030 horizon.

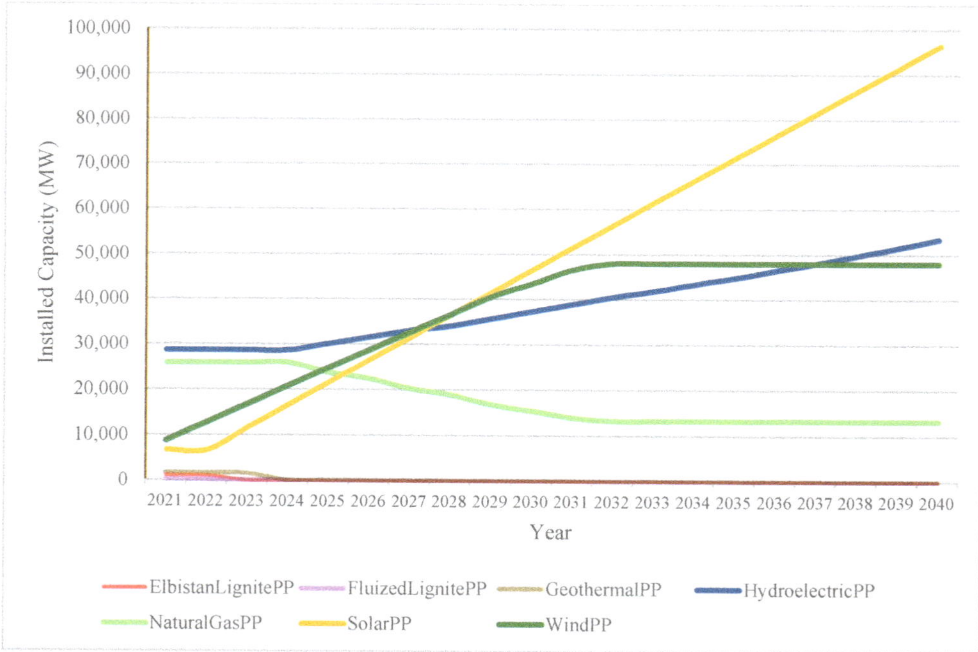

Figure 5. Installed capacities for the first scenario in the 2021–2040 horizon.

The starting capacity of each power plant type was supplied in the base scenario, and the new power plant requirements were estimated based on this initial capacity. In the second scenario, we want to explore what happens when the model determines all power plant types and their capacities. Furthermore, as in the first scenario, it is anticipated that no power plant openings are planned. To summarize, the model determines all power plant types and capacities in this scenario. Since there is no available capacity at time zero, the requisite number of power plants should be opened to meet demand throughout the first period. As a result, we abandoned the power plant building schedule limits. Otherwise, because no power plants can be operational in the early stages, demand cannot be met, and the model becomes unsustainable. Furthermore, we remove the renewable capacity restriction constraint to check if the model selects only renewable resources or not. Installed capacities for 2021–2030 and 2021–2040 planning horizons under the assumptions of the second scenario are visualized in Figures 6 and 7, respectively. The main power plant types chosen in the second scenario for the period 2021–2030 are solar, wind, and natural gas power plants. In comparison to the base scenario, solar share climbs to 39.9% (7.4% higher), wind share decreases to 22.5% (9.3% lower), and natural gas share remains same at around 32%. Hydroelectric and geothermal power plants are also used, but their contributions are less than 4%. When compared to the baseline scenario, the geothermal potential is fully utilized, while hydroelectric capacity is significantly reduced. Wind and solar power plant shares are close to the base scenario for the period 2021–2040. Natural gas capacity remains unchanged, but hydropower capacity drops by 17% towards the end of the planning horizon compared to the base scenario, and geothermal potential is utilized to maximum capacity.

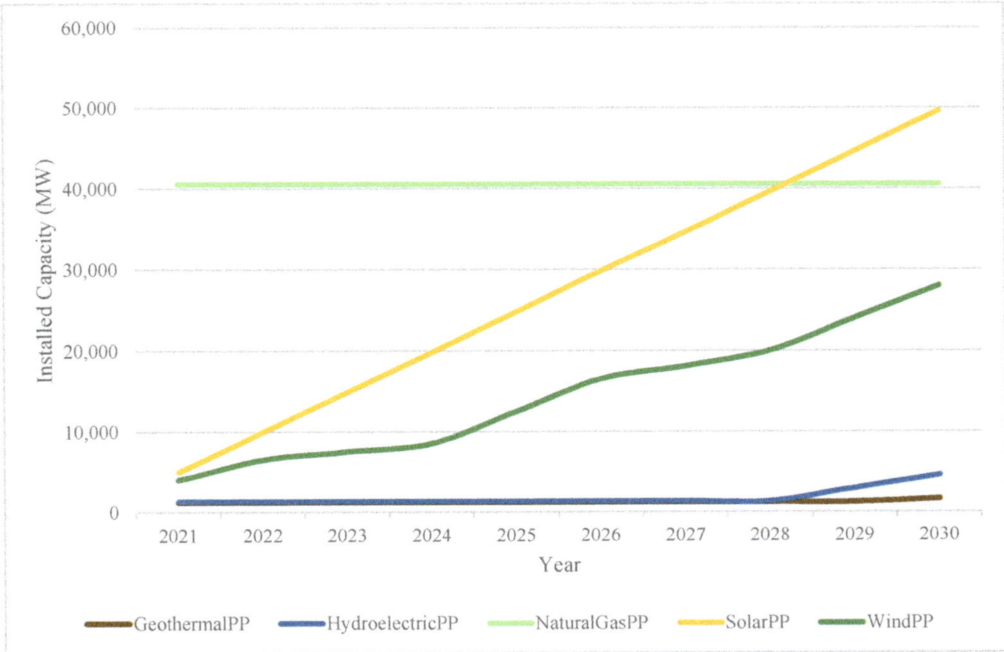

Figure 6. Installed capacities for the second scenario in the 2021–2030 horizon.

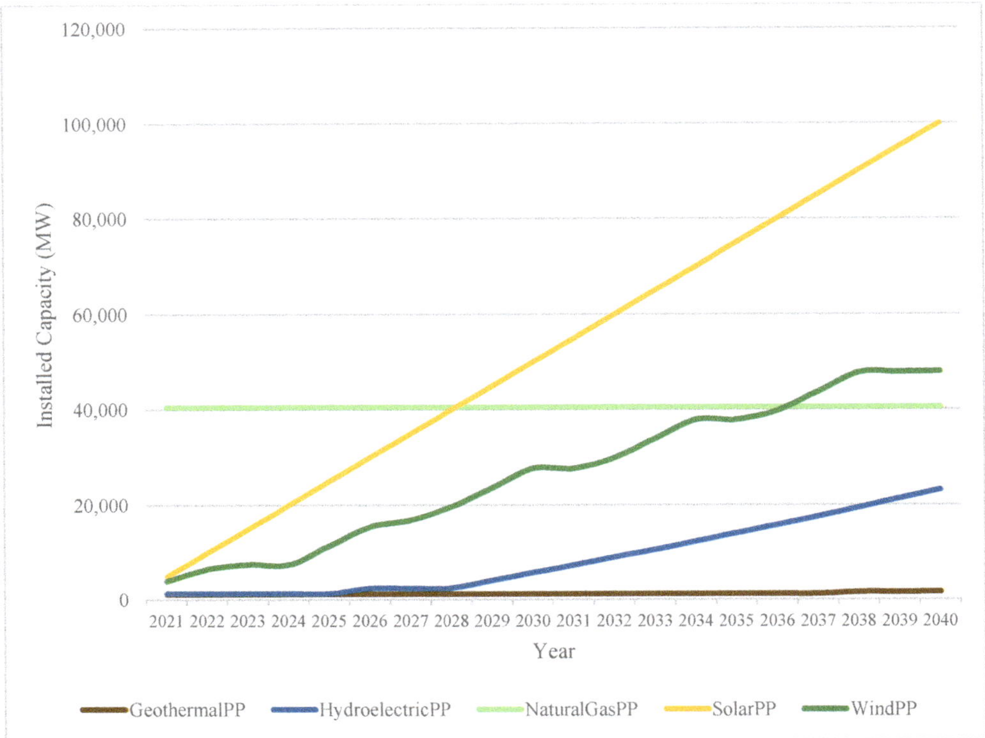

Figure 7. Installed capacities for the second scenario in the 2021–2040 horizon.

In the third scenario, we considered the case that all power plants are renewable. The goal of this scenario is to examine what occurs when all capacity is constrained to renewable sources. We will also be able to determine whether the current renewable potential is sufficient to meet the available demand. In this scenario, the available capacity of renewables is stated as the beginning capacity, and additional power plants other than renewables are not included. Furthermore, prospective nonrenewable power projects are not considered. Construction time limits are dropped, as in the second scenario. In addition to satisfy demands, import limits are removed. Findings under the assumptions of the third scenario are shown in Figures 8 and 9 for 2021–2030 and 2021–2040 planning horizons, respectively. All electricity demand is met by renewable power plants and imports. As existing renewable potential is insufficient to fulfill demand, the import option is adopted. The majority of the demand is met by solar, wind, and hydroelectric power sources. Furthermore, geothermal power plants are utilized to their utmost capability. Solar, hydroelectric, wind, and geothermal power plants, and imports, provide the demand. The average import contribution is around 19%. Solar power plants supply around 49% of energy, 28% of hydropower, 22% of wind, and 1% of geothermal energy.

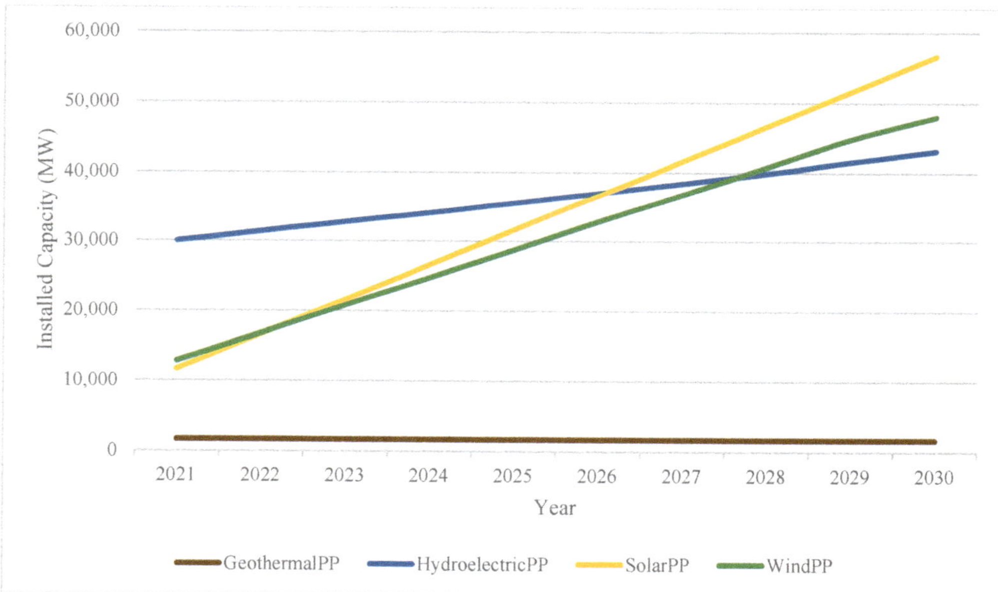

Figure 8. Installed capacities for the third scenario in the 2021–2030 horizon.

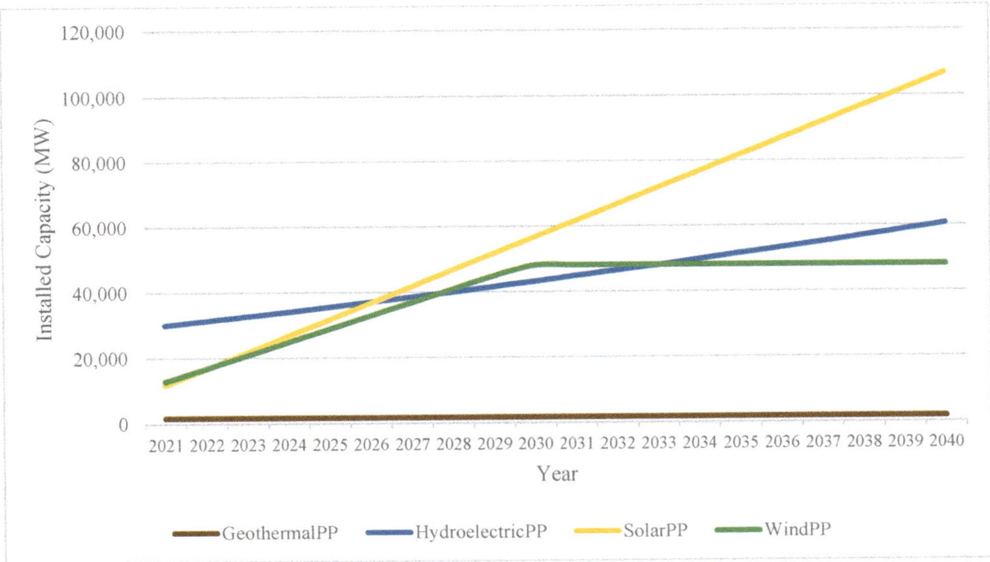

Figure 9. Installed capacities for the third scenario in the 2021–2040 horizon.

5. Conclusions

This study examines Turkey's strategic level electrical energy planning challenge from various perspectives, including the strategic plan, energy policy, capacity planning, and environmental policies of the government. To tackle the given problem, a mixed integer mathematical programming model that takes into account alternative power plant categories, such as fossil fuels, renewable energy, nuclear energy, and so on, is proposed. As several energy resources are included, the defined problem is characterized as a "general electricity planning problem". In addition to lowering electricity generation costs, a variety of alternative policies, such as lowering CO_2 emissions, limiting energy resource share regulations (such as limiting the use of fossil fuels), and promoting renewable energy, are taken into account in this study. This study is also falling under the "energy policy analysis" category in this regard. Two different planning horizons are considered, namely 2021–2030 and 2021–2040, and it is observed that the share of renewable resources increases while the share of fossil fuels declines with time.

As a result of this research, various key insights and outcomes involving power investment and production planning have been achieved. Due to the highest levelized costs of all choices, the first nuclear energy option is not chosen if the model is not required to do so. If nuclear energy is required by government regulations, all fossil fueled power facilities must be shut down during the planning horizon. Otherwise, coal power facilities are shut down, while natural gas power plants are up and running. Hydroelectric power plants are the least appealing renewable energy source because they have a higher levelized cost than wind and solar power plants and a lower availability factor than geothermal power plants.

The findings of this study indicate the trend toward renewable energy. Although nuclear energy is perceived as an effective energy resource, it is shown that renewable energy resources are more cost effective under the determination of CO_2 emissions and generation capabilities. These results can be used as a guide to update strategic energy generation plans to improve the long-term effectiveness of future investments in power plants. It is advocated in this research to steadily boost renewable energy expenditures (particularly solar, wind, and geothermal) and eventually replace fossil fuel alternatives. The proposed energy plan not only saves investment, operation, and maintenance expenses, but also cuts emissions. Nuclear energy can also be used as an alternate and reliable source of energy, but the possible risks and greater costs must be addressed. Additionally, more renewable energy resources, such as hydrogen power can be included in the analysis, and minimization of total emission can be introduced as an additional goal in the objective function in future research. Additionally, the suggested model is deterministic, and it is assumed that the parameter values are known precisely. However, in reality, this is not the case, and the values of various factors may fluctuate based on economic, political environmental, and strategic aims. The renewable shares, import and export limits, demand, and levelized cost parameters can also be modeled as stochastic variables in future studies. Moreover, as in [53] microstructure of Turkey's renewable electricity sources can be studied in the future to create cost efficiency and reduce carbon emissions. This type of study may provide insights into developing countries. In conclusion, ARIMA is used as a statistical time series model. Although statistical estimation methods are used in optimization algorithms in some studies, the integration of forecasting results from the ARIMA model to the mixed integer linear programming is a new and recently evolving area of interest [54,55]. In addition to that, the contribution to energy planning is especially appropriate for developing countries, such as Turkey, in which switching to renewable energy resources is in the early phases. The findings of this study may serve as a guideline to prioritize energy resource preferences in developing countries during the planning phase since it is shown that optimization is required before the preparation of any regulations since these regulations have a considerable effect on the distribution of plants used.

The results show that the use of renewable energy generation options is the most preferable source of energy as expected. However, it is found that some regulations and not optimized plans may prevent the effective use of these resources. As seen in the results of the scenario analyses geothermal and hydroelectric alternatives are found as better options when compared to nuclear power plants when current plans are neglected which causes the shutdown of geothermal plants and inefficient increase in hydropower facilities. In this context, it should be added that if there are not any existing hydropower facilities it is found that building many new ones is not a feasible option. Furthermore, it is observed that natural gas power plants are preferred to nuclear power plants and even as an alternative to hydropower facilities. Even though natural gas is a fossil fuel-based resource, this alternative is used to support renewable energy plants in optimal scenarios since CO_2 emission rates can still be fulfilled. Finally, solar power plants are found to be the best energy generation option, especially in long term plans as they become more feasible than wind power plants, whereas geothermal resources are found to be used at full capacity even though they are much scarcer than solar and wind options.

Author Contributions: Conceptualization, G.Y., U.B. and K.D.Ü.; methodology, G.Y., U.B. and F.Y.-Ö.; software, G.Y., U.B. and F.Y.-Ö.; validation, G.Y., U.B., F.Y.-Ö. and K.D.Ü.; formal analysis, G.Y., U.B. and F.Y.-Ö.; investigation, G.Y., U.B., F.Y.-Ö. and K.D.Ü.; resources, G.Y.; data curation, G.Y.; writing—original draft preparation, K.D.Ü.; writing—review and editing, U.B., F.Y.-Ö. and K.D.Ü.; visualization, G.Y.; supervision, U.B. and F.Y.-Ö. All authors have read and agreed to the published version of the manuscript.

Funding: This research received no external funding.

Data Availability Statement: Publicly available datasets were analyzed in this study. This data can be found here: https://www.teias.gov.tr/tr-TR/turkiye-elektrik-uretim-iletim-istatistikleri (accessed on 15 May 2020).

Acknowledgments: We appreciate the associate editor's and four anonymous reviewers' helpful comments and revisions.

Conflicts of Interest: The authors declare no conflict of interest.

References

1. IEA. World Energy Outlook. Available online: https://www.iea.org/reports/world-energy-outlook-2019 (accessed on 1 December 2022).
2. Zhou, P.; Ang, B.W.; Poh, K.L. Decision analysis in energy and environmental modeling: An update. *Energy* **2006**, *31*, 2604–2622. [CrossRef]
3. Lazard. Lazard's Annual Levelized Cost of Energy Analysis (LCOE 12.0). Available online: https://www.lazard.com/ (accessed on 1 December 2022).
4. Baños, R.; Manzano-Agugliaro, F.; Montoya, F.G.; Gil, C.; Alcayde, A.; Gómez, J. Optimization methods applied to renewable and sustainable energy: A review. *Renew. Sustain. Energy Rev.* **2011**, *15*, 1753–1766. [CrossRef]
5. Yörük, G. Strategic Energy Production Planning of Turkey Using Mixed Integer Programming Based on Electricity Demand Forecasting. Master's Thesis, Atilim University, Ankara, Turkey, 2021.
6. Ramanathan, R.; Ganesh, L.S. Energy alternatives for lighting in households: An evaluation using an integrated goal programming-AHP model. *Energy* **1995**, *20*, 63–72. [CrossRef]
7. Pokharel, S.; Chandrashekar, M. A multiobjective approach to rural energy policy analysis. *Energy* **1998**, *23*, 325–336. [CrossRef]
8. Mavrotas, G.; Diakoulaki, D.; Papayannakis, L. An energy planning approach based on mixed 0-1 multiple objective linear programming. *Int. Trans. Oper. Res.* **1999**, *6*, 231–244. [CrossRef]
9. Agrawal, R.K.; Singh, S.P. Energy allocations for cooking in UP households (India): A fuzzy multi-objective analysis. *Energy Convers. Manag.* **2001**, *42*, 2139–2154. [CrossRef]
10. Antunes, C.H.; Martins, A.G.; Brito, I.S. A multiple objective mixed integer linear programming model for power generation expansion planning. *Energy* **2004**, *29*, 613–627. [CrossRef]
11. San Cristóbal, J.R. A goal programming model for the optimal mix and location of renewable energy plants in the north of Spain. *Renew. Sustain. Energy Rev.* **2012**, *16*, 4461–4464. [CrossRef]
12. Özcan, E.C.; Erol, S. A multi-objective mixed integer linear programming model for energy resource allocation problem: The case of Turkey. *Gazi Univ. J. Sci.* **2014**, *27*, 1157–1168.
13. Şengül, Ü.; Eren, M.; Eslamian Shiraz, S.; Gezder, V.; Sengül, A.B. Fuzzy TOPSIS method for ranking renewable energy supply systems in Turkey. *Renew. Energy* **2015**, *75*, 617–625. [CrossRef]
14. Cormio, C.; Dicorato, M.; Minoia, A.; Trovato, M. A regional energy planning methodology including renewable energy sources and environmental constraints. *Renew. Sustain. Energy Rev.* **2003**, *7*, 99–130. [CrossRef]
15. Ai, Y.P.; Huang, G.H.; Yang, Z.F.; Lin, Q.G.; Bass, B.; Tan, Q. Development of an optimization model for energy systems planning in the Region of Waterloo. *Int. J. Energy Res.* **2008**, *32*, 988–1005.
16. Huang, Y.H.; Wu, J.H. A portfolio risk analysis on electricity supply planning. *Energy Policy* **2008**, *36*, 627–641. [CrossRef]
17. Daniel, J.; Dicorato, M.; Forte, G.; Iniyan, S.; Trovato, M. A methodology for the electrical energy system planning of Tamil Nadu state (India). *Energy Policy* **2009**, *37*, 904–914. [CrossRef]
18. Arnette, A.; Zobel, C.W. An optimization model for regional renewable energy development. *Renew. Sustain. Energy Rev.* **2012**, *16*, 4606–4615. [CrossRef]
19. Pina, A.; Silva, C.A.; Ferrão, P. High-resolution modeling framework for planning electricity systems with high penetration of renewables. *Appl. Energy* **2013**, *112*, 215–223. [CrossRef]
20. Zhang, Q.; Mclellan, B.C.; Tezuka, T.; Ishihara, K.N. An integrated model for long-term power generation planning toward future smart electricity systems. *Appl. Energy* **2013**, *112*, 1424–1437. [CrossRef]
21. Habib, M.A.; Chungpaibulpatana, S. Electricity generation expansion planning with environmental impact abatement: Case study of Bangladesh. *Energy Procedia* **2014**, *52*, 410–420. [CrossRef]
22. Moret, S.; Bierlaire, M.; Maréchal, F. Strategic Energy Planning under Uncertainty: A Mixed-Integer Linear Programming Modeling Framework for Large-Scale Energy Systems. *Comput. Aided Chem. Eng.* **2016**, *38*, 1899–1904.
23. Santos, M.J.; Ferreira, P.; Araújo, M. A methodology to incorporate risk and uncertainty in electricity power planning. *Energy* **2016**, *115*, 1400–1411. [CrossRef]
24. Kachoee, M.S.; Salimi, M.; Amidpour, M. The long-term scenario and greenhouse gas effects cost-benefit analysis of Iran's electricity sector. *Energy* **2018**, *143*, 585–596. [CrossRef]
25. Shen, W.; Qiu, J.; Dong, Z. Electricity network planning targeting Low-Carbon energy transition. *Glob. Energy Interconnect.* **2018**, *1*, 487–499.
26. Walmsley, M.R.W.; Walmsley, T.G.; Atkins, M.J. Linking greenhouse gas emissions footprint and energy return on investment in electricity generation planning. *J. Clean. Prod.* **2018**, *200*, 911–921. [CrossRef]
27. Senatla, M.; Nchake, M.; Taele, B.M.; Hapazari, I. Electricity capacity expansion plan for Lesotho—Implications on energy policy. *Energy Policy* **2018**, *120*, 622–634. [CrossRef]
28. Levin, T.; Kwon, J.; Botterud, A. The long-term impacts of carbon and variable renewable energy policies on electricity markets. *Energy Policy* **2019**, *131*, 53–71. [CrossRef]
29. Anderson, D. Models for Determining Least-Cost Investments in Electricity Supply. *Bell J. Econ.* **2008**, *3*, 267–299. [CrossRef]
30. Trotter, P.A.; McManus, M.C.; Maconachie, R. Electricity planning and implementation in sub-Saharan Africa: A systematic review. *Renew. Sustain. Energy Rev.* **2017**, *74*, 1189–1209. [CrossRef]

31. Melikoglu, M. Vision 2023: Scrutinizing achievability of Turkey's electricity capacity targets and generating scenario base nationwide electricity demand forecasts. *Energy Strateg. Rev.* **2018**, *22*, 188–195. [CrossRef]
32. Calikoglu, U.; Koksal, M.A. Green electricity and Renewable Energy Guarantees of Origin demand analysis for Türkiye. *Energ Policy* **2022**, *170*, 113229. [CrossRef]
33. Ünal, B.B.; Onaygil, S.; Acuner, E.; Cin, R. Application of energy efficiency obligation scheme for electricity distribution companie in Turkey. *Energy Policy* **2022**, *163*, 112851. [CrossRef]
34. Selçuklu, S.B.; Coit, D.W.; Felder, F.A. Electricity generation portfolio planning and policy implications of Turkish power systen considering cost, emission, and uncertainty. *Energy Policy* **2023**, *173*, 113393. [CrossRef]
35. Singh, A.K.; Khatoon, S. An Overview of Electricity Demand Forecasting Techniques. *Natl. Conf. Emerg. Trends Electr. Instrum Commun. Eng.* **2013**, *3*, 38–48.
36. Srinivasan, D.; Lee, M.A. Survey of hybrid fuzzy neural approaches to electric load forecasting. *Proc. IEEE Int. Conf. Syst. Man Cybern.* **1995**, *5*, 4004–4008.
37. Jiang, W.; Wu, X.; Gong, Y.; Yu, W.; Zhong, X. Holt–Winters smoothing enhanced by fruit fly optimization algorithm to forecas monthly electricity consumption. *Energy* **2020**, *193*, 116779. [CrossRef]
38. Al-Musaylh, M.S.; Deo, R.C.; Adamowski, J.F.; Li, Y. Short-term electricity demand forecasting with MARS, SVR and ARIMA models using aggregated demand data in Queensland, Australia. *Adv. Eng. Inform.* **2017**, *35*, 1–16. [CrossRef]
39. Yukseltan, E.; Yucekaya, A.; Bilge, A.H. Forecasting electricity demand for Turkey: Modeling periodic variations and demand segregation. *Appl. Energy* **2017**, *193*, 287–296. [CrossRef]
40. Abd Jalil, N.A.; Ahmad, M.H.; Mohamed, N. Electricity load demand forecasting using exponential smoothing methods. *World Appl. Sci. J.* **2013**, *22*, 1540–1543.
41. Lee, C.M.; Ko, C.N. Short-term load forecasting using lifting scheme and ARIMA models. *Expert Syst. Appl.* **2011**, *38*, 5902–5911 [CrossRef]
42. Dilaver, Z.; Hunt, L.C. Modelling and forecasting Turkish residential electricity demand. *Energy Policy* **2011**, *39*, 3117–3127 [CrossRef]
43. Taylor, J.W. Triple seasonal methods for short-term electricity demand forecasting. *Eur. J. Oper. Res.* **2010**, *204*, 139–152. [CrossRef]
44. Shin, S.-Y.; Woo, H.-G. Energy Consumption Forecasting in Korea Using Machine Learning Algorithms. *Energies* **2022**, *15*, 4880 [CrossRef]
45. Leite Coelho da Silva, F.; da Costa, K.; Canas Rodrigues, P.; Salas, R.; López-Gonzales, J.L. Statistical and Artificial Neural Networks Models for Electricity Consumption Forecasting in the Brazilian Industrial Sector. *Energies* **2022**, *15*, 588. [CrossRef]
46. Ünlü, K.D. A Data-Driven Model to Forecast Multi-Step Ahead Time Series of Turkish Daily Electricity Load. *Electronics* **2022**, *11*, 1524. [CrossRef]
47. Akbal, Y.; Ünlü, K.D. A univariate time series methodology based on sequence-to-sequence learning for short to midterm wind power production. *Renew. Energy* **2022**, *200*, 832–844. [CrossRef]
48. Akdi, Y.; Gölveren, E.; Okkaoğlu, Y. Daily electrical energy consumption: Periodicity, harmonic regression method and forecasting. *Energy* **2020**, *191*, 116524. [CrossRef]
49. Kazancık, L.B. Türkiye Elektrik Enerji Sektöründe Robust Optimizasyon Modeli. Ph.D. Thesis, Department of Statistics, Ankara Üniversitesi, Ankara, Turkey, 2016.
50. İncekara, Ç.Ö. Türkiye'nin Sürdürülebilir Stratejik Enerji Politikalarının Oluşturulması için Optimizasyon Modellerinin Geliştirilmesi ve Uygulama Adımları. Master's Thesis, Department of Ind. Eng., Çukurova Üniversitesi, Adana, Turkey, 2007.
51. Box, G.; Jenkins, G. *Time Series Analysis*; Forecasting and Control; Holden-Day: San Francisco, CA, USA, 1970.
52. TEİAŞ. Electric Generation and Transfer Statistics 2018, Electric Statistics. 2020. Available online: https://www.teias.gov.tr/tr-TR/turkiye-elektrik-uretim-iletim-istatistikleri (accessed on 29 January 2021).
53. Cosic, A.; Stadler, M.; Mansoor, M.; Zellinger, M. Mixed-integer linear programming based optimization strategies for renewable energy communities. *Energy* **2021**, *237*, 121559. [CrossRef]
54. Meier, A.; Kramer, O. Recurrent neural network-predictions for PSO in dynamic optimization. In Proceedings of the Genetic and Evolutionary Computation Conference, Kyoto, Japan, 15–19 July 2018; Volume 29, p. 36.
55. Ordu, M.; Demir, E.; Tofallis, C.; Gunal, M.M. A novel healthcare resource allocation decision support tool: A forecasting-simulation-optimization approach. *J. Oper. Res. Soc.* **2021**, *72*, 485–500. [CrossRef]

mathematics

Article

Performance Evaluation of Grid-Connected DFIG-Based WECS with Battery Energy Storage System under Wind Alterations Using FOPID Controller for RSC

Pradeep Singh [1], Krishan Arora [1], Umesh C. Rathore [2], Eunmok Yang [3], Gyanendra Prasad Joshi [4,*] and Kwang Chul Son [5,*]

1 School of Electronics and Electrical Engineering, Lovely Professional University, Phagwara 144411, India; pradeep.42000074@lpu.in (P.S.); krishan.12252@lpu.co.in (K.A.)
2 Department of Electrical Engineering, Atal Bihari Vajpayee Government Institute of Engineering & Technology-Pragtinagar, Shimla 171202, India
3 Department of Financial Information Security, Kookmin University, Seoul 02707, Republic of Korea; emyang@kookmin.ac.kr
4 Department of Computer Science and Engineering, Sejong University, Seoul 05006, Republic of Korea
5 Department of Information Contents, Kwangwoon University, Seoul 01897, Republic of Korea
* Correspondence: joshi@sejong.ac.kr (G.P.J.); kcson@kw.ac.kr (K.C.S.); Tel.: +82-2-940-8100 (K.C.S.)

Abstract: In the present energy scenario, wind energy is the fastest-growing renewable energy resource on the globe. However, wind-energy-based generation systems are also associated with increasing demands for power quality and active power control in the power network. With the advancements in power-electronics-based technology and its use in non-conventional energy conversion systems, it has witnessed tremendous growth in wind energy conversion systems (WECSs). At the same time, integrating wind farms into the grid system also results in many power quality issues in the power system that involve these renewable energy sources feeding power networks. This paper reports the effectiveness of grid-connected doubly fed induction generator (DFIG)-based WECS with a battery energy storage system (BESS) under variable wind conditions. In this study, a rotor side converter (RSC) is controlled to achieve the optimal torque for a given maximal wind power. The control scheme is simulated using MATLAB for a 2 MW-rated DFIG used in a WECS. Additionally, in this paper, a new fraction order proportional integral derivative (FOPID) controller is introduced into the system's RSC, and its performance is also observed. The BESS technique is used with a DC link to improve the overall performance of the DFIG-based WECS under different wind conditions. To control the BESS, a proportional integral (PI) controller is introduced to increase the charging and discharging rates. Two models are developed in MATLAB/Simulink: one model is a basic model, and other model is equipped with a BESS and a PI controller in the BESS. The results validate the effectiveness of the proposed PI-controller-equipped BESS at improving the overall performance of the WECS system under study.

Keywords: renewable energy; WECS; DFIG; BESS; PI controller; FOPID-controller; RSC

MSC: 68W99; 65K10

Citation: Singh, P.; Arora, K.; Rathore, U.C.; Yang, E.; Joshi, G.P.; Son, K.C. Performance Evaluation of Grid-Connected DFIG-Based WECS with Battery Energy Storage System under Wind Alterations Using FOPID Controller for RSC. *Mathematics* **2023**, *11*, 2100. https://doi.org/10.3390/math11092100

Academic Editors: Jinfeng Liu, Atanda Raji and Khaled M. Abo-Al-Ez

Received: 16 March 2023
Revised: 16 April 2023
Accepted: 26 April 2023
Published: 28 April 2023

1. Introduction

One third of global greenhouse gas (GHG) releases are attributed to the source materials responsible for producing power, such as fossil fuels, petroleum, and gasses. India's present attempts at commercial improvement are leading to an increase in the nation's energy requirements. Indeed, the maximal extraction of natural resources is a necessary component of the expansion of a country's economic growth [1]. The changing climate could result in a shift in the planet's natural balance. The UNFCCC (UN Framework for Convention on Climatic Changes) and the PT (Paris Treaty) have both received submissions of INDCs (intended national voluntarily determined contributions), with the latter aimed at limiting the rise in the global mean temp. to far below 2 °C.

Among the alternatives mentioned above, the most well-known and prevalent sources of clean energy in India are wind and solar [2]. The power scenario in India up to 2022 is shown in Figure 1.

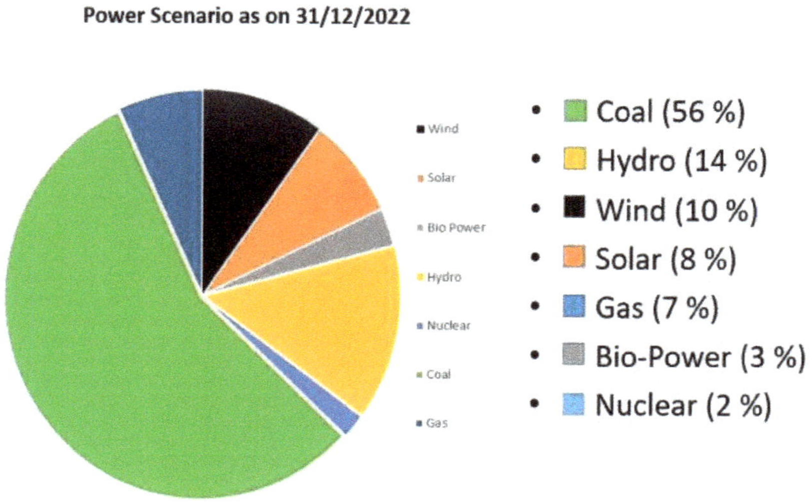

Figure 1. Power scenario in India as of 31 December 2022.

By 2023, India aims to produce 180 GW of power in the form of renewable sources. Of the 180 GW, 5 GW will come from small hydroelectric plants, 10 GW from biofuels, 100 GW from photovoltaic cells, and 65 GW from wind farms [3]. The government must provide 330,000 new jobs in order to meet the challenging goal of 180 GW of sustainable power generation by 2023 [4].

Disabling an RSC results in a loss of power control and the occurrence of significant transients immediately after the fixation of the fault. In some cases, this may necessitate disconnecting the machine from the grid [5]. The power scenario expected by 2039–2040 is in Figure 2.

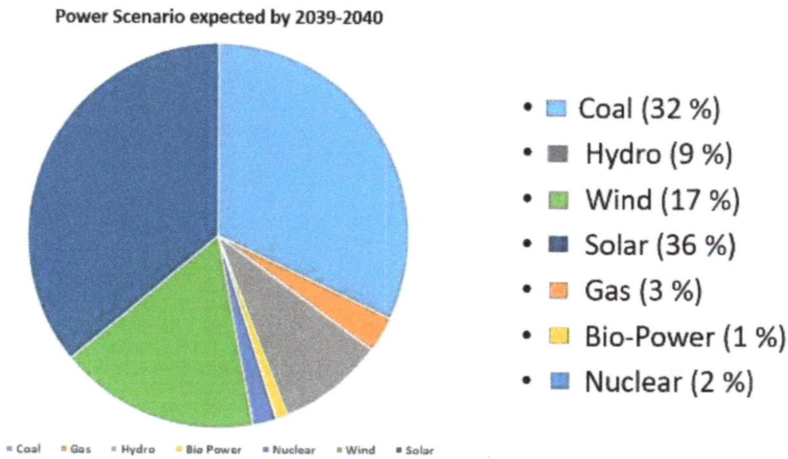

Figure 2. Power scenario expected by 2039–2040.

However, the operation limit of a changing-velocity WECS may be split into four main zones. In first zone, low wind speeds do not produce enough energy. In second zone, the structure is optimized to produce maximum power while adjusting to changes in wind velocity using the maximum/peak power point tracking (MPPT) approach, in which the angle of the blade pitch is fixed to its perfect value. In third zone, when the wind velocity goes above the predefined values, the angle of the pitch is adjusted to control the electrical energy generated around its predefined value. In the fourth zone, high wind speeds can cause damage to the wind turbine (WT), requiring the devices that work in emergency situations to stop the turbine so that the damage can be prevented [6]. A. Santhi Mary Antony et al. [7] presented simulation results for different types of WECS generators, including the permanent magnet synchronous generator (PMSG), squirrel cage induction generator (SCIG), and DFIGs with bridge and bridgeless configurations, all of which utilize a WT with a horizontal axis. The conventional system couples the RSC and GSC through a direct current capacitor link, which helps in to maintain a steady direct current voltage and reduce voltage ripple but also has the disadvantage of being bulky [8].

However, with a direct connection to the power grid, the DFIG can struggle with more variations in torque and harmonic content in the grid, leading to shaft vibration, increased mechanical wear and tear, and higher maintenance costs [9]. To mitigate these issues, ESSs can be employed to store excess wind energy during high-production periods. Additionally, integrating the system of energy storage directly at the DC link terminals of the DFIG to maintain the variability of the WE on the grid side has been proposed by Singh et al. [10].

According to the reviewed literature, adaptation control has been employed less extensively than vector control with DFIG power transformers. This same application of adaption control to grid-connected photo voltaic arrays has shown outstanding performance. To exploit its advantages, Singh et al. [11] used adaption control to effectively operate the GSC of an induction generator for a WT. It has also been discovered that the use of adaption control methods leads to increased weighted computation efficiency, less fluctuations, and a quicker settling time, which are critical factors in determining power balancing, PQ, resilience, and the maximum supply of energy fed into network in rotary-machine-integrated systems. Different types of energy storage systems are shown in Figure 3, some of which are briefly explained herein. Batteries are the most commonly used energy storage devices and store electrical energy in a chemical form. Some common types of batteries are lithium-ion, lead–acid, nickel–cadmium, and sodium-ion, etc. Flywheels store energy in the form of rotational kinetic energy. A flywheel consists of a rotor that rotates at a high speed in a vacuum, and energy is extracted by slowing down the rotor. In the case of pumped hydroelectric storage, energy is stored by pumping water uphill to a reservoir during periods of low demand and then releasing it through a turbine to generate electricity during high-demand periods. In the case of compressed air energy storage, energy is stored by compressing air in an underground reservoir during low-demand periods and then releasing it through a turbine to generate electricity during high-demand periods. Thermal energy storage involves storing energy in the form of heat; common types of thermal energy storage are water-based systems, phase-change materials, molten salt systems, etc. Superconducting magnetic energy storage stores energy in the magnetic field generated by a superconducting coil, which can be used to generate electricity during periods of peak demand. Hydrogen energy storage involves using excess electricity to produce hydrogen through electrolysis, which can then be stored and used to generate electricity during periods of peak demand. Capacitors store energy in an electric field and are commonly used in electronics and power electronics.

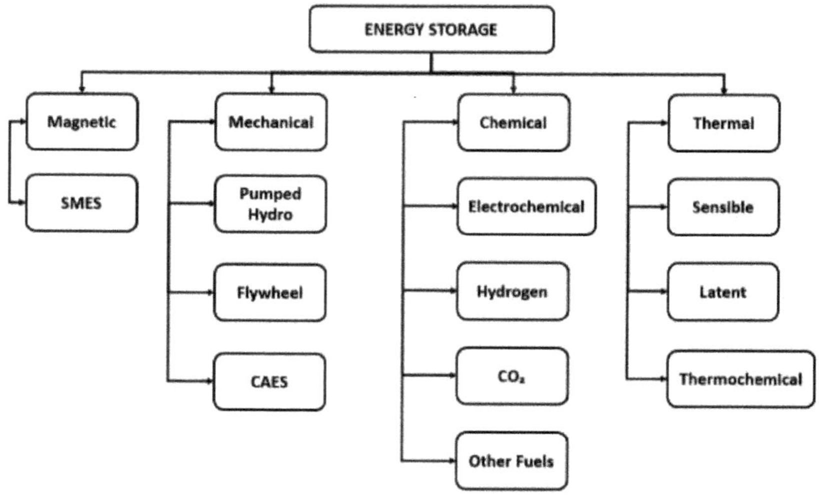

Figure 3. Classification. of energy storage system (ESSs).

To achieve these objectives, the proposed control scheme integrates a BESS with a direct current link through a power conversion device which compensates for wind alterations and employs a high-order adaption control for the GSC of the DFIG, supplying the system with a consistent amount of high energy [12]. An application scenario of an energy storage system is shown in Figure 4.

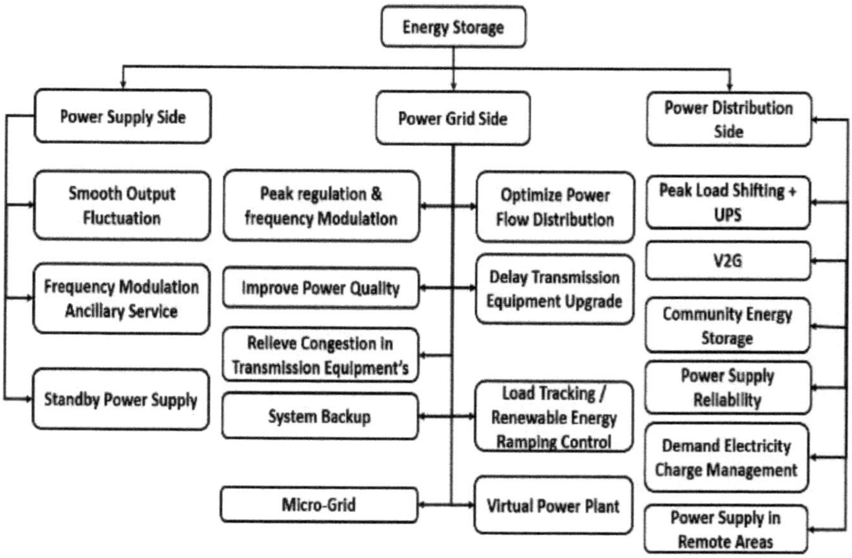

Figure 4. Application scenario of an ESS.

The control architecture's purpose is to maintain a constant V and f at the terminal of the DFIG in response to wind velocity alterations and additional load variations. However, developing a control system for a grid-integrated WECS is a challenging task due to the erratic patterns of wind velocity. However, the high switching frequency associated with DTC leads to increased current harmonic distortion, elevated electromagnetic torque fluctuations, and elevated heating in IGBT components [13]. These controllers, including back stepping control, feedback linearization control, H-infinity control, adaptive control, model predictive control, and sliding model control, have the ability to handle nonlinearities in a wide range of operations [14].

The use of sliding mode control (SMC) is advantageous in grid-tied DFIG systems due to its quick calculation time and straightforward installation. Fuzzy logic (FL) and artificial neural networks (ANNs) are examples of AI (artificial intelligence) approaches that are employed to improve the effectiveness of SMC and to lessen the buzzing influence. In HOSMCs (higher-order sliding mode controllers), the super twisting technique (STT) is frequently employed [15]. Several state of charge (SOC) restrictions have also been taken into consideration in distinct experiments towards the fuzzy-logic-based charging and discharging control of Li-ion battery systems [16]. An FL-based BESS is mainly proposed to regulate the condition of a battery's SOC. This SOC restriction was set in the range of 50% to 100% [17]. Martinez et al. [18] also employed a similar approach, utilizing 5 MF to create 21 FL rules to restrict the battery's SOC.

In the coastal and Himalayan regions of Himachal Pradesh, wind input conditions vary greatly, with some areas experiencing high wind speeds that are not constant. Data for an entire year from Dharamshala in the Kangra district of Himachal Pradesh are analyzed and depicted in Figure 5. Twin capacitors are charged using a DC link voltage thanks the R and L series connection of the BESS. Quick velocity resolution, power control, and the regulation of the DC link voltage via the GSC are all ensured by the recommended converters and control techniques applied to the RSC. Using MATLAB Simulation, the overall efficiency of the suggested technique was confirmed, has and it was shown to work superbly under typical operational settings.

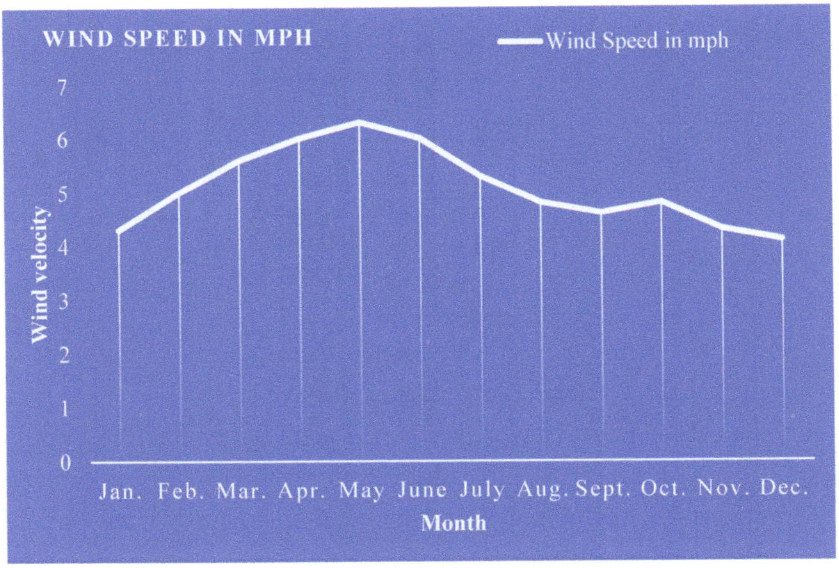

Figure 5. Continuous average velocity of nearby Kangra in Himachal Pradesh for 1 year.

The contributions of this paper are as follows: first, we recommend various energy storage devices for direct current links in place of capacitors and provide scenarios of their application in a wind energy conversion system. We observed the wind speeds at Dharamshala in HP for an entire year. Section 2 provides a model and control for a doubly-fed-induction-generator-based wind energy conversion system; Section 3 provides the design of a battery energy storage system; Section 4 explains the design of a fractional order proportional integral derivative controller used in the converter. In Section 5, we provide an analysis of the proposed technique in MATLAB. In Section 6, the results from the simulation performed in this study are presented and analyzed. Section 7 concludes the paper. This paper investigates the impact of a BESS on the performance and stability of the system under varying wind conditions. This research problem statement could contribute to the development of more efficient and reliable DFIG-based WECSs with BESSs, which can help increase the prevalence of renewable energy sources in the power grid. The performance of the proposed algorithms is discussed, and potential areas for future research are outlined.

2. Modeling and Control

2.1. Modeling of WT

The core component of a WECS is the WT, which converts the wind energy into mechanical energy. A gearbox is employed to link the WT to the DFIG. The generated power is then integrated into an electrical grid through a control system with the aim of minimizing disruptions, enhancing stability, and optimizing the quality of the system. The power output (o/p) of a WT is a nonlinear function that is dependent on the wind velocity; therefore, the amount of power o/p is described by Equation (1), as indicated in [19]. The main components of a WECS are shown in Figure 6.

$$P_m = 0.5\rho A C_p(\lambda, \beta) V^3 \tag{1}$$

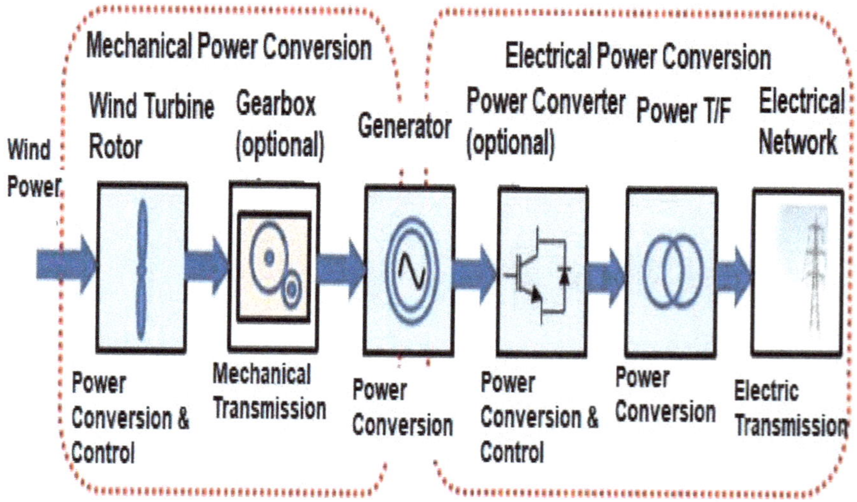

Figure 6. Main components of a WECS.

Moreover, λ is outlined as the ratio of the linear velocity at the tip of the wind rotor blade to the wind velocity and may be written as in Equation (2):

$$\lambda = \frac{\omega R}{V} \tag{2}$$

where ρ represents the density present in the wind (kg/m^3), A refers to the area covered by the WT blades (m^2), **V** denotes the wind velocity (m/s), R stands for the sweep of turbine blades in (rad/s), and C_p represents the performance coefficient of the WT.

The evolution of torque is as follows:

$$T_a = \frac{\rho}{\omega_t} = \frac{1}{2\lambda}\rho\pi R^3 V_w^3 C_p(\lambda, \beta) \tag{3}$$

where R is the radius of the WT and ω_t signifies the rotational velocity of the WT. The role of the gearbox is to link the WT shaft with the DFIG's rotor, enabling an increase in the low velocity of the WT to increase the external velocity of the DFIG. The equation that governs this gearbox is as follows:

$$\omega_t = \frac{\omega_m}{G} \text{ and } T_m = \frac{T_a}{G}$$

The expression for the velocity of the WT is provided through the use of fundamental dynamic relationship and is shown in Equation (4) as:

$$J\frac{d\omega_m}{dt} = T_g - T_{em} - f\omega_m \quad (4)$$

J = Moment of inertia (MOI) on the generator side (GS) of the WECS;
f = Coefficient of viscous friction (VFC) on the GS;
T_g = Torque of the gearbox;
T_{em} = Electro-magnetic torque of the generator;
ω_m = Mechanical velocity of generator.

The output power of the WT is dependent on the rotor velocity and the wind velocity. Coefficients **λ and β**, which are functions of the rotor velocity, determine the performance coefficient C_p. Equation (2) provides the optimal speed, which is obtained by considering the power o/p as a function of the rotor velocity for various wind speeds (whenever the angle of pitch **β = 0°**). A performance factor diagram for different values of **β** is shown in Figure 7.

$$\omega_{opt} = \frac{\lambda_{opt} V}{R} \quad (5)$$

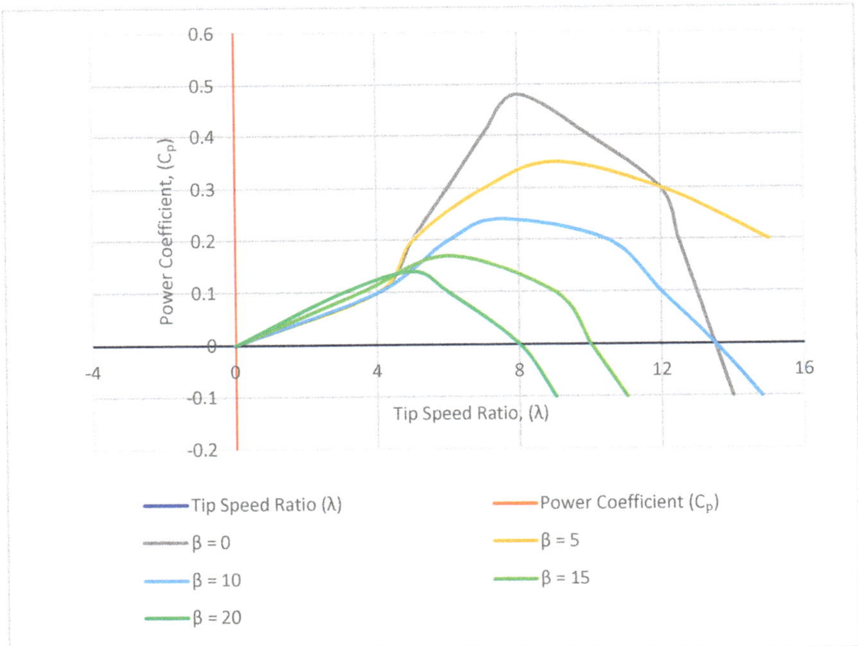

Figure 7. Performance factor diagram for different values of β.

2.2. DFIG Modeling and Control

The dynamic voltages of the rotor **(V_{dr},V_{qr})** and stator **(V_{ds}, V_{qs})** in the dq frame of reference are determined using Equations (6)–(9) as follows:

$$V_{dr} = R_r I_{dr} + P\Phi_{dr} - n_r \Phi_{qr} \quad (6)$$

$$V_{qr} = R_r I_{qr} + P\Phi_{qr} + n_r \Phi_{qr} \tag{7}$$

$$V_{ds} = R_s I_{ds} + P\Phi_{ds} - n_s \Phi_{qs} \tag{8}$$

$$V_{qs} = R_s I_{qs} + P\Phi_{qs} + n_s \Phi_{ds} \tag{9}$$

The fluxes on the rotor side (Φ_{dr}, Φ_{qr}) and the fluxes on the stator side (Φ_{ds}, Φ_{qs}) are in proportion to their identical d and q currents of the rotor and stator, respectively, i.e., I_{dr}, I_{qr}, I_{ds} and I_{qs}, as shown in Equations (10)–(13) as:

$$\Phi_{dr} = L_r I_{dr} + M I_{ds} \tag{10}$$

$$\Phi_{qr} = L_r I_{qr} + M I_{qs} \tag{11}$$

$$\Phi_{ds} = L_s I_{ds} + M I_{dr} \tag{12}$$

$$\Phi_{qs} = L_s I_{qs} + M I_{qr} \tag{13}$$

The electro-magnetic torque T_{em} is provided in Equation (14) is shown as:

$$T_{em} = \frac{PM}{L_s}(\Phi_{qs} I_{dr} - \Phi_{ds} I_{qr}) \tag{14}$$

where:

R_r = Resistance of each phase in the rotor;
R_s = Resistance of each phase in the stator;
L_r = Inductance of each phase in the rotor;
L_s = Inductance of each phase in the stator;
M = Magnetizing inductance;
n_r = Speed of rotor;
n_s = Speed at the stator terminals.

The voltage equations of the generator in the dq rotating reference frame are shown in Equations (15) and (16), as in [20]:

$$V_{gq} = -R_r i_q = PL_q i_{q_{\omega_e L_d i_d}} + \omega_e \lambda_m \tag{15}$$

$$V_{gd} = -R_r i_d - PL_d i_d - \omega_e L_q i_q \tag{16}$$

The generator's voltages in the dq axis (V_{gd} and V_{gq}) and the generator's currents in the dq axis (i_d and i_q) are also characterized by the generator's inductances in the d-q axis $(L_d$ and $L_q)$, stator resistance (R_g), electrical rotating speed (ω_e), and magnetic flux (λ_m). The electro-magnetic torque is defined by Equations (17) and (18) as:

$$T_e = \frac{3}{2}P\left[\lambda_m i_q - (L_d - I_q) i_d i_q\right] \tag{17}$$

$$T_e = \frac{3}{2}P i_q \lambda_m \tag{18}$$

Here, P shows the no. of pole pairs. The direct axis of frame is in line with the magnetic flux through the application of field-oriented control (FOC), and as a result of considering $i_d = 0$, the EM torque can be explained by the equation shown above. The graph for the maximum power output of a wind turbine is shown in Figure 8.

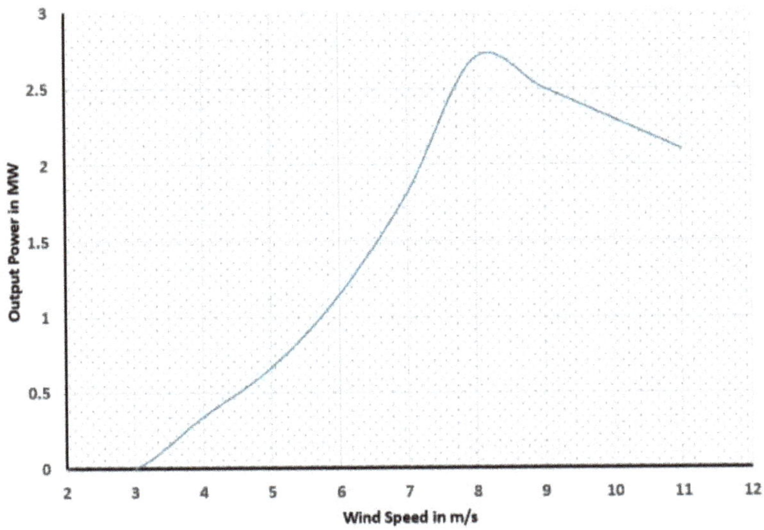

Figure 8. Graph for maximum power output from a wind turbine.

The q-axis current can be utilized for generator speed regulation through the implementation of MPPT, while the d-axis current is kept at zero.

2.3. MPPT Control

The MPPT approach seeks to maximize the amount of electricity that can be obtained from the WT by tracking power curves, as shown in Figure 9. This can be achieved via the optimal power and can be defined by Equation (18).

$$P_{opt} = 0.5\rho AC_{p-Max}(\lambda_{opt})V^3 \tag{19}$$

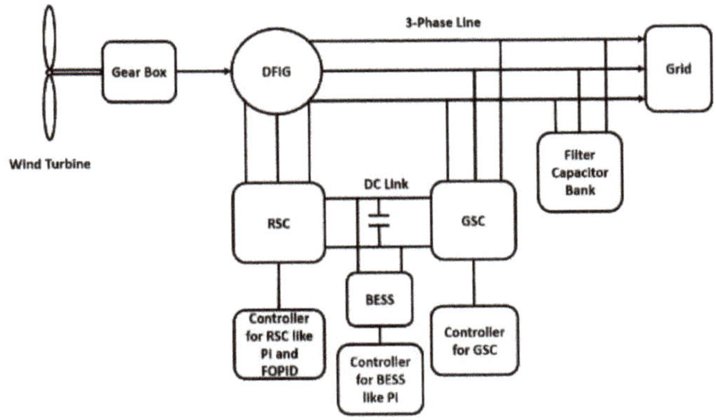

Figure 9. Schematic of DFIG WT for PQ improvement using a BESS and controllers.

The speed controller adjusts the rotor speed to the optimal value for each wind speed, thus controlling the electrical power generated by the generator. By adjusting the rotor speed ω_r,λ can be maintained at its optimum value, λ_{opt}, thereby maximizing the power output that can be obtained from the WT. The defined Equations (20) and (21) show the correlation between the maximum power and the corresponding torque at the optimum speed of the generator.

$$P_{MPPT} = K\omega_{opt}^3 \tag{20}$$

$$T_{MPPT} = K\omega_{opt}^2 \tag{21}$$

3. Design of BESS (Battery Energy Storage System)

Various energy storage devices can also be used in DFIG-based WECS, depending upon the specific requirements of the system and the available technologies. The major advantage of this is that the BESS has a high power density and fast response time, making it well-suited for use in applications that require a high power output and rapid load response, such as wind energy systems. It has a long life cycle and can operate in harsh environments, making it a reliable and durable option for energy storage in wind energy systems. For this reason, many authors use a BESS rather than another energy storage device. The characteristics of different ESS devices are provided by [21].

Such a model analyses the overall power transfers across a battery bank and updates the device's SOC across the specified time period while considering its steady-state performance [22–24]. This approach does not take into account factors such as lifetime deterioration, temperature changes, capacity fluctuations that are dependent upon operational voltage, and particular electrical circuits [25–27]. The battery model obtains the intended 1PSET (specified in units of nominal capacity) from its regulation model and updates the SOC of the battery. The BESS model and parameters are shown in Figure 10. According to generator conventions, the calculation of the actual power, 1PB, needed from or inserted into the battery is described in Equations (22) and (23) as:

$$\Delta P_B = \eta_{CHARGE} \Delta P_{SET}, \Delta P_{SET} < 0 \tag{22}$$

$$P_B = \frac{\Delta P_{SET}}{\eta_{DISCHARGE}}, \Delta P_{SET} \geq 0 \tag{23}$$

where $\eta_{DISCHARGE}$ and η_{CHARGE} are the discharge and charge efficiencies of the BESS, respectively. It should be noted that the outcome of the BESS is calculated in the form of Δ. This implies that a portion of the BESS's usable capacity may be put to use for other tasks, such as energy storage or buffering.

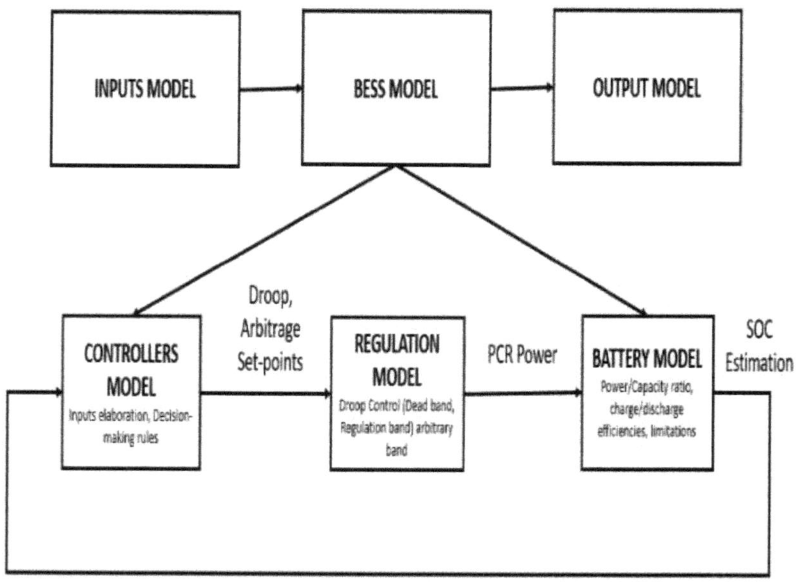

Figure 10. BESS Model and Parameters.

At each time step, the SOC variation can be computed by Equation (24) as follows:

$$\Delta SOC = \frac{\int_t^{t+1} \Delta P_B dt}{EPR} \tag{24}$$

where EPR stands for the energy–power ratio of the BESS.

4. Tuning of FOPID Controller

The main limitation of the PI controller is that it cannot provide satisfactory control performance for systems with complex dynamics and non-linearities. In DFIG-based WECSs, the system has complex dynamics due to the presence of wind turbine dynamics, power electronics converter dynamics, and grid dynamics. The non-linear behavior of the wind turbine and converter also affect the control performance.

On the other hand, an FOPI controller is a variant of the traditional PI controller that uses fractional calculus. FOPI controllers have been shown to achieve better performance than traditional PI controllers in some applications, and they do not have the additional derivative term that is present in an FOPID. The derivative term in an FOPID controller can help provide a faster and more accurate response to changes in the system, especially in cases of non-linear systems. As the wind speed and load conditions can change rapidly in the case of a DFIG-based WECS, the system must respond quickly to maintain stability and optimal power generation.

The FOPID controller is a more advanced and flexible version of the traditional PID controller. It uses fractional calculus to improve the performance of traditional PID controllers. In the context of DFIG-based WECSs, the FOPID controller can be used in the RSC to improve the power quality of the system. It introduces three additional parameters (alpha, beta, and gamma) that allow for better tuning and control of the system. These additional parameters include the fractional order derivative and fractional order integral terms, which provide more flexibility in controlling the system dynamics. These regulate the rotor currents in a better way, which improves the power quality of the system. This helps in reducing the impact of disturbances, such as wind speed variations, on the system. In the RSC of a DFIG-based WECS, this controller can provide better control over the system's dynamic response, transient stability and output performance. Compared with the PID controller, it has additional degree of freedom that allows it to better handle nonlinearities and uncertainties in the system. This is important in DFIG-based WECSs in which the wind turbine's operating conditions can vary widely and quickly. It also provides better robustness against disturbances, noise, and other factors that can affect the system's performance. The results with the FOPID controller are provided in the Results section.

Although fractional calculus is an ancient branch of mathematics, it has only recently been applied in variety of scientific and technical fields. The differential equations for the $PI^{\lambda}D^{\mu}$ controller in the time and frequency domains are presented in Equations (25) and (26) as:

$$\psi(t) = K_p\delta(t) + K_i H_t^{-\lambda}\delta(t) + K_d H_t^{\mu}\delta(t) \tag{25}$$

$$\psi(s) = K_p + K_i s^{-\lambda} + K_d s^{\mu} \tag{26}$$

Here, $K_i, K_d,$ and K_q represent the gains of the integral, derivative, and proportional components, respectively; μ and λ are the real numbers, with $0 < \mu < 2$ and $0 < \lambda < 2$. The block diagram of an FOPID controller is shown in Figure 11. The settings are changed to maximize or lower the index of performance to achieve the best CS, which is, in this case, the integral of time absolute error (ITAE), a measure of error that includes time weighting. This is defined by Equation (27) as:

$$ITAE = \int_0^t t(|\Delta V| + |\Delta I| + |\Delta P|)dt \tag{27}$$

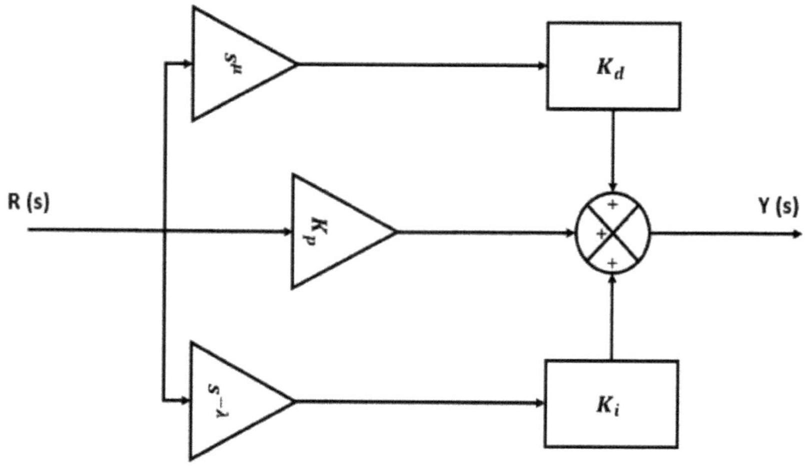

Figure 11. Block diagram of a FOPID controller.

The errors between the actual parameters (voltage, current, and power) and the reference parameters are represented by $\Delta V = V_i^* - V_t, \Delta I = I^* - I,$ and $\Delta P = P^* - P,$ respectively. The FOPID controller parameters were calculated by considering a 0.1 step change in the reference and minimizing the performance index through the use of algorithms [28–31]. The design of the fractional order PID (FOPID) controller was achieved through the minimization of the performance index, which has the property of weighing initial large errors lightly and penalizing late-occurring errors heavily. The classical tuning method was employed, which is based on trial-and-error or heuristic methods, and it was used to adjust the parameters of the controller until satisfactory performance was achieved. For an FOPID controller in the RSC of a DFIG-based WECS, the classical tuning method can be carried out using the following steps: setting the integral and derivative gains to zero, setting the proportional gain and low value, increasing the proportional gain until the system oscillates continuously, tuning the integral gain to reduce the steady-state error, tuning the derivative gain to reduce the overshoot and settling time, and finally, tuning the fractional order of the controller to improve the overall system response. The fopid_optim tool was utilized to carry out the design of the FOPID controller in which the LTI system contains the workspace name of the plant and could either be in state space or transfer function form.

The classical tuning method is used in the FOPID and PI controllers to find the values of $K_p, K_i,$ and K_d gains. This involves determining the values of these gains using mathematical formulas that are based on the system parameters. The perturb and observe (P and O) algorithm is used for MPPT in DFIG-based WECS.

5. Analysis of the Proposed and Established Control Technique in MATLAB

The PI controller for the basic MATLAB model is shown below (Figure 12). The classical tuning approach was used for the PI controller. The various waveforms for the same controller are shown in Figures 15–25.

The proposed FOPID controller for the DFIG-based WECS is shown in Figure 13.

The main purpose of the FOPID controller in the RSC of DFIG-based WECS is to regulate the rotor current and voltage, which, in turn, control the speed of the generator. It improves the response of the system. By using a FOPID in the RSC, the system can achieve better control of the rotor speed and ultimately increase the energy conversion efficiency of the system. It also improves the stability of the system by regulating the reactive power of the generator, ensuring that the generator operates at a stable voltage and maintaining the stability of the grid.

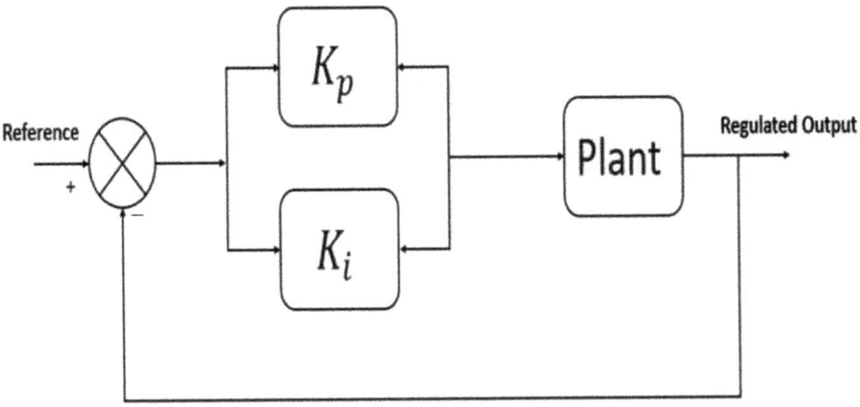

Figure 12. PI controller.

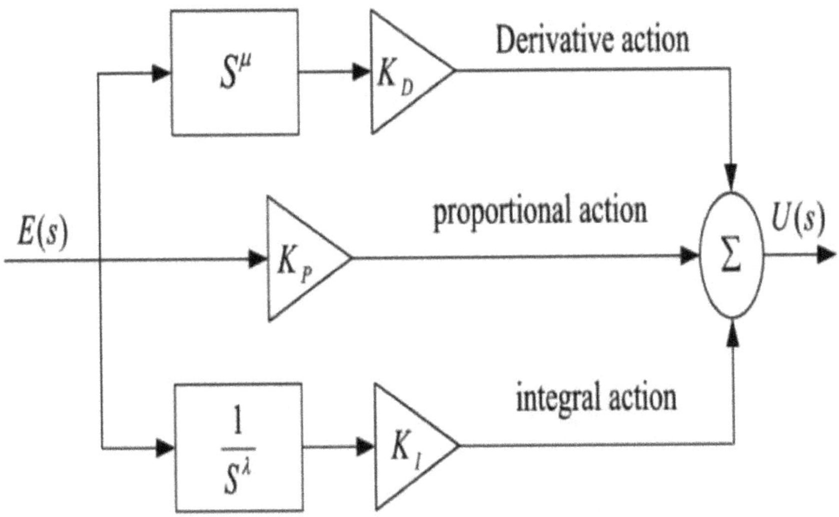

Figure 13. Proposed FOPID controller for RSC.

The proposed BESS for the DFIG-based WECS with a PI controller is as shown in Figure 14.

The parameters of the BESS in a DFIG-based WECS may vary depending on the system's specific applications and requirements. Some common parameters that can be considered in a DFIG-based WECS are battery capacity, battery voltage, battery chemistry, battery management system, and charging and discharging strategies.

Here, the function of the BESS is to mitigate the effects of the intermittency and variability of wind power by storing excess energy during periods of high wind speed and supplying energy to the grid during periods of low wind speed. The controller in the BESS can regulate the charging and discharging of the battery, ensuring that the battery operates within safe and optimal operating conditions. The function of the controller in a BESS is to provide or improve the power quality of the system by reducing the total harmonic distortion (THD) of the output voltage. By using a controller in the BESS, the output voltage waveform can be made more sinusoidal, reducing THD and improving the quality of power. It also provide power quality support by regulating the voltage and frequency of the output power. The controller can respond quicky to changes in voltage and frequency, injecting and absorbing power from the grid as needed to maintain the stability of the system.

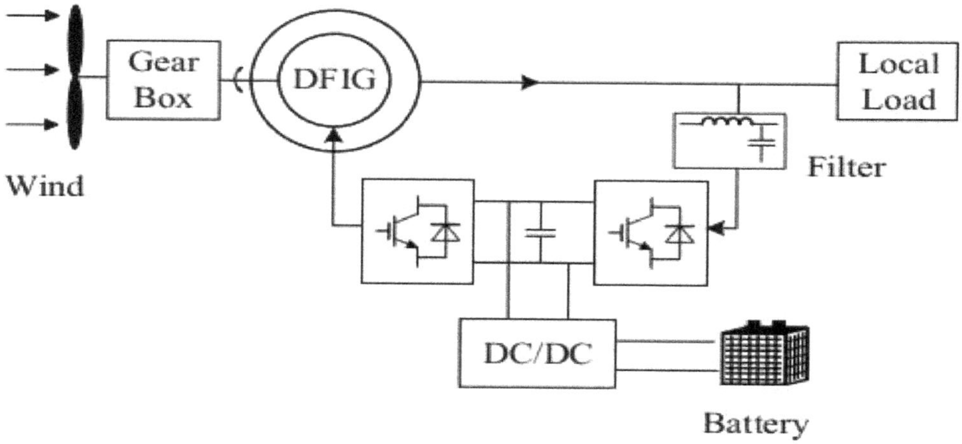

Figure 14. Proposed battery energy storage system (BESS) for DFIG-based WECS.

6. Results and Discussions

The various waveforms for the coefficient of performance (C_p), rotor current, rotor speed, output power, i.e., stator and rotor power, bus voltage (V_{bus}), and torque are analyzed at different wind velocities in the MATLAB/Simulink and are provided below. We have taken three wind velocities of 3 m/s (less than minimum), 11 m/s (within limits), and 15 m/s (greater than maximum).

The performance coefficient (C_p) in vertical axis with respect to time (t) in the horizontal axis for a wind turbine at different wind velocities with an FOPID and a PI controller (i) at 3 m/s (ii), at 11 m/s (iii), and at 15 m/s are presented in Figure 15a–f. The performance coefficient (C_p) is the ratio of the power captured by the rotor to the total power available in the wind (P) just before it interacts with the rotor of the wind turbine. As can be seen in Figure 15a,b, (C_p) fluctuates when the WECS is started and has a constant value of −0.43 after some time with a wind velocity of 3 m/s. When the wind speed is further increased to 11 m/s and 15 m/s, we can then see that in Figure 15c–f, it is around 0.44.

The waveforms for rotor currents in the vertical axis for three phases with respect to time in the horizontal axis for a wind turbine at different wind velocities with an FOPID and PI controller for (i) 3 m/s, (ii) 11 m/s, and (iii) 15 m/s are shown in Figure 16a–f. In Figure 16a,b, we can see that in the initial rotor current is not sinusoidal; however, after some time, it has a sinusoidal waveform. The magnitude of the current is significantly lower with a PI controller but increases slightly with the FOPID controller, whereas in Figure 16c–f, the rotor currents are, once again, initially not sinusoidal, but the magnitude of the rotor currents is high with the FOPID controller.

The waveforms for rotor speed in the vertical axis with respect to time in the horizontal axis for a wind turbine at different wind velocities with an FOPID and PI controller for (i) 3 m/s, (ii) 11 m/s, and (iii) 15 m/s are shown in Figure 17a–f. In Figure 17a,b, we can see that at a wind velocity of 3 m/s, the speed of the rotor slowly increases and reaches up to 950 rpm after some time; in Figure 17c,d, at a wind velocity of 11 m/s, the rotor speed slowly increases and reaches up to 1750 rpm after some time; in Figure 17e,f, at a wind velocity of 15 m/s, the rotor speed slowly increases and reaches up to 2250 rpm.

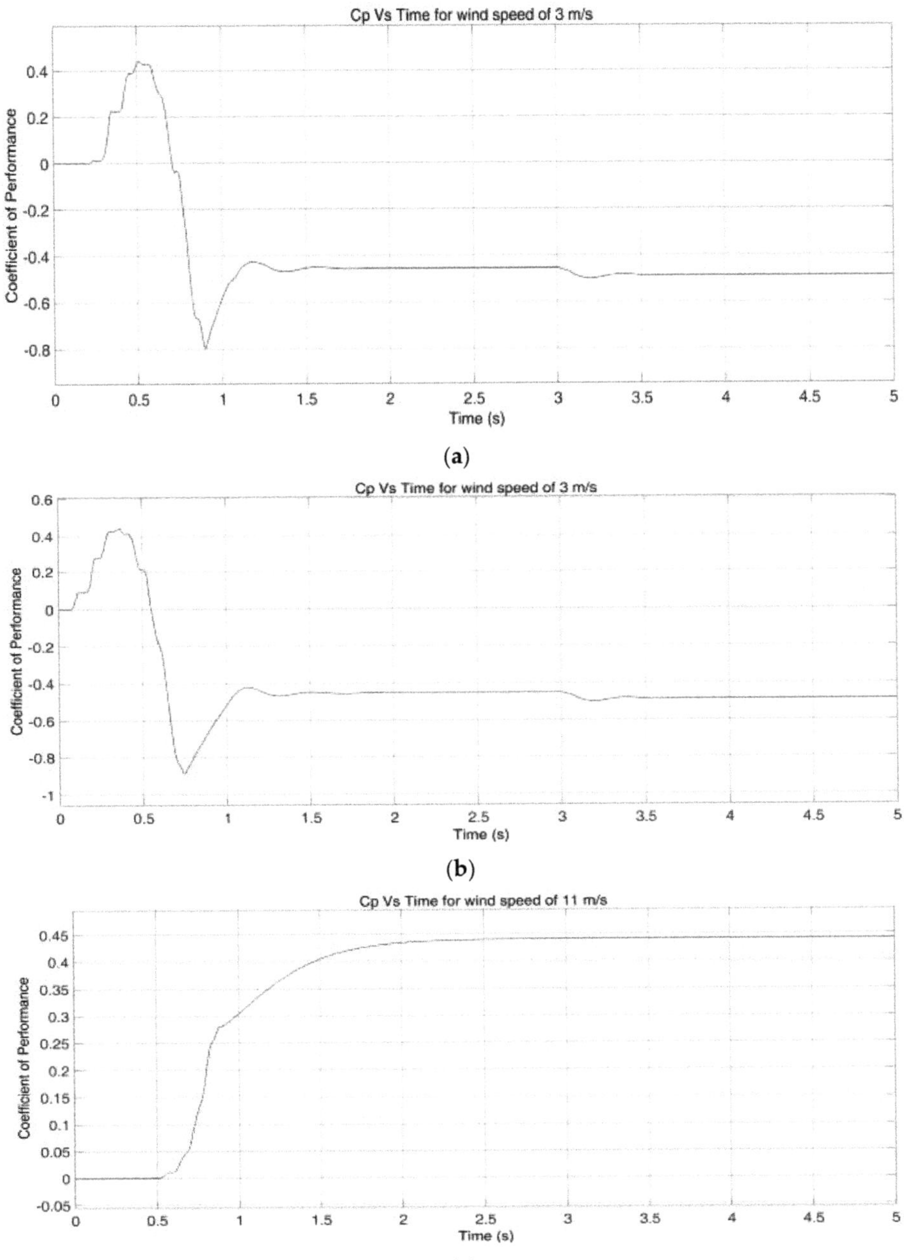

Figure 15. *Cont*.

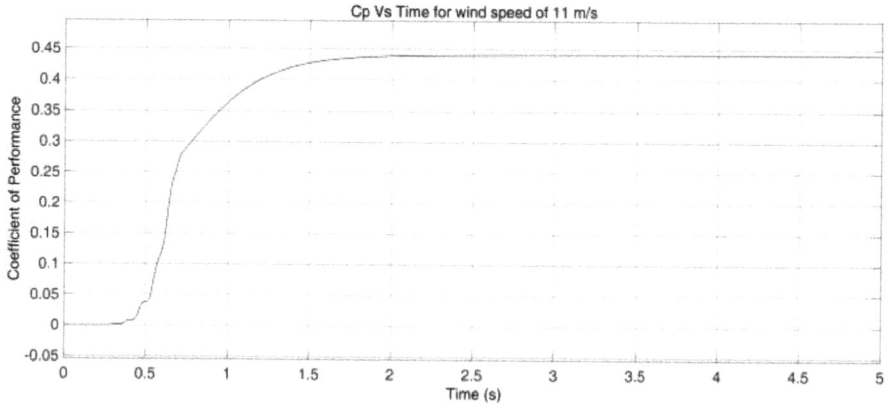

(**d**)

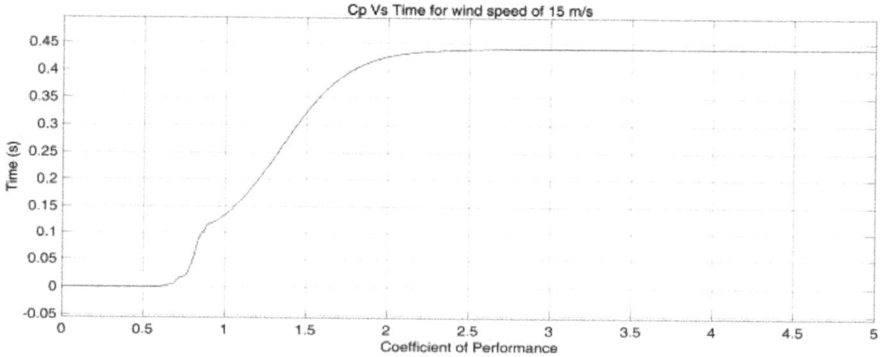

(**e**)

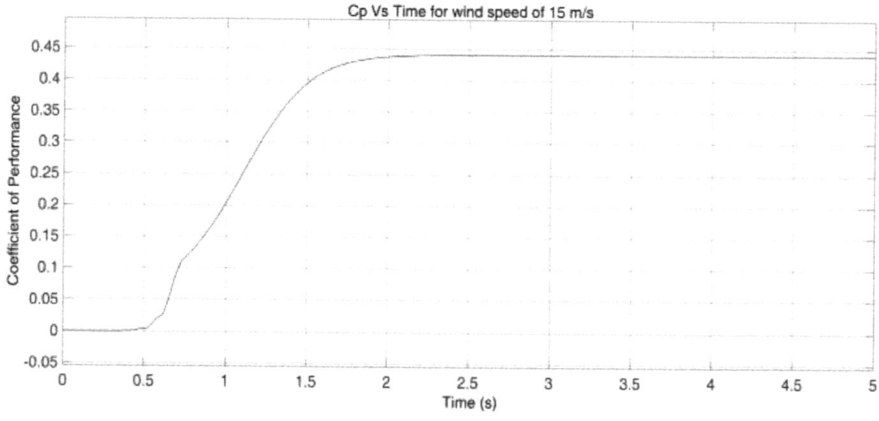

(**f**)

Figure 15. (**a**–**f**). Performance coefficient of a wind turbine at different velocities. (**a**) Performance coefficient with FOPID controller for wind speed of 3 m/s. (**b**) Performance coefficient with PI controller for wind speed of 3 m/s. (**c**) Performance coefficient with FOPID controller for wind speed of 11 m/s. (**d**) Performance coefficient with PI Controller for wind speed of 11 m/s. (**e**) Performance coefficient with FOPID controller for wind speed of 15 m/s. (**f**) Performance coefficient with PI controller for wind speed of 15 m/s.

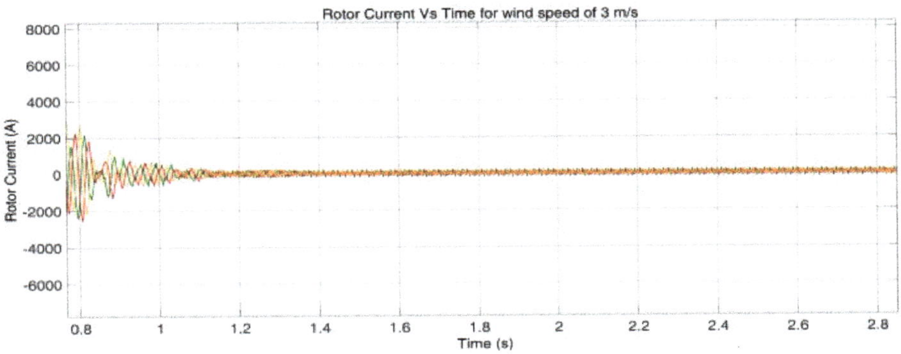

(a)

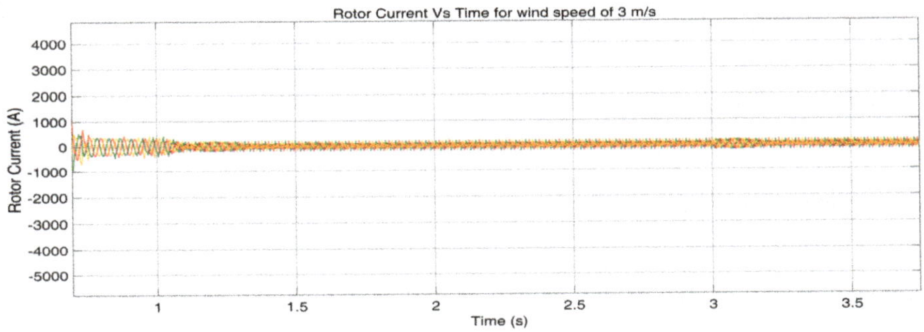

(b)

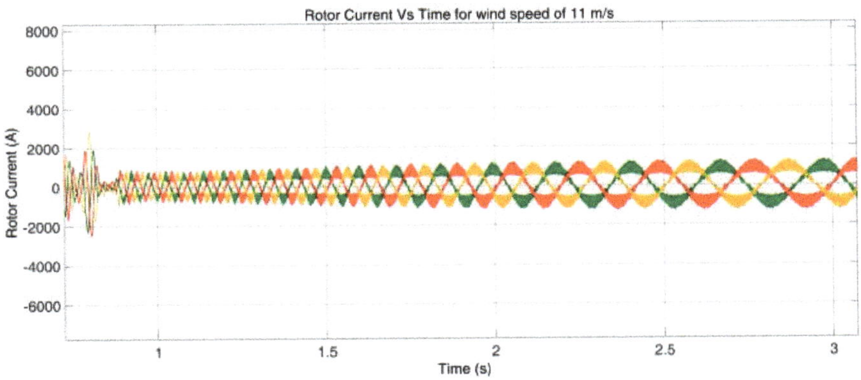

(c)

Figure 16. *Cont.*

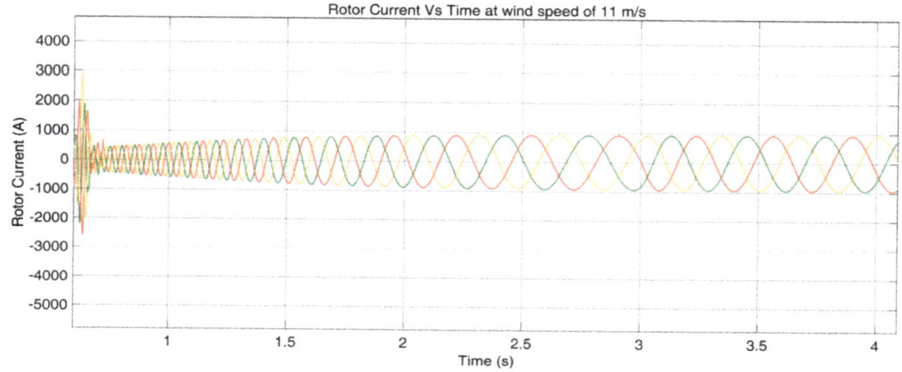

(**d**)

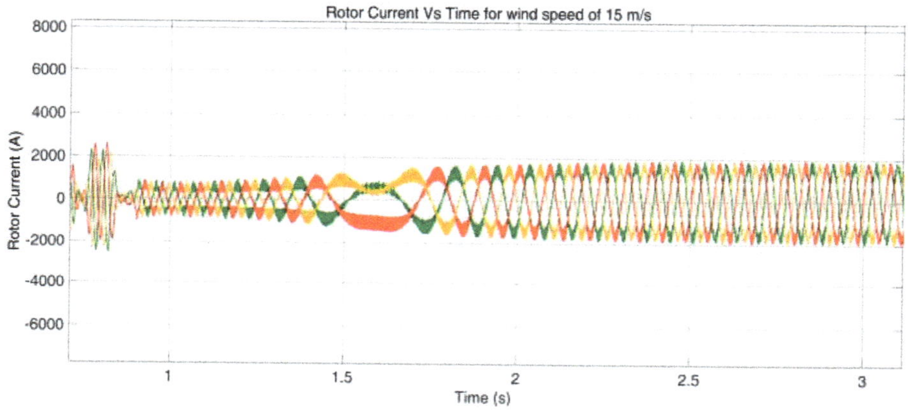

(**e**)

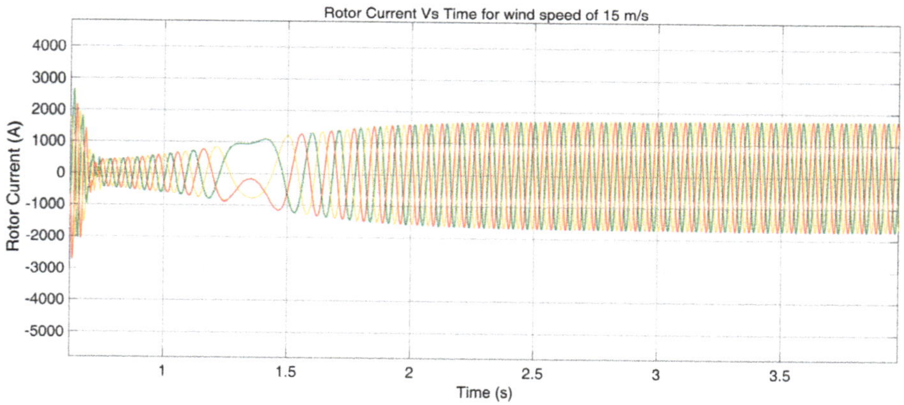

(**f**)

Figure 16. (**a**–**f**). Waveforms of rotor current of a wind turbine at different wind velocities. (**a**) Rotor current with FOPID controller for wind speed of 3 m/s. (**b**) Rotor current with PI controller for wind speed of 3 m/s. (**c**) Rotor current with FOPID controller for wind speed of 11 m/s. (**d**) Rotor current with FOPID controller for wind speed of 11 m/s. (**e**) Rotor current with FOPID controller for wind speed of 15 m/s. (**f**) Rotor current with PI controller for wind speed of 15 m/s.

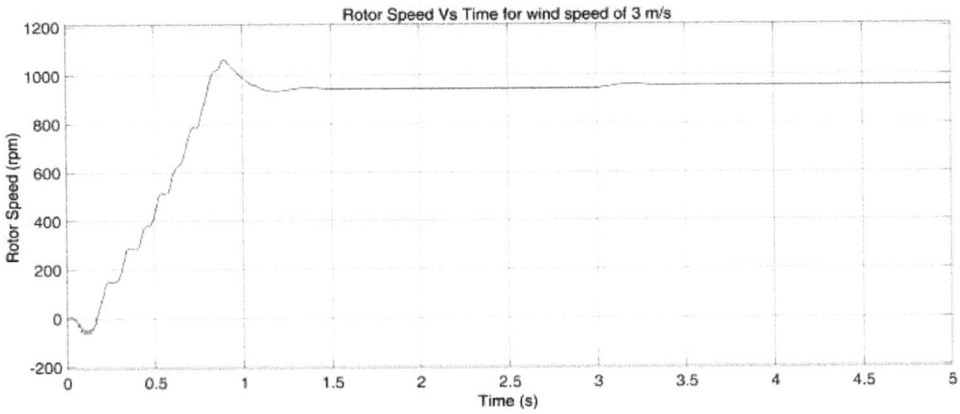

(a)

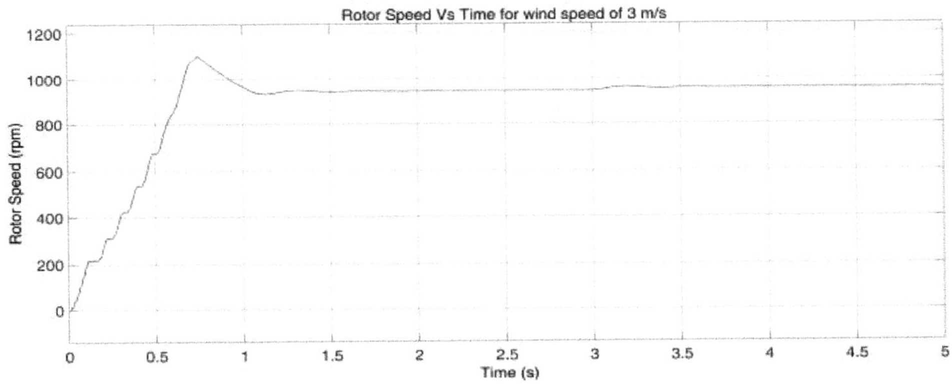

(b)

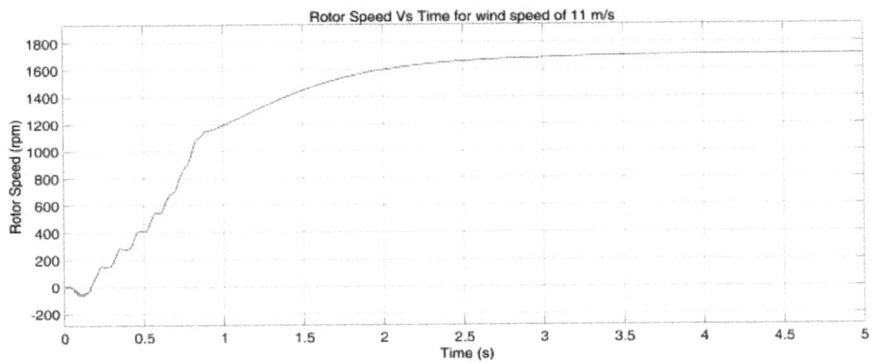

(c)

Figure 17. *Cont.*

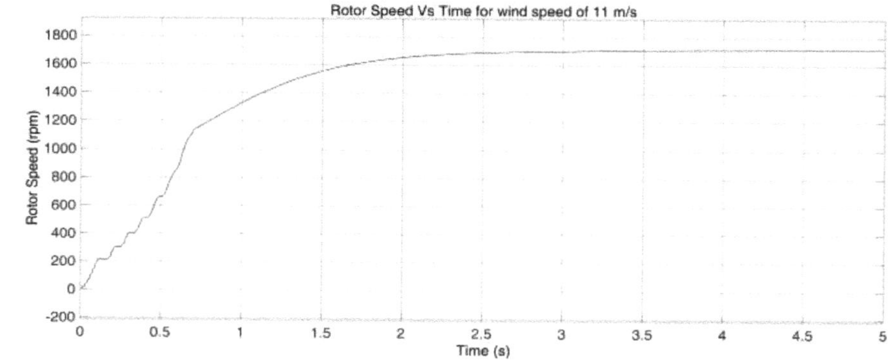

(**d**)

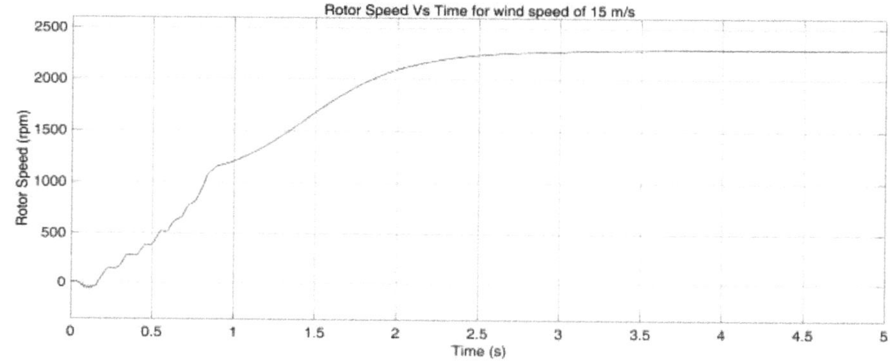

(**e**)

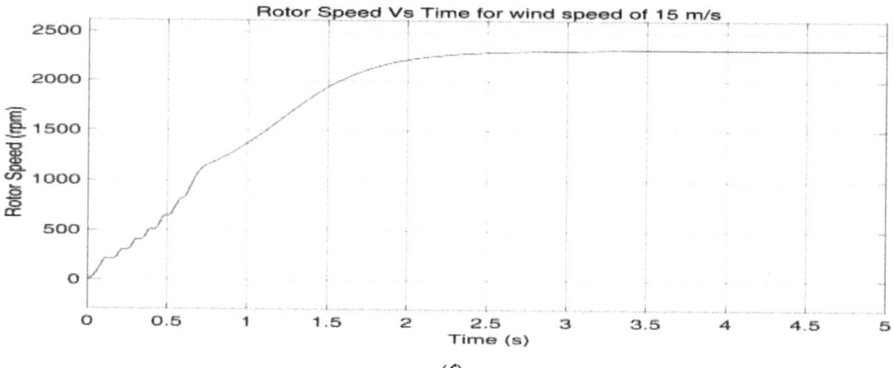

(**f**)

Figure 17. (**a**–**f**). Rotor speed of a wind turbine at different wind velocities. (**a**) Rotor speed with FOPID controller for wind speed of 3 m/s. (**b**) Rotor speed with PI controller for wind speed of 3 m/s. (**c**) Rotor speed with FOPID controller for wind speed of 11 m/s. (**d**) Rotor speed with PI controller for wind speed of 11 m/s. (**e**) Rotor speed with FOPID controller for wind speed of 15 m/s. (**f**) Rotor speed with PI controller for wind speed of 15 m/s.

The stator and rotor power in the vertical axis with respect to time in the horizontal axis for a wind turbine at different wind velocities with an FOPID and PI controllers for (i) 3 m/s, (ii) 11 m/s, and (iii) 15 m/s are shown in Figure 18a–f. The red color indicates the real stator power and pink color indicates the real rotor power. As can be seen in Figure 18a,b, at a wind velocity of 3 m/s, the stator power and rotor power delivered by the wind turbine are very low, of the orders of 0.1 MW and 0.2 MW, respectively, because the wind speed is low; in the initial phase, it fluctuates and begins to increase as it achieves its speed. The rotor power is high because the BESS supplies power to the grid during low-wind-speed periods. In Figure 18c,d, at a wind velocity of 11 m/s, we can see that the power developed by the stator and the power developed by the rotor are around 2 MW and zero, respectively, because at this point, the rotor is not supplying any power and all the real power is received across the stator. In Figure 18e,f, the power developed is around 2.5 MW; as the wind velocity is 15 m/s at this time, the excess power is stored in the BESS.

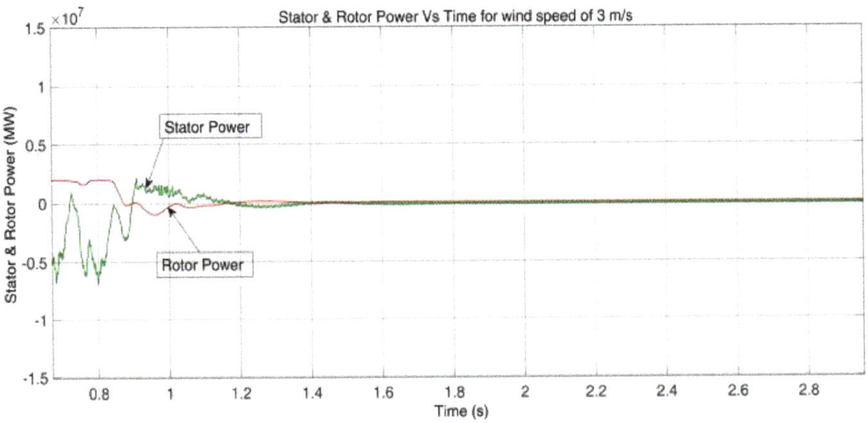

(a)

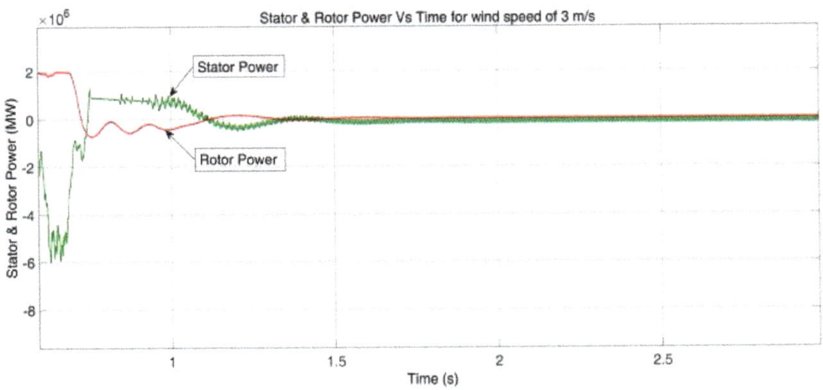

(b)

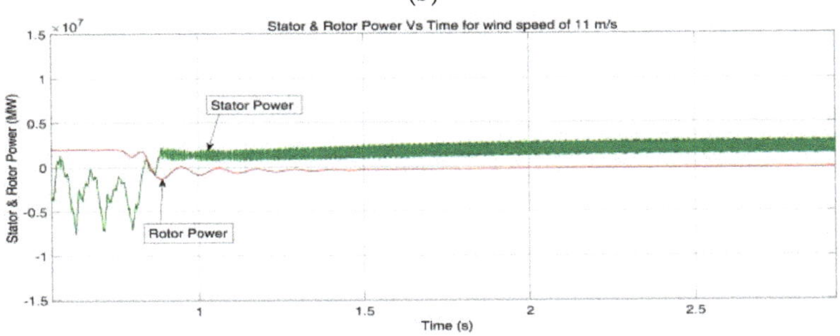

(c)

Figure 18. *Cont.*

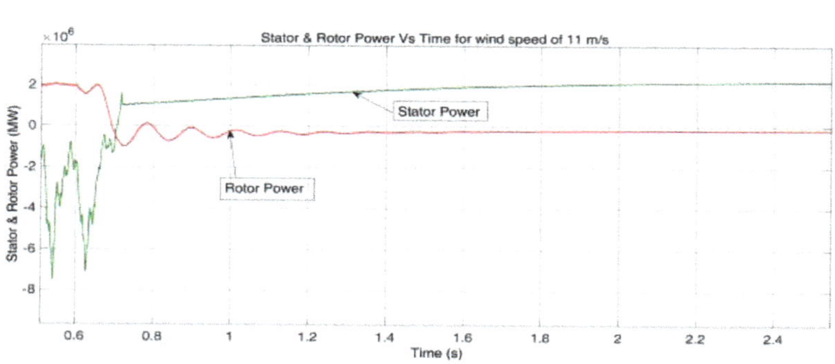

(d)

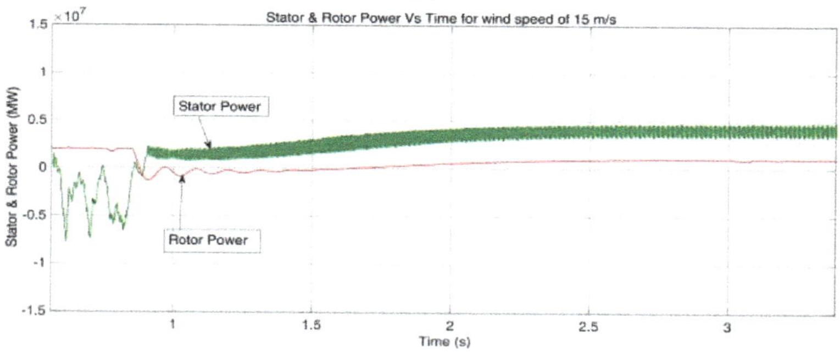

(e)

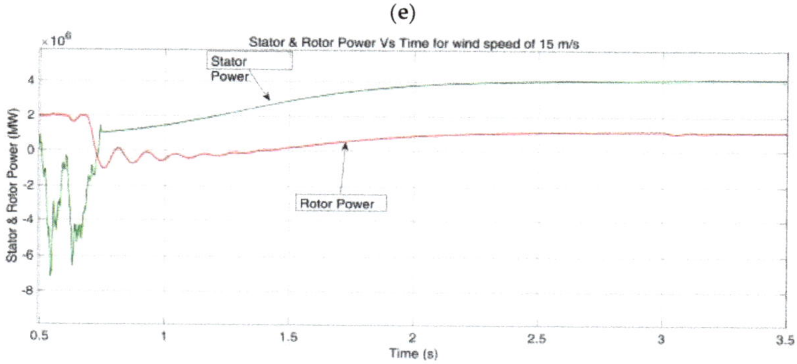

(f)

Figure 18. (**a**–**f**). Stator and rotor power of a wind turbine at different wind velocities. (**a**) Stator and rotor power with FOPID controller for wind speed of 3 m/s. (**b**) Stator and rotor power with PI controller for wind speed of 3 m/s. (**c**) Stator and rotor power with FOPID controller for wind speed of 11 m/s. (**d**) Stator and rotor power with PI controller for wind speed of 11 m/s. (**e**) Stator and rotor power with FOPID controller for wind speed of 15 m/s. (**f**) Stator and rotor power with PI controller for wind speed of 15 m/s.

The bus voltage (V_{bus}) in the vertical axis at the DC link with respect to time (t) in the horizontal axis for a wind turbine at different wind velocities with an FOPID and PI controllers for (i) 3 m/s, (ii) 11 m/s, and (iii) 15 m/s are shown in Figure 19a–f. As we can see in Figure 19a–f, the bus voltage is received at the DC link and is almost constant for all wind velocities. This is because the BESS is always connected in the system, and it supplies power to the grid during periods of low wind velocity. When there are wind gusts, it stores the excess power in the BESS.

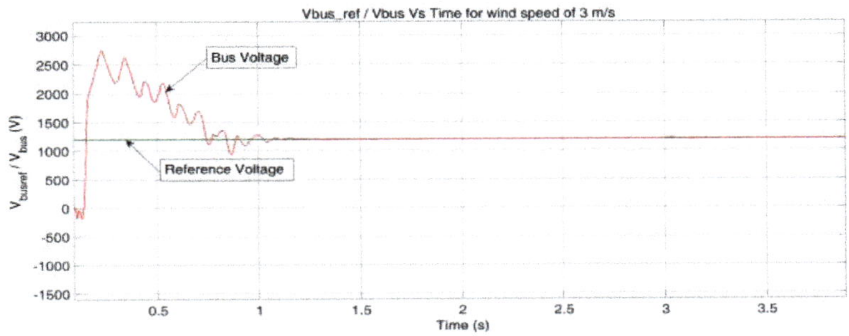

(a)

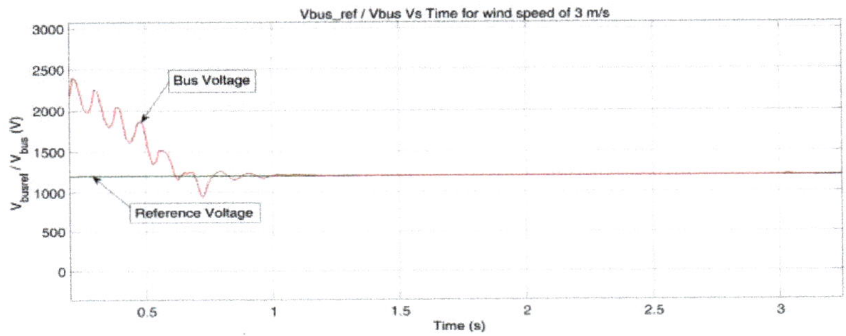

(b)

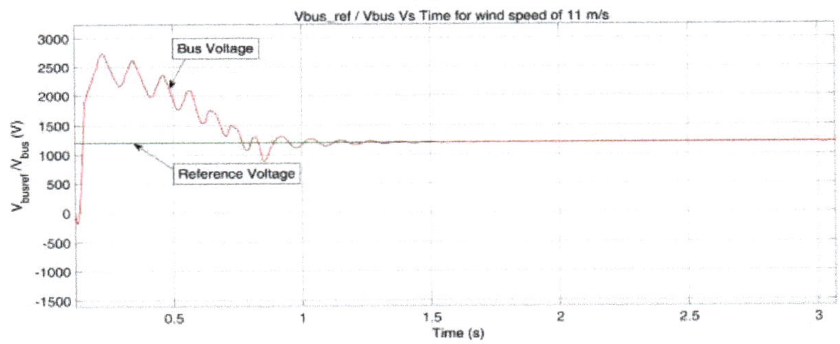

(c)

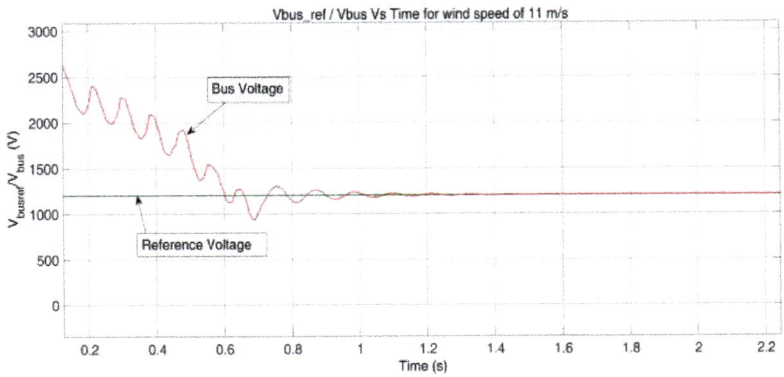

(d)

Figure 19. *Cont.*

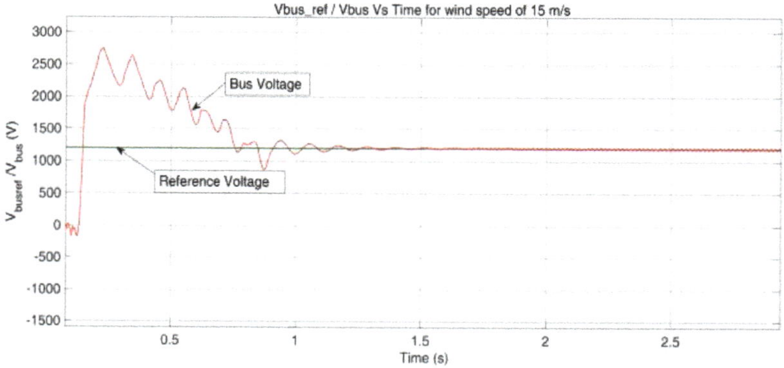

(e)

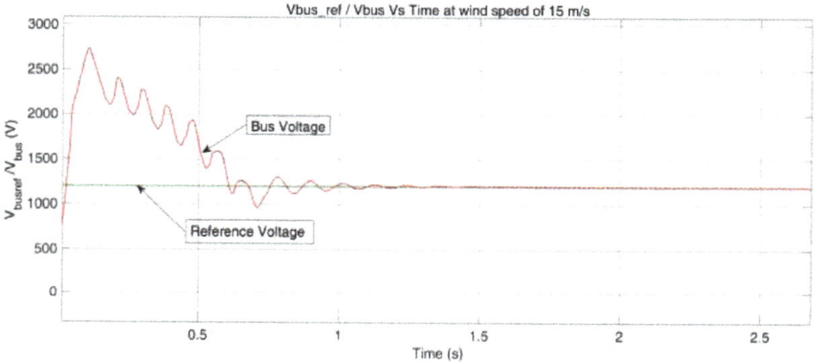

(f)

Figure 19. (**a–f**). Waveform for bus voltage (V_{bus}) of a wind turbine at different velocities. (**a**) Bus voltage with FOPID controller for wind speed of 3 m/s. (**b**) Bus voltage with PI controller for wind speed of 3 m/s. (**c**) Bus voltage with FOPID controller for wind speed of 11 m/s. (**d**) Bus voltage with PI controller for wind speed of 11 m/s. (**e**) Bus Voltage with FOPID controller for wind speed of 15 m/s. (**f**) Bus voltage with PI controller for wind speed of 15 m/s.

From Figures 20–25, it can be seen that the FFT analysis was performed for the rotor current with an FOPID controller and PI controller. From the FFT analysis, it can be observed that with the FOPID, the THD was improved when compared with the PI controller at same wind speeds. The FOPID provides a major contribution to the power quality in a DFIG-based WECS. We can see that the with the FOPID controller, performance was improved at all wind speeds, and it can further be improved by providing the BESS with some new control algorithms.

Implementing the model in a practical way involves designing the control algorithms (PI, FOPID, FL, and ANFIS) and tuning the controller parameters to achieve the desired performance, hardware selection, and simulation; this involves obtaining appropriate hardware components for the BESS and RSC, i.e., selecting the battery type and capacity for the BESS and selecting the power electronics components (IGBTs, capacitors, and inductors) for the RSC. The next step is to design the circuitry for the BESS and RSC, which involves designing a battery management system (BMS) for the BESS and designing the power electronics circuitry for the RSC. The next step is the development of a prototype, which involves assembling hardware components and circuitry according to the design. The last step is to test and validate the prototype, which involves conducting experiments and taking measurements to verify that the BESS and RSC are functioning as expected and that the prototype meets the desired performance criteria. For the low-power scale, it has a lower efficiency, which results in increased losses and decreased power generation. It also requires larger generators and power electronics converters, and it has a high cost and complex grid integration, which makes it less suitable for low-power applications.

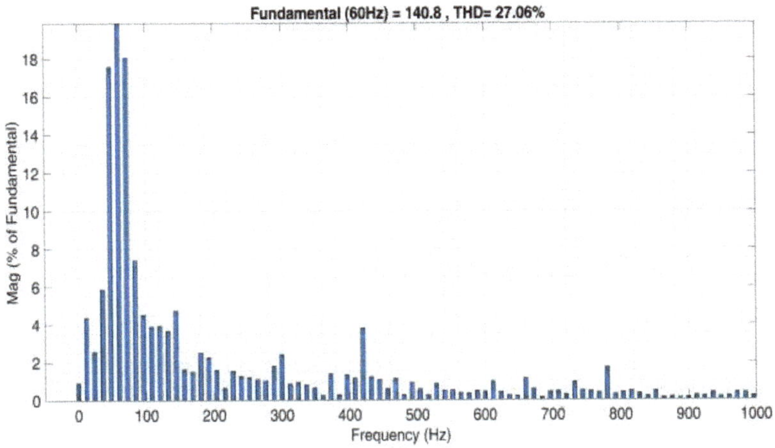

Figure 20. FFTT analysis of rotor current for wind speed of 3 m/s with FOPID controller.

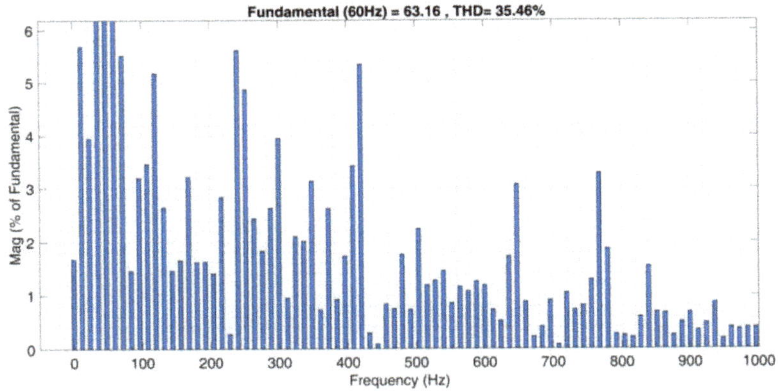

Figure 21. FFT analysis of rotor current at wind speed of 3 m/s with PI controller.

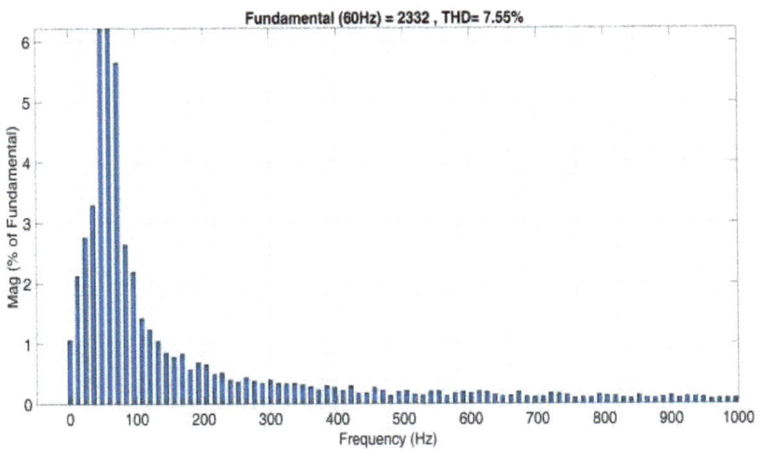

Figure 22. FFT analysis of rotor current for wind speed of 11 m/s with FOPID controller.

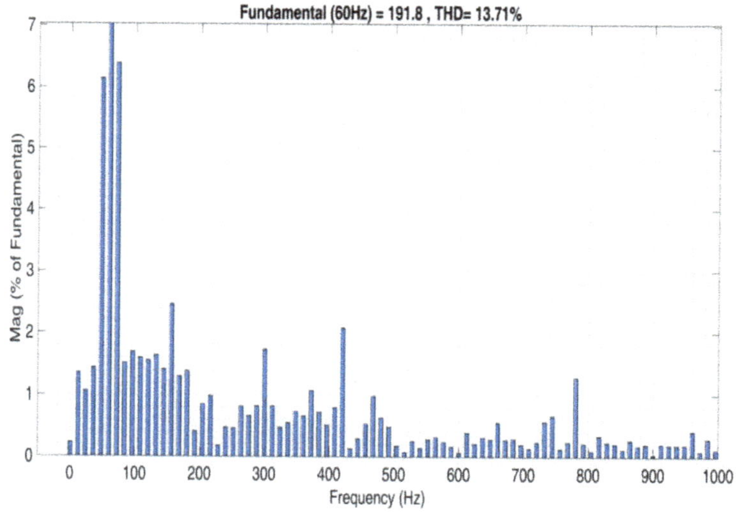

Figure 23. FFT analysis of rotor current for wind speed of 11 m/s with PI controller.

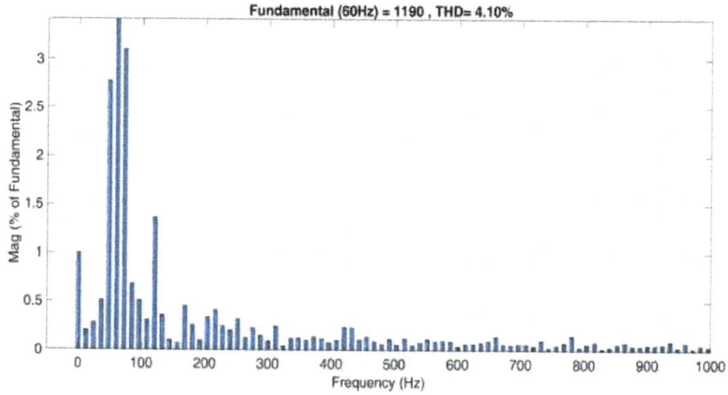

Figure 24. FFT analysis of rotor current for wind speed of 15 m/s with FOPID controller.

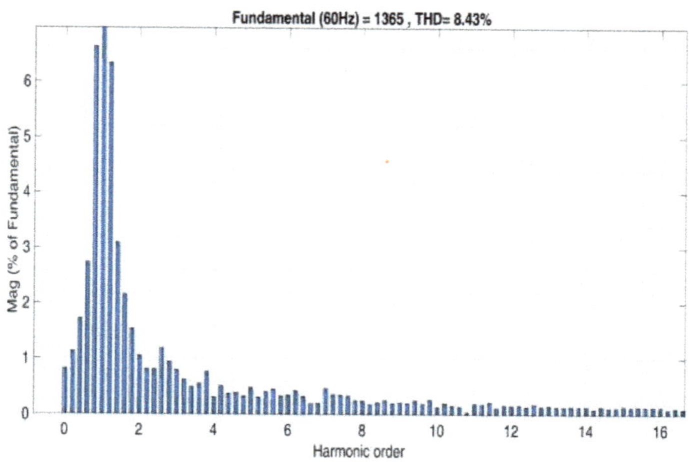

Figure 25. FFT analysis of rotor current for wind speed of 15 m/s with PI controller.

7. Conclusions

In the proposed work, we have introduced the utilization of an FOPID controller with five system variables which has been shown to be more efficient than conventional controllers in several areas. The FOPID which was used to execute the suggested controllers within the RSC, whichwas implemented using the FOPID and the fractional order PID controller block from the Fractional-Order Modeling and Control (FOMCON) toolbox, which was used for simulation purposes. fopid_optim is a Simulink toolbox that was used to optimize the FOPID controller. The purpose of the controller is to automatically regulate the rated power of a 2 MW wind turbine generation system in response to varying wind speeds. The WECS consisted of a WT and a DFIG with a BESS incorporated into the DC link with an initial SOC of 71.2%. The BESS stores energy when the wind speed exceeds the MPPT threshold and supplies voltage to the grid through the GSC when the wind speed falls below the MPPT. A simple PI controller was used instead of a conventional controller to increase the storing and releasing charge rate of the battery. The simulation findings were developed in a MATLAB/Simulink environment with validated parameters for the WT and generator. The various waveforms shown in Figures 14–18 are from MATLAB/Simulink software, such as the performance coefficient of the WT, stator power, rotor power, rotor current, DC bus voltage through the DC link, and the speed of the rotor at multiple wind velocities of 5 m/s, 11.2 m/s, and 15 m/s. The performance coefficient of the WT exhibits fluctuations during start-up and periods of low wind speeds, while the rotor speed and the output power at the rotor and stator increase. There is a slight increase in the voltage at the DC link terminal when the wind velocity is high. The study was conducted using the parameters shown in Table 1, with a fluctuating wind velocity applied to the WECS, and various parameters for the BESS are shown in Table 2.

Table 1. Parameters of Wind Turbine.

Total O/P Power at Generator Terminals	2 MW
Pole Pair	2
Frequency	50 Hz
Voltage	690 V (L–L)
Current	1760 A
Type of Machine	Wound Rotor IM
Stator Resistance, (R_s)	2.6 mΩ
Resistance of the Rotor, (R_r)	2.9 mΩ
Inductance of the Stator, (L_s)	2.587 mH
Stator Leakage Inductance and Rotor Leakage Inductance, $(L_{si}$ and $L_{sr})$	87 µH
Rotor Inductance, (L_r)	2.587 mH
Mutual Inductance, L_m	2.5 mH
Inertia of the M/C (J)	130 kg.m^2
Friction Factor for Damping	0.001 N.m.s.
Rated Torque, (T_{em})	12,732 N.m.
Reference Frame	Rotor
Stator/Rotor Turns Ratio	0.34
Stator Flux	1.8
Gearbox Ratio	100
Air Density	1.225 kg/m^3
Switching Frequency	4000 Hz

Table 2. Battery characteristics, wind profile, and control parameters.

Name of Parameter	Rating and Control Strategy Used
Battery Capacity	100 Ah
Battery Voltage	1200 V
Battery Chemistry	Li-Ion Battery
Battery Management System, such asSOC	71.4%
Charging and Discharging strategy	PI Controller with Buck-Boost Converter
Wind Profile	3 m/s, 11 m/s and 15 m/s

Every controller we used has its own advantages and limitations. The FO-PID controller has a complex control strategy, higher computational requirements, non-linear behavior, a high sensitivity to parameter variations, and difficulties in physical implementation. On the other hand, we used a PI controller in the BESS, which had certain limitations such as its limited control, poor dynamic control, sensitivity to load changes, fixed tuning parameters, sensitivity to model uncertainties, difficulty in tuning, and limited capability.

In the future, to address the limitations of FOPID controllers, we can improve the design and tuning processes of the FOPID controllers and reduce their computational requirements. Some new control techniques are being explored that may offer improved performance in specific situations. For BESS control, we can also use advanced control techniques that can provide better control over the BESS. On the other hand, the use of advanced sensors and communication systems can help improve the accuracy and responsiveness of the control system.

Author Contributions: Conceptualization, P.S.; data curation, P.S. and K.C.S.; formal analysis, P.S., U.C.R. and E.Y.; funding acquisition, G.P.J. and K.C.S.; investigation, U.C.R. and G.P.J.; methodology, P.S., K.A. and E.Y.; project administration, E.Y., G.P.J. and K.C.S.; resources, E.Y., G.P.J. and K.C.S.; software, P.S., K.A. and U.C.R.; supervision, G.P.J. and K.C.S.; validation, K.A., E.Y. and K.C.S.; visualization, U.C.R. ; writing—original draft, P.S.; writing—review and editing, G.P.J. All authors have read and agreed to the published version of the manuscript.

Funding: The present research was conducted with a Research Grant from Kwangwoon University in 2023.

Data Availability Statement: Data sharing is not applicable to this article as no datasets were generated during the current study.

Conflicts of Interest: The authors declare no conflict of interest.

References

1. Kumar, J.C.R.; Kumar, D.V.; Majid, M.A. Wind energy programme in India: Emerging energy alternatives for sustainable growth. *Energy Environ.* **2019**, *30*, 1135–1189.
2. Ganthia, P.B.; Barik, S.K. Steady-state and dynamic comparative analysis of pi and fuzzy logic controller in stator voltage oriented controlled DFIG fed wind energy conversion system. *J. Inst. Eng. India Ser. B* **2020**, *101*, 273–286. [CrossRef]
3. Kumar, J.C.R.; Majid, M.A. Renewable energy for sustainable development in India: Current status, future prospects, challenges, employment, and investment opportunities. *Energy Sustain. Soc.* **2020**, *10*, 2. [CrossRef]
4. Hamid, C.; Derouich, A.; Taoussi, M.; Zamzoum, O.; Hanafi, A. An improved performance variable speed wind turbine driving a doubly fed induction generator using sliding mode strategy. In Proceedings of the 2020 IEEE 2nd International Conference on Electronics, Control, Optimization and Computer Science (ICECOCS), Kenitra, Morocco, 2–3 December 2020; pp. 1–8.
5. Paulraj, C.R.K.J.; Bernard, M.A.; Raju, J.; Abdulmajid, M. Sustainable waste management through waste to energy technologies in India—Opportunities and environmental impacts. *Int. J. Renew. Energy Res.* **2019**, *9*, 309–342.
6. Sholapurkar, R.B.; Mahajan, Y.S. Review of wind energy development and policy in India. *Energy Technol. Policy* **2015**, *2*, 122–132. [CrossRef]
7. Antony, A.S.M.; Immanuel, G. A Novel Stage AC/DC PFC converter for renewable energy sources. *J. Green Eng.* **2020**, *10*, 483–490.
8. Xiao, S.; Yang, G.; Zhou, H.; Geng, H. Analysis of the control limit for rotor-side converter of doubly fed induction generator-based wind energy conversion system under various voltage dips. *IET Renew. Power Gener.* **2013**, *7*, 71–81. [CrossRef]

9. Kiani, M.M.; Wang, W.; Lee, W.-j. Elimination of System-Induced Torque Pulsations in Doubly-Fed Induction Generators via Field Reconstruction Method. *IEEE Trans. Energy Convers.* **2015**, *30*, 1228–1236. [CrossRef]
10. Singh, P.; Arora, K.; Rathore, U.C. Performance Evaluation of a Grid Connected Variable Speed 3-Φ Doubly Fed Induction Generator in Wind Energy Conversion System Using Battery Energy Storage System. In Proceedings of the 10th International Conference on Reliability, Infocom Technologies and Optimization (Trends and Future Directions) (ICRITO), Noida, India, 13–14 October 2022.
11. Singh, P.; Arora, K.; Rathore, U.C. Control Strategies for Improvement of Power Quality in Grid Connected Variable Speed WECS with DFIG—An Overview. In Proceedings of the 4th International Conference on Intelligent Circuits and Systems, Online, 7–9 April 2022; IOP Publishing: Bristol, UK, 2022; Volume 2327, p. 012008.
12. Kumar, N.; Singh, B.; Panigrahi, B.K.; Chakraborty, C.; Suryawanshi, H.M.; Verma, V. Integration of Solar PV with Low-Voltage Weak Grid System: Using Normalized Laplacian Kernel Adaptive Kalman Filter and Learning Based InC Algorithm. *IEEE Trans. Power Electron.* **2019**, *34*, 10746–10758. [CrossRef]
13. Kashkooli, M.A.; Madani, S.M.; Lipo, T.A. Improved direct torque control for a DFIG under symmetrical voltage dip with transient flux damping. *IEEE Trans. Ind. Electron.* **2019**, *67*, 28–37. [CrossRef]
14. Mohamed, M.A.; Diab, A.A.Z.; Rezk, H.; Jin, T. A novel adaptive model predictive controller for load frequency control of power systems integrated with DFIG wind turbines. *Neural Comput. Appl.* **2020**, *32*, 7171–7181. [CrossRef]
15. Sami, I.; Ullah, S.; Ali, Z.; Ullah, N.; Ro, J.-S. A super twisting fractional order terminal sliding mode control for DFIG-based wind energy conversion system. *Energies* **2020**, *13*, 2158. [CrossRef]
16. Chen, Y.-K.; Wu, Y.-C.; Song, C.-C.; Chen, Y.-S. Design and implementation of energy management system with fuzzy control for DC microgrid systems. *IEEE Trans. Power Electron.* **2013**, *28*, 1563–1570. [CrossRef]
17. Arcos-Aviles, D.; Pascual, J.; Marroyo, L.; Sanchis, P.; Guinjoan, F. Fuzzy logic-based energy management system design for residential grid-connected microgrids. *IEEE Trans. Smart Grid* **2018**, *9*, 530–543. [CrossRef]
18. Li, J.; Zhang, M.; Yang, Q.; Zhang, Z.; Yuan, W. SMES/battery hybrid energy storage system for electric buses. *IEEE Trans. Appl. Supercond.* **2016**, *26*, 5700305. [CrossRef]
19. Kusiak, A.; Zheng, H.; Song, Z. On-line monitoring of power curves. *Renew. Energy* **2009**, *34*, 1487–1493. [CrossRef]
20. Golnary, F.; Moradi, H. Dynamic modelling and design of various robust sliding mode controls for the wind turbine with estimation of wind speed. *Appl. Math. Model.* **2019**, *65*, 566–585. [CrossRef]
21. Panhwar, I.H.; Ahmed, K.; Seyedmahmoudian, M.; Stojcevski, A.; Horan, B.; Mekhilef, S.; Aslam, A.; Asghar, M. Mitigating Power Fluctuations for Energy Storage in Wind Energy Conversion System Using Supercapacitors, Special Section on Evolving Technologies in Energy Storage Systems for Energy Storage Applications. *IEEE Access* **2020**, *8*, 189747–189760. [CrossRef]
22. Rekioua, D.; Matagne, E. *Optimization of Photovoltaic Power Systems Modelization, Simulation and Control*; Bentham Science Publishers: Sharjah, United Arab Emirates, 2012.
23. Dufo-López, R.; Lujano-Rojas, J.M.; Bernal-Agustín, J.L. Comparison of different lead–acid battery lifetime prediction models for use in simulation of stand-alone photovoltaic systems. *Appl. Energy* **2014**, *115*, 242–253. [CrossRef]
24. Sauer, D.U.; Wenzl, H. Comparison of different approaches for lifetime prediction of electrochemical systems—Using lead–acid batteries as example. *J. Power Sources* **2008**, *176*, 534–546. [CrossRef]
25. Sun, H.Y.; Jou, H.L.; Wu, J.C. Aging estimation method for lead–acid battery. *IEEE Trans. Energy Convers.* **2011**, *26*, 264–271. [CrossRef]
26. Barré, A.; Deguilhem, B.; Grolleau, S.; Gérard, M.; Suard, F.; Riu, D. A review on lithium-ion battery ageing mechanisms and estimations for automotive applications. *J. Power Sources* **2013**, *241*, 680–689. [CrossRef]
27. Schuster, S.F.; Bach, T.; Fleder, E.; Müller, J.; Brand, M.; Sextl, G.; Jossen, A. Nonlinear ageing characteristics of lithium-ion cells under different operational conditions. *J. Energy Storage* **2015**, *1*, 44–53. [CrossRef]
28. Yamparala, S.; Lakshminarasimman, L.; Rao, G.S. Improvement of LVRT Capability for DFIG based WECS by Optimal Design of FoPID Controller using SLnO + GWO Algorithm. *Int. J. Intell. Eng. Syst.* **2023**, *16*, 202–213.
29. Karad, S.G.; Thakur, R. Enhanced control of doubly fed induction generator based wind turbine system using soft computing assisted fractional order controller. *Renew. Energy Focus* **2022**, *43*, 291–308. [CrossRef]
30. Kasbi, A.; Rahali, A. Fractional order PI controller for highly efficient power regulation of DFIG based wind power conversion system. *Int. J. Energy* **2020**, *14*, 66–82. [CrossRef]
31. Mahvash, H.; Taher, S.A.; Rahimi, M.; Shahidehpour, M. DFIG performance improvement in gridconnected mode by using fractional order [PI] controller. *Int. J. Electr. Power Energy Syst.* **2018**, *96*, 398–411. [CrossRef]

mathematics

Article

Optimizing Power Exchange Cost Considering Behavioral Intervention in Local Energy Community

Pratik Mochi [1], Kartik Pandya [2], Joao Soares [3,*] and Zita Vale [4]

[1] Department of Electrical Engineering, Chandubhai S Patel Institute of Technology (CSPIT), Charotar University of Science and Technology, CHARUSAT Campus, Changa 388421, India; pratikmochi.ee@charusat.ac.in
[2] Independent Researcher, Anand 388001, India; kartikpandya75@gmail.com
[3] GECAD—Research Group on Intelligent Engineering and Computing for Advanced Innovation and Development, LASI—Intelligent Systems Associate Laboratory, School of Engineering (ISEP)—Polytechnic of Porto, 4200-072 Porto, Portugal
[4] School of Engineering, Polytechnic of Porto, 4249-015 Porto, Portugal; zav@isep.ipp.pt
* Correspondence: jan@isep.ipp.pt

Abstract: In order to encourage energy saving and the adoption of renewable sources, this study provides a comprehensive experimental framework that integrates socioeconomic and behavioral objectives for the local energy community. The experiment aims to find out how successfully using behavioral interventions might encourage customers to save electrical energy and encourage them to adopt renewable energy, e.g., solar photovoltaic energy, in the present case. Using this method, we can calculate the causal impact of the intervention on consumer participation in the local electricity sector. The study uses consumer data on the import and export of electrical power from retailer electricity utilities at a predetermined power exchange price and a midmarket price for local energy community power transactions. The local energy community model simulates the consumption, storage, and export of 20 residential customers who, in different scenarios, are the test subjects of an empirical experiment and embrace electricity conservation and renewable energy. We address the optimization issue of calculating the power exchange cost and revenue in various scenarios and comparing them with the base case cost. The cases are built on the customers' behavioral interventions' empirical response. The findings demonstrate that the interaction of socioeconomic and behavioral objectives leads to impressive cost savings of up to 19.26% for energy utility customers. The policy implication is suggested for local energy utilities.

Keywords: behavioral economics; cost optimization; energy community; energy conservation; energy economics; energy policy; local electricity market; renewable energy; social nudge

MSC: 90C90

Citation: Mochi, P.; Pandya, K.; Soares, J.; Vale, Z. Optimizing Power Exchange Cost Considering Behavioral Intervention in Local Energy Community. *Mathematics* **2023**, *11*, 2367. https://doi.org/10.3390/math11102367

Academic Editor: Jinfeng Liu

Received: 10 April 2023
Revised: 6 May 2023
Accepted: 14 May 2023
Published: 19 May 2023

1. Introduction

Reducing the world's energy needs through efficient electricity consumption is now more crucial than ever. This situation calls for challenging the current efficiency policies' investment-focused concepts and the involvement of behavioral interventions [1] that encourage electricity conservation and utilize renewable energy. Our present culture requires energy conservation and appropriate energy resource management to address significant issues like global warming mitigation and heat regulation within a particular range [2]. The renewable energy revolution cannot be governed as a purely technical or commercial endeavor as it was in the past due to its extensive societal impacts [3]; it should be understood as developing a sophisticated sociotechnical system that necessitates novel types of cooperation. The ability to test innovative concepts such as energy auctions [4] and energy-saving behavior [5] should be provided in order to develop relevant contextual factors for business and technology models that aid in the achievement of the environment and energy policy priorities [6]. The research in [7] suggests three energy policies using solar PV and battery storage; the results showed that self-usage of generated power and energy strategy focused on capital expenditure or subsidies would give customers more financial benefits. The study in [8] suggests that a behavioral intervention method for energy conservation can increase the willingness of energy users to adopt and engage in lowering their energy consumption.

A case study is used in the research [9] to identify the best electrification options based on the unique characteristics and needs of an isolated energy community without electricity. Several specific system behavioral findings based on yearly, monthly, and hourly energy patterns with seasonal fluctuations were also displayed to demonstrate how well the ideal microgrid solution performs. In this energy community, it was possible to construct load consumption, power loss, voltage profile, and storage as optimization issues in order to reduce system costs. This was acheived by taking into account the technological uncertainties associated with the generation of renewable energy [10]. As solar photovoltaic power generation [11] is frequently uncertain, it may affect the planning and operational performance of the system [12], yet, research in [13] proved that by taking an optimization approach in uncertain power generation, a cost saving up to 15% could be achieved. In the proposed optimization model of our paper, any uncertainty in solar power generation would result in higher energy storage in the battery and so power export will be optimized accordingly. We introduced a power balance equation to minimize the effect of such uncertainty.

Scholars in various disciplines have begun to pay more attention to electricity conservation and have presented novel ideas, theories, and methods. However, only a few studies try to capture the various aspects of electricity conservation in a more comprehensive manner. The findings of the study [14] imply that by taking into account the role of cognitive mechanisms that underlie the implementation of electricity-saving activities, our comprehension of these behaviors may be enhanced. A field experiment [15] conducted on 237 individuals confirmed that 6% of electricity was achieved by sending behavioral interventions. The authors of [16] conducted a social comparison treatment for electricity saving on 525 households and found a 6.7% reduction in electricity use. A study on the same objective of electricity conservation in 2927 households showed electricity savings of 8.6% using the social nudge approach [17]. An energy-saving nudge approach reported a 10% reduction in electricity usage in the field experiment [18] conducted for 528 households. The possibility of reducing home energy usage with non-fiscal rewards that respond to consumer environmental values was looked at in [19]; the results showed a 5% monthly reduction in energy consumption. In research conducted on a government workplace, electricity savings of up to 14% resulted from social nudges that contained comparable energy consumption facts [20]. In the quarter year from the intervention, an 8.5% electricity saving was achieved, while social power participants sustained saving more electricity in an experiment conducted [21] in Switzerland. A meta-analysis [22] found that electricity consumption has been reduced by about 3.91% from the most recent experimental results of publications. The study suggests a few points for policy implications, e.g., the inclusion of control groups with and without incentives, collection of sociodemographic information, focus on individual incentives, and more experimental analysis. In order to help regulators, utilities, and politicians use energy efficiency as a resource, the study [23] offers insights into the economics of consumer-subsidized efficiency initiatives. Conclusive proof for energy saving was discovered by the authors, suggesting that regions with lower energy savings compared to retail sales can expand the scope of their energy efficiency plans without significantly increasing the electricity costs saved. The authors describe cases of energy efficiency predictions and prospective modeling and pinpoint technical advancements essential to utilities and energy providers.

Interventions in behavior that rely on social pressure might be potential tools for changing people's preferences to opt for renewables. In contrast to other actions, it is important to be more circumspect before implementing a norm-based intervention since different people may interpret the same information differently. Consequently, it is essential to identify the type of behavior that has to be altered. It is challenging to draw in customers when they have little knowledge or engagement. [24]. It is possible that people will not pick renewable energy since they are worried about the high purchase prices associated with it because it demands an initial investment. The fact that humans have a propensity to myopically overrate things that are close in time and underrate those that are far in the future is the root cause of this phenomenon [25]. Individuals may also choose not to utilize renewable energy since they overestimate the cost reductions in the future owing to a lack of knowledge needed to make optimal judgments, as is believed by rational decision-making. It may be useful to change attitudes toward renewables by letting them know that their peers have made the same decision and have already adopted solar electricity. Making the potential cost reductions from renewables more visible may prove to be a successful method for assisting the local electricity community in appreciating the benefits of renewables in the future. Additionally, heterogeneity in behavior may influence a variety of personal decisions. Considering the behaviors practiced by others in the relevant social environment is another important consideration when determining whether to embrace new technology [26]. A research study in [27] suggests that customers have the chance to choose low-emission electrical retailers inside the liberalized market, helping to reduce pollutants from power-producing facilities. The study [28] looks at how customer aspirations to adopt smart appliances are impacted by both interpersonal and technology-specific views. The results show that there are differences in the proportional weight of personal views and particular technologies across different smart energy offers.

Service quality and behavioral intention have a large and favorable impact on a person's decision to purchase smart appliances. In a laboratory experiment [29] involving 300 participants, researchers looked at how social norms and decision observability affect acceptability for renewable electricity, even at one's own financial cost. According to the findings, when requested to adhere to pro-environmental public standards, individuals contributed 35% more to a running renewable energy program compared to their control groups. This suggests that the government might enlighten people more about the renewable energy sector so they may make better purchase decisions. Information does not, however, guarantee a change in energy consumption behavior [30] or a movement in the preferences of the electricity community. Instead, when information is presented in a way that respects community members' poor information processing abilities, it may be more successful in altering behavior. The authors of the work [31] combined studies on societal support for energy policy with the idea of vocal partisanship for energy conservation. An optimization approach was proposed in [32] to model the local energy market by considering a mid-market rate tariff for prosumers. A demand response program [33] was suggested to motivate customers to participate in the energy transition.

By suggesting behavioral economics as a path from the root to determine how behavioral aspects can be used to understand energy costs and to supplement conventional interventions aimed at addressing them, this research involves an idea that emphasizes the significance of human initiative in affecting energy conservation and the adoption of renewable energy sources. The goal is to emphasize how behavioral economics might offer an enhancing competencies framework to analyze and handle the challenging problem of lowering the cost of energy rather than to give an exhaustive overview of the pertinent literature. We provide a local energy community model that illustrates and includes socioeconomic elements and investigates the effects of these factors on operational costs. Beyond its theoretical appeal, the study's justification is to investigate the notion that, in order to enhance the current policy, policymakers should take advantage of the many components and be conscious of the results of their interactions.

The structure of the paper is as follows: Section 2 describes the synergy of behavioral interventions and the energy community; Section 3 provides the details of the mathematical model, case study, and result details; We provide a general discussion and policy implication suggestions in Section 4; and the conclusion is discussed in Section 5.

2. Behavioral Intervention and Energy Community

2.1. Behavioral Economics and Intervention

Behavioral economics acknowledges the enormous effect that situation has on conduct. The specifics of the choice issue, in particular, impact the chance that decisions may display implicit dissonance. Individuals are also more likely to regard things as more important when they are closer in time or when they may provide greater benefits if delayed from now on [1]. This method focuses on the short-term costs and long-term benefits of energy measures such as energy conservation and renewable energy adoption. Behavioral intervention methods, such as energy conservation measures, may provide monetary benefits to residents in a nearby local energy region. Giving advice on the most effective approach to save energy, for example, might encourage improved energy consumption habits and, as a result, save money. To modify behavior that is recognized as a situational social practice, a more detailed investigation of behavior determinants beyond the focus on people is required. People also demonstrate persuasive departures from prudent choice suspicions in behavioral economics, in addition to demonstrating mental abnormalities. People differ not just in their preferences but also in their levels of personal responsibility and inspiration [34]. People are considered sane leaders with limited mental assets, such as bounded rational individuals, in the behavioral science perspective, and as a result, when making decisions under limited self-assuredness, they choose different methods.

The ethical consequences of legislation and technology interventions have recently received much attention throughout the world [35]. The hypothesis [36] anticipates that people's behavior is unaffected by their surroundings and results in decisions based on a scientific connection of the costs and benefits associated with alternative choices, which can be changed simply by adjusting financial motivators and providing additional data. Using commitment devices [37] is one technique for overcoming the challenge of energy-saving behavior. It is a system that pushes people to follow through on their goals by laying out negative consequences, such as financial or social repercussions, if they do not.

2.2. Energy Community

An energy community would be prepared to engage in renewable energy even in the presence of advantages or government subsidies that are greater than costs is explained by a variety of motives and levels of self-interest. In a consumer-centric energy community [38], significant advantages are provided by the electricity savings brought about by energy efficiency [39]. These include lowering regional emissions, enhancing business efficiency, lowering home energy costs, increasing productivity, enhancing resident wellbeing, and helping to lower energy scarcity. Energy conservation strategies may have various other benefits for the community in addition to lowering energy costs. Living in disadvantaged areas has a negative impact on a community's ability to build the social capital required to impose desirable behaviors via socialization.

From an economic viewpoint, the choice to invest in renewable power is generally portrayed as being driven by energy and cost reserves. This suggests that the energy community may decide to put resources into environmentally friendly energy, considering that this is monetarily ideal. Adopting renewable energy does not just increase the probability that the energy objectives are accomplished; it likewise improves individuals' self-achievement and mental self-portrait discernments by giving them the option to accomplish the presented objectives. Saving electricity or making investments in renewable energy is a choice to advance the common good and is thus a form of ethical behavior.

Given that the local energy community may make various assumptions from the sort of information presented, social interventions should be built with extra care. When people are simply informed about other people's electricity consumption habits, for instance, a boomerang effect could occur. This would have a positive impact on people who were previously using more electricity than recommended and a negative impact on people who were previously using less electricity than recommended. As a result, for behavioral interventions to be successful, it is crucial to accurately assess the behavior that needs to change. The social problem that results from encouraging electricity conservation in a community is caused by the conflict between each family's individual and communal goals, which is to consume energy as they see fit. It is important to understand the barriers to energy-efficient technology, which may lead to insight into locally based solutions for adaptability so that distributed electricity production from renewable sources may be incorporated [40]. Strategies for information dissemination account for the influence of energy community behavior. These actions significantly contribute to strengthening the energy community's knowledge of their rights and market rates and their comprehension of common electricity issues and energy conservation consciousness. By being encouraged to carry out beneficial activities, the energy community may improve their willingness to do so and, in turn, improve their level of self-efficacy, which is often lower due to the social isolation brought on by their status on the periphery.

3. Empirical Study and Analysis

3.1. Mathematical Model

The power cost minimization is considered an optimization problem and is solved by a mixed integer linear programming method. The objective of cost minimization is shown in Equation (1):

$$\text{Minimize}\left(\sum_{t=1}^{N_t}\sum_{x=1}^{N_x}\left(\mathrm{C}_{x,t}^{import} - \mathrm{R}_{x,t}^{export}\right) + \mathrm{FC}_{x,t}\right) \quad (1)$$

where $\mathrm{C}_{x,t}^{import}$ is the cost associated with the import of electrical power and $\mathrm{R}_{x,t}^{export}$ is the revenue made by customers while exporting power, and $\mathrm{FC}_{x,t}$ is the fixed charge paid by x customers for utilizing the energy community resources.

Equations (2) and (3) signify the cost and revenue for buying and selling electricity to the utility by all players in time t, subject to constraints (4)–(8):

$$\mathrm{C}_{x,t}^{import} = \left(l_{x,t}^{buy} \times \mathrm{E}_{x,t}^{buy} \times \mathrm{P}_{x,t}^{buy}\right) \times dt \quad \forall x \in \mathrm{N}_x,\ \forall t \in \mathrm{N}_t \quad (2)$$

$$\mathrm{R}_{x,t}^{export} = \left(l_{x,t}^{sell} \times \mathrm{E}_{x,t}^{sell} \times \mathrm{P}_{x,t}^{sell}\right) \times dt \quad \forall x \in \mathrm{N}_x,\ \forall t \in \mathrm{N}_t \quad (3)$$

where $\mathrm{E}_{x,t}^{buy}$ and $\mathrm{E}_{x,t}^{sell}$ denote the electric power purchase and sell to the community, respectively. $\mathrm{P}_{x,t}^{buy}$ and $\mathrm{P}_{x,t}^{sell}$ are the buy and sell price of electric power, and dt is the time period adjustment factor. The power loss multipliers for power purchase and sell are $l_{x,t}^{buy}$ and $l_{x,t}^{sell}$, correspondingly. The sum of the power loss multipliers is taken as 5% [41].

In this local energy community model, the customers are limited to buy or sell electric power within their upper limits of power import and export, as per Equations (4) and (5). They are also limited to buy or sell the power in the community at the same time, as per constraints Equations (6)–(8):

$$0 \le \mathrm{E}_{x,t}^{buy} \le \mathrm{E}_{x,t}^{buy\ max} \times \mathrm{X}_{x,t}^{\mathrm{EB}} \quad \forall x \in \mathrm{N}_x,\ \forall t \in \mathrm{N}_t \quad (4)$$

$$0 \le \mathrm{E}_{x,t}^{sell} \le \mathrm{E}_{x,t}^{sell\ max} \times \mathrm{X}_{x,t}^{\mathrm{ES}} \quad \forall x \in \mathrm{N}_x,\ \forall t \in \mathrm{N}_t \quad (5)$$

where the upper limits for power purchase and sell are designated as $\mathrm{E}_{x,t}^{buy\ max}$ and $\mathrm{E}_{x,t}^{sell\ max}$, respectively. The binary variables for power purchase are $\mathrm{X}_{x,t}^{\mathrm{EB}}$ and for power sale is $\mathrm{X}_{x,t}^{\mathrm{ES}}$. They are introduced to limit the power trading simultaneously:

$$0 \le \mathrm{X}_{x,t}^{\mathrm{EB}} \le 1 \quad \forall x \in \mathrm{N}_x,\ \forall t \in \mathrm{N}_t \quad (6)$$

$$0 \leq X_{x,t}^{ES} \leq 1 \quad \forall x \in N_x, \ \forall t \in N_t \tag{7}$$

$$X_{x,t}^{EB} + X_{x,t}^{ES} \leq 1 \quad \forall x \in N_x, \ \forall t \in N_t \tag{8}$$

In some cases, the customers are permitted to exchange power at a mid-market price. A mid-market price allows customers to buy power from the local community at a lower price and sell power at a higher price compared to utility grid prices. The condition for this power transaction in the local community, as per Equation (9), is the total electric power buy and total electric power sell should be the same. The mid-market price is calculated as Equation (10) [42]:

$$\sum_{x=1}^{N_x} E_{x,t}^{buy\ mmp} = \sum_{x=1}^{N_x} E_{x,t}^{sell\ mmp} \quad \forall x \in N_x, \ \forall t \in N_t \tag{9}$$

$$P_{x,t}^{mmp} = \frac{\min(P_{x,t}^{buy}) + P_{x,t}^{sell}}{2} \quad \forall x \in N_x, \ \forall t \in N_t \tag{10}$$

where $E_{x,t}^{buy\ mmp}$ and $E_{x,t}^{sell\ mmp}$ are electric power buy and sell by the customer at a mid-market price $P_{x,t}^{mmp}$. It should be noted that power trading at mid-market price will also follow a similar power exchange constraint as per Equations (11)–(15):

$$0 \leq E_{x,t}^{buy\ mmp} \leq E_{x,t}^{buy\ max} \times X_{x,t}^{EB\ mpp} \quad \forall x \in N_x, \ \forall t \in N_t \tag{11}$$

$$0 \leq E_{x,t}^{sell\ mpp} \leq E_{x,t}^{sell\ max} \times X_{x,t}^{ES\ mmp} \quad \forall x \in N_x, \ \forall t \in N_t \tag{12}$$

The binary variables, at a mid-market price, for power purchase is $X_{x,t}^{EB\ mmp}$ and the same for power sale is $X_{x,t}^{ES\ mmp}$. They are limiting factor power exchange simultaneously as mid-market price:

$$0 \leq X_{x,t}^{EB\ mmp} \leq 1 \quad \forall x \in N_x, \ \forall t \in N_t \tag{13}$$

$$0 \leq X_{x,t}^{ES\ mmp} \leq 1 \quad \forall x \in N_x, \ \forall t \in N_t \tag{14}$$

$$X_{x,t}^{EB\ mmp} + X_{x,t}^{ES\ mmp} \leq 1 \quad \forall x \in N_x, \ \forall t \in N_t \tag{15}$$

Further, power purchase from the utility grid and from the local energy community is limited as per Equation (16), and the same goes for power sell, as per Equation (17):

$$X_{x,t}^{EB} + X_{x,t}^{EB\ mmp} \leq 1 \quad \forall x \in N_x, \ \forall t \in N_t \tag{16}$$

$$X_{x,t}^{ES} + X_{x,t}^{ES\ mmp} \leq 1 \quad \forall x \in N_x, \ \forall t \in N_t \tag{17}$$

The customers can store the electrical power generated by their solar PV in a battery; the battery power is estimated by Equation (18):

$$E_{x,t}^{bat} = E_{x,ini}^{bat} + E_{x,1}^{bat\ ch} \times \eta_{x,ch} - \frac{E_{x,1}^{bat\ dch}}{\eta_{x,dch}} \quad \forall x \in N_x \tag{18}$$

where $E_{x,t}^{bat}$ is the electric energy content of the battery, $E_{x,ini}^{bat}$ is the initial electric energy of the battery, $\eta_{x,ch}$ and $\eta_{x,dch}$ are the charging and discharging efficiencies of the battery, respectively. The battery charging power $E_{x,1}^{bat\ ch}$ and discharging power $E_{x,1}^{bat\ dch}$ are limited to staying within the maximum limit of the battery, and customers can either charge or discharge the battery at the same time. Equation (19) represents power balance:

$$E_{x,t}^{buy} + E_{x,t}^{bat\ dch} + E_{x,t}^{gen} = E_{x,t}^{load} + E_{x,t}^{sell} + E_{x,t}^{bat\ ch} \quad \forall x \in N_x\,,\ \forall t \in N_t \tag{19}$$

where $E_{h,t}^{gen}$ is power generated by the customers and $E_{h,t}^{load}$ is the electric load.

3.2. Case Study and Result Analysis

The study is carried out for a total of five cases to determine the operating cost of a small local energy community consisting of 20 household customers. All these customers are assumed to be prosumers, i.e., they own the solar PV power generation and the battery storage according to their contracted power limits with the power grid. The first case is a base case where the customers buy and sell electric power without energy conservation and without adopting higher PV installation. In case 01, it is assumed that half of the customers respond positively to the energy conservation interventions and reduce their electric power consumption by 3–5%. However, this case does not include behavioral interventions for adopting a higher PV installation. Case 02, on the other hand, considers a 3–5% higher PV installation by half of the total customers collectively, resulting from responding positively to behavioral interventions. Case 03 is formulated as all customers being nudged by energy conservation and higher PV installation requests. In this case, it is assumed that half of the customers reduce their power consumption by 3–5% and the other half increase their PV installation capacity by 3–5% collectively. Case 04 is similar to case 03, but in this case, the customers are allowed to exchange their power at the mid-market price in their local community. All these assumptions are based on the studies and proof provided in the literature [14–20]. As this is an empirical study, the authors have also taken liberty with assumptions, yet at a level acceptable based on the literature. The various cases are depicted in Table 1.

Table 1. Cases under study.

Case	Energy Conservation	Higher PV Installation	Mid-Market Price
Base case	No	No	No
Case 01	Yes	No	No
Case 02	No	Yes	No
Case 03	Yes	Yes	No
Case 04	Yes	Yes	Yes

The system under study consists of a total of 20 customers, which is a part of the system used in research [42,43] and is available publicly to download and use [44]. All the community participants have a contract with a retailer about the power buy/sell limits, power exchange rate, fixed cost to use resources, and storage of power. The power consumption, generation, and storage are recorded by a smart meter, and aggregators or system operators may use it to forecast future values. In the proposed energy community model, forecasted data can also be used to estimate the system cost and revenue. The local energy community system specifications are listed in Table 2. The power capacity of the battery and solar PV installation for each customer is shown in Figure 1. The power trade prices are shown in Figure 2.

The simulation platform is MATLAB, and a mixed integer linear programming optimization problem is solved by using the TOMLAB toolbox. The results of system cost $C_{x,t}^{import}$ and customer revenue $R_{x,t}^{export}$ were obtained for five cases, as stated in Table 3.

Table 2. System specification.

Parameter	Symbol	Value		Unit
		Min	Max	
Number of customers	N_x	20		-
Customers' buy limit	-	4.6	10.35	kW
Customers' sell limit	-	2.3	5.175	kW
Fixed cost	$FC_{x,t}$	0.32	0.62	EUR/day
Power buy price	$p^{buy}_{x,t}$	0.0922	0.1836	EUR/kWh
Power sell price	$p^{sell}_{x,t}$	0.045		EUR/kWh
Mid-market price	$P^{mmp}_{x,t}$	0.0686	0.0937	EUR/kWh
Electric load of customer	$E^{load}_{x,t}$	0	7.07	kW
Generation of customer	$E^{gen}_{x,t}$	0	7.75	kW
Initial battery power	$E^{bat}_{x,ini}$	0		kW
Battery capacity of customers	$E^{bat}_{x,t}$	13.5	15	kWh
Charging efficiency of battery	$\eta_{x,ch}$	90%		-
Discharging efficiency of battery	$\eta_{x,dch}$	90%		-

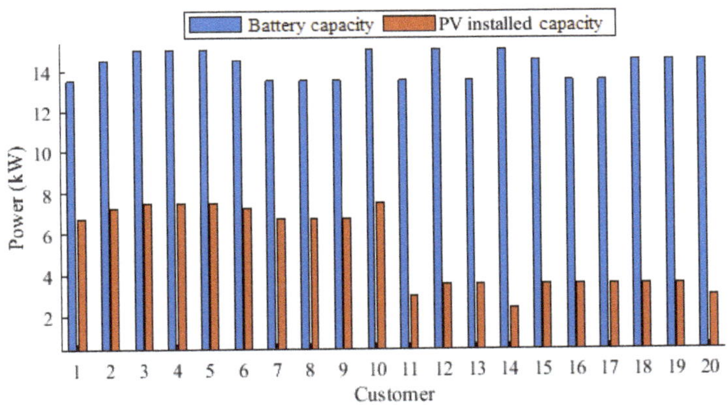

Figure 1. Battery capacity and PV installed capacity.

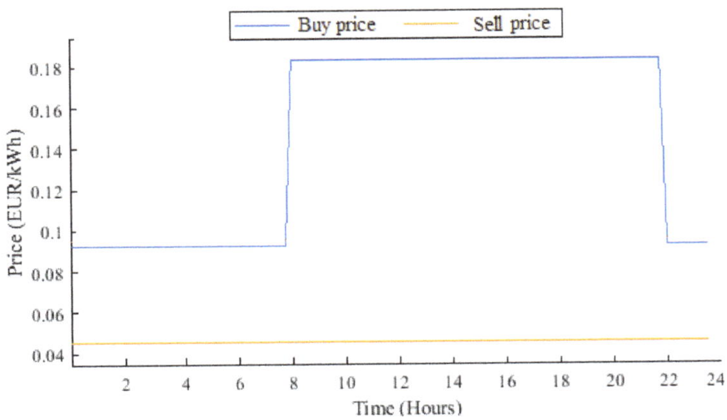

Figure 2. Buy and sell energy prices.

Table 3. System cost results, cost saving, and revenue.

Case	System Cost (EUR)	% Cost Saving	Revenue (EUR)
Base case	42.25	-	2.09
Case 01	39.88	5.60%	2.13
Case 02	41.04	2.86%	2.42
Case 03	38.70	8.40%	2.46
Case 04	34.11	19.26%	5.48

The system cost was highest in the base case and lowest in case 4. The cost reduction was achieved in each case with respect to the base case. The economic savings were in the range of 2.86–19.26%. This result indicates that when the behavioral interventions work positively for energy conservation and increase solar power adoption in the mid-market price scenario, the local energy community achieved an economic benefit of 19.26%. While considering the revenue of customers for selling the power, the lowest revenue was observed in the base case. If the customers responded positively to interventions and changed their behavior for energy conservation and adoption of solar power at the same time, i.e., comparing the base case with case 03, an economic benefit of 17.7% was obtained by the local energy community. The same behavior with the mid-market price increased the revenue from EUR 2.09 to EUR 5.48. In overall comparison, case 04 was the most economical way of operating the system, as it corresponded to the lowest system cost and the highest revenue.

Figure 3 depicts the peak power export of all twenty customers. As the optimization model for reducing the economic cost of the community ran, it was discovered that only a few customers exported a negligible quantity of energy to the community. Therefore, the most cost-effective option for these consumers was to generate their own electricity. On the other hand, exporting excess electricity to a utility or community was the best method to reduce system costs, so few customers had a substantial quantity of excess electricity. Comparing the average power exported in all five cases, the base case exported the least power, while case 04 exported the most.

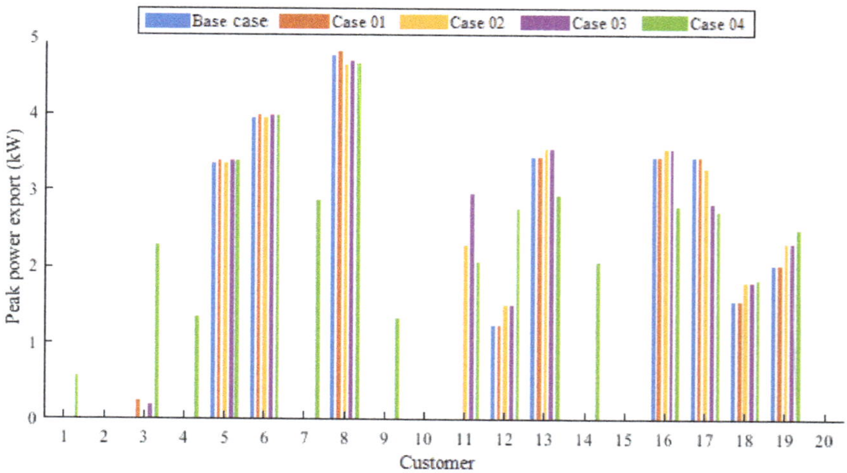

Figure 3. Peak power export by customers.

Table 4 and Figure 4 present the statistics for the local cost of purchasing electricity for five cases. The case-by-case results presented in Table 4 depict the minimum and maximum power purchase costs incurred by customers, the average power purchase cost, and the reduction in power purchase costs compared to the base case. In this context, average power buy cost refers to the ratio of the sum of power buy costs for all customers in a specific case to the total number of customers. Figure 4 depicts the comprehensive results of the power purchase cost for all consumers in all circumstances.

Table 4. Power buy cost values.

Case	Cost Range (Min–Max) (EUR)	Average Cost (EUR)	% Reduction in avg. Power Buy Cost
Base case	0.46–3.13	1.788	-
Case 01	0.40–3.13	1.671	6.53%
Case 02	0.46–2.99	1.745	2.40%
Case 03	0.40–2.93	1.628	8.95%
Case 04	0.66–2.49	1.262	29.41%

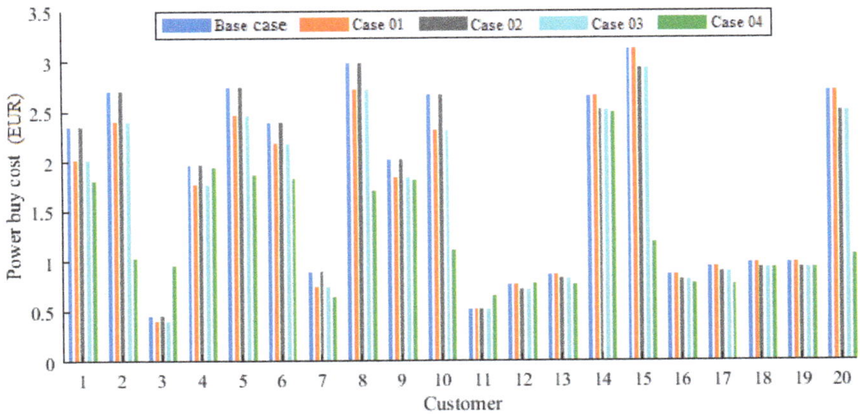

Figure 4. Cost of power bought in community.

Case 04 was the exception; in all other instances, consumers purchased power from the utility power grid at a price indicated in Figure 2's price chart. In case 04, consumers were permitted to purchase electricity at the calculated mid-market price (5). Considering battery energy storage, case 04 gave optimum results compared to the other cases. Battery power trading was the most economical, as the lowest system cost and highest revenue were reported. According to the results, it is evident that the base case had the maximum power purchase cost, and case 04 had the lowest.

Table 5 and Figure 5 detail the revenue derived from the sale of energy to the community. Comparing all cases, it is evident that the base case had the lowest economic benefit for customers, while case 04 had the highest. Case 04 has an average revenue of 0.274 EUR, which was 163.46% higher than the base case. Customers were permitted to sell power in their local community at a mid-market price that was less than the utility grid export price, and they were also permitted to purchase power at a mid-market price that was less than the utility grid import price.

Table 5. Power sold revenue values.

Case	Revenue Range (Min–Max) (EUR)	Average Revenue (EUR)	% Increase in Avg. Power Sell Revenue
Base case	0–0.398	0.104	-
Case 01	0–0.559	0.106	1.92%
Case 02	0–0.553	0.121	16.34%
Case 03	0–0.561	0.123	18.26%
Case 04	0–0.874	0.274	163.46%

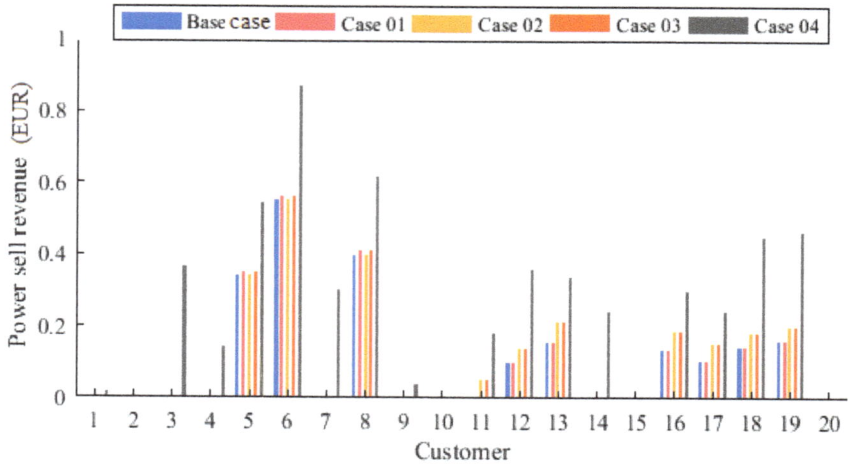

Figure 5. Revenue of power sold in the community.

4. Discussion and Policy Implication

We will now go through how the experiment's findings could be relevant to policymakers who are debating whether to execute a behavioral intervention or a cost-based intervention aimed at encouraging consumers to use less energy and switch to renewable sources. When people are unable to charge higher premiums, social intervention may be a particularly attractive policy instrument to urge them to conserve electricity. We contend that motivating factors and social position, in addition to economic reasons, also play a role in making judgment calls and that policymakers may take advantage of these elements to boost policy effectiveness. We conducted an experiment based on the local energy community members' heterogeneity, a characteristic common to social computational science, to examine how a policy will progress after the socioeconomic drivers of electricity savings and renewable energy adoption are taken into consideration. The research we used to develop an effective approach to electricity community simulation is a good illustration of how behavioral elements, including ecological issues, are becoming more and more important to policymakers. Additionally, our findings imply that a further step is required: details on objectives should be collected in a way that enables communicating their relative importance within the energy community member's decision-making strategy in order to obtain a complete picture of the policy receivers.

Overall, by experimentally examining the effect of local energy communities on costs, our results add to studies on energy conservation and the uptake of renewable energy sources. Despite this, there are certain limitations to our study. First, this study evaluated a hypothetical decision between adopting renewable energy sources and energy conservation. Even so, we think that this analysis offers policymakers an important empirical understanding of how to deploy interventions for local power cost reduction, despite some evidence to the contrary. Secondly, we assume the infrastructure required for communication is in a ready condition in the local electric community. For instance, it gave decision-makers new instruments for influencing social behavior as well as new perspectives for more precisely forecasting the effects of current policies. Social interventions are commands to do or not do something that are supported by the acceptance or rejection of others. Social influencing frequently promotes collaboration and has a significant impact on behavior in the local community. They may be successful, for instance, in lowering energy use. Because of this, social interventions may be used to address significant social problems. This is particularly vital for the energy local area because, by saving more energy costs, they will have more monetary assets accessible for other important merchandise that they generally can not manage.

While we have reproduced intervention-based cases independently, further examination ought to explore the possibility in future that behaviorally informed interventions could be utilized mixed with conventional ones, for example, how and what mixtures of traditional, behavioral, and socially informed instruments are powerful at advancing energy viability choices. One method for evaluating its viability would be by investigating the instruments on which such combinations work. An incentive-based policy proposal might be interesting research work in the future.

Due to its explicit behavior expectations and good mathematical representation, the projected model can be used as a reference for policymaking. Policymakers can address the social drivers of energy saving and embracing renewables by planning interventions that recognize that people are not generally judicious leaders, particularly when the unfortunate circumstances where they live drain the mental assets important for reasonable direction. Only behavioral interventions that are fair to the target population, that are designed to meet people's demands and aspirations, and that stop the private sector from creating deceptive interventions should be supported by policymakers. Nevertheless, relying solely on behavioral economic perspectives is insufficient for policymakers to determine if the behavioral intervention has been successful in encouraging better choices and results at a wider level, if it should be improved, whether it can be scaled up, or if it is reproduced in other locations. They must make use of the insights from impact evaluations in order to rely on the strongest evidence. According to this branch of study, interventions can produce the best evidence because they make it possible to establish the ideal circumstances to take into account when developing an intervention's effects.

The proposed model would also work in the urban population. However, if the local community is larger, i.e., customers exchanging the power are located at a far distance, the internal power losses would significantly affect the system cost. In such cases, a sophisticated optimization model with customer groups in large quantities would be an option to get the optimum system cost.

5. Conclusions

Behavioral economics emphasizes economic incentives and offers a wider view of the problem that considers the diversity of people. We also offer an additional explanation that involves the social framework in which contact takes place via behavioral intervention. Most people agree that social involvement encourages technology adoption through emulation. To encourage more consumers to modify their energy behavior and foster bottom-up initiatives, an energy policy that provides more opportunities for investment in renewable power and behavioral interventions for energy conservation is required.

To the best of our knowledge, this is the first empirical study experiment that uses behavioral intervention results to calculate the system cost of a local energy community. In this study, we looked at how much behavioral changes like electricity conservation and renewable energy adoption may lower the cost of the local electrical grid. Based on the efficacy of interventions for a locally controlled power system, we examined a variety of instances. While adopting energy conservation only and while adopting higher PV generation individually, the system cost was reduced by 5.60% and 2.86%, respectively. The lowest system cost of EUR 34.11 for case 04 and the highest revenue of EUR 5.48 in the same case were achieved. The average cost reduction of up to 29.41% was achieved in case 04 compared to the base case, and also the average power sell revenue increased by up to 163.46% when energy conservation and higher renewable power generation were considered. The results point to favorable effects and give policymakers proof to use when adopting behavioral intervention strategies for energy conservation and preference for renewable energy sources. Outcomes of an empirical study show that when socioeconomic and behavioral objectives are taken into consideration, policy interventions may result in paradoxical results, notwithstanding a few constraints. The findings also imply that the path followed is worthwhile continuing as long as this kind of modeling is improved, and perhaps as quantitative social science develops by gathering additional qualitative data on the judgment process of the energy community.

Author Contributions: Conceptualization, P.M.; methodology, P.M. and J.S.; software, P.M. and J.S.; validation, P.M., J.S. and K.P.; formal analysis, K.P.; investigation, P.M.; data curation, J.S.; writing—original draft preparation, P.M.; writing—review and editing, K.P., J.S. and Z.V.; supervision, Z.V. All authors have read and agreed to the published version of the manuscript.

Funding: This article is a result of the project RETINA (NORTE-01-0145-FEDER-000062), supported by Norte Portugal Regional Operational Programme (NORTE 2020), under the PORTUGAL 2020 Partnership Agreement, through the European Regional Development Fund (ERDF). João Soares has received funding from FCT, namely CEECIND/00420/2022. The authors acknowledge the work facilities and equipment provided by GECAD research center (UIDB/00760/2020).

Data Availability Statement: The dataset utilized in our publication can be accessed at the following link: https://zenodo.org/record/4737293#.ZCxOUXZBxPY, accessed on 15 November 2022.

Conflicts of Interest: The authors declare no conflict of interest.

Nomenclature

Indices	
x	customers
t	periods
Parameters	
N_x	number of customers
N_t	number of periods
$E_{x,t}^{buy\ max}$	power purchase upper limit
$E_{x,t}^{sell\ max}$	power sell upper limit
$E_{x,ini}^{bat}$	initial battery energy
dt	time period adjustment factor
$FC_{x,t}$	fixed cost
$E_{h,t}^{load}$	electric load of customer
$E_{h,t}^{gen}$	power generation of customer
$\eta_{x,ch}$	charging efficiency of battery
$\eta_{x,dch}$	discharging efficiency of battery
$P_{x,t}^{mmp}$	mid-market price
$P_{x,t}^{sell}$	power sell price
$P_{x,t}^{buy}$	power buy price
$l_{x,t}^{sell}$	power loss multiplier for power sell
$l_{x,t}^{buy}$	power loss multiplier for power buy
Variables	
$E_{x,t}^{buy}$	power purchased by community members
$X_{x,t}^{EB}$	binary variable for power purchase
$E_{x,t}^{buy\ mmp}$	power buy at mid-market price
$X_{x,t}^{EB\ mpp}$	binary variable for power buy at mid-market price
$E_{x,t}^{sell}$	power sell by community member
$X_{x,t}^{ES}$	binary variable for power sell
$E_{x,t}^{sell\ mmp}$	power sell at mid-market price
$X_{x,t}^{ES\ mmp}$	binary variable for power sell at mid-market price
$E_{x,t}^{bat}$	electric energy content of the battery
$E_{x,t}^{bat\ ch}$	battery charging power
$E_{x,t}^{bat\ dch}$	battery dis-charging power
$C_{x,t}^{import}$	cost of power import from grid
$C_{h,t}^{lem}$	cost of power buy in LEM
$R_{x,t}^{export}$	revenue from power export to grid
$R_{h,t}^{lem}$	revenue from power sell in LEM

References

1. DellaValle, N.; Sareen, S. Nudging and boosting for equity? Towards a behavioural economics of energy justice. *Energy Res. Soc. Sci.* **2020**, *68*, 101589. [CrossRef]
2. Mochi, P.; Pandya, K.; Dabhi, D.; Rajput, V. Risk Based Day-ahead Energy Resource Management with Renewables via Computational Intelligence. *Int. J. Renew. Energy Res.* **2022**, *12*, 921–929. [CrossRef]
3. Genc, T.S.; Kosempel, S. Energy Transition and the Economy: A Review Article. *Energies* **2023**, *16*, 2965. [CrossRef]
4. Teixeira, D.; Gomes, L.; Vale, Z. Single-unit and multi-unit auction framework for peer-to-peer transactions. *Int. J. Electr. Power Energy Syst.* **2021**, *133*, 107235. [CrossRef]
5. Guibentif, T.; Patel, M.; Yilmaz, S. Using energy saving deficit distributions to assess calculated, deemed and metered electricity savings estimations. *Appl. Energy* **2021**, *304*, 117721. [CrossRef]
6. Peyerl, D.; Barbosa, M.O.; Ciotta, M.; Pelissari, M.R.; Moretto, E.M. Linkages between the Promotion of Renewable Energy Policies and Low-Carbon Transition Trends in South America's Electricity Sector. *Energies* **2022**, *15*, 4293. [CrossRef]
7. Zakeri, B.; Cross, S.; Dodds, P.; Gissey, G. Policy options for enhancing economic profitability of residential solar photovoltaic with battery energy storage. *Appl. Energy* **2021**, *290*, 116697. [CrossRef]
8. d'Adda, G.; Galliera, A.; Tavoni, M. Urgency and engagement: Empirical evidence from a large-scale intervention on energy use awareness. *J. Econ. Psychol.* **2020**, *81*, 102275. [CrossRef]
9. Kumar, A.; He, X.; Deng, Y.; Singh, A.; Sah, B.; Kumar, P.; Bansal, R.; Bettayeb, M.; Rayudu, R. A sustainable rural electrification based on a socio-techno-economic-environmental-political microgrid design framework. *Energy Environ. Sci.* **2022**, *15*, 4213–4246. [CrossRef]
10. Kumar, A.; Deng, Y.; He, H.; Singh, A.; Kumar, P.; Bansal, R.; Bettayeb, M.; Ghenai, C.; Naidoo, R. Impact of demand side management approaches for the enhancement of voltage stability loadability and customer satisfaction index. *Appl. Energy* **2023**, *339*, 120949. [CrossRef]
11. Chow, T. A review on photovoltaic/thermal hybrid solar technology. *Appl. Energy* **2010**, *87*, 2. [CrossRef]
12. Fu, X. Statistical machine learning model for capacitor planning considering uncertainties in photovoltaic power. *Prot. Control Mod. Power Syst.* **2022**, *7*, 5. [CrossRef]
13. Fu, X.; Zhou, Y. Collaborative Optimization of PV Greenhouses and Clean Energy Systems in Rural Areas. *IEEE Trans. Sustain. Energy* **2023**, *14*, 642–656. [CrossRef]
14. Corradi, N.; Priftis, K.; Jacucci, G.; Gamberini, L. Oops, I forgot the light on! The cognitive mechanisms supporting the execution of energy saving behaviors. *J. Econ. Psychol.* **2013**, *34*, 88–96. [CrossRef]
15. Callery, P.; Goodwin, C.; Moncayo, D. Norm proximity and optimal social comparisons for energy conservation behavior. *J. Environ. Manag.* **2021**, *296*, 113332. [CrossRef] [PubMed]
16. Kažukauskas, A.; Broberg, T.; Jaraitė, J. Social comparisons in real time: A field experiment of residential electricity and water use. *Scand. J. Econ.* **2021**, *123*, 558–592. [CrossRef]
17. Asmare, F.; Jaraitė, J.; Kažukauskas, A. The effect of descriptive information provision on electricity consumption: Experimental evidence from Lithuania. *Energy Econ.* **2021**, *104*, 105687. [CrossRef]
18. Ruokamo, E.; Meriläinen, T.; Karhinen, S.; Räihä, J.; Suur-Uski, P.; Timonen, L.; Svento, R. The effect of information nudges on energy saving: Observations from a randomized field experiment in Finland. *Energy Policy* **2022**, *161*, 112731. [CrossRef]
19. Ghesla, C.; Grieder, M.; Schmitz, J.; Stadelmann, M. Pro-environmental incentives and loss aversion: A field experiment on electricity saving behavior. *Energy Policy* **2020**, *137*, 111131. [CrossRef]
20. Klege, R.; Visser, M.; Datu, S.; Darling, M. The Effectiveness of Competition and Responsibility Assignment in Saving Energy: A Non-Residential Example of the Power of the "Nudge". 2018. Available online: https://www.researchgate.net/publication/323258372_The_Effectiveness_of_Competition_and_Responsibility_Assignment_in_Saving_Energy_A_Non-Residential_example_of_the_Power_of_the_Nudge (accessed on 15 November 2022). [CrossRef]
21. Wemyss, D.; Cellina, F.; Lobsiger-Kägi, E.; de Luca, V.; Castri, R. Does it last? Long-term impacts of an app-based behavior change intervention on household electricity savings in Switzerland. *Energy Res. Soc. Sci.* **2019**, *47*, 16–27. [CrossRef]
22. Buckley, P. Prices, information and nudges for residential electricity conservation: A meta-analysis. *Ecol. Econ.* **2020**, *172*, 106635. [CrossRef]
23. Murphy, S.; Frick, N.M. Estimating the Drivers of the Cost of Saved Electricity in Utility Customer-Funded Energy Efficiency Programs. *Energies* **2023**, *16*, 2177. [CrossRef]
24. de Almeida, L.; Esposito, F.; van Zeben, J. When indicators fail electricity policies: Pitfalls of the EU's retail energy market Barrier Index. *Energy Policy* **2022**, *165*, 112892. [CrossRef]
25. Frederick, S.; Loewenstein, G.; O'Donoghue, T. Time Discounting and Time Preference: A Critical Review. *J. Econ. Lit.* **2002**, *40*, 351–401. Available online: https://www.jstor.org/stable/2698382 (accessed on 15 November 2022). [CrossRef]
26. Wilson, C.; Crane, L.; Chryssochoidis, G. Why do homeowners renovate energy efficiently? Contrasting perspectives and implications for policy. *Energy Res. Soc. Sci.* **2015**, *7*, 12–22. [CrossRef]
27. Gajdzik, B.; Jaciow, M.; Wolniak, R.; Wolny, R.; Grebski, W.W. Energy Behaviors of Prosumers in Example of Polish Households. *Energies* **2023**, *16*, 3186. [CrossRef]
28. Jittayasotorn, T.; Sadidah, M.; Yoshida, T.; Kobashi, T. On the Adoption of Rooftop Photovoltaics Integrated with Electric Vehicles toward Sustainable Bangkok City, Thailand. *Energies* **2023**, *16*, 3011. [CrossRef]

29. Vesely, S.; Klöckner, C.; Carrus, G.; Chokrai, P.; Fritsche, I.; Masson, T.; Panno, A.; Tiberio, L.; Udall, A. Donations to renewable energy projects: The role of social norms and donor anonymity. *Ecol. Econ.* **2022**, *193*, 107277. [CrossRef]
30. Guo, Z.; O'Hanley, J.; Gibson, S. Predicting residential electricity consumption patterns based on smart meter and household data: A case study from the Republic of Ireland. *Util. Policy* **2022**, *79*, 101446. [CrossRef]
31. Mayer, A. Social support for de-carbonizing the energy system: The role of expressive partisanship. *Environ. Sci. Policy* **2020**, *109*, 83–94. [CrossRef]
32. Mochi, P.; Pandya, K. The Impact of Social Nudge on System Cost & Revenue Optimization in Local Electricity Market. In Proceedings of the 2023 International Conference on Power Electronics and Energy (ICPEE), Bhubaneswar, India, 3–5 January 2023. [CrossRef]
33. Ivanov, O.; Chattopadhyay, S.; Banerjee, S.; Neagu, B.; Grigoras, G.; Gavrilas, M. A Novel Algorithm with Multiple Consumer Demand Response Priorities in Residential Unbalanced LV Electricity Distribution Networks. *Mathematics* **2020**, *8*, 1220. [CrossRef]
34. Chersoni, G.; DellaValle, N.; Fontana, M. Modelling thermal insulation investment choice in the EU via a behaviourally informed agent-based model. *Energy Policy* **2022**, *163*, 112823. [CrossRef]
35. Della Valle, N.; Poderi, G. What works for consumer engagement in the energy transition: Experimenting with a behavioural-sociological approach. In *Control, Change and Capacity-Building in Eneryg Systems: SHAPE ENERGY—Research Design Challenge*; Shape Energy: Cambridge, UK, 2018; pp. 104–113.
36. Loewestein, G.; Chater, N. Putting nudges in perspective. *Behav. Public Policy* **2017**, *1*, 26–53. [CrossRef]
37. Bryan, G.; Karlan, D.; Nelson, S. Commitment devices. *Annu. Rev. Econ.* **2010**, *2*, 671–698. [CrossRef]
38. Takeda, Y.; Nakai, Y.; Senoo, T.; Tanaka, K. Designing a User-Centric P2P Energy Trading Platform: A Case Study—Higashi-Fuji Demonstration. *Energies* **2021**, *14*, 7289. [CrossRef]
39. Dzwigol, H.; Kwilinski, A.; Lyulyov, O.; Pimonenko, T. The Role of Environmental Regulations, Renewable Energy, and Energy Efficiency in Finding the Path to Green Economic Growth. *Energies* **2023**, *16*, 3090. [CrossRef]
40. Kachirayil, F.; Weinand, J.; Scheller, F.; McKenna, R. Reviewing local and integrated energy system models: Insights into flexibility and robustness challenges. *Appl. Energy* **2022**, *324*, 119666. [CrossRef]
41. Azizi, A.; Aminifar, F.; Moeini-Aghtaiet, M.; Alizadeh, A. Transactive Energy Market Mechanism with Loss Implication. *IEEE Trans. Smart Grid* **2021**, *12*, 1215–1223. [CrossRef]
42. Faia, R.; Soares, J.; Vale, Z.; Corchado, J.M. An Optimization Model for Energy Community Costs Minimization Considering a Local Electricity Market between Prosumers and Electric Vehicles. *Electronics* **2021**, *10*, 129. [CrossRef]
43. Mochi, P.; Pandya, K. Two Stage Optimization for Customer Fiscal Incentive in Local Electricity Market. In Proceedings of the 2022 IEEE International Power and Renewable Energy Conference, Kerala, India, 16–18 December 2022. [CrossRef]
44. Faia, R.; Soares, J.; Ghazvini, M.A.F.; Franco, J.F.; Vale, Z. Energy Consumption and PV Generation Data of 50 Prosumers and Energy Consumption of 40 Electric vehicles—15-Minute Resolution [Data Set]. *Zenodo* **2021**. [CrossRef]

mathematics

Article

Efficient Red Kite Optimization Algorithm for Integrating the Renewable Sources and Electric Vehicle Fast Charging Stations in Radial Distribution Networks

Sami M. Alshareef [1,*] and Ahmed Fathy [1,2,*]

1 Department of Electrical Engineering, College of Engineering, Jouf University, Sakaka 72388, Saudi Arabia
2 Department of Electric Power and Machines, Faculty of Engineering, Zagazig University, Zagazig 44519, Egypt
* Correspondence: smalshareef@ju.edu.sa (S.M.A.); afali@zu.edu.eg (A.F.)

Abstract: The high penetration of renewable energy resources' (RESs) and electric vehicles' (EVs) demands to power systems can stress the network reliability due to their stochastic natures. This can reduce the power quality in addition to increasing the network power losses and voltage deviations. This problem can be solved by allocating RESs and EV fast charging stations (FCSs) in suitable locations on the grid. So, this paper proposes a new approach using the red kite optimization algorithm (ROA) for integrating RESs and FCSs to the distribution network through identifying their best sizes and locations. The fitness functions considered in this work are: reducing the network loss and minimizing the voltage violation for 24 h. Moreover, a new version of the multi-objective red kite optimization algorithm (MOROA) is proposed to achieve both considered fitness functions. The study is performed on two standard distribution networks of IEEE-33 bus and IEEE-69 bus. The proposed ROA is compared to dung beetle optimizer (DBO), African vultures optimization algorithm (AVOA), bald eagle search (BES) algorithm, bonobo optimizer (BO), grey wolf optimizer (GWO), multi-objective multi-verse optimizer (MOMVO), multi-objective grey wolf optimizer (MOGWO), and multi-objective artificial hummingbird algorithm (MOAHA). For the IEEE-33 bus network, the proposed ROA succeeded in reducing the power loss and voltage deviation by 58.24% and 90.47%, respectively, while in the IEEE-69 bus it minimized the power loss and voltage deviation by 68.39% and 93.22%, respectively. The fetched results proved the competence and robustness of the proposed ROA in solving the problem of integrating RESs and FCSs to the electrical networks.

Keywords: electric vehicles; charging stations; renewable energy; red kite optimization algorithm

MSC: 90C31

Citation: Alshareef, S.M.; Fathy, A. Efficient Red Kite Optimization Algorithm for Integrating the Renewable Sources and Electric Vehicle Fast Charging Stations in Radial Distribution Networks. *Mathematics* **2023**, *11*, 3305. https://doi.org/10.3390/math11153305

Academic Editor: Ioannis G. Tsoulos

Received: 12 June 2023
Revised: 20 July 2023
Accepted: 24 July 2023
Published: 27 July 2023

1. Introduction

Recently, there has been a rapid growth in the use of fossil fuel sources, especially in electric power generation plants and the transportation sector. These sources increase environmental pollution as they emit greenhouse gas; they also cause global warming [1]. Therefore, many countries are looking to replace gasoline vehicles with clean energy cars, known as electric vehicles (EVs), to reduce the amount of pollution [2]. EVs are environmentally friendly, but have different economic costs than gasoline ones. EVs have advanced batteries and power electronic devices that enable them to be installed to grids as controllable loads. The integration of EVs to power systems faces great challenges like violation of transmission line thermal constraints due to overload; this may cause a voltage drop in some sensitive buses. Also, the uncertainties associated with these vehicles represent challenges to: the distribution network operator, as the sources of uncertainties are time rounding; the amount of daily energy consumption; the range of driving; and the EV battery capacity [3]. When EVs charge from public charging stations, mostly fast charging stations (FCSs) are utilized by demanding high power from the grid to reduce the required charging duration to meet the required battery state-of-charge (SOC). However, the high demand required by these stations causes negative effects on the distribution network, as they increase the network power losses and voltage deviations. However, to reduce the demand on the grid, renewable energy sources (RESs) can be installed to supply the excess loads during peak time. Identifying the optimal allocations of RESs and FCSs in the distribution network is mandatory to minimize their associated negative effects. Improvement of the power generation from RESs is essential; various technologies have been presented to improve the power quality of renewable energy sources integrated in the microgrid [4,5].

Many reported approaches have been implemented to identify the best allocations and sizes on both RESs and FCSs. Amer et al. [6] presented a planning model to evaluate the sizes and sites of FCSs in addition to wind turbines in distribution networks. The authors considered the stochastic features of RESs, FCSs, and residential EV loads. In [7], parking lots and capacitor allocations have been identified in the electrical distribution network via a biogeography-based optimizer (BBO) to compensate the system reactive power. A multi-objective problem was introduced and solved via hybrid and grey wolf optimizers (GWO) and the particle swarm optimizer (PSO) to allocate FCSs, and shunt capacitors and distributed generators (DGs) [8]. Reducing the cost of power loss, minimizing the voltage fluctuation, reducing the development costs of FCSs, minimizing the costs of EV energy consumption, and reducing the costs of DGs have been considered as targets. Bayram et al. [9] determined the allocations of parking lots via a combinatorial optimization algorithm and the two-stages stochastic programming model. A comprehensive review of allocating the EV rapid charging stations based on economic benefits has been conducted by Gupta et al. [10]. A planning method for penetrating FCSs in the electrical distribution network has been presented in [11] to find the optimal operators, traffic conditions, vehicles, power grids, and drivers. Moreover, the authors used real-time data for the practical third ring of Beijing. In [12], the non-dominated sorting genetic algorithm II (NSGA-II) was presented to evaluate the places and sizes of FCSs and DGs installed in the electrical distribution network. The considered targets are mitigating the EV user loss, minimizing the power loss, reducing the cost of FCS development, and enhancing the voltage shape. A FCS connected to the grid has been simulated in [13] such that the harmonic currents were minimized. Moreover, an energy management strategy has been presented via integrating the photovoltaic (PV) generation system. Pal et al. [14] identified the best allocations of FCSs and solar DGs in addition to battery storage system in the electrical distribution network using the hybrid Harris hawks optimizer and GWO. The targets are mitigating the energy loss, investment costs, operating and maintenance costs, and the voltage violation index. Moreover, the number of charging ports, FCSs' capacities, and the captured power via EVs have been evaluated. A planning model of FCS has been introduced and solved by binary PSO to minimize the costs of construction, operating and maintenance, trips to the station, and power loss [15]. In [16], optimal places and capacities of FCSs and RESs integrated to the distribution network have been identified when considering the uncertainties of renewable-based generators. Also, the capacitated deviation flow refueling location-based model has been presented to cover the EVs' charging demands on transportation network. Amer et al. [17] developed a stochastic program to evaluate the optimal locations and sizes of small wind turbines connected to FCSs in urban and suburban areas. Moreover, a worthiness metric has been employed to classify the FCS candidate sites according to the EV drivers' attractiveness. The EV charging station place was evaluated in the electrical network and covered by the transportation network using HHO and differential evolution (DE) [18]. The targets are mitigating the voltage fluctuation, minimizing the energy loss, and reducing the cost of land for maximizing the service to EV with minimized founding costs. A quantum-behaved Gaussian mutational dragonfly algorithm (QGDA) has been employed to conduct the best planning of capacitors and EV charging stations in the distribution network [19]. The authors in [20] reviewed various configurations of charging station designs, and the different modes of renewable DGs were summarized. Ahmad et al. [21] solved the problem of sitting solar-operated charging stations integrated to the grid using an improved chicken swarm optimizer where the network voltage profile was enhanced while the power loss and operating cost were minimized. Moreover, the authors used a stochastic method to forecast the EV demand and the neural network to predict the power generated from the solar PV plant. The places of charging stations integrated to the grid have been evaluated via the hybrid metaheuristic approach comprising the chicken swarm optimizer (CSO) and the teaching learning-based optimizer (TLBO) [22]. A solar hybrid EV charging station has been presented to mitigate the dependence on the main grid [23]. The authors used a stochastic model to forecast the EV arrival time, battery SOC, and charging demand. Moreover, a stochastic firefly algorithm (SFA) has been used as a maximum power point tracker (MPPT) for the solar system to extract its maximum power. Furthermore, SFA has been employed to solve multi-objective planning to mitigate the investment cost and enhance the charging profit. In [24], the sizing problem of EV charging stations has been solved via optimization frameworks to reduce the charging station investment cost and provide a certain quality of service to the client. The locations of FCSs in the distribution network have been identified via solving multi-objective problems using a transient search optimizer (TSO) to mitigate the active and reactive losses and enhance the network voltage stability [25]. A model with two stages for optimizing charging stations and charging schedules has been presented by Yi et al. [26] to achieve complete satisfaction among members of society. The load demand and starting point of the trip have been predicted via Monte Carlo simulations in the first phase while a binary PSO has been employed in the second one to find the optimal path of the trip. In [27], the state-of-the-art features for many design approaches of FCSs have been reviewed in addition to the future challenges of each one. Zhou et al. [28] presented a model to calculate the charging station's total operating cost which is divided into economic and environmental costs. Also, the locations of these stations in Irish regions have been identified using a genetic algorithm (GA) where the total cost is reduced. A black widow optimizer (BWO) has been used to identify the optimal places of charging stations and renewable DGs in the distribution system with the aid of model predictive control (MPC) that simulates the actual SOC of storage batteries [29]. Many methods conducted in optimizing the charging station have been reviewed and categorized according to the fitness functions, algorithms, constraints, modeling of EV uncertainties, and DG integration [30]. In [31], FCSs, solar PV, and storage batteries have been installed in distribution networks with sizes and locations decided by hybrid NSGA-II and Fuzzy satisfaction. The authors considered many targets, like system power loss, voltage violation, flow of served EVs, costs of investment, operation and maintenance of PV, and charging stations. The planning of charging station location and battery-swapping stations have been presented as multi-objective problems to mitigate the total cost, enhance the satisfaction of user, and reduce the EV's consumed energy [32]. A hybrid approach combining the student psychology optimizer and the AdaBoost algorithm has been introduced to allocate the EV charging station linked to distribution generation such that the peak power and voltage regulation are mitigated [33]. Al Wahedi et al. [34] implemented a techno-economic analysis via the HOMER software for renewable-based charging stations to evaluate its optimal configuration in different cities in Qatar. Excessive review of different nature-inspired optimizers employed in solving the problem of FCS placement has been presented in [35]. The optimal planning of FCSs has been expressed as a multi-objective problem with multi-criteria decision-making [36]. Minimizing the total charging time and cost is the main target of the work presented in [37] to model the vehicle charging via a bi-level optimizer. The optimal locations of EV parking lots in smart distribution systems have been identified using a hybrid metaheuristic algorithm to reduce the network loss and voltage fluctuation [38]. Moreover, the cost of EV charging/discharging and the cost of purchased power from the grid are considered in the presented problem. Fathy et al. [39] presented a competition over resource (COR) approach to determine the optimal sites and sizes of EV parking lots in the electrical distribution system. The authors considered the cost of reliability enhancement, investment cost, and the cost of power loss improvement as the targets to be minimized. Table 1 outlines most of the reported methods to integrate RESs and FCSs in electrical networks, the reader can observe the following items:

- Most of the reported works used metaheuristic optimization algorithms to integrate RESs and FCSs to the network.
- Many of these methods lack accuracy due to the fall in local optimal solution in addition to the slow convergence rate of some approaches.
- Also, the reported hybrid algorithms were complicated to implement and required excessive effort and time.
- Many researchers ignored the installation of distributed generators (DGs) and they relied mainly on the grid as the source of energy.

The authors considered all these shortages in the reported methods and covered them via the following contributions:

- A new methodology incorporating the simple and efficient red kite optimization algorithm (ROA) is proposed to evaluate the optimal capacities and places of RESs and FCSs in distribution networks.
- The considered fitness functions are: reducing the network active loss and minimizing the voltage deviation.
- A multi-objective red kite optimization algorithm (MOROA) is proposed to reduce both targets.
- The proposed approach competency is proved through the obtained results.

The paper is outlined as follows: Section 2 explains the model of th considered system; the form of the optimization problem is presented in Section 3; the basics of ROA are introduced in Section 4; the proposed ROA-based methodology is explained in Section 5; the results and discussions are presented in Section 6; and Section 7 handles the conclusions.

Table 1. Outlines of most reported methods to integrate the RESs and FCSs in electrical networks.

Author	Year	DG	Type	Objective	Algorithm	Metaheuristic	Remarks
Amer et al. [6]	2020	√	wind	- Maximize the revenues of both wind DGs and FCS. - Minimize the total associated costs.	Genetic algorithm	√	The genetic algorithm has a slow convergence rate
Sachan et al. [7]	2020	×	--	- Minimize total losses.	Biogeography-based optimizer	√	The installation of DGs is ignored
Mohanty et al. [8]	2022	√	NA	- Minimize the cost of power loss. - Minimize the voltage violations. - Minimize the EV energy consumption cost. - Minimize the cost of DGs.	GWO-PSO	√	The presented hybrid algorithm is complicated and requires excessive effort for implementation
Zeng et al. [9]	2020	√	wind	Maximize the overall profit of parking lots	Genetic algorithm	√	The genetic algorithm has a slow convergence rate
Kong et al. [11]	2019	√	NA	Minimize the costs of construction and operation	Iterative optimization algorithm	×	The authors considered fast charging stations for all EVs
Battapothula et al. [12]	2019	√	NA	- Minimize the EV user loss - Minimize the power loss - Minimize the cost of FCS development - Enhance the network voltage profile	NSGA-II	√	NSGA-II is very complicated in construction and implementation
Khan et al. [13]	2019	√	PV	Minimize the net power exchange between the charging station and grid	Constant current-constant voltage	×	An energy management strategy between DG, charging station, and grid has been implemented

Table 1. *Cont.*

Author	Year	DG	Type	Objective	Algorithm	Metaheuristic	Remarks
Pal et al. [14]	2021	√	Solar	- Minimize the energy loss - Minimize voltage violations - Minimize investment, operation, and maintenance costs	Harris hawks optimizer and GWO	√	Both HHO and GWO are easy to trap in local optima
Wu et al. [15]	2021	×	--	- Minimize the costs of construction, operation, and maintenance - Minimize the EV travel cost - Minimize the power loss cost	Binary PSO	√	PSO can not avoid the local optima and has a low convergence rate
aSa'adati et al. [16]	2021	√	Wind and solar	Minimize the costs of investment and energy losses	CFRLM	×	The authors ignored the driving range uncertainty and the EV's SOC during arrival at the transportation network
Amer et al. [17]	2021	√	wind	- Maximizing the FCS and DG profits gained from FCSs - Minimize the network energy losses	Genetic algorithm	√	The genetic algorithm has a slow convergence rate
Pal et al. [18]	2021	×	--	- Minimize of the energy loss - Minimize the voltage violation - Minimize the land cost	DE and HHO	√	The authors ignored the installation of DGs
Rajesh et al. [19]	2021	√	NA	- Minimize the total losses - Minimize the network voltage violation	Quantum-behaved Gaussian mutational dragonfly algorithm	√	The employed approach is difficult and requires large computational time

Table 1. *Cont.*

Author	Year	DG	Type	Objective	Algorithm	Metaheuristic	Remarks
Ahmad et al. [21]	2021	√	Solar	- Improve the voltage profile - Mitigate the power loss	Improved chicken swarm optimizer	√	Many steps are followed in the presented approach that make it complicated in implementation
Deb et al. [22]	2021	×	--	Minimize the overall cost of FSC construction	CSO-TLBO	√	The optimal sites of swapping stations and EV charging have not been considered
Goswami et al. [23]	2021	√	Solar	- Minimize the investment cost - Maximize the charging station profit	Stochastic firefly algorithm	√	Firefly suffers from high complexity, computational time, and slow convergence speed
Khaksari et al. [24]	2021	×	--	Minimize the investment cost of FCS	Gurobi optimization	×	Gurobi optimization is limited to the complexity of the handled problem. Also, it is not fast enough to solve complex problems
Bhadoriya et al. [25]	2022	√	NA	Mitigate the total active power loss	Transient search optimizer	√	TDO may trap in local optima during handling complex problems with high dimensions
Yi et al. [26]	2022	×	--	Minimize the annual cost paid by the car owners and investors	Binary PSO	√	PSO falls in local optima and has a slow convergence rate
Zhou et al. [28]	2022	×	--	Minimize the total social cost	Genetic algorithm	√	The genetic algorithm has a slow convergence rate
Aljehane et al. [29]	2022	√	RESs	Reduce the charging time and cost	Black widow optimizer	√	BWO can not avoid the local optima and has a slow rate of convergence
Kumar et al. [31]	2022	×	--	Mitigate the investment cost, power loss, and voltage deviation	Fuzzy optimized via NSGA-II	√	NSGA-II is very complicated in construction and implementation

Table 1. *Cont.*

Author	Year	DG	Type	Objective	Algorithm	Metaheuristic	Remarks
Zu et al. [32]	2022	×	--	- Reduce the total cost - Enhance the user satisfaction - Mitigate the EV's consumed energy	CPLEX and YALMIP languages	×	The solver needs high memory for solving complex problems
Thangaraju [33]	2022	√	NA	Minimize the annualized costs	Student psychology optimizer and AdaBoost algorithm	√	Large consumed time is required for implementing the student psychology optimizer
Al Wahedi et al. [34]	2022	√	Wind and PV	Minimize investment and operating costs	HOMER	×	Detailed inputs, data, and time are mandatory to obtain adequate results from HOMER
Erdogan et al. [36]	2021	×	--	Minimize the overall cost of charging station	Multi-objective optimization (MOO)	NA	The presented MOO method is not clear, also the authors did not consider DG installation
Ma et al. [37]	2021	×	--	Minimize the daily charging time	Surrogate optimization algorithm	√	The surrogate optimization algorithm has a slow convergence rate
Ahmadi et al. [38]	2021	√	Wind and PV	Minimize the loss and voltage fluctuation	- Genetic PSO - Genetic imperialist competitive algorithm (GICA)	√	Excessive computational time is required by both employed hybrid approaches
Fathy et al. [39]	2020	×	--	- Enhance the network reliability - Reduce the investment cost - Minimize the cost of power loss	Competition over resource	√	The authors did not consider the installation of DGs

2. The Considered System Model

The integration of renewable-based DGs and EVs in the distribution network is considered in this work. This section presents the models of the photovoltaic (PV) system, wind turbines (WT), and electric vehicles (EVs).

2.1. Model of the PV System

The PV system transforms light to electrical energy, and has many methods employed in producing electrical energy from sunlight irradiance. It composes a series of cells to produce the required voltage. The generation of the PV system relies on temperature and solar radiation striking its surface, so it is essential to consider both terms while studying the PV system's behavior. Normal operating cell temperature (NOCT) is used as an indicator of cell temperature that can be computed as follows [40]:

$$T_c = T_a + \frac{NOCT - 20^\circ}{0.8} \cdot G \tag{1}$$

where T_a is ambient temperature and G is irradiance.

The cell generated power can be expressed as

$$P_c = P \times [\eta \times (T_c - 25^\circ)] \tag{2}$$

where P is the cell rated power and η is the efficiency of the solar cell. The PV panel generated power can be obtained via multiplying the cell output power by the number of cells as follows:

$$P_{Panel} = n_{cell} \times P_c \tag{3}$$

where n_{cell} denotes the number of cells in the panel. The used temperature and solar radiation daily profiles are shown in Figure 1 [41].

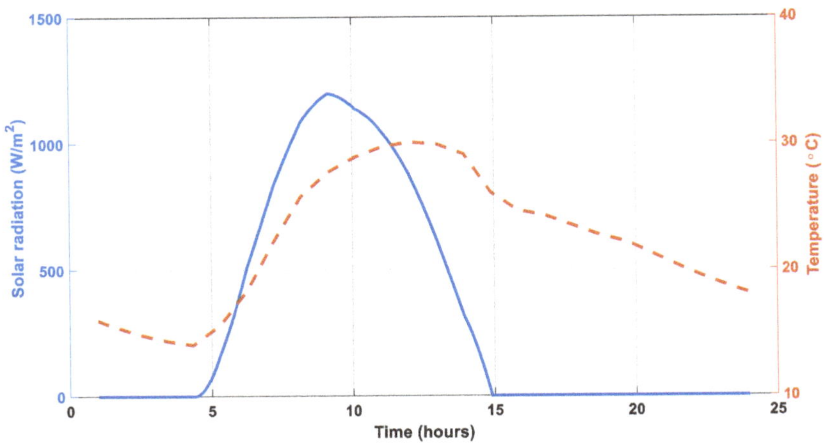

Figure 1. Daily profiles of temperature and solar radiation.

2.2. Model of a Wind Turbine

The wind turbine (WT) output power depends on the wind speed and wind direction in addition to the geography site of installation and wind density. The extracted power from WT can be written as follows:

$$P_{WT}(t) = \begin{cases} 0 & V < V_{cut-in}\ or\ V > V_{cut-off} \\ P^r_{WT}\left(\frac{P_{WT0}-P^r_{WT}}{V_{cut-off}-V_r}\right)(V(t) - V_r) & V_r < V \leq V_{cut-off} \\ P^r_{WT}\left(\frac{V(t)-V_{cut-in}}{V_r-V_{cut-in}}\right)^3 & V_{cut-in} \leq V \leq V_{cut-off} \end{cases} \tag{4}$$

where P^r_{WT} is the WT rated power, V_{cut-in}, V_r, and $V_{cut-off}$ are the cut-in, rated, and cut-off speeds of the turbine, respectively, V is wind speed, and P_{WT0} is WT power at the cut-off speed. The WT output power can be calculated as [42]

$$P_{Wind-total} = n_{WT} \times P_{WT} \tag{5}$$

where n_{WT} denotes wind turbine number. The wind speed daily profile is given in Figure 2 [41].

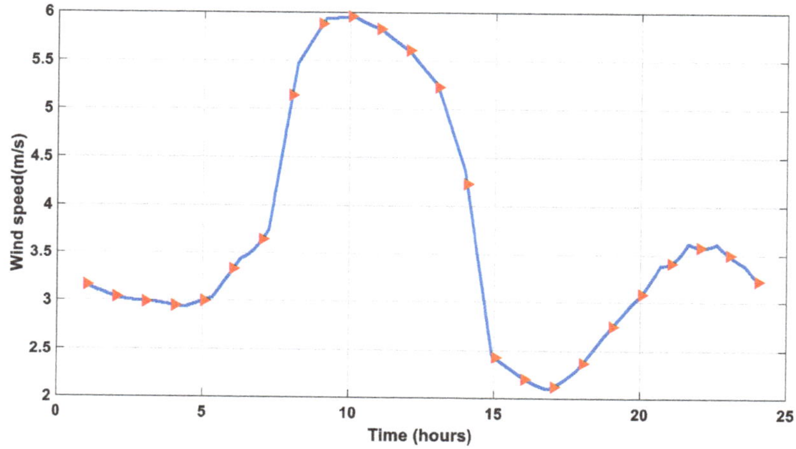

Figure 2. Wind speed daily profile.

2.3. Model of an Electric Vehicle

In order to model the EVs, three elements should be considered which are the expected mileage per day, the consumed energy per mile, and the wait time spent in the station. The first one can be simulated through lognormal distribution [43]; the lognormal distribution probability density function (PDF) can be computed as

$$f(x) = \frac{1}{\sqrt{2\pi}\sigma x} exp\left(-\frac{(ln(x) - \mu)^2}{2\sigma^2}\right),\ x > 0 \tag{6}$$

where x is a random number with one variance and zero mean, μ and σ denote the location and scaling parameters, respectively, and they can be calculated as follows:

$$\mu = ln\left(\frac{m}{\sqrt{1 + \frac{v}{m^2}}}\right),\ \sigma = \sqrt{ln\left(1 + \frac{v}{m^2}\right)} \tag{7}$$

where m and v represent the standard deviation and mean created via historical data. The expected mileage per day can be expressed as follows:

$$M_d = e^{(\mu_m + \sigma_m \times \sqrt{-2 \times lnc_1} \times cos(2\pi c_2))} \tag{8}$$

where σ_m and μ_m are the parameters of lognormal probability distribution, respectively, and c_1 and c_2 are random variables that follow the normal distribution; they are in the range of [0, 1]. The values of σ_m and μ_m can be calculated with the aid of the standard deviation (σ_{md}) and mean (μ_{md}) of EV mileage statistical data as follows:

$$\mu_m = ln\left(\frac{\mu_{md}^2}{\sqrt{\mu_{md}^2+\sigma_{md}^2}}\right), \sigma_m = \sqrt{ln\left(1+\frac{\sigma_{md}^2}{\mu_{md}^2}\right)} \tag{9}$$

The second important parameter that should be considered while modeling the EV is the consumed energy per mile, it can be computed as [44]

$$E_m = \alpha \times K_{EV}^b \tag{10}$$

where α and b represent the EV model constant coefficients and K_{EV} is the total energy supplied via battery. The EV can travel the maximum mileage (M_{dMax}) with a fully charged battery through the following formula:

$$M_{dMax} = \frac{B_{Cap}}{E_m} \tag{11}$$

where B_{Cap} is the battery capacity, and the charging demand can be computed as follows:

$$E_d = \begin{cases} B_{Cap} & M_d \geq M_{dMax} \\ M_d \times E_m & M_d < M_{dMax} \end{cases} \tag{12}$$

A Gaussian distribution can be used to calculate the waiting time spent in the station as follows [45]:

$$t_a = \mu_a + \sigma_a \cdot x_1, t_d = \mu_d + \sigma_d \cdot x_2 \tag{13}$$

$$t_{dur} = t_d - t_a \tag{14}$$

where t_a, t_d, and t_{dur} are arrival, departure, and charging duration times, respectively, σ_a, σ_d, μ_a, μ_d are standard deviations and means of entrance/leaving of EV to/from the station, and x_1 and x_2 are random numbers with one variance and zero mean.

The required state of charge ($SOC^{desired}$) of EV battery can be calculated as follows [44]:

$$SOC^{desired} = min\left\{\left(SOC^{init} + \frac{E_d}{B_{Cap}}\right), \left(SOC^{init} + \frac{t_{dur}}{B_{Cap}} \cdot r_{ch}\right)\right\} \tag{15}$$

where SOC^{init} and r_{ch} are the battery's initial state of charge and charging rate, respectively. In this study, four EVs with specifications given in Table 2 are considered [44]. The layout of the considered model is shown in Figure 3.

Table 2. The specifications of the four considered EVs.

Vehicle Model	Honda Accord	Toyota Prius	Chevrolet Volt	Ford Fusion
Consumed power	29 kW/mile	29 kW/mile	36 kW/mile	34 kW/mile
Distance with battery capacity	13 miles	11 miles	37 miles	21 miles
Capacity of battery	6.6 kWh	4.4 kWh	16 kWh	7.6 kWh
Maximum rate of charge	6.6 kW	3.5 kW	3.5 kW	3.5 kW

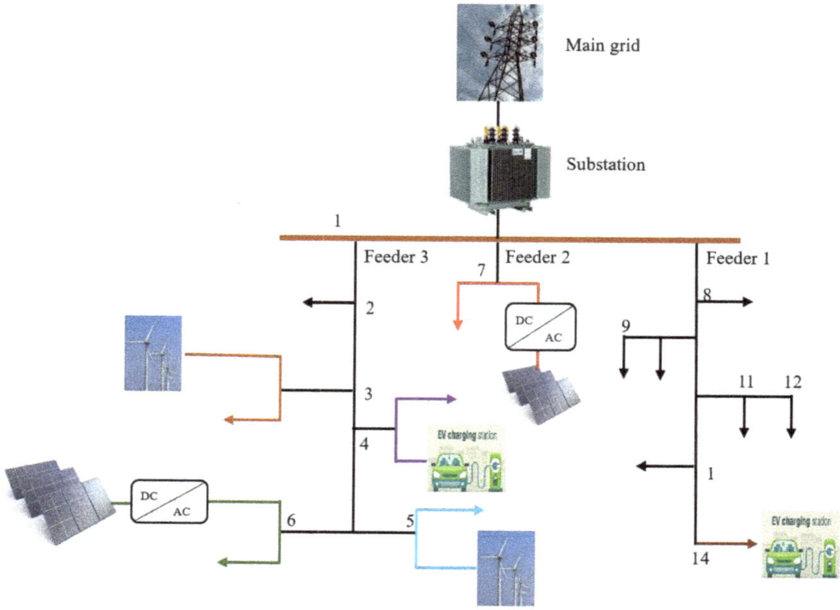

Figure 3. Layout of the considered model.

3. Form of Optimization Problem

In this section, the objective function and constraints are established to allocate and seize the renewable energy DGs and EV charging stations. Minimizing the total power loss of the network and reducing the voltage violation are the two considered objective functions. The accompanied constraints are supply-demand balance, limits of bus voltage, thermal constraint, generation limits, and constraints related to EVs.

3.1. Network Power Loss

The active power loss of the network is considered as the first target, it can be written as follows:

$$f_1 = Minimize\ P_{loss} \tag{16}$$

$$P_{loss} = \sum_{t=1}^{24} \sum_{i=1}^{n_b} \sum_{j>1}^{n_b} Y_{ij} \left(V_{i,t}^2 + V_{j,t}^2 + 2V_{i,t}V_{j,t}cos\left(\delta_{i,t} - \delta_{j,t}\right) \right) \tag{17}$$

where n_b is the number of branches, $V_{i,t}$ and $V_{j,t}$ are the magnitudes of voltage at buses i and j during time t, respectively, $\delta_{i,t}$ and $\delta_{j,t}$ are the voltages' angles at buses i and j, respectively, and Y_{ij} is the admittance of feeder $i-j$. The mitigation of the network losses is the required target from the distribution system operator perspective.

3.2. Network Voltage Violation

The second target is reducing the network voltage violation; the penetrations of RESs and FCSs to the grid may increase the capacity of supply part, this helps in reducing the losses and enhancing the voltage violation. This can be expressed as follows:

$$f_2 = Minimize \sum_{t=1}^{24} \sum_{i=1}^{n_b} |1 - V_{i,t}| \tag{18}$$

3.3. Constraints

Balance of supply-demand, limits of bus voltage, thermal limits, generation limits, and constraints related to EVs are five constraints considered in the formulated problem.

3.3.1. Supply-Demand Balance

This constraint is given by load flow analysis, the supplied power at each bus should be equal to the demand power plus the power losses of the branches connected to this bus. This can be written as follows:

$$P_{gi,t} - P_{di,t} = P_{chi,t} + |V_{i,t}| \sum_{j=1}^{n_b} |Y_{ij}| |V_{j,t}| cos(\delta_{i,t} - \delta_{j,t} - \theta_{ij}) \quad (19)$$

$$Q_{gi,t} - Q_{di,t} = |V_{i,t}| \sum_{j=1}^{n_b} |Y_{ij}| |V_{j,t}| sin(\delta_{i,t} - \delta_{j,t} - \theta_{ij}) \quad (20)$$

where $P_{gi,t}$, $P_{di,t}$, and $P_{chi,t}$ are the generated, demand, and EV charged active powers at bus i during time t, respectively, $Q_{gi,t}$ and $Q_{di,t}$ are the generated and demand reactive powers at bus i during time t, respectively, and θ_{ij} is the angle of Y_{ij}.

3.3.2. Bus Voltage Constraint

During integrating charging station and RESs, the bus voltage should be kept inside its normal limits as follows:

$$V^{min} \le V_{i,t} \le V^{min} \quad (21)$$

$$\delta^{min} \le \delta_{i,t} \le \delta^{min} \quad (22)$$

where min and max denote minimum and maximum values.

3.3.3. Thermal Constraint

Integrating the EV to the grid increases the transmission line power flow, therefore the temperature of lines will raise, the power flow should not exceed the allowable range, this can be written as follows:

$$|S_{i,t}| \le |S_i^{max}|,\ i = 1,\ 2, \ldots,\ n_b \quad (23)$$

where $S_{i,t}$ is the power flow in line i at time t while S_i^{max} is the maximum allowable flow in line i.

3.3.4. Generation Limit

The generated power from renewable energy DGs should be in its normal limits as follows:

$$P_{RES}{}^{min} \le P_{RESi,t} \le P_{RES}{}^{max} \quad (24)$$

where $P_{RESi,t}$ is the output power from RES installed at bus i during time t, and $P_{RES}{}^{min}$ and $P_{RES}{}^{max}$ denote the minimum and maximum generated powers from RES, respectively.

3.3.5. EV Constraint

The power required by the EV should be inside min and max limits as follows:

$$P_{EV}{}^{min} \le P_{EVi,t} \le P_{EV}{}^{max} \quad (25)$$

where $P_{EVi,t}$ is the output power from EV connected to bus i at time t, and $P_{EV}{}^{min}$ and $P_{EV}{}^{max}$ represent the minimum and maximum required powers by EV, respectively.

4. The Basics of the Red Kite Optimization Algorithm

The red kite optimization algorithm (ROA) is a novel metaheuristic approach introduced by Gahruei et al. [46]; it was inspired by the red kites' social life. The red kites usually build nests near lakes and wooded areas that are suitable for hunting. They live together, with random movements, and are affected by each other's positions during flight, and they use high speed while hunting. They have voices, called the sound of unity, that have been generated in times like finding good bait, water source, migration, and birth. Also, the sounds that occur in times of danger such as enemy attack, death of another animal, earthquake, and storm are known as the sound of danger. To simulate the behavior of a red kite in finding food, each bird can be defined through its position, value of evaluation function, amount of displacement of points, sound of danger (in the direction of the individual component), sound of unity (in the direction of the social component), new position of the bird, and new evaluation function. In order to obtain good results, the metaheuristic algorithm must first navigate the problem search space well to prevent trapping in local optima. Then it gradually moves from the exploration to exploitation phase and exploits the best solution in the last iterations. ROA has three main stages which are explained as follows:

1. **The first stage—the initial position of the birds:** In this stage, according to Equation (26), the position of red kites can be initialized randomly as,

$$Pos_{i,j}(t) = lb + rand \times (ub - lb),\ i = 1,\ 2, \ldots, n \ and \ j = 1, 2, \ldots, d \tag{26}$$

   where $Pos_{i,j}(t)$ is i^{th} red kite's position at iteration t, lb and ub are lower and upper boundaries, respectively, n is size of population, d denotes problem dimension, and $rand$ is a random number in [0, 1].

2. **The second stage—selection of the leader:** Selecting the leader is obtained according to Equation (27):

$$\overrightarrow{Best(t)} = \overrightarrow{Pos_i(t)} \ if \ f_i(t) < f_{best}(t) \tag{27}$$

   where $Best(t)$ denotes position of the best bird in iteration t, $Pos_i(t)$ denotes the position of ith red kite in iteration t, $f_i(t)$ is value of the bird evaluation function in iteration t, and $f_{best}(t)$ is the value of the evaluation function of the best bird in iteration t.

3. **The third stage—the movement of the birds:** It is considered that red kites must gradually move from exploration phase to exploitation stage through considering decreasing coefficient (D) according to Equation (28).

$$D = \left(exp\left(\frac{t}{t_{\_max}} \right) - \frac{t}{t_{\_max}} \right)^{-10} \tag{28}$$

   where t is the current iteration and $t_{\_max}$ denotes the maximum iteration.

   The birds update their positions through Equations (29) and (30):

$$\overrightarrow{pos_i^{new}(t+1)} = \overrightarrow{Pos_i(t)} + \overrightarrow{P_{mi}(t+1)} \tag{29}$$

$$\overrightarrow{P_{mi}(t+1)} = D(t) \times \overrightarrow{P_{mi}(t)} + \overrightarrow{SC(t)} \odot \left(\overrightarrow{Pos_{rws}(t)} - \overrightarrow{Pos_i(t)} \right) + \overrightarrow{UC(t)} \odot \left(\overrightarrow{Best(t)} - \overrightarrow{Pos_i(t)} \right) \tag{30}$$

where $Pos_{rws}(t)$ is the bird position selected by roulette wheel in iteration t, $pos_i^{new}(t+1)$ denotes the new position of the bird, and SC and UC are random vectors of social and individual components, respectively. After updating the position, it is important to check the search space boundaries, this can be conducted using Equation (31) as,

$$\overrightarrow{pos_i^{new}(t+1)} = max\left(min\left(\overrightarrow{pos_i^{new}(t+1)} + ub \right), lb \right) \tag{31}$$

The new temporary position will be replaced if the evaluation function is improved. In such case, $Pos_i(t+1)$ is equal to $pos_i^{new}(t+1)$. As mentioned, SC and UC are random vectors of social and individual components, they represent the voice of unity and danger of each bird, and they are obtained according to the following relation:

$$\begin{cases} \overrightarrow{SC(t+1)} = \vec{r_1} \\ \overrightarrow{UC(t+1)} = \vec{r_2} \end{cases} if\ rand \leq 0.5$$

$$\begin{cases} \overrightarrow{SC(t+1)} = \vec{r_3} \\ \overrightarrow{UC(t+1)} = \vec{r_1} \end{cases} Otherwise \tag{32}$$

where $\vec{r_1}$ is a random vector in [1, 2], $\vec{r_2}$ is a random vector in [1, 3], and $\vec{r_3}$ is a random vector in [0, 1].

In the ROA, based on the current position of each bird, the position of a neighbor is randomly chosen via a roulette wheel and the best solution found so far. In the early iterations, the value of $D(t)$ is close to one for exploring and searching new spaces. In the movement based on the individual component, the red kite explores new spaces based on its position and that of a randomly selected neighbor. The social component also leads the algorithm to global optimum. Gradually, as the algorithm moves from the initial iterations to intermediate iterations, the coefficient $D(t)$ decreases to achieve balance between the exploration and exploitation phases. In the final iterations, this coefficient tends to zero and the algorithm exploits searching for the best solution among the obtained good solutions. The ROA is characterized by its ease in structure and execution, also it has few controlling parameters and a high convergence rate. The flowchart of the ROA is shown in Figure 4.

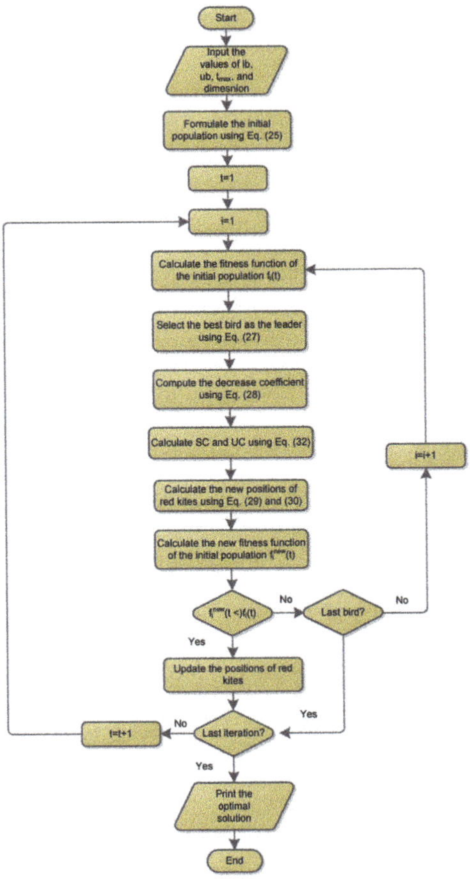

Figure 4. ROA flowchart.

5. The Proposed ROA-Based Methodology

This work proposes a new approach of the ROA to identify the optimal sites and sizes of RESs and FCSs in a radial distribution network. The considered objective functions are selected based on the distributor operator's point of view, and are mitigating the network power loss and minimizing the voltage fluctuation. Also, a new multi-objective ROA is proposed to achieve both targets. The assigned memory of the problem is divided into four vectors that represent the locations and sizes of RESs and FCSs as shown in Figure 5. The process of updating followed in the ROA using Equations (29) and (30) is adapted such that the first and third vectors of the variables have integer numbers assigned to the best sites of both RESs and FCSs. The most key features that characterize the proposed ROA are simplicity of the construction, need of few controlling parameters, and balance between exploration and exploitation phases. These features enhance the convergence rate of the algorithm and prevent falling in local optima. All these merits encourage the authors to apply the ROA in solving the presented problem. The proposed ROA pseudo code assigned to solve the single objective problem is given in Algorithm 1.

Algorithm 1 The proposed ROA pseudo code to solve the single objective optimization problem.

```
Define the ROA parameters like max iteration (t_max), size of population (n), d, lb, ub, and
number of runs (n_run).
Input the load data and line data of the network under study.
Conduct load flow analysis and keep the voltage fluctuation and power loss.
Formulate the initial population using Equation (26).
    for i = 1: n
    Integrate Pos_i in the network, where Pos_i is the probable solution from the population.
    Conduct power flow for the network with integrating Pos_i.
    Compute the initial evaluation function (f_i(Pos_i)).
    end for
while k > n_run do
  for t > t_max do
    for i = 1: n
        Calculate the values of SC, UC, and D using Equations (28) and (32).
        Calculate the red kites' new positions using Equations (19) and (30).
        Check the positions' limits using Equation (31).
        Compute the new objective function (f_i^t(pos_i^new).
     if f_i^t(pos_i^new) > (f_i^(t-1)(Pos_i)
      Update Pos_i bu pos_i^new
      end if
         i = i + 1
     end for
       t = t + 1
    end for
 end for
   k = k + 1
end while
Save the optimal places and sizes of RESs and FCSs.
```

RESs' sites			RESs' sizes (kW)			FCSs' sites			FCSs' sizes (kW)		
L_{r1}	...	L_{rn}	P_{r1}	...	P_{rn}	L_{c1}	...	L_{cm}	P_{c1}	...	P_{cm}

Figure 5. The proposed ROA memory.

A multi-objective red kite optimization algorithm (MOROA) is proposed to minimize both power loss and voltage violation, two components of archiving and hunting the food are proposed in MOROA. The first one saves the nondominant solutions achieved so far while the other component selects the best one from the obtained archive. Moreover, the solution entrance to the archive is controlled via considering the archive controller.

When the new solution is governed by one archived solution, it should be excluded from the archive entering. On the other hand, if the new solution is governed via one or more archived solutions, it will be included in the archive and the governed solutions are ignored. Also, when recent solutions and archive members have no control between them, it must be included in the archive. The top solution is chosen from the archive using the roulette wheel method as follows:

$$P_i = \frac{C}{N_i} \tag{33}$$

where C is a constant with a value greater than unity and N_i is the number of pareto solutions.

6. Numerical Analysis and Discussions

The analysis was performed on two standard distribution systems, which are the IEEE-33 bus network and the IEEE-69 bus network; the proposed ROA was simulated for 100 iterations, 50 population sizes, and 10 independent runs [47]. The maximum generation of RESs (PV and WT) was 1000 kW, and 1500 kW for FCS [44]. Three cases were studied in each network, the first one was minimizing power loss whereas the second one was mitigating network voltage fluctuation. The last case was a multi-objective to reduce both power loss and voltage fluctuation.

6.1. IEEE-33 Bus Network

The network single line diagram is shown in Figure 6; the network had 32 branches and 33 nodes, its nominal voltage was 12.66 kV while 100 MVA was the base power. In this network, it was assumed that two RESs were required to be installed, the first one was PV and the second one was WT. Also, two FCSs were integrated to the network to serve 200 vehicles selected randomly from Table 2 over 24 h. The base loads were 3.715 MW and 2.3 MVar while the network losses were 3905.628 kW and 2604.031 kVar. Figure 7 shows the demand level as a percentage of the base demand during each hour. The proposed ROA was implemented, and the fetched results were compared to other approaches of the dung beetle optimizer (DBO) [48], the African vultures optimization algorithm (AVOA) [49], the bald eagle search (BES) algorithm [50], the bonobo optimizer (BO) [51], and the grey wolf optimizer (GWO) [52] which are programmed. The analysis was performed using a laptop with specifications of 11th Gen Intel(R) Core(TM) i7-11370 @3.30 GHz processor, and 16.00 GB RAM.

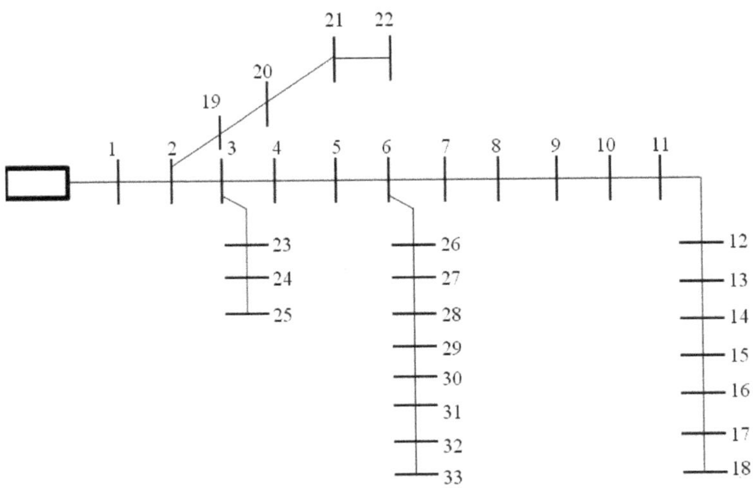

Figure 6. IEEE-33 bus network single line diagram.

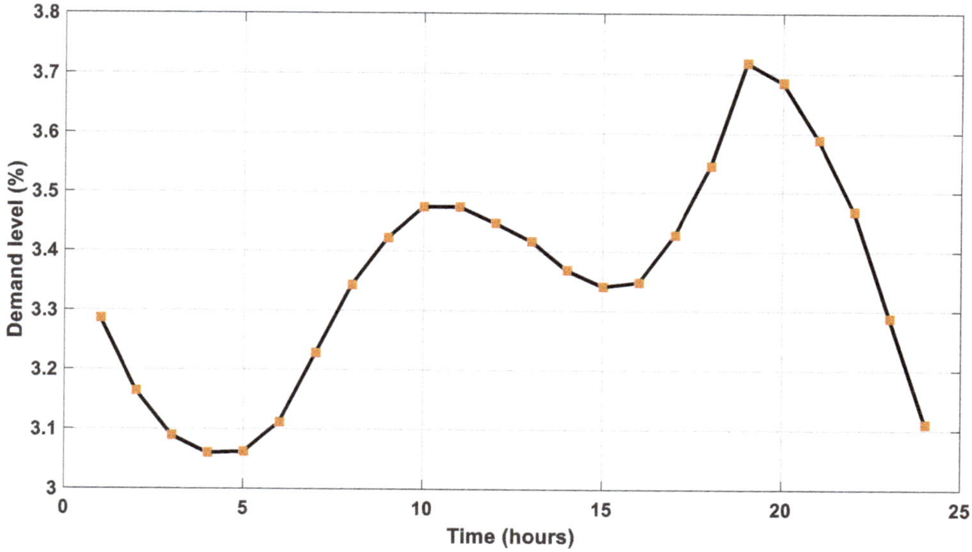

Figure 7. The daily load demand of the IEEE-33 bus network.

The fetched results during minimizing the power loss are given in Table 3. The proposed ROA achieved the best power loss over 24 h with a value of 1631.1189 kW obtained with integrating a RES of 770.3162 kW and 1126.969 kW on buses 13 and 30, respectively, and a FCS of 63.34486 kW and 50.23074 kW on buses 30 and 2, respectively. In such case, the voltage violation of the network was 14.5663 pu. On the other hand, BO came second with a power loss of 1633.4916 kW through integrating RESs and FCSs of 1072.105 kW, 870.675 kW, 103.400 kW, and 103.950 kW on buses 30, 13, 2, and 13, respectively. The highest power loss was 1716.946 kW, obtained via AVOA. The results proved the preference of the proposed ROA in such case. Another important item that is considered in comparison is the computational time required to implement one run, it is clear that the proposed ROA is the fastest one as it consumed 64.569 s., whereas the slowest one is BES with 131.489 s. The power loss versus number of iterations is shown in Figure 8. The voltage profiles of the network throughout minimizing the power loss are shown in Figure 9. The proposed ROA achieved good improvement in the voltage profile, being better than the original network.

Table 3. The optimal results throughout minimizing the first objective function of the IEEE-33 bus network.

	DBO	AVOA	BES	BO	GWO	ROA
RES (kW)/location	1172.455/(30) 768.3386/(13)	1415.316/(8) 841.6776/(30)	767.7161/(13) 1073.831/(30)	1072.105/(30) 870.675/(13)	796.258/(13) 1179.83/(30)	770.3162/(13) 1126.969/(30)
FCS (kW)/location	99.000/(33) 107.05/(2)	109.800/(26) 107.625/(19)	104.8500/(2) 112.6125/(19)	103.400/(2) 103.950/(13)	79.4101/(6) 95.9678/(30)	63.34486/(30) 50.23074/(2)
Active power loss (kW)	1650.078	1716.946	1641.0623	1633.4916	1652.1002	1631.1189
Reactive power loss (kVar)	1067.1	1104.5	1060.6	1067.3	956.48116	947.36830
Vmin (pu)/location	0.9718/(33)	0.9609/(18)	0.9724/(33)	0.9692/(18)	0.9707/(33)	0.9708/(33)
Vmax (pu)/location	1.000/(1)	1.000/(1)	1.000/(1)	1.000/(1)	1.000/(1)	1.000/(1)
Voltage deviation (pu)	14.1639	15.2245	14.2351	15.2438	14.5806	14.5663
Time (s)	92.287	89.6085	247.461	66.547	131.489	64.569

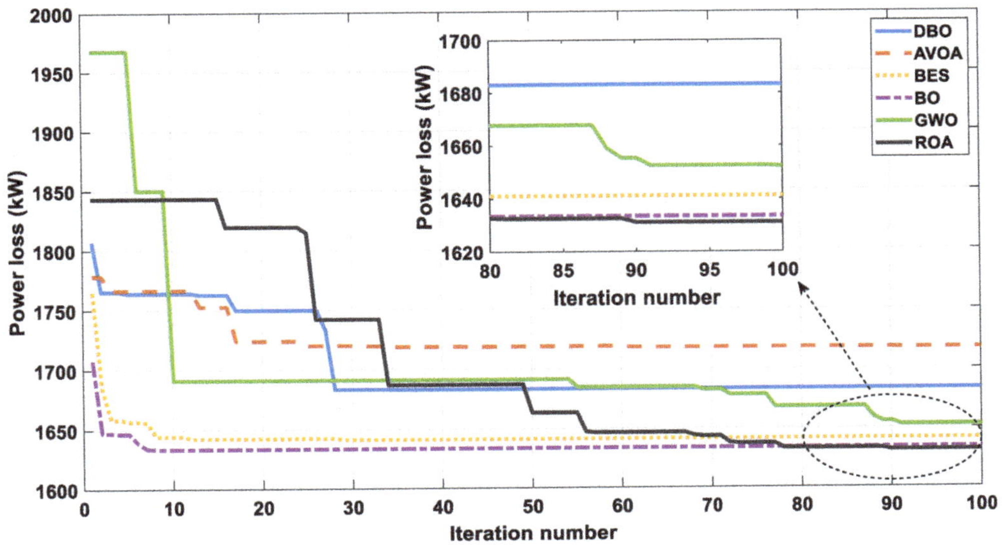

Figure 8. Power loss variation throughout minimizing the first objective function of the IEEE-33 network.

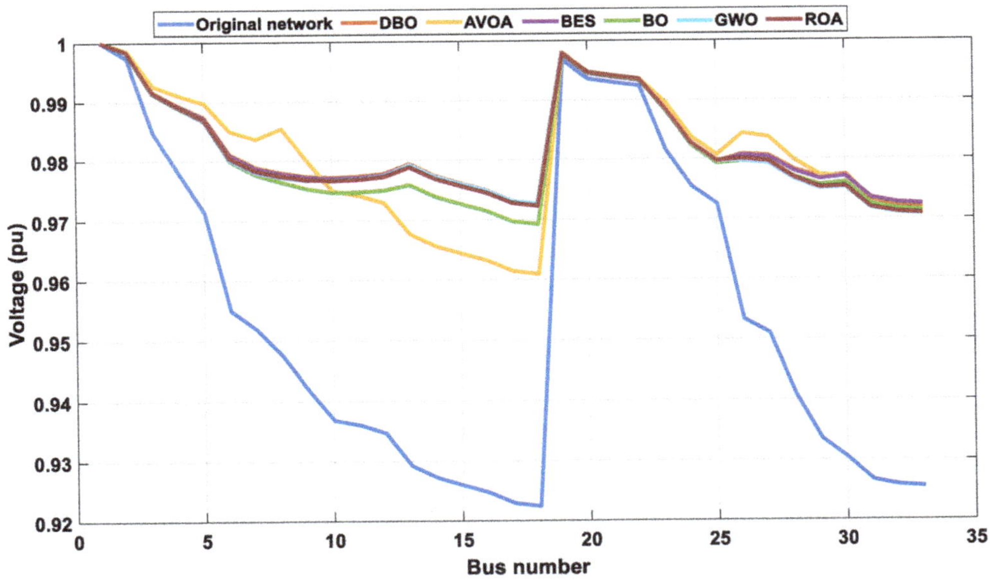

Figure 9. The network voltage profile throughout minimizing the first objective function of the IEEE-33 network.

The second fitness function minimized the voltage violation; the original network had a voltage violation of 36.477 pu. The fetched results in such case are given in Table 4, the proposed ROA accomplished the best voltage violation of 3.4762 pu, about a 90.47% enhancement of the original network, by installing RESs of 1499.994 kW and 1500 kW on buses 10 and 30, respectively, as well as 50 kW and 79.68228 kW FCSs on buses 2 and 10, respectively. This integration resulted in an active power loss of 1643.5811 kW. AVOA was the worst optimizer with a voltage deviation of 3.7866 pu. The voltage fluctuation versus number of iterations is shown in Figure 10, while the voltage patterns of the network are displayed in Figure 11, the profile clarified significant improvement of the network voltages. The proposed ROA was the best optimizer compared to the others in achieving the least voltage deviation.

Table 4. The optimal results throughout minimizing the second objective function of the IEEE-33 bus network.

	DBO	AVOA	BES	BO	GWO	ROA
RES (kW)/location	1498.299/(10) 1500/(31)	1449.13/(11) 1500/(31)	1485.9678/(10) 1500/(30)	1500/(10) 1380.22/(31)	1500/(10) 1500(20)	1499.994/(10) 1500/(30)
FCS (kW)/location	50/(2) 181.6571/(33)	89.58208/(30) 82.92387/(33)	50/(33) 50/(26)	50/(2) 52.7957/(23)	74.7382/(10) 121.3876/(3)	50/(2) 79.68228/(10)
Active power loss (kW)	1838.1837	1748.6105	1743.0893	1840.8733	1654.1723	1643.5811
Reactive power loss (kVar)	1307.9386	1246.4018	1223.3889	1307.5806	1154.3989	1148.2444
Vmin (pu)/location	0.9833/(25)	0.9833/(25)	0.9837/(25)	0.9833/(25)	0.9828/(25)	0.9831/(25)
Vmax (pu)/location	1.0024/(10)	1.0026/(11)	1.0036/(10)	1.0025/(10)	1.0/(1)	1.0/(1)
Voltage deviation (pu)	3.7273	3.7866	3.5632	3.7378	3.5399	3.4762

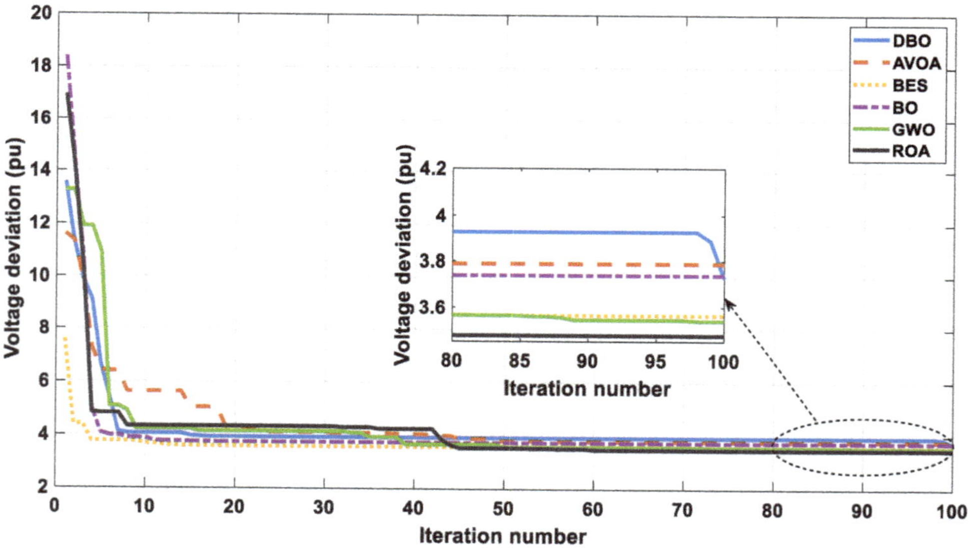

Figure 10. The voltage deviation versus the number of iterations throughout minimizing the second objective function of the IEEE-33 network.

The third case involved solving a multi-objective problem to minimize power loss and volage fluctuation; the proposed MOROA was compared to the multi-objective grey wolf optimizer (MOGWO), the multi-objective multi-verse optimizer (MOMVO), and the multi-objective artificial hummingbird algorithm (MOAHA) [47]. The optimal results are given in Table 5, RESs with 994.2378 kW and 1472.334 kW, and FCSs with 128.1094 kW and 165.3984 kW are recommended to be installed via the proposed MOROA on buses 13, 30, 2, and 30, respectively. This integration achieved active power loss and voltage violation of 1763.93 kW and 6.6547 pu, respectively, while MOAHA achieved the worst power loss and voltage violation of 1829.26 kW and 6.7704 pu, respectively. The results demonstrated the superiority of the proposed MOROA over the others. Moreover, the variations of both targets with number of iterations obtained via the proposed approach are given in Figure 12. Furthermore, the network voltage profiles before and after installing RESs and

FCSs are shown in Figure 13. The curves revealed that there is significant improvement in the network voltage profile.

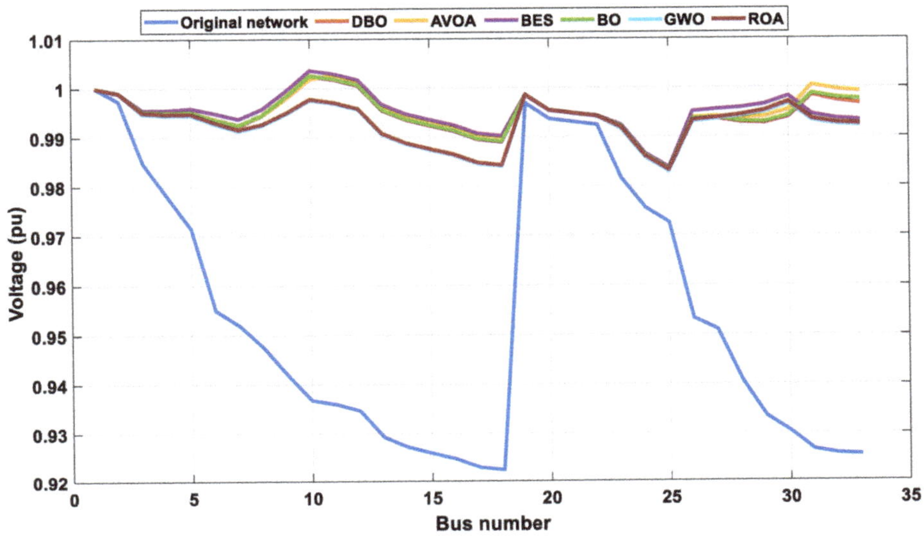

Figure 11. The network voltage profile throughout minimizing the second objective function of the IEEE-33 network.

Table 5. The optimal results throughout solving the multi-objective problem for the IEEE-33 bus network.

	MOAHA [47]	MOMVO	MOGWO	MOROA
RES (kW)/location	1475.0424/(30) 1073.4234/(15)	885.715/(14) 1465.69/(30)	1283.274/(11) 1264.667/(30)	994.2378/(13) 1472.334/(30)
FCS (kW)/location	63.763105/(14) 188.82647/(17)	51.3188/(3) 92.0779/(2)	190.1696/(2) 96.09850/(10)	128.1094/(2) 165.3984/(30)
Active power loss (kW)	1829.26	1810.31	1801.96	1763.93
Voltage deviation (pu)	6.7704	6.2317	6.4819	6.6547
Reactive power loss (kVar)	1152.29	1122.65	1108.20	1025.81
Vmin (pu)/location	0.9813/(25)	0.9813/(25)	0.9802/(18)	0.9798/(33)
Vmax (pu)/location	1.0/(1)	1.0/(1)	1.0/(1)	1.0/(1)

The obtained results confirmed the efficiency of ROA as it achieved the best fitness values for all scenarios investigated on the IEEE-33 bus network.

6.2. IEEE-69 Bus Network

The proposed ROA was also applied on the IEEE-69 bus system, it consisted of 68 branches and 69 nodes, the network nominal voltage was 12.66 kV and the base power was 100 MVA. The single line diagram of the IEEE-69 bus system is shown in Figure 14. The demand and branch data of the network were given in [53], the network is loaded by 24 h demand level given in Figure 7, the active power loss was 8665.356 kW whereas the reactive power loss was 3938.366 kVar. It was assumed that three renewable DGs were integrated in addition to three FCSs.

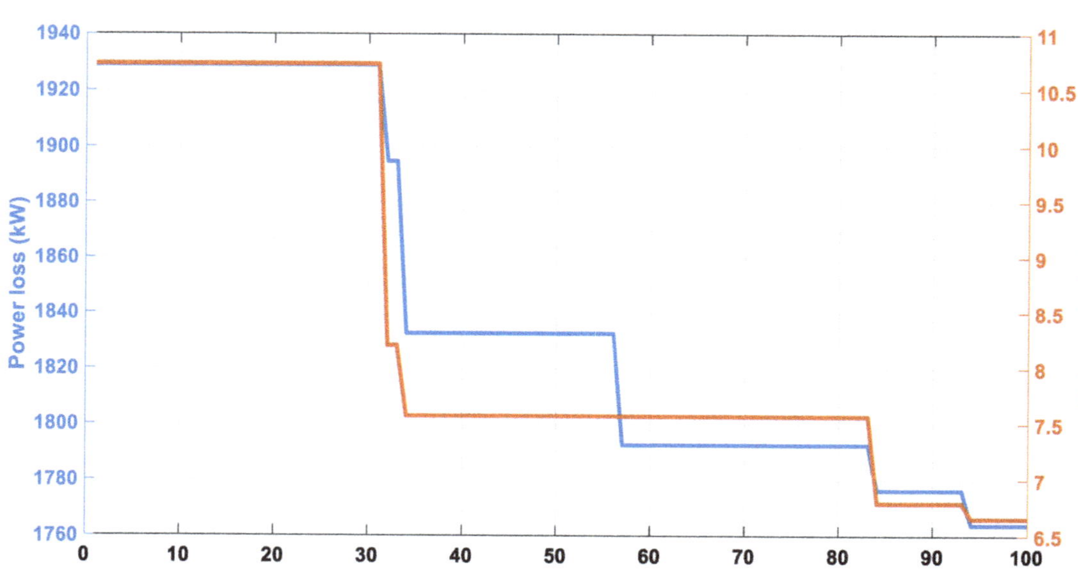

Figure 12. Variations of power loss and voltage violation throughout solving the multi-objective problem for the IEEE-33 network.

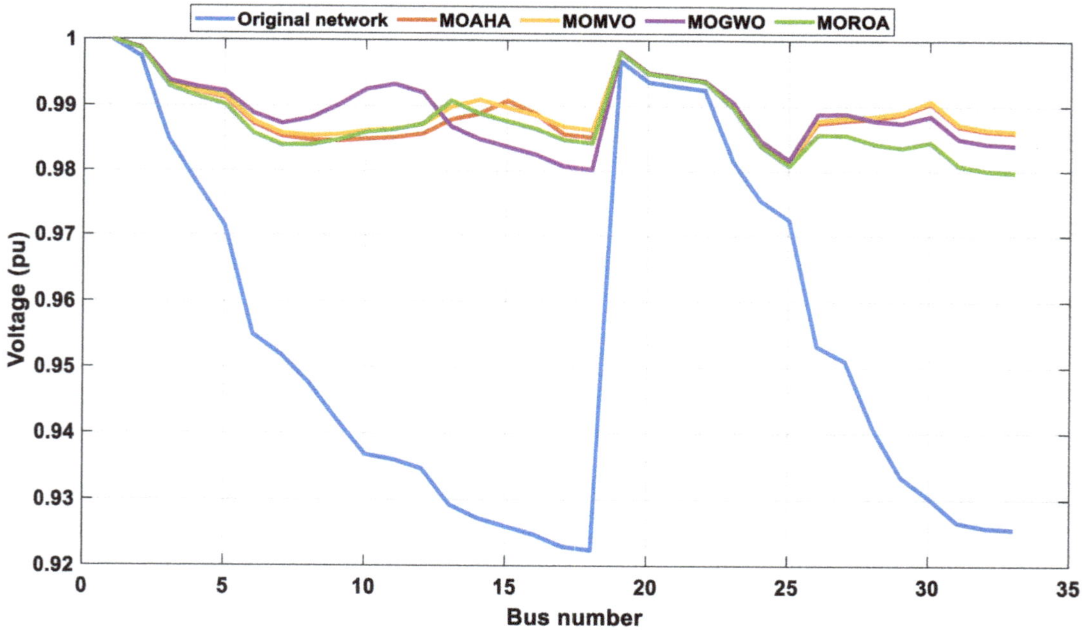

Figure 13. The network voltage profile throughout solving the multi-objective problem for the IEEE-33 network.

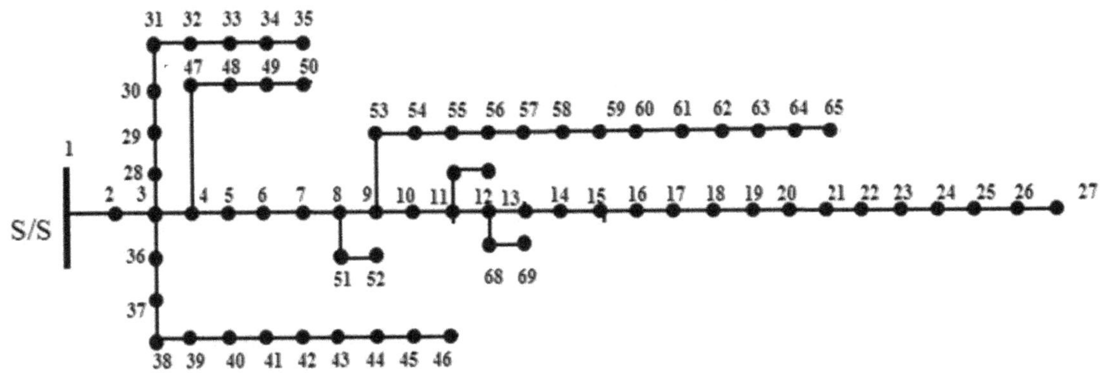

Figure 14. IEEE-69 bus network single line diagram.

The fetched results obtained by ROA and others throughout minimizing the network power loss are given in Table 6. The best loss was 2738.731 kW, achieved through the proposed ROA via installing RESs of 1500 kW, 663.715 kW, and 502.0473 kW on buses 61, 69, and 19, respectively, in addition to FCSs of 156.5290 kW, 50 kW, and 346.3399 kW on buses 19, 4, and 69, respectively. This integration reduced the active power loss by 68.39% compared to the original network. BO came in the second rank with a power loss of 2742.766 kW, while the worst one was 2905.728 kW, obtained by AVOA. Moreover, the proposed ROA required 62.078 s. to implement one run, this was the best obtained time. Figure 15 shows the variations of power losses during iterative process followed in each optimizer. Moreover, the voltage profile of the network is shown in Figure 16, it is clear that the voltage pattern is improved after integrating RESs and FCSs with sizes and sites obtained via the proposed ROA. The fetched results proved the superiority of ROA in minimizing the IEEE-69 bus system power losses.

Table 6. The optimal results throughout minimizing the first objective function of the IEEE-69 bus network.

	DBO	AVOA	BES	BO	GWO	ROA
RES (kW)/location	1500/(61) 518.641/(17) 54.7547/(14)	1500/(61) 295.224/(6) 813.401/(10)	713.821/(17) 584.281/(62) 1036.10/(61)	1500/(61) 426.56/(17) 485.21/(53)	1500/(61) 47.971/(23) 575.167/(12)	1500/(61) 663.715/(69) 502.0473/(19)
FCS (kW)/location	51.0967/(18) 50/(69) 50.6284/(5)	67.2109/(4) 53.2949/(51) 54.3663/(29)	58.7486/(18) 82.7298/(47) 191.153/(17)	50.019/(53) 50/(2) 50/(47)	64.7808/(35) 196.489/(47) 173.715/(29)	156.5290/(19) 50/(4) 346.3399/(69)
Active power loss (kW)	2810.358	2905.728	2775.538	2742.766	2819.619	2738.731
Reactive power loss (kVar)	1291.968	1311.428	1274.645	1259.164	1296.032	1276.838
Vmin (pu)/location	0.9796/(65)	0.9798/(27)	0.9836/(65)	0.9819/(65)	0.9804/(65)	0.9800/(65)
Vmax (pu)/location	1.0/(1)	1.0/(1)	1.0/(1)	1.0/(1)	1.0/(1)	1.0/(1)
Voltage deviation (pu)	11.4739	13.3273	11.5038	10.4405	13.3135	12.1014
Time (s)	159.893	104.103	347.468	62.493	145.821	62.078

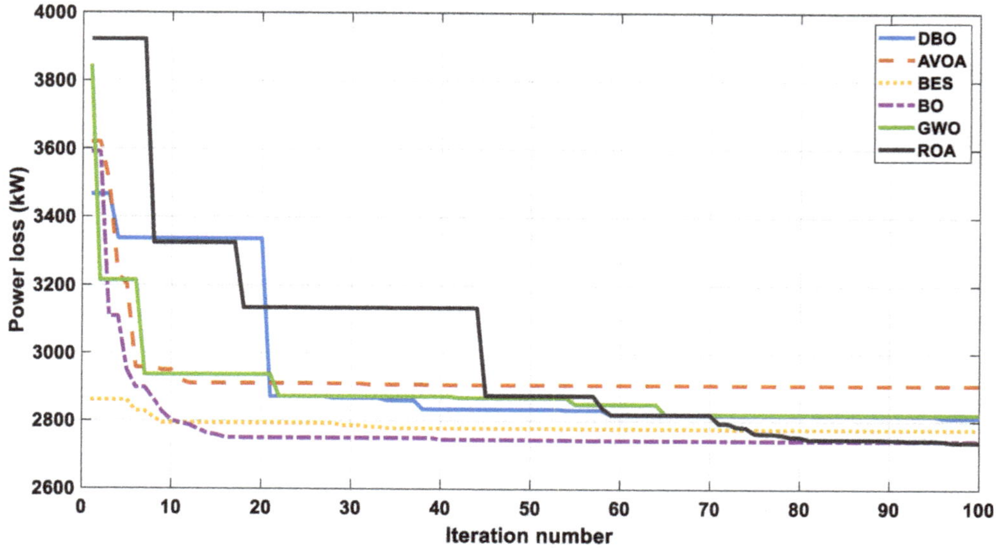

Figure 15. The power loss versus the number of iterations throughout minimizing the first objective function of the IEEE-69 network.

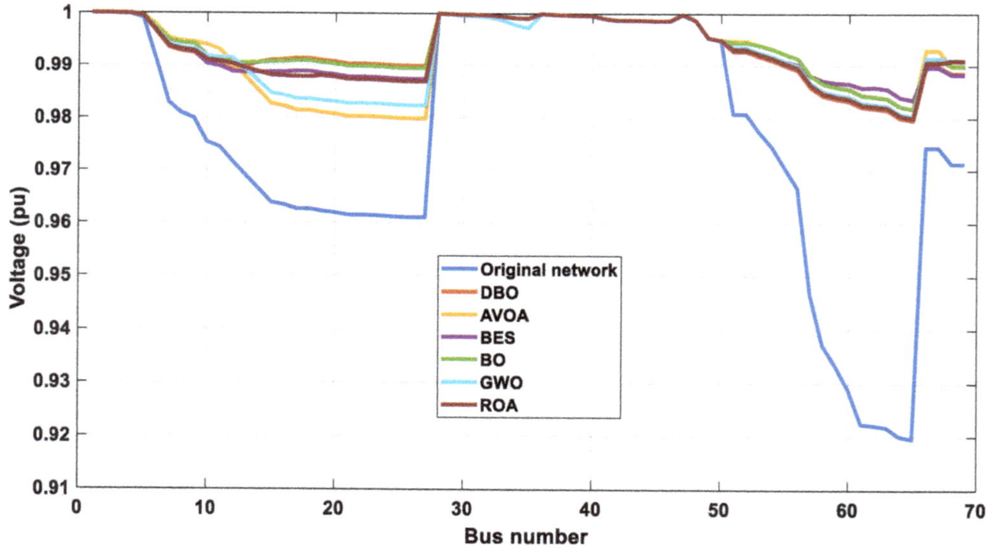

Figure 16. The network voltage profile throughout minimizing the first objective function of the IEEE-69 network.

The original network had a voltage violation of 39.229 pu; minimization of the voltage fluctuation was the second target, Table 7 tabulates the optimal fetched results in such case. The proposed ROA succeeded in mitigating the network voltage deviation to 2.6607 pu, about 93.22% enhancement of the original value, via installing RESs of 1464.69 kW, 1495.45 kW, and 891.503 kW on buses 63, 56, and 15, respectively, in addition to FCSs of 311.624 kW, 337.688 kW, and 231.826 kW on buses 7, 16, and 5, respectively. AVOA was still in the last rank, achieving a voltage deviation of 4.2701 pu. The performances of the optimizers considered are given in Figure 17, while the voltage patterns with/without the installed DGs and stations are given in Figure 18. The results demonstrated the excellence of the proposed method while reducing the voltage fluctuation of the IEEE-69 bus network.

Table 7. The optimal results throughout minimizing the second objective function of the IEEE-69 bus network.

	DBO	AVOA	BES	BO	GWO	ROA
RES (kW)/location	987.669/(63) 1152.20/(13) 1416.90/(63)	1500/(12) 1500/(64) 607.458/(54)	1499.996/(63) 1499.999/(68) 469.3114/(58)	801.0936/(64) 1500/(59) 1500/(69)	549.582/(19) 1500/(62) 1500/(55)	1464.69/(63) 1495.45/(56) 891.503/(15)
FCS (kW)/location	245.784/(43) 253.716/(64) 308.648/(56)	350/(7) 234.083/(46) 333.914/(31)	53.66301/(58) 50/(2) 163.3407/(67)	50/(36) 350/(58) 192.210/(66)	162.323/(29) 58.5327/(21) 149.570/(47)	311.624/(7) 337.688/(16) 231.826/(5)
Active power loss (kW)	3316.866	3518.4419	3534.8244	3636.6704	3176.7778	3284.3169
Reactive power loss (kVar)	1626.321	1740.1091	1659.1212	1696.3122	1602.5207	1645.6244
Vmin (pu)/location	0.9941/(56)	0.9870/(61)	0.9912/(65)	0.9922/(61)	0.9903/(65)	0.9915/(65)
Vmax (pu)/location	1.0040/(13)	1.0055/(12)	1.0096/(68)	1.0100/(69)	1.0024/(55)	1.0021/(56)
Voltage deviation (pu)	3.748	4.2701	3.4861	3.5533	2.7751	2.6607

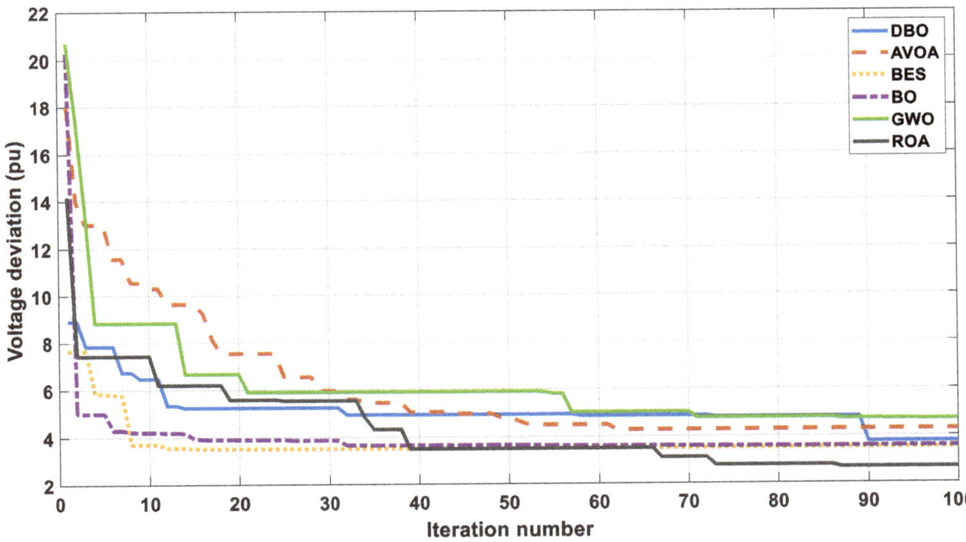

Figure 17. Voltage deviation variation throughout minimizing the second objective function of the IEEE-69 network.

Finally, the multi-objective problem for the IEEE-69 bus network was solved via the proposed MOROA in comparison to others, the fetched results are tabulated in Table 8. The best power loss and voltage violation were 2929.075 kW and 4.3347 pu, respectively, obtained via the proposed algorithm. The MOGWO achieved the worst power loss with a value of 3351.509 kW, whereas the largest voltage deviation was 6.832 pu, obtained via MOMVO. Also, the power loss and voltage fluctuation versus the number of iterations are shown in Figure 19, while the network voltage patterns are shown in Figure 20. The proposed approach proved its preference in finding the best locations and sizes of RESs and charging stations while solving the multi-objective problem.

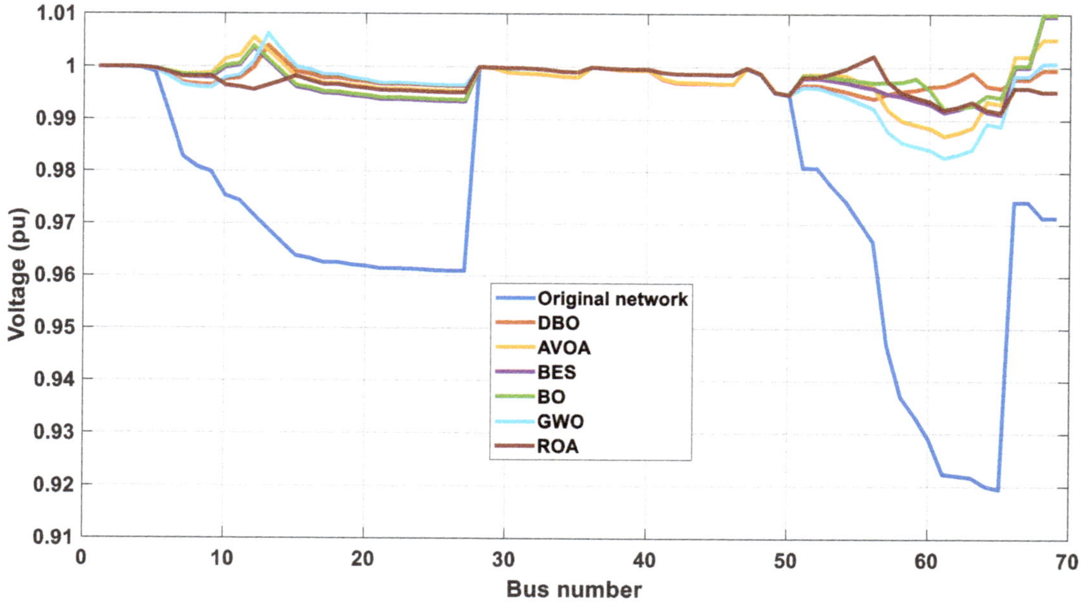

Figure 18. The network voltage profile throughout minimizing the second objective function of the IEEE-69 network.

Table 8. The optimal results throughout solving the multi-objective problem for the IEEE-69 bus network.

	MOAHA [47]	MOMVO	MOGWO	MOROA
RES (kW)/location	1469.52/(61) 916.516/(18) 301.015/(59)	1249.069/(49) 1500/(61) 855.449/(14)	1260.96/(68) 1395.23/(51) 1470.41/(62)	917.811/(9) 647.6445/(15) 1500/(61)
FCS (kW)/location	307.607/(47) 60.7376/(52) 207.741/(26)	98.1616/(34) 80.8805/(43) 207.898/(31)	305.866/(68) 307.682/(6) 296.051/(51)	248.222/(28) 135.164/(2) 336.811/(37)
Active power loss (kW)	2974.105	2960.5889	3351.509	2929.075
Voltage deviation (pu)	5.1243	6.832	5.4632	4.3347
Reactive power loss (kVar)	1294.045	1232.638	1501.993	1303.072
Vmin (pu)/location	0.9881/(65)	0.9817/(65)	0.9812/(65)	0.9856/(65)
Vmax (pu)/location	1.0002/(18)	1.0/(1)	1.0008/(68)	1.0005/(15)

A new methodology incorporating the ROA is proposed to find the best locations and capacities of RESs and FCSs in distribution systems. Power loss and voltage fluctuation are the considered targets to be minimized. Both single objective and multi-objective problems are formulated and solved via the proposed ROA. The power loss and voltage deviation of the IEEE-33 bus were reduced by 58.24% and 90.47%, respectively, with the aid of the proposed ROA. While benefits of 68.39% for losses and 93.22% for voltage deviation were achieved for the IEEE-69 bus network. Finally, the proposed ROA can be recommended as an effective tool to solve the problem of integrating RESs and EV FCSs in a radial distribution network.

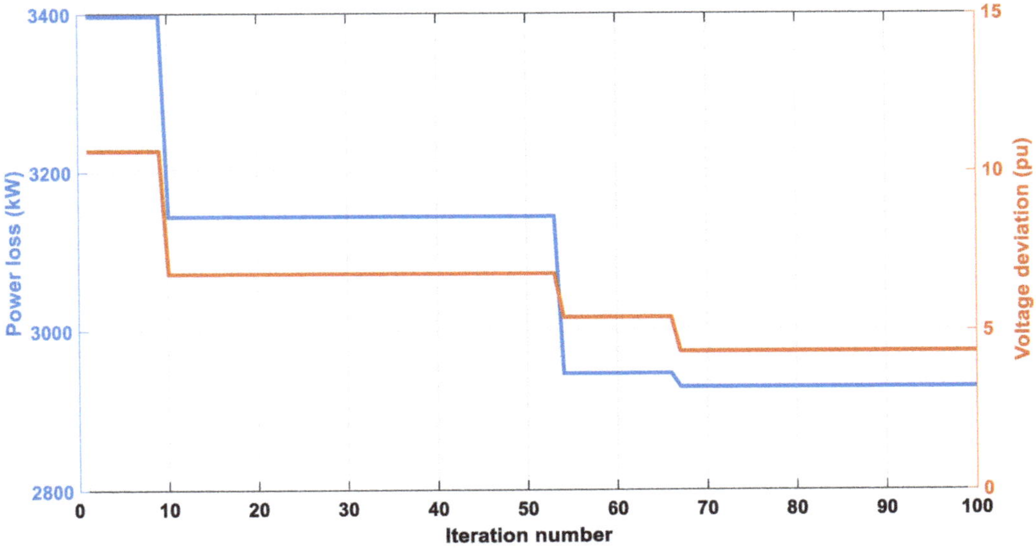

Figure 19. Variations of power loss and voltage violation throughout solving the multi-objective problem for the IEEE-69 network.

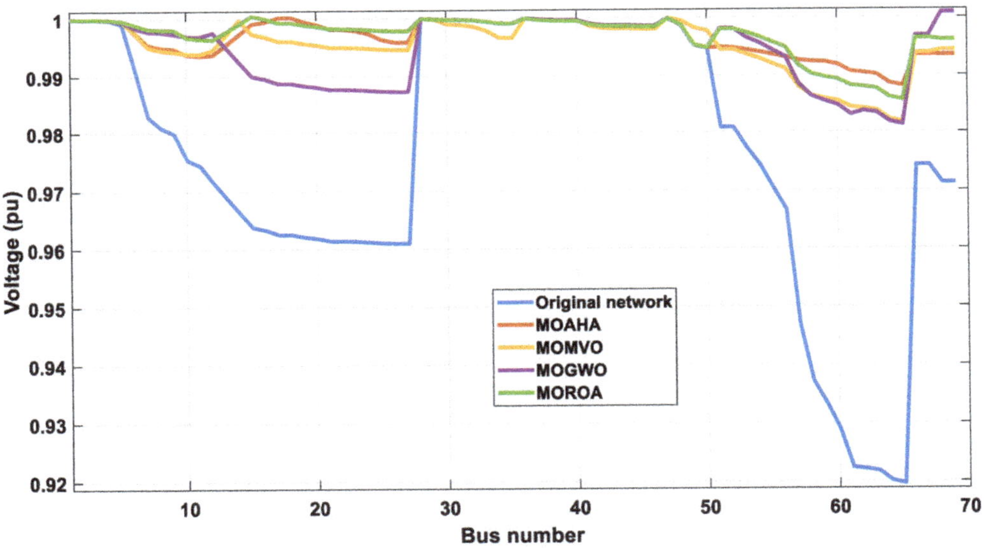

Figure 20. The network voltage profile throughout solving the multi-objective problem for the IEEE-69 network.

7. Conclusions

This paper proposed a new metaheuristic approach of the red kite optimization algorithm (ROA) to identify the best sites and sizes of RESs and FCSs in distribution networks. The ROA was selected due to its simplicity, requirement of less controlling parameters, high convergence rate, and balance between exploration and exploitation phases that enabled the algorithm to escape from local optima. The targets were minimizing the network active power loss and voltage fluctuation. Also, a multi-objective red kite optimization algorithm (MOROA) was proposed to mitigate both targets. Two standard radial distribution networks, the IEEE-33 bus and the IEEE-69 bus, were analyzed. In the first network, two RESs and two FCSs were installed while in the second network three RESs and three FCSs were integrated. Comparisons to DBO, AVOA, BES, BO, and GWO in the single objective problem in addition to MOGWO, MOMVO, and MOAHA in the multi-objective problem were conducted. The proposed ROA gave the best solution in the IEEE-33 bus, reducing the network loss and voltage violation by 58.24% and 90.47% respectively, whereas it achieved benefits of 68.39% for losses and 93.22% for voltage deviation of the IEEE-69 bus network. The obtained results revealed the robustness and competence of the proposed ROA in achieving the best results. Minimizing the period of charging for EVs will be considered as the target in future works. Also, the investigation of a real distribution network will be conducted in the next works.

Author Contributions: Conceptualization, S.M.A. and A.F.; methodology, A.F.; software, A.F.; validation, S.M.A. and A.F.; formal analysis, S.M.A.; investigation, A.F.; resources, A.F.; data curation, S.M.A.; writing—original draft preparation, A.F.; writing—review and editing, S.M.A.; visualization, S.M.A.; supervision, S.M.A.; project administration, A.F.; funding acquisition, S.M.A. All authors have read and agreed to the published version of the manuscript.

Funding: The authors extend their appreciation to the Deputyship for Research & Innovation, Ministry of Education in Saudi Arabia for funding this research through project number 223202.

Data Availability Statement: Not applicable.

Conflicts of Interest: The authors declare no conflict of interest.

References

1. Fazelpour, F.; Vafaeipour, M.; Rahbari, O.; Rosen, M.A. Intelligent optimization to integrate a plug-in hybrid electric vehicle smart parking lot with renewable energy resources and enhance grid characteristics. *Energy Convers. Manag.* **2014**, *77*, 250–261. [CrossRef]
2. Amini, M.H.; Boroojeni, K.G.; Wang, C.J.; Nejadpak, A.; Iyengar, S.S.; Karabasoglu, O. Effect of electric vehicle parking lots' charging demand as dispatchable loads on power systems loss. In Proceedings of the 2016 IEEE International Conference on Electro Information Technology (EIT), Grand Forks, ND, USA, 19–21 May 2016; pp. 499–503. [CrossRef]
3. Xu, N.Z.; Chung, C.Y. Uncertainties of EV charging and effects on well-being analysis of generating systems. *IEEE Trans. Power Syst.* **2014**, *30*, 2547–2557. [CrossRef]
4. Kumar, Y.V.P.; Rao, S.N.V.B.; Padma, K.; Reddy, C.P.; Pradeep, D.J.; Flah, A.; Kraiem, H.; Jasiński, M.; Nikolovski, S. Fuzzy hysteresis current controller for power quality enhancement in renewable energy integrated clusters. *Sustainability* **2022**, *14*, 4851. [CrossRef]
5. Rao, S.B.; Kumar, Y.P.; Amir, M.; Ahmad, F. An adaptive neuro-fuzzy control strategy for improved power quality in multi-microgrid clusters. *IEEE Access* **2022**, *10*, 128007–128021. [CrossRef]
6. Amer, A.; Azzouz, M.A.; Azab, A.; Awad, A.S. Stochastic planning for optimal allocation of fast charging stations and wind-based DGs. *IEEE Syst. J.* **2020**, *15*, 4589–4599. [CrossRef]
7. Sachan, S.; Amini, M.H. Optimal allocation of EV charging spots along with capacitors in smart distribution network for congestion management. *Int. Trans. Electr. Energy Syst.* **2020**, *30*, 12507. [CrossRef]
8. Mohanty, A.K.; Suresh Babu, P.; Salkuti, S.R. Optimal allocation of fast charging station for integrated electric-transportation system using multi-objective approach. *Sustainability* **2022**, *14*, 14731. [CrossRef]
9. Zeng, B.; Zhu, Z.; Xu, H.; Dong, H. Optimal public parking lot allocation and management for efficient PEV accommodation in distribution systems. *IEEE Trans. Ind. Appl.* **2020**, *56*, 5984–5994. [CrossRef]
10. Gupta, R.S.; Tyagi, A.; Anand, S. Optimal allocation of electric vehicles charging infrastructure, policies and future trends. *J. Energy Storage* **2021**, *43*, 103291. [CrossRef]
11. Kong, W.; Luo, Y.; Feng, G.; Li, K.; Peng, H. Optimal location planning method of fast charging station for electric vehicles considering operators, drivers, vehicles, traffic flow and power grid. *Energy* **2019**, *186*, 115826. [CrossRef]
12. Battapothula, G.; Yammani, C.; Maheswarapu, S. Multi-objective simultaneous optimal planning of electrical vehicle fast charging stations and DGs in distribution system. *J. Mod. Power Syst. Clean Energy* **2019**, *7*, 923–934. [CrossRef]
13. Khan, W.; Ahmad, F.; Alam, M.S. Fast EV charging station integration with grid ensuring optimal and quality power exchange. *Eng. Sci. Technol. Int. J.* **2019**, *22*, 143–152. [CrossRef]
14. Pal, A.; Bhattacharya, A.; Chakraborty, A.K. Placement of public fast-charging station and solar distributed generation with battery energy storage in distribution network considering uncertainties and traffic congestion. *J. Energy Storage* **2021**, *41*, 102939. [CrossRef]

15. Wu, X.; Feng, Q.; Bai, C.; Lai, C.S.; Jia, Y.; Lai, L.L. A novel fast-charging stations locational planning model for electric bus transit system. *Energy* **2021**, *224*, 120106. [CrossRef]
16. Sa'adati, R.; Jafari-Nokandi, M.; Saebi, J. Allocation of RESs and PEV fast-charging station on coupled transportation and distribution networks. *Sustain. Cities Soc.* **2021**, *65*, 102527. [CrossRef]
17. Amer, A.; Azab, A.; Azzouz, M.A.; Awad, A.S. A stochastic program for siting and sizing fast charging stations and small wind turbines in urban areas. *IEEE Trans. Sustain. Energy* **2021**, *12*, 1217–1228. [CrossRef]
18. Pal, A.; Bhattacharya, A.; Chakraborty, A.K. Allocation of electric vehicle charging station considering uncertainties. *Sustain. Energy Grids Netw.* **2021**, *25*, 100422. [CrossRef]
19. Rajesh, P.; Shajin, F.H. Optimal allocation of EV charging spots and capacitors in distribution network improving voltage and power loss by Quantum-Behaved and Gaussian Mutational Dragonfly Algorithm (QGDA). *Electr. Power Syst. Res.* **2021**, *194*, 107049. [CrossRef]
20. Narasipuram, R.P.; Mopidevi, S. A technological overview & design considerations for developing electric vehicle charging stations. *J. Energy Storage* **2021**, *43*, 103225. [CrossRef]
21. Ahmad, F.; Khalid, M.; Panigrahi, B.K. An enhanced approach to optimally place the solar powered electric vehicle charging station in distribution network. *J. Energy Storage* **2021**, *42*, 103090. [CrossRef]
22. Deb, S.; Gao, X.Z.; Tammi, K.; Kalita, K.; Mahanta, P. A novel chicken swarm and teaching learning based algorithm for electric vehicle charging station placement problem. *Energy* **2021**, *220*, 119645. [CrossRef]
23. Goswami, A.; Sadhu, P.K. Stochastic firefly algorithm enabled fast charging of solar hybrid electric vehicles. *Ain Shams Eng. J.* **2021**, *12*, 529–539. [CrossRef]
24. Khaksari, A.; Tsaousoglou, G.; Makris, P.; Steriotis, K.; Efthymiopoulos, N.; Varvarigos, E. Sizing of electric vehicle charging stations with smart charging capabilities and quality of service requirements. *Sustain. Cities Soc.* **2021**, *70*, 102872. [CrossRef]
25. Bhadoriya, J.S.; Gupta, A.R.; Zellagui, M.; Saxena, N.K.; Arya, A.K.; Bohre, A.K. Optimal allocation of electric vehicles charging station in distribution network beside DG using TSO. In *Planning of Hybrid Renewable Energy Systems, Electric Vehicles and Microgrid: Modeling, Control and Optimization*; Springer Nature: Singapore, 2022; pp. 785–808. [CrossRef]
26. Yi, T.; Cheng, X.; Peng, P. Two-stage optimal allocation of charging stations based on spatiotemporal complementarity and demand response: A framework based on MCS and DBPSO. *Energy* **2022**, *239*, 122261. [CrossRef]
27. Ghasemi-Marzbali, A. Fast-charging station for electric vehicles, challenges and issues: A comprehensive review. *J. Energy Storage* **2022**, *49*, 104136. [CrossRef]
28. Zhou, G.; Zhu, Z.; Luo, S. Location optimization of electric vehicle charging stations: Based on cost model and genetic algorithm. *Energy* **2022**, *247*, 123437. [CrossRef]
29. Aljehane, N.O.; Mansour, R.F. Optimal allocation of renewable energy source and charging station for PHEVs. *Sustain. Energy Technol. Assess.* **2022**, *49*, 101669. [CrossRef]
30. Ahmad, F.; Iqbal, A.; Ashraf, I.; Marzband, M. Optimal location of electric vehicle charging station and its impact on distribution network: A review. *Energy Rep.* **2022**, *8*, 2314–2333. [CrossRef]
31. Kumar, N.; Kumar, T.; Nema, S.; Thakur, T. A comprehensive planning framework for electric vehicles fast charging station assisted by solar and battery based on Queueing theory and non-dominated sorting genetic algorithm-II in a co-ordinated transportation and power network. *J. Energy Storage* **2022**, *49*, 104180. [CrossRef]
32. Zu, S.; Sun, L. Research on location planning of urban charging stations and battery-swapping stations for electric vehicles. *Energy Rep.* **2022**, *8*, 508–522. [CrossRef]
33. Thangaraju, I. Optimal allocation of distributed generation and electric vehicle charging stations-based SPOA2B approach. *Int. J. Intell. Syst.* **2022**, *37*, 2061–2088. [CrossRef]
34. Al Wahedi, A.; Bicer, Y. Techno-economic optimization of novel stand-alone renewables-based electric vehicle charging stations in Qatar. *Energy* **2022**, *243*, 123008. [CrossRef]
35. Deb, S.; Gao, X.Z.; Tammi, K.; Kalita, K.; Mahanta, P. Nature-inspired optimization algorithms applied for solving charging station placement problem: Overview and comparison. *Arch. Comput. Methods Eng.* **2021**, *28*, 91–106. [CrossRef]
36. Erdogan, N.; Pamucar, D.; Kucuksari, S.; Deveci, M. An integrated multi-objective optimization and multi-criteria decision-making model for optimal planning of workplace charging stations. *Appl. Energy* **2021**, *304*, 117866. [CrossRef]
37. Ma, T.Y.; Xie, S. Optimal fast charging station locations for electric ridesharing with vehicle-charging station assignment. *Transp. Res. Part D Transp. Environ.* **2021**, *90*, 102682. [CrossRef]
38. Ahmadi, M.; Hosseini, S.H.; Farsadi, M. Optimal allocation of electric vehicles parking lots and optimal charging and discharging scheduling using hybrid metaheuristic algorithms. *J. Electr. Eng. Technol.* **2021**, *16*, 759–770. [CrossRef]
39. Fathy, A.; Abdelaziz, A.Y. Competition over resource optimization algorithm for optimal allocating and sizing parking lots in radial distribution network. *J. Clean. Prod.* **2020**, *264*, 121397. [CrossRef]
40. Khan, M.J.; Iqbal, M.T. Pre-feasibility study of stand-alone hybrid energy systems for applications in Newfoundland. *Renew. Energy* **2005**, *30*, 835–854. [CrossRef]
41. Moghaddam, A.A.; Seifi, A.; Niknam, T. Multi-operation management of a typical micro-grids using Particle Swarm Optimization: A comparative study. *Renew. Sustain. Energy Rev.* **2012**, *16*, 1268–1281. [CrossRef]
42. Fathy, A.; Kaaniche, K.; Alanazi, T.M. Recent approach based social spider optimizer for optimal sizing of hybrid PV/wind/battery/diesel integrated microgrid in aljouf region. *IEEE Access* **2020**, *8*, 57630–57645. [CrossRef]

43. Roe, C.; Meliopoulos, A.P.; Meisel, J.; Overbye, T. November. Power system level impacts of plug-in hybrid electric vehicles using simulation data. In Proceedings of the 2008 IEEE Energy 2030 Conference, Atlanta, GA, USA, 17–18 November 2008; pp. 1–6 [CrossRef]
44. Mozafar, M.R.; Moradi, M.H.; Amini, M.H. A simultaneous approach for optimal allocation of renewable energy sources and electric vehicle charging stations in smart grids based on improved GA-PSO algorithm. *Sustain. Cities Soc.* **2017**, *32*, 627–637 [CrossRef]
45. Li, Z.; Chowdhury, M.; Bhavsar, P.; He, Y. Optimizing the performance of vehicle-to-grid (V2G) enabled battery electric vehicles through a smart charge scheduling model. *Int. J. Automot. Technol.* **2015**, *16*, 827–837. [CrossRef]
46. Gahruei, J.R.; Beheshti, Z. The Electricity Consumption Prediction using Hybrid Red kite Optimization Algorithm with Multi-Layer Perceptron Neural Network. *J. Intell. Proced. Electr. Technol.* **2022**, *15*, 1–22.
47. Fathy, A. A novel artificial hummingbird algorithm for integrating renewable based biomass distributed generators in radial distribution systems. *Appl. Energy* **2022**, *323*, 119605. [CrossRef]
48. Xue, J.; Shen, B. Dung beetle optimizer: A new meta-heuristic algorithm for global optimization. *J. Supercomput.* **2023**, *79*, 7305–7336. [CrossRef]
49. Abdollahzadeh, B.; Gharehchopogh, F.S.; Mirjalili, S. African vultures optimization algorithm: A new nature-inspired metaheuristic algorithm for global optimization problems. *Comput. Ind. Eng.* **2021**, *158*, 107408. [CrossRef]
50. Alsattar, H.A.; Zaidan, A.A.; Zaidan, B.B. Novel meta-heuristic bald eagle search optimisation algorithm. *Artif. Intell. Rev.* **2020**, *53*, 2237–2264. [CrossRef]
51. Das, A.K.; Pratihar, D.K. Bonobo optimizer (BO): An intelligent heuristic with self-adjusting parameters over continuous spaces and its applications to engineering problems. *Appl. Intell.* **2022**, *52*, 2942–2974. [CrossRef]
52. Mirjalili, S.; Mirjalili, S.M.; Lewis, A. Grey wolf optimizer. *Adv. Eng. Softw.* **2014**, *69*, 46–61. [CrossRef]
53. Fathy, A.; Abdelaziz, A.Y. Grey wolf optimizer for optimal sizing and siting of energy storage system in electric distribution network. *Electr. Power Compon. Syst.* **2017**, *45*, 601–614. [CrossRef]

mathematics

Article

Real-Time Management for an EV Hybrid Storage System Based on Fuzzy Control

Dimitrios Rimpas *, Stavros D. Kaminaris, Dimitrios D. Piromalis and George Vokas

Department of Electrical & Electronics Engineering, University of West Attica, 250 Thivon Av., 122-44 Egaleo, Greece; skamin@uniwa.gr (S.D.K.); piromali@uniwa.gr (D.D.P.); gvokas@uniwa.gr (G.V.)
* Correspondence: drimpas@uniwa.gr; Tel.: +30-210-5381581

Abstract: Following the European Climate Law of 2021 and the climate neutrality goal for zero-emission transportation by 2050, electric vehicles continue to gain market share, reaching 2.5 million vehicles in Q1 of 2023. Electric vehicles utilize an electric motor for propulsion powered by lithium batteries, which suffer from high temperatures caused by peak operation conditions and rapid charging, so hybridization with supercapacitors is implemented. In this paper, a fuzzy logic controller is employed based on a rule-based scheme and the Mamdani model to control the power distribution of the hybrid system, driven by the state of charge and duty cycle parameters. An active topology with one bi-directional DC-to-DC converter at each source is exploited in the MATLAB/Simulink environment, and five power states like acceleration and coasting are identified. Results show that the ideal duty cycle is within 0.40–0.50 as a universal value for all power states, which may vary depending on the available state of charge. Total efficiency is enhanced by 6%, sizing is increased by 22%, leading to a more compact layout, and battery life is extended by 20%. Future work includes testing with larger energy sources and the application of this management strategy in real-time operations.

Keywords: battery; ultracapacitor; hybrid; energy; management; fuzzy; control

MSC: 93C42

Citation: Rimpas, D.; Kaminaris, S.D.; Piromalis, D.D.; Vokas, G. Real-Time Management for an EV Hybrid Storage System Based on Fuzzy Control. *Mathematics* **2023**, *11*, 4429. https://doi.org/10.3390/math11214429

Academic Editor: Nicu Bizon

Received: 21 September 2023
Revised: 24 October 2023
Accepted: 24 October 2023
Published: 25 October 2023

1. Introduction

The need for a cleaner environment and the reduction of greenhouse gases has been identified as the primary target for the global population. The Paris Agreement of 2015 and the European Climate Law of 2021 set the need for a reduction of mean global temperature by two degrees Celsius via green transportation adoption by 2050 [1]. With those regulations, a 55% reduction in greenhouse gases is expected to be achieved. This margin is mainly focused on the road transport sector and especially light-duty passenger cars powered by internal combustion engines (ICE), which represented over 25% of total carbon dioxide emissions over the last five years, and are expected to increase until 2030 [2]. Therefore, decarbonizing the transportation sector and increasing energy efficiency in all other sectors is imminent for the protection of the environment [3].

Over the last 20 years, the trend of green transportation and improving fuel efficiency has led to hybrid (HEV) and battery electric vehicles (BEV). HEVs utilize an electric motor to assist the ICE or even power the car fully in conditions like traffic or low-speed driving. Hence, fuel consumption can be limited up to 30% in certain conditions without any significant changes in driving habits [4]. Due to their manufacturing simplicity, low emissions, and public demand for a sustainable price, a market share of 28% for EVs was achieved in 2021. However, the road to emission-free transportation can only be fully accomplished with the adoption of pure electric vehicles (EVs). An electric motor is responsible for the propulsion of the vehicle powered by lithium batteries and an advanced battery management system [5]. There are major benefits to the utilization of EVs like:

- No greenhouse gases emissions;
- Noiseless operation;
- Energy recovery through regenerative braking;
- Ability to charge at home via renewable sources.

These parameters played a major role in the recent sales of EVs. Up to 12% of the market share for 2022 is pure electric, with huge waiting lists due to the limited resources that manufacturers face, with diesel cars declining [6]. This trend is expected to continue strongly in 2023, with almost 2.5 million EVs sold in the first quarter, about to reach 15 million units in total, resulting in 35% year-on-year escalation and 18% of the total market share. By following this direction, 5 million crude oil barrels can be saved per day by 2030, ensuring the utilization of green environmental transportation means. Additional policies and measures are shown in Figure 1.

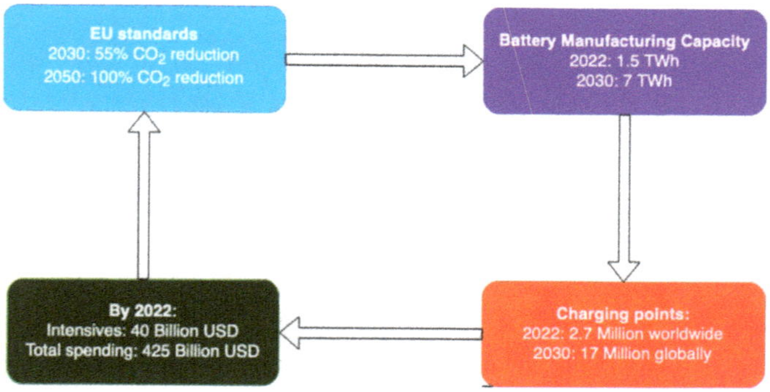

Figure 1. Policies and actions required for full electrification of the light duty vehicles [6].

However, there are still certain limitations regarding the operation of EVs, as summarized in previous work [5]. High charging time, a lack of chargers, low mileage causing range anxiety, and increased purchase costs as rare minerals are required for current battery and motor manufacturing are still considered major and practical limitations. It is evident that the main issues refer to the battery pack's limited operational conditions and sensitivity to extreme setups. The temperature window of a typical lithium battery is within 20 to 65 degrees Celsius, while functioning outside this range can cause lithium deposition, dendrites, or even total breakdown [7]. Proper cooling is crucial for protecting the cells from overheating, and while unified with the climate control system, it can decrease total energy consumed while offering maximum thermal comfort [8]. Additionally, charging and discharging sequences are supervised and controlled by complex power electronics to ensure safety and protection, avoiding peak conditions. As the battery is a part of the vehicle chassis for better weight management, even maintenance is practically unfeasible, so the layout has to be precisely designed and sized.

There are certain parameters affecting battery life [9–13]. State of charge (Soc) and state of health (SoH), which reveal the degradation of the cells, are crucial to calculating battery aging and cannot be measured directly on the pack. State of Power (SoP) supplies information about the battery's peak current capabilities, while state of voltage refers to the ratio of current-voltage divided by the nominal value and can be connected to the SoC. Lastly, Depth of Discharge (DoD) represents the percentage that the battery pack is discharged and proved to be valuable as a low DoD massively affects battery degradation, limiting stress on the cells and increasing its viability. It is essential to monitor all values to ensure battery-safe operation and nominal performance within the projected time applied by the manufacturer, as depicted in Figure 2.

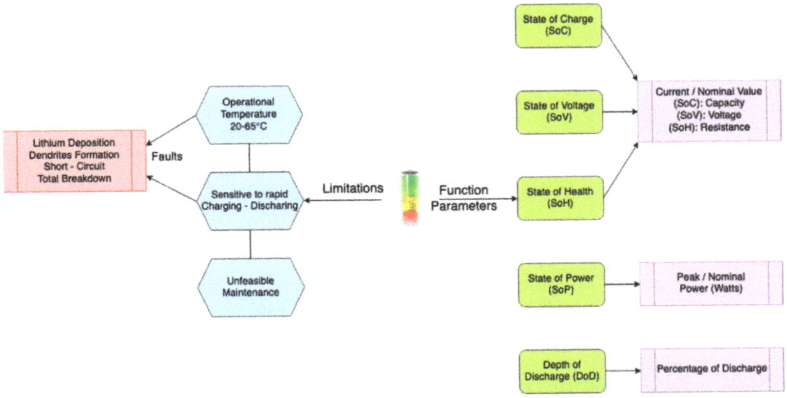

Figure 2. Limitations and operational parameters for a lithium battery.

To address those restrictions, a hybrid energy storage system (HESS), consisting of batteries and supercapacitors, is utilized. In everyday use, battery life is shortened due to significant heat generation caused by the need to cover the high power requirements of electrically powered vehicles [14]. Supercapacitors (SC) or ultracapacitors (UC) solve this problem by providing both average and peak power loads, reducing the maximum current of the battery by up to 55% and ensuring lower voltage drops, hence fewer fluctuations and capacity costs [15,16]. SC also inherits a lifespan of over 1 million operating cycles, making it the perfect auxiliary source. However, since the energy density is low, they cannot operate as a standalone source for an EV.

The combination of these different systems enables electric vehicles to operate over a longer range using high-energy-density batteries while increasing performance with ultracapacitors acting as power buffer units. In addition, the supercapacitor takes advantage of regenerative braking harvesting, providing better control and up to 16.2% more range with less wear [17]. The system performs greatly in low-temperature conditions where conventional batteries are outside their safety zone. However, the connection of power modes, as well as the duty cycle indication are not studied widely in the literature. The power demand has to be classified into different states for better understanding, and the operation cycle or duty cycle, namely the energy used in each time interval compared to the maximum, e.g., acceleration, are important parameters of proper power distribution.

The purpose of this work Is the implementation of a fuzzy logic simulation model for the operation of HESS via MATLAB/Simulink. This model calculates the various parameters affecting the hybrid system states (SOC, SOH, etc.) to manage the power split between the two sources connected in active topology. The introduction of duty cycle and power mode sequences is also a major step in designing the ideal scheme for power distribution, which is the contribution of this paper. Additionally, the energy management strategy (EMS) will be validated for enhanced efficiency of the system with constant protection of the battery cells.

The manuscript is structured into five main sections. Section 2 describes the equations and the methodology used to conduct the experiment, along with an explanation of duty cycles term and classification of the distinct power modes. In Section 3, the results of the simulation are presented. In Section 4, discussion about the outcome and comparison with other works is thoroughly examined. In the last section, conclusions about the model are included along with suggestions about future work.

2. Materials and Methods

2.1. Methodology

2.1.1. Hybrid Storage System Bank

The topology of the applied HESS is a key parameter in the power-split and dynamic performance of the whole setup. There are three different topologies: Passive, semi-active, and active [5]. In passive configuration, the two sources are simply connected in parallel, reducing peak currents adequately, but the ultracapacitor voltage is limited by the battery bank [18]. Semi-active format, which requires a DC-to-DC converter plugged into the battery or the SC while the other is connected in parallel with the converter, offers better cell protection and HESS sizing, but the layout is restricted by a low performance-to-cost ratio and increased converter sizing. Lastly, active topology implies two bi-directional converters directly connected to both the energy sources (battery and SC) and then toward the DC bus powering the motor and other electronics. Both sources can be coupled or decoupled from the load simultaneously. This arrangement was chosen for this project as performance, control, efficiency, and voltage regulation are the main priorities, even though it is by far the most complex setup. Power distribution is vastly configurable, offering the ability to test different energy management strategies [19]. The projected topology is depicted in Figure 3 below.

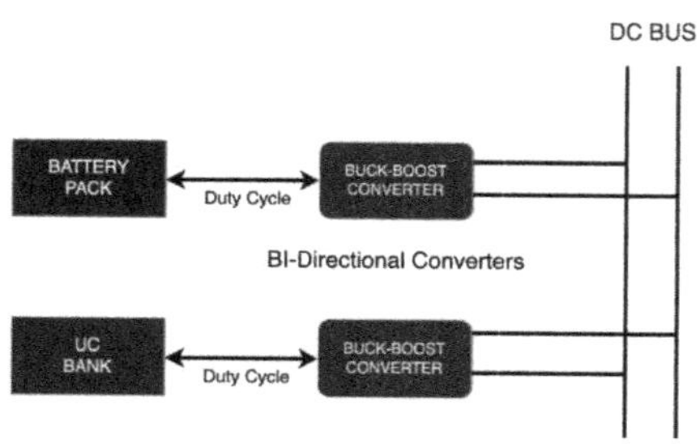

Figure 3. Experimental layout with two buck-boost converters connected directly to the DC bus with power output directly connected to power mode and duty cycle.

For calculations, the power loss model is considered. Since the internal parameters of both converters cannot be neglected, like inductor resistance, voltage drop, and switch on-resistance, the assumption that both models are identical, as found in [20], will be utilized. The power loss of the DC/DC components is related to the term known as the duty cycle. This describes the percentage that each source operates until it is recharged again, for example, 80% to 30% SoC and off to recharging, which complies with 50% duty cycle. As it is suggested, the duty cycle is expressed as the energy supplied by each source to cover the projected power demand within specified time interval. A 30% drop in the ultracapacitor SoC required to cover the vehicle acceleration from stop is also indicated as 30% duty cycle. It is renowned that small duty cycles and low charging times or rates (total amperage drawn), are essential for prolonging battery lifetime or prevent any possible breakdown as well as increasing charging capacity [21]. To calculate the power loss, the following equations are required [20]:

$$\text{Boost mode}: \begin{cases} P_{VEHICLE} - V_H \times I_{UC} = V_L \times I_B - P_{CON,LOSS} \\ \frac{P_{VEHICLE}}{V_H} = I_{UC} + I_B \times (1 - D_{BOOST}) \end{cases} \tag{1}$$

$$\text{Buck mode}: \begin{cases} P_{VEHICLE} - V_H \times I_{UC} = V_L \times I_B + P_{CON,LOSS} \\ \frac{P_{VEHICLE}}{V_H} = I_{UC} + I_B \times D_{BUCK} \end{cases} \tag{2}$$

where P_D is the power demand of the vehicle, $P_{CON,LOSS}$ refers to the power loss of the converter, V_H equals to voltage of the low voltage side, V_L is the high voltage value,

I_B is the DC/DC inductor current, mainly affecting the battery, I_{UC} denotes the current of the supercapacitor, and D_{BOOST} and D_{BUCK} represent the duty cycles at boost and buck modes, respectively.

Accordingly, further processing with the conversation of power and Kirchhoff's current law shows that the power loss of the total system is equal to the square of the inductor's current multiplied by internal resistances, the duty cycles at each state, and the voltages of the switches of the converters, which cannot be neglected [20]:

$$P_{CON,LOSS} = \begin{cases} I_B \times [\,(V_S + V_D) \times D_{BOOST}\,] + \times I_B^2 \times [R_L + (R_S + R_D) \times D_{BOOST}] \\ I_B \times [\,(V_S + V_D) \times D_{BUCK}\,] + \times I_B^2 \times [R_L + (R_S + R_D) \times D_{BUCK}] \end{cases} \quad (3)$$

where V_D and R_D are the voltage drop and resistance of the diode, V_S and R_S denote the voltage drop and resistance of the switch, and R_L refer to the inductor resistance [20].

2.1.2. Energy Sources Models

As every component and energy source, battery and supercapacitor can be simplified following Thevenin modeling for better explanation, ignoring losses that are infinitesimal. So, the battery model consists of the internal resistance R_B, the polarization voltage V_P, polarization capacitance C_P, polarization resistance R_P, and Terminal voltage V_T. V_{OC} accounts as the open circuit voltage, thus the maximum voltage the battery can provide with no load. So, the final model is presented in Figure 4 below [20].

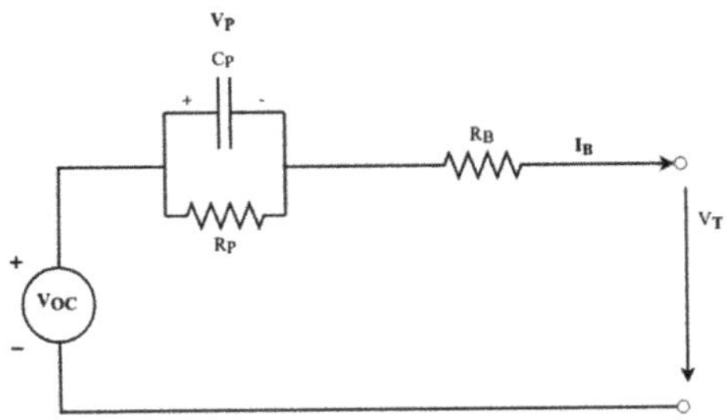

Figure 4. Lithium Battery Thevenin model by [20].

Depending on the state of charge and temperature at each specific time k, the polarization values are different and the terminal voltage has to be calculated from equations by [20,21]. It is obvious that these parameters are always in a dynamic state and must be supervised by the energy management system for further adjustments, improving efficiency and control.

$$V_{T(k)} = V_{OC} \times SOC_{(k)} - I_{B(k)} \times R_B \times SOC_{(k)} - V_{P(k)} \quad (4)$$

State of charge, at the time k is expressed at the next equation.

$$SOC(k) = \left(\frac{V_{P(k)} + V_{T(k)}}{V_{OC}} \right) \quad (5)$$

Finally, the lithium battery losses $P_{B,LOSS}$ can be calculated at any specific time k including all losses induced by polarization and internal resistance [20].

$$P_{B,LOSS} = I_B^2 \times R_B \times SOC + \frac{V_P}{R_P \times SOC} \quad (6)$$

The ultracapacitor can be simplified as an RC model or a electrical model containing the equivalent series resistance (ESR) or heat losses, the equivalent parallel resistance (EPR), representing leakage current and capacitance [22]. However, this model adds complexity to the simulation, so the RC model is preferred. For simplicity, both rated capacitance

(C_R) and internal resistance (R_{UC}) will be accounted consistently even though they diffe during charging-discharging due to the effects of ESR and EPR. V_{UC} and I_{UC} represent th ultracapacitor voltage and current, with V_T as the voltage terminal, similar to the batter layout. Figure 5 illustrates the final model of the supercapacitor.

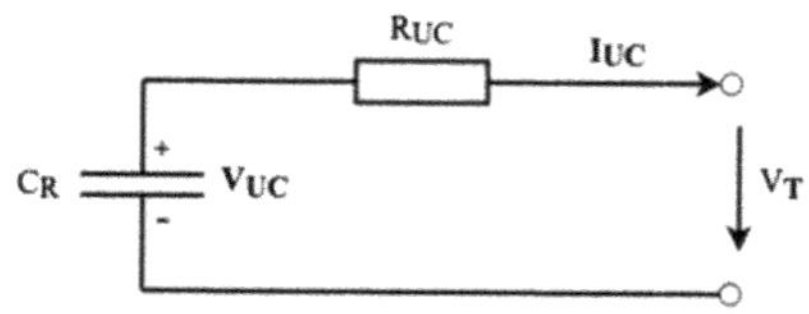

Figure 5. Supercapacitor RC model [20–22].

Since the model has been simplified, it is now possible to calculate all values, along with the power losses, by utilizing Kirchhoff's laws. The next Equations (7)–(11) modified from [20–22] are expressing these computations.

The Open circuit voltage of the ultracapacitor at a certain n time:

$$V_{UC(n)} = VUC(n-1) - \frac{I_{UC(n)}}{C_R} \quad (7)$$

Next is the equation for the terminal voltage of the SC at n time:

$$V_{T(n)} = V_{T(n-1)} - I_{UC(n)} \times R_{UC} \quad (8)$$

The total power delivered by the ultracapacitor is expressed as:

$$P_{UC(n)} = V_{T(n)} \times I_{UC(n)} \quad (9)$$

The current flowing in and out of the capacitor is calculated with Equation (10). The same equation applies for charging or discharging with a minus when the capacitor is being charged:

$$I_{UC} = \frac{V_{UC} - \sqrt{V_{UC}^2 - 4 \times R_{UC} \times P_{UC}}}{2 \times R_{UC}} \quad (10)$$

Equation (11) shows the power losses due to internal resistance and leakage current:

$$P_{UC,LOSS} = R_{UC} \times I_{UC}{}^2 \quad (11)$$

The state of voltage, similar to SOC for supercapacitors equals to the terminal voltage divided by the maximum rated voltage by the manufacturer as presented in Equation (12) [23].

$$SOV = \frac{V_T}{V_{RATED}} \quad (12)$$

Equation (13) is applied to calculate the state of health parameter revealing battery lifetime where charge is expressed as capacity, measured in ampere-hours [12].

$$\mathbf{SOH} = \frac{\boldsymbol{Qmax}}{\boldsymbol{Qrated}} \quad (13)$$

Finally, the total power losses of HESS plus the converter are calculated with Equation (13) below [14,20].

$$P_{LOSS,TOTAL} = P_{CON,LOSS} + P_{B,LOSS} + P_{UC,LOSS} \quad (14)$$

2.1.3. Duty Cycles

As mentioned previously, the duty cycle is an essential term for proper energy management and power distribution. The SOC, SOV, and other parameters should be maintained high to cover the necessary power for the electric motor. At fractional levels, battery over-charging or discharging will lead to the end of the service life or even a failure. It is crucial to keep the frequency fixed without major fluctuations while keeping the state of charge of both sources above 25% as a lower threshold for protection and efficiency [22–24]. The topology chosen offers great control through buck-boost mode by the converters, along with fast power switching for adjustments in different power modes, minimizing voltage ripple and losses. Since the efficiency of the DC/DC can reach over 94% at average power/current outputs, the duty cycle should be low and focused on maintaining the SC charge high, thus keeping the converter at its rated power area [23–25].

The hybrid energy system Is required to cover the driving power demand $P_{VEHICLE}$ of the vehicle which is described in Equation (15) below [23].

$$P_{VEHICLE} = \frac{V_A}{3600 * \eta} \times (M \times g \times f \times \cos(\beta) + M \times g \times \sin(\beta) + \frac{C_{ARC} \times A}{21.15} * {V_A}^2 + \frac{\delta \times M}{3.6} \times \frac{dV_A}{dt}) \quad (15)$$

where V_A is the car speed, η denotes the powertrain efficiency, M is the total mass, g is the gravitational acceleration, f and C_{ARC} represent the rolling assistance and air resistance coefficients, β indicates the road incline, A is the wind exposed area and δ designates the rotation mass coefficient correction factor.

Based on Equation (14), the size of HESS utilized must be able to meet and overcome the power demand of the vehicle. For safety reasons, the battery power alone should also be able to meet this requirement, so the total number of cells, needed capacitance, and voltage are expressed at Equation (16) [23].

$$U_R \times C_R \times n_P \times n_S \geq P_{VEHICLE} \quad (16)$$

where U_R is the rated voltage, C_R denotes the rated capacitance of each cell, n_P indicates the battery cells required in parallel and n_S represents the cells in series.

2.1.4. Power Modes

There are various power modes for an EV based on the motor, and HESS states [24–26]: Initialization: When the vehicle starts, HESS needs to provide high torque quickly. At this time, the power is only provided by the SC, taking advantage of its high-power density. During acceleration, high current is required, so the battery should assist the SC in covering the motor power demand. The battery voltage is stepped up by a DC-DC converter to obtain a high DC bus voltage; hence, the HESS provides the energy required. The output power ratio of these sources should be reasonably distributed, with the supercapacitor providing most of the power and minimizing losses. However, as the voltage of SC decreases, the duty cycle lessens, so power losses can be minimized by discharging the lithium-ion battery and the ultracapacitor simultaneously. As a result, HESS can perfectly adapt to future instantaneous power demand.

The next mode is regenerative braking. If the vehicle is decelerating or coasting at high speeds, the engine is running as a generator and is in a power-generating state. The electrical energy generated is sent back to the HESS, almost entirely on the SC, to avoid high-frequency current on the battery. Then, depending on the state of charge of each source, energy is strategically distributed. During the braking phase, the voltage of the DC busbar is low and the DC/DC duty cycle is small. When the voltage of the ultracapacitor is higher, the DC/DC duty cycle is smaller, so losses are minimal.

Finally, in coasting, or cruising mode, the motor is running at constant speed due to a low load or road slope; thus, only a small current may be required. Therefore, SC is used to supply all load currents, as it is sufficient. The battery DC-DC converter is temporarily switched off to reduce circuit losses. However, if the SC voltage drops below the battery voltage, the battery can supply energy to the drive and charge the SC.

2.2. Fuzzy Model

For this project, the deployment of a fuzzy logic model via MATLAB/Simulink is selected. Fuzzy logic is a math model leading to a non-binary state of mind and a sophisticated control system [27]. Fuzzy control systems consist of four main parts: fuzzification, rule-based, fuzzy reasoning, and defuzzification. The power requirement is a valuable operating parameter and a key factor in the project. Even though increased processing power is required to run the models, there are multiple benefits regarding the use of FL strategy and controllers [28,29]. They are as follows:

1. Voltage regulation of the DC bus during load variations;
2. Limited battery current and frequency fluctuations hence lower temperatures;
3. High performance, simplicity, reliability, and enhanced lifetime;
4. The linearity pattern of data acquisition is a crucial parameter for a fuzzy model. A weighted memory H controller is utilized at [30] for increased precision in non-linear monitoring and identification of memory states based on Takagi-Sugeno fuzzy systems. The Wavelet packet decomposition method can be utilized for time sequence determination for consistent data retrieval, identification of charging or discharging states, and even power mode classification [31]. In addition, fuzzy logic modeling can be applied not only to EVs but also to smaller applications like electric scooters or three wheelers as well, limiting fluctuations and enhancing range [32].
5. The developed fuzzy set-based controller has five inputs: Power demand, UC, and battery voltages, and SoC. The controller uses a list of fuzzy rules to manipulate the correlation between inputs and outputs. According to the principles, efficient energy management must be provided with respect to the operating conditions of the energy source [33]. The supply is based on the SoC of each source and the power modes described before, which may also indicate the driving style [34]. The basic relationship between input and output, as defined by the fuzzy rules, is that the battery acts as the main energy source for the vehicle as long as the SOC is above 25% and the voltage of the ultracapacitor is below 50%.
6. At constant speeds with low energy consumption, energy is provided by the lithium battery, while the SC should be saved for peak demand loads like high slopes and rapid accelerations. Additionally, supercapacitors are efficient in regenerative braking exploitation since their high power density allows swift charging, even with high-frequency currents. Afterwards, SC can charge the battery or utilize a low-pass filter for that case to avoid high frequencies. The rules that comprise the fuzzy model are the following [35,36]:
7. Power demand and SOC_{SC} are M to H:SC will provide the demand entirely.
8. Power demand is H and SOC_{SC} is ML: SC will provide its available power.
9. Power demand is M to H and SOC_{SC} is L: Both sources output will be balanced.
10. Power demand is L to ML and SOC_{SC} is L: Again, sources are balanced.
11. Power demand is ML and SOC_{SC} is L: Battery will cover the power mostly.
12. Power demand is L to ML and SOC_{SC} is M to H: SC will provide most of the power.

Where L is low, ML stands for Medium Low, M for Medium and H for High. The scale is represented as a percentage as follows:

13. Low: 0–25%
14. Medium Low: 25–50%
15. Medium: 50–75%
16. High: >75%

The fuzzy control aims to generate commands for charging and discharging the two energy sources as a function of the fuzzy logic controller, presented in Table 1.

Table 1. Rules for the proposed fuzzy logic controller.

Power State	Power Output	
	SC	Battery
Regenerative braking	Charging *	Charging *
Coasting	Charging	Charging
Initialization	Maximum	Low
Slow Acceleration	Medium	Medium
Fast Acceleration	High	Medium

* Depending on SoC and duty cycle.

The scheme selected for the EV is presented at Figure 6 below.

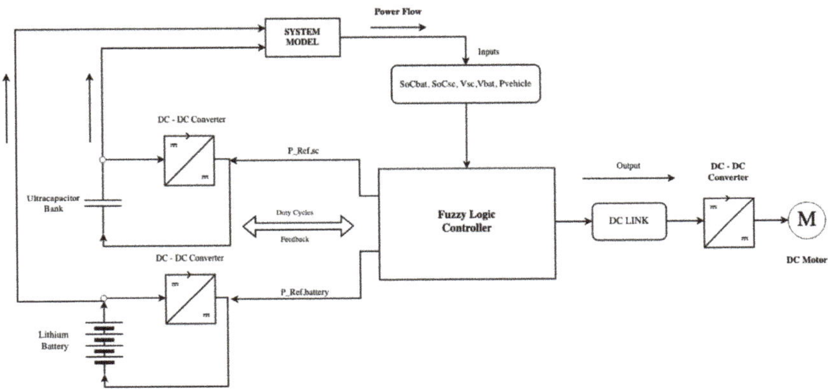

Figure 6. Experimental layout highlighting the power flow to the motor through the converter and the importance of the fuzzy logic controller.

Next, the power management flowchart shows the strategy followed by the FL controller to cover the power demand and handle the output of both sources. Membership functions are shown in previous works [32–38], based on the Mamdani model, and are projected in this paper as illustrated in Figure 7. The fuzzy model incorporates a small difference in rule definition (low to high) to specify the value of temperature and duty cycle in the power distribution scenarios implemented.

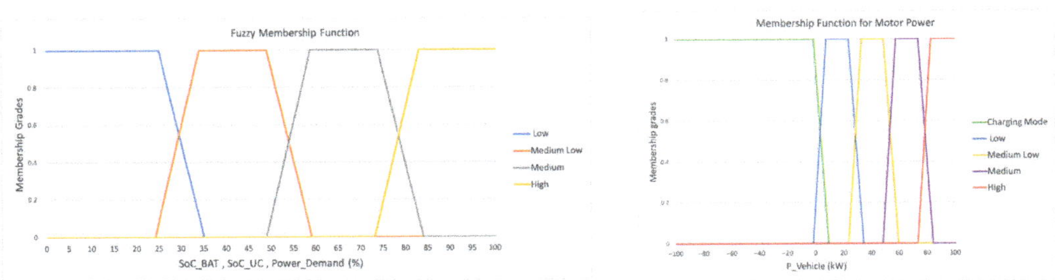

(a) **(b)**

Figure 7. Membership functions utilized for this fuzzy model (**a**) Distinct energy sources SoC to cover the power demand; (**b**) Motor power ($P_{VEHICLE}$ parameter) function depending on the power state.

The regulation of charging and discharging as well as energy demand coverage are priorities shown in Figure 8 below:

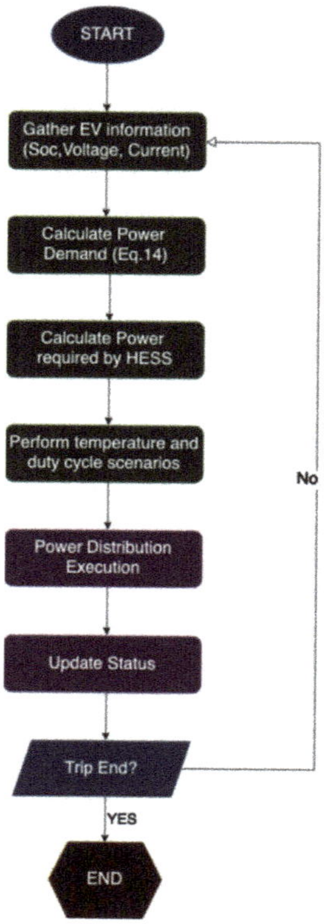

Figure 8. Fuzzy logic controller scheme as utilized in this work.

The power distribution ratios will vary depending on the state of charge of each source, as previously mentioned. In previous work, a specific ratio of 70% battery and 30% supercapacitor was presented to offer better sizing, cost, and weight, but with a modestly lower range [39]. Over 100,000 values validated the result, while temperature management and prevention of high-frequency currents provided increased battery life. This dataset will be utilized in this study, as well as a more advanced and precise control pattern is followed. However, control and charging were conducted manually at [39] so the limitations were escalated. The modelling of the fuzzy logic controller is implemented through MATLAB/Simulink via the ready-to-use libraries. Therefore, the diagram of this model in the application environment is projected in Figure 9.

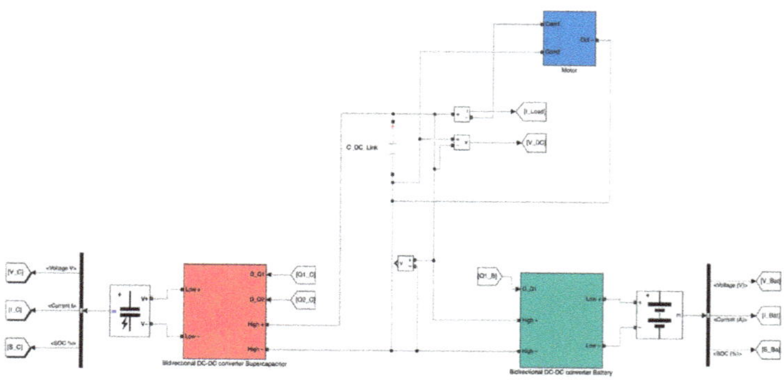

Figure 9. Fuzzy logic controller simulation using MATLAB/Simulink.

3. Results and Discussion

3.1. HESS Power Variations under Different Duty Cycles

Duty cycles play a major role in the power distribution of HESS. At low values, below 40%, the majority of power is covered by the supercapacitors due to their high-power density, so the battery pack is protected. As the duty cycles increase, batteries undertake more and more of the power demand while the ultracapacitor output becomes minimum, as depicted in Figure 10. Therefore, it is crucial to keep the duty cycle ratio as low as possible, ideally at 0.50, depending on the power state shown in Table 1.

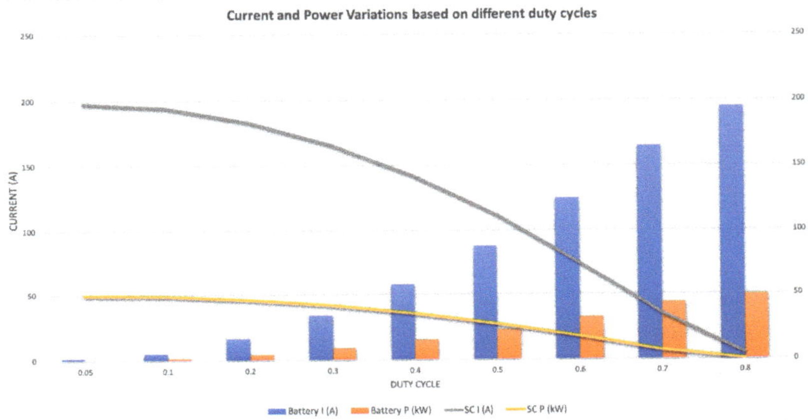

Figure 10. Power output variations of HESS based on powertrain demand and duty cycle.

3.2. Power Distribution of HESS Depending on Total Demand

The power required from the powertrain varies as the vehicle overcomes different power states; thus, the fuzzy logic controller prompts the HESS to properly distribute the power, with the rule of $SoC_{min} \geq 25\%$ without depleting any source. At the start of the simulation, after defuzzification, the motor power state is high so the ultracapacitor covers that demand while the battery supply is gradually increasing to level the power split, as represented in Figure 11. Then, after 500 ms, the motor enters a regenerative braking state so both sources are recharged quickly. Afterward, at T = 1000 ms, the acceleration state is engaged, but since the power demand is not rapid, the battery can sufficiently handle the energy needed without overheating, and the supercapacitor remains with high SoC for high-density operation occurring at T = 1250 ms. Similarly, the battery will provide power in a steady function, leading to T = 2000 ms, where the supercapacitor is reaching a lower limit based on the rules applied, so the battery has to cover the load and charge the ultracapacitor as well since demand is medium. Finally, UC is now recharged and available to cover peak loads.

Figure 12 presents the load and power distribution fluctuations shown before. Fuzzy logic controllers based on utilized rules, keep the state of charge at safe levels not leaving any source depleted, a situation that can lead to major breakdown or damage to the EV electrical systems.

The battery temperature operating range is maintained within 20–40 °C, the nominal power of the motor is 50 kW, the battery energy capacity is 40 kWh running at 360 V, and the ultracapacitor bank has a total capacitance of 5000 F at 320 V. The SoC range as stated previously, should be maintained at 20–80% effective range for both sources, but due to DC/DC converter limitations for cut-off voltage in real applications, testing mainly refers to 30% and more available SoV for the entire HESS. The dataset mainly used is available from previous work conducted [39]. However, for the previous figure, the data sequence is summarized in Table 2 below.

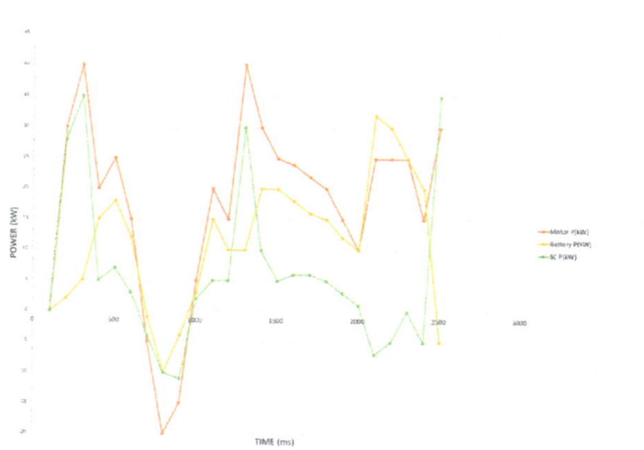

Figure 11. Power distribution of the hybrid storage system under different loads.

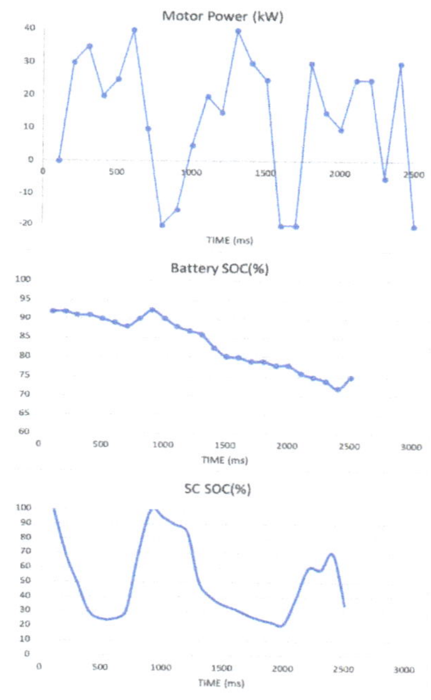

Figure 12. SOC fluctuations of the two sources separately according to the powertrain output.

Compared to a typical HESS with fuzzy rules or other management system, this work shows safe and high operation time, sufficiently prolonging battery life. With the introduction of duty cycles and constant parameter calculation, each energy source is obliged to function within a specified power, temperature, and voltage range to cover the demand of the motor or even auxiliary loads, like climate control. Based on the simulation, efficiency increases by 6% as regenerative braking utilization is highly achievable through the UC bank. The range is 2.5% boosted as HESS can cover the demand for more time, hence the distance traveled is increased. Sizing is improved by 22% as fewer batteries are required if the distance is planned as consistent and weight minimization is the requisite, while the energy source's lifetime is enhanced by 20% as battery last longer, since less stress is applied due to UC. This gain is further explained at Figure 13 below. As the charge rate increases, an energy source cannot operate for adequate amount of time. Sizing must be sufficient, but dimensions are a key factor associated with duty cycle and useful range. Supercapacitors have an elevated ability to handle high charge rates as the left diagram explained, compared to the lithium discharge curve by [40], where the battery cells show low tolerance to high currents and charge rates, causing immense temperatures.

Table 2. Data sequence employed for Figure 12.

Time (ms)	$P_{VEHICLE}$ (kW)	SoC_{BAT} (%)	SoC_{UC} (%)
100	0	92	100
200	30	92	70
300	35	91	50
400	20	91	30
500	25	90	25
600	40	89	25
700	10	88	30
800	−20	90.2	70
900	−15	92.4	100
1000	5	90.2	95
1100	20	88	90
1200	15	86.9	85
1300	40	85.8	50
1400	30	82.5	40
1500	25	80.3	35
1600	−20	80	32
1700	−20	79	28
1800	30	79	25
1900	15	78	23
2000	10	78	22
2100	25	76	40
2200	25	75	60
2300	−5	74	59
2400	−30	72	70
2500	−20	75	35

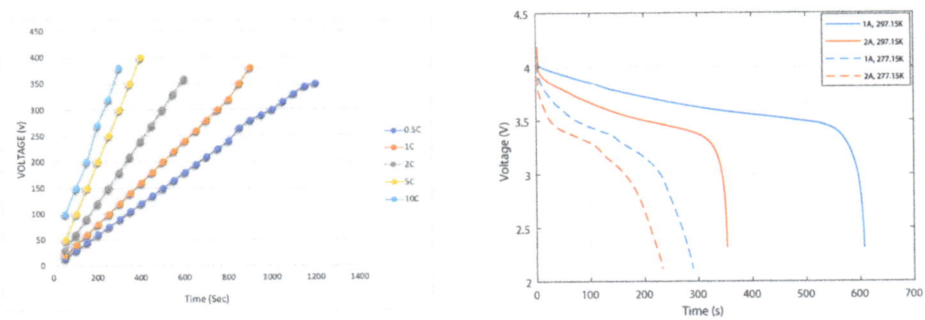

(a) (b)

Figure 13. Difference of the two energy sources behavior and voltage output at various charge rates (**a**) Ultracapacitor operating time and voltage at five distinct rates; (**b**) Battery voltage and operating time drop at low–high charging rates and temperatures by [40].

Furthermore, the efficiency chart of the proposed method compared to the reference scheme tested in [39] is presented in Figure 14. A mean 3% efficiency improvement is available at the entire power range due to enhanced power distribution and regenerative braking utilization, except at minimal loads where the HESS is still slow in providing the supply due to DC/DC switching lag. Efficiency peaks at medium power demand 20–30 kW, reaching 90% efficiency, including all losses from Equation (14). Motor losses are not incorporated as the HESS and converter are the main components studied in this project.

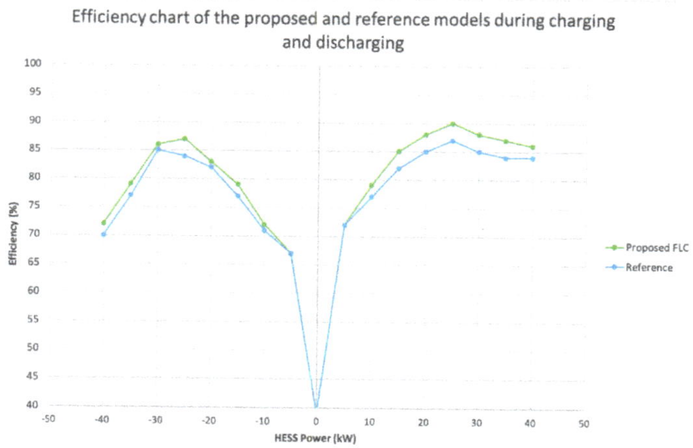

Figure 14. Efficiency chart of the experimental layout followed in [39] and this paper.

Compared to previous literature this work's novelty is the introduction of battery-safe operational zone indication, that is 20 to 80% State of charge, along with the connection of the power mode criteria with duty cycles. Depending on the power state of the engine, each source has to work in a specified duty cycle so depletion is prevented while covering the power demand completely without lag or unsafe operation. This assessment is summarized in Table 3.

Table 3. Contribution of this work supported by the latest literature.

A/A	Paper	Layout	Outcome
1	[7]	Sustainable Fuzzy Model	Power distribution through duty cycles, Utilization for lower battery stress
2	[8]	Fuzzy PID Controller	Smaller battery aging
3	[11]	Model Predictive Control	Lower battery current fluctuations, SOV indication, Single power mode
4	[13]	Fractional Order Modelling	Optimized SoC and HESS discharge time calculation
5	[19]	Semi-Active Topology	Smooth distribution for 2.5 C operation rates
6	[20]	Optimization Control for Trams	Power mode differentiation utilize better efficiency and safe operation
7	[21]	Fine tuning resistance and duty cycle method	Safe HESS Operation, battery life prolonging but higher charging time
8	[23]	Rule based Strategy with CanBUS	Power modes approach minimize Battery current and high UC employment
9	[24]	Multi-Layer Strategy on HESS	20–80% SoC indication for safe operation Single power mode

Table 3. *Cont.*

A/A	Paper	Layout	Outcome
10	[25]	Buck-Boost Converter modelling	Two power modes implementation for lower operational temperature
11	[26]	Particle Filter	Accurate state of charge calculation, with power mode introduction
12	[27]	Genetic Algorithm Optimization	Better efficiency at higher temperatures, no power modes
13	[29]	Semi-Active adaptive controller	High recovery through regenerating braking at all drive states
14	[33]	Fuzzy logic controller for electric bus	UC exploitation as primary source for better sizing, no power modes
15	[35]	Fuzzy Logic Control Optimization	Power demand-based distribution achieving 50% lower fluctuations
16	[36]	Fuzzy Logic rule-based Strategy	Duty cycle and sources SoC Consideration for better control
17	This paper	Fuzzy Logic Rule-Based Strategy	Duty cycle and power modes connection for optimized power distribution within safe Source operational range

4. Conclusions

This paper describes an energy management strategy for electric vehicles using a fuzzy logic controller for a hybrid storage system. Different power modes of the propulsion system are stated with the introduction of duty cycle importance for power distribution. Battery should be maintained within safe SoC limits (20–80%), while the ultracapacitors bank supplies the power demand as an auxiliary source and further exploits regenerative braking. The novelty of this work is the introduction and classification of power modes, providing better power distribution possibilities as duty cycles can be constantly modified, leading to better and more accurate control in real time. Power modes also indicate the driving style of the user and can be easily monitored with certain sensors, like the throttle potentiometer output. The results show that the FLC is able to regulate the power supply between the battery and the supercapacitor by changing the duty cycle value of the buck-boost converter. The proposed system is then able to adjust the power distribution according to the current power state. When the vehicle is in the acceleration state, the supercapacitor provides more or all of the power to the powertrain while the battery output gradually increases. When the vehicle is operating under normal conditions or coasting, the battery acts as the main energy source, and the ultracapacitor is utilized supportively or charging if the SoC is low. At regenerative breaking, both sources are charged depending on the converter voltage. It has been validated that duty cycles from 0.40 to 0.50 are the ideal ratios for most of the states without stressing the battery or depleting any source. This scheme achieves 6% better efficiency and 2.5% range, with 22% enhanced sizing plus 20% higher battery longevity. Future work implies testing with larger energy sources and the implementation of this management strategy in real-life applications.

Author Contributions: Conceptualization, D.R. and S.D.K.; methodology, D.R.; software, D.D.P.; validation, D.R. and S.D.K.; formal analysis, G.V.; investigation, D.R. and D.D.P.; resources, D.R.; data curation, S.D.K. and G.V.; writing—original draft preparation, D.R.; writing—review and editing, D.R. and S.D.K.; visualization, D.D.P.; supervision, S.D.K.; project administration, S.D.K. and G.V.; funding acquisition, S.D.K. All authors have read and agreed to the published version of the manuscript.

Funding: This research received no external funding.

Data Availability Statement: The data that support the findings of this study are available from the corresponding author upon reasonable request.

Conflicts of Interest: The authors declare no conflict of interest.

Abbreviations

$P_{VEHICLE}$	Power demand of the Vehicle
V_H	Buck Boost converter high voltage side
I_{UC}	Ultracapacitor current
V_L	Buck-Boost converter high voltage side
I_B	Battery Current
$P_{CON,LOSS}$	Buck-Boost converter power losses
D_{BOOST}	Duty cycle on the boost mode
D_{BUCK}	Duty cycle on the buck mode
V_S	Voltage drop on the converter switch
V_D	Voltage drop on the converter diode
R_S	Resistance on the converter switch
R_L	Inductor resistance of the converter
R_D	Resistance on the converter diode
V_P	Battery polarization voltage
R_P	Battery polarization resistance
R_B	Battery internal resistance
C_P	Battery polarization capacitance
V_{OC}	Open circuit voltage
V_T	Terminal voltage
k	Time interval for the battery pack
$P_{B,LOSS}$	Battery power losses
R_{UC}	Ultracapacitor internal resistance
I_{UC}	Ultracapacitor current
C_R	Ultracapacitor rated capacitance
n	Time interval for the ultracapacitor bank
P_{UC}	Ultracapacitor power
$P_{UC,LOSS}$	Ultracapacitor power losses
V_{RATED}	Rated voltage by the manufacturer
$P_{LOSS,TOTAL}$	Total power losses
η	Powertrain efficiency
M	Vehicle mass
g	Gravitational acceleration
f	Rolling assistance coefficient
β	Road incline
C_{ARC}	Air assistance coefficient
A	Wind exposed area of the vehicle
V_A	Vehicle speed
δ	Road mass coefficient correction factor
U_R	Battery rated voltage
C_R	Battery rated voltage
n_P	Battery cells in parallel
n_S	Battery cells in series
SoC	State of Charge
SoH	State of Health

References

1. Climate Action and the Green Deal. Available online: https://commission.europa.eu/strategy-and-policy/priorities-2019-2024/european-green-deal/climate-action-and-green-deal_en (accessed on 18 August 2023).
2. Tsemekidi Tzeiranaki, S.; Economidou, M.; Bertoldi, P.; Thiel, C.; Fontaras, G.; Clementi, E.L.; Franco De Los Rios, C. The Impact of Energy Efficiency and Decarbonisation Policies on the European Road Transport Sector. *Transp. Res. Part A Policy Pract.* **2023**, *170*, 103623. [CrossRef]
3. Komnos, D.; Tsiakmakis, S.; Pavlovic, J.; Ntziachristos, L.; Fontaras, G. Analysing the Real-World Fuel and Energy Consumption of Conventional and Electric Cars in Europe. *Energy Convers. Manag.* **2022**, *270*, 116161. [CrossRef]
4. Tansini, A.; Fontaras, G.; Millo, F. A Multipurpose Simulation Approach for Hybrid Electric Vehicles to Support the European CO_2 Emissions Framework. *Atmosphere* **2023**, *14*, 587. [CrossRef]

5. Rimpas, D.; Kaminaris, S.D.; Aldarraji, I.; Piromalis, D.; Vokas, G.; Papageorgas, P.G.; Tsaramirsis, G. Energy Management and Storage Systems on Electric Vehicles: A Comprehensive Review. *Mater. Today Proc.* **2022**, *61*, 813–819. [CrossRef]
6. Electric Vehicles. Available online: https://www.iea.org/energy-system/transport/electric-vehicles (accessed on 18 August 2023).
7. Hartani, M.A.; Hamouda, M.; Abdelkhalek, O.; Mekhilef, S. Sustainable Energy Assessment of Multi-Type Energy Storage System in Direct-Current-Microgrids Adopting Mamdani with Sugeno Fuzzy Logic-Based Energy Management Strategy. *J. Energy Storage* **2022**, *56*, 106037. [CrossRef]
8. Zhao, Y.; Dan, D.; Zheng, S.; Wei, M.; Xie, Y. A Two-Stage Eco-Cooling Control Strategy for Electric Vehicle Thermal Management System Considering Multi-Source Information Fusion. *Energy* **2023**, *267*, 126606. [CrossRef]
9. Tran, M.-K.; Panchal, S.; Khang, T.D.; Panchal, K.; Fraser, R.; Fowler, M. Concept Review of a Cloud-Based Smart Battery Management System for Lithium-Ion Batteries: Feasibility, Logistics, and Functionality. *Batteries* **2022**, *8*, 19. [CrossRef] [PubMed]
10. Lai, X.; Yuan, M.; Tang, X.; Yao, Y.; Weng, J.; Gao, F.; Ma, W.; Zheng, Y. Co-Estimation of State-of-Charge and State-of-Health for Lithium-Ion Batteries Considering Temperature and Ageing. *Energies* **2022**, *15*, 7416. [CrossRef]
11. Chen, H.; Xiong, R.; Lin, C.; Shen, W. Model Predictive Control Based Real-Time Energy Management for a Hybrid Energy Storage System. *CSEE J. Power Energy Syst.* **2020**, *7*, 862–874. [CrossRef]
12. Liu, K.; Kang, L.; Xie, D. Online State of Health Estimation of Lithium-Ion Batteries Based on Charging Process and Long Short-Term Memory Recurrent Neural Network. *Batteries* **2023**, *9*, 94. [CrossRef]
13. Wang, Y.; Gao, G.; Li, X.; Chen, Z. A Fractional-Order Model-Based State Estimation Approach for Lithium-Ion Battery and Ultra-Capacitor Hybrid Power Source System Considering Load Trajectory. *J. Power Sources* **2020**, *449*, 227543. [CrossRef]
14. Lemian, D.; Bode, F. Battery-Supercapacitor Energy Storage Systems for Electrical Vehicles: A Review. *Energies* **2022**, *15*, 5683. [CrossRef]
15. Einan, M.; Torkaman, H.; Pourgholi, M. Optimized Fuzzy-Cuckoo Controller for Active Power Control of Battery Energy Storage System, Photovoltaic, Fuel Cell and Wind Turbine in an Isolated Micro-Grid. *Batteries* **2017**, *3*, 23. [CrossRef]
16. Zhang, Q.; Li, G. Experimental Study on a Semi-Active Battery-Supercapacitor Hybrid Energy Storage System for Electric Vehicle Application. *IEEE Trans. Power Electron.* **2020**, *35*, 1014–1021. [CrossRef]
17. Bhurse, S.S.; Bhole, A.A. A Review of Regenerative Braking in Electric Vehicles. In Proceedings of the 2018 International Conference on Computation of Power, Energy, Information and Communication (ICCPEIC), Chennai, India, 28–29 March 2018; pp. 363–367.
18. Pipicelli, M.; Sessa, B.; De Nola, F.; Gimelli, A.; Di Blasio, G. Assessment of Battery–Supercapacitor Topologies of an Electric Vehicle under Real Driving Conditions. *Vehicles* **2023**, *5*, 424–445. [CrossRef]
19. Chmielewski, A.; Piórkowski, P.; Bogdziński, K.; Możaryn, J. Application of a Bidirectional DC/DC Converter to Control the Power Distribution in the Battery–Ultracapacitor System. *Energies* **2023**, *16*, 3687. [CrossRef]
20. Lin, H.; Jiang, J.; Wei, S.; Cheng, L. Optimization Control for the Efficiency of an On-Board Hybrid Energy Storage System in Tramway Based on Fuzzy Control. In Proceedings of the 2017 11th IEEE International Conference on Compatibility, Power Electronics and Power Engineering (CPE-POWERENG), Cadiz, Spain, 4–6 April 2017; pp. 454–459.
21. Dong, A.; Ma, R.; Deng, Y. Optimization on Charging of the Direct Hybrid Lithium-Ion Battery and Supercapacitor for High Power Application through Resistance Balancing. *Energy* **2023**, *273*, 127233. [CrossRef]
22. Ghoulam, Y.; Pavot, T.; Mamouri, L.; Mesbahi, T.; Durand, S.; Lallement, C.; Kiefer, R.; Laroche, E. Energy Management Strategy with Adaptive Cut-off Frequency for Hybrid Energy Storage System in Electric Vehicles. In Proceedings of the 2022 IEEE Vehicle Power and Propulsion Conference (VPPC), Merced, CA, USA, 1–4 November 2022; pp. 1–6.
23. Xiong, R.; Duan, Y.; Cao, J.; Yu, Q. Battery and Ultracapacitor In-the-Loop Approach to Validate a Real-Time Power Management Method for an All-Climate Electric Vehicle. *Appl. Energy* **2018**, *217*, 153–165. [CrossRef]
24. Xu, W.; Liu, M.; Xu, L.; Zhang, S. Energy Management Strategy of Hydrogen Fuel Cell/Battery/Ultracapacitor Hybrid Tractor Based on Efficiency Optimization. *Appl. Sci.* **2023**, *13*, 151. [CrossRef]
25. Pai, F.S. Design and Control Method of a Battery/Ultra-Capacitor Energy Storage System for EVs. *Int. J. Electr. Electron. Eng. Telecommun.* **2023**, *12*, 203–208. [CrossRef]
26. Feng, N.; Ma, T.; Chen, C. Fuzzy Energy Management Strategy for Hybrid Electric Vehicles on Battery State-of-Charge Estimation by Particle Filter. *SN Appl. Sci.* **2022**, *4*, 256. [CrossRef]
27. Wang, C.; Liu, R.; Tang, A. Energy Management Strategy of Hybrid Energy Storage System for Electric Vehicles Based on Genetic Algorithm Optimization and Temperature Effect. *J. Energy Storage* **2022**, *51*, 104314. [CrossRef]
28. Podder, A.K.; Chakraborty, O.; Islam, S.; Manoj Kumar, N.; Alhelou, H.H. Control Strategies of Different Hybrid Energy Storage Systems for Electric Vehicles Applications. *IEEE Access* **2021**, *9*, 51865–51895. [CrossRef]
29. Hussain, S.; Ali, M.U.; Park, G.-S.; Nengroo, S.H.; Khan, M.A.; Kim, H.-J. A Real-Time Bi-Adaptive Controller-Based Energy Management System for Battery–Supercapacitor Hybrid Electric Vehicles. *Energies* **2019**, *12*, 4662. [CrossRef]
30. Yan, S.; Gu, Z.; Ding, L.; Park, J.H.; Xie, X. Weighted Memory H_∞ Stabilization of Time-Varying Delayed Takagi-Sugeno Fuzzy Systems. *IEEE Trans. Fuzzy Syst.* **2023**, 1–6. [CrossRef]
31. Zhao, X.; Zhang, Y.; Cui, X.; Wan, L.; Qiu, J.; Shang, E.; Zhang, Y.; Zhao, H. Wavelet Packet-Fuzzy Optimization Control Strategy of Hybrid Energy Storage Considering Charge–Discharge Time Sequence. *Sustainability* **2023**, *15*, 10412. [CrossRef]
32. Boumediene, S.; Nasri, A.; Hamza, T.; Hicham, C.; Kayisli, K.; Garg, H. Fuzzy Logic-Based Energy Management System (EMS) of Hybrid Power Sources: Battery/Super Capacitor for Electric Scooter Supply. *J. Eng. Res.* **2023**, S2307187723001682. [CrossRef]

33. Rocha, S.P.D.; Silva, S.M.D.; Ekel, P.I. Fuzzy Set-Based Approach for Grid Integration and Operation of Ultra-Fast Chargin Electric Buses. *Int. J. Electr. Power Energy Syst.* **2022**, *138*, 107919. [CrossRef]
34. Shelma, G.; Rajeev, T. Fuzzy-Based Control Strategy for Supercapacitor Assisted Battery Powered EV. In Proceedings of the 202 International Conference on Control, Communication and Computing (ICCC), Thiruvananthapuram, India, 19–21 May 202 pp. 1–6.
35. Angundjaja, C.Y.; Wang, Y.; Jiang, W. Power Management for Connected EVs Using a Fuzzy Logic Controller and Artificia Neural Network. *Appl. Sci.* **2022**, *12*, 52. [CrossRef]
36. Van Jaarsveld, M.J.; Gouws, R. An Active Hybrid Energy Storage System Utilising a Fuzzy Logic Rule-Based Control Strateg *World Electr. Veh. J.* **2020**, *11*, 34. [CrossRef]
37. Khan, M.A.; Zeb, K.; Sathishkumar, P.; Ali, M.U.; Uddin, W.; Hussain, S.; Ishfaq, M.; Khan, I.; Cho, H.-G.; Kim, H.-J. A Novel Supercapacitor/Lithium-Ion Hybrid Energy System with a Fuzzy Logic-Controlled Fast Charging and Intelligent Energ Management System. *Electronics* **2018**, *7*, 63. [CrossRef]
38. Zakzouk, N.E.; Lotfi, R.A. Power Flow Control of a Hybrid Battery/Supercapacitor Standalone PV System under Irradiance and Load Variations. In Proceedings of the 2020 10th International Conference on Power and Energy Systems (ICPES), Chengdu, China, 25–27 December 2020; pp. 469–474.
39. Rimpas, D.; Kaminaris, S.D.; Piromalis, D.; Vokas, G.; Papageorgas, P. *Design and Implementation of a Small-Scaled Hybrid Storage System for Optimal Sizing in Electric Vehicles*; AIP Publishing: Metz, France, 2023; p. 020007.
40. Sheikh, S.S.; Anjum, M.; Khan, M.A.; Hassan, S.A.; Khalid, H.A.; Gastli, A.; Ben-Brahim, L. A Battery Health Monitoring Method Using Machine Learning: A Data-Driven Approach. *Energies* **2020**, *13*, 3658. [CrossRef]

MDPI
St. Alban-Anlage 66
4052 Basel
Switzerland
www.mdpi.com

Mathematics Editorial Office
E-mail: mathematics@mdpi.com
www.mdpi.com/journal/mathematics

www.ingramcontent.com/pod-product-compliance
Lightning Source LLC
Chambersburg PA
CBHW061620220926
51850CB00005B/558
9783725807758